Topics in Applied Physics Volume 63

Topics in Applied Physics Founded by Helmut K. V. Lotsch

Volume 57 **Strong and Ultrastrong Magnetic Fields and Their Applications**
Editor: F. Herlach

Volume 58 **Hot-Electron Transport in Semiconductors**
Editor: L. Reggiani

Volume 59 **Tunable Lasers**
Editors: L. F. Mollenauer and J. C. White

Volume 60 **Ultrashort Laser Pulses** and Applications
Editor: W. Kaiser

Volume 61 **Photorefractive Materials and Their Applications I**
Fundamental Phenomena
Editors: P. Günter and J.-P. Huignard

Volume 62 **Photorefractive Materials and Their Applications II**
Applications
Editors: P. Günter and J.-P. Huignard

Volume 63 **Hydrogen in Intermetallic Compounds I**
Electronic, Thermodynamic, and Crystallographic Properties, Preparation
Editor: L. Schlapbach

Volume 64 **Hydrogen in Intermetallic Compounds II**
Surface and Dynamic Properties, Applications
Editor: L. Schlapbach

Volumes 1–56 are listed on the back inside cover

Hydrogen in Intermetallic Compounds I

Electronic, Thermodynamic, and Crystallographic Properties, Preparation

Edited by L. Schlapbach

With Contributions by
P. Fischer T. B. Flanagan R. Griessen M. Gupta
G. Hilscher W.A. Oates A. Percheron-Guégan
T. Riesterer L. Schlapbach J.-M. Welter
G. Wiesinger K. Yvon

With 118 Figures and 12 Tables

Springer-Verlag Berlin Heidelberg GmbH

Dr. Louis Schlapbach
Laboratorium für Festkörperphysik, ETH Zürich, Hönggerberg,
CH-8093 Zürich, Switzerland and
Institut de Physique, Université de Fribourg,
CH-1700 Fribourg, Switzerland

ISBN 978-3-662-30925-4 ISBN 978-3-540-47882-9 (eBook)
DOI 10.1007/978-3-540-47882-9

Library of Congress Cataloging-in-Publication Data. Hydrogen in intermetallic compounds/edited by L. Schlapbach; with contributions by P. Fischer... [et al.]. p. cm. – (Topics in applied physics; v. 63) Includes index. Contents: v. 1. Electronic, thermodynamic, and crystallographic properties, preparation. 1. Intermetallic compounds – Hydrogen content. I. Schlapbach, L. (Louis), 1944–. II. Fischer, P. (Peter) III. Series. QD171.H94 1988 546'.3–dc 19 87-32257

Originally published by Springer-Verlag Berlin Heidelberg New York in 1988
Softcover reprint of the hardcover 1st edition 1988

2153/3150-543210

Preface

Phenomena related to the topic "Hydrogen in Metals" and the physics behind them as they were known at the end of the seventies are reviewed in the books *Hydrogen in Metals*, Vols. I and II, edited by G. Alefeld and J. Völkl (Vols. 28 and 29 of this series). These books, which treat mostly hydrogen in elemental metals, are still very valuable and will continue to be important in the coming years.

Since their publication, a lot of new results have been obtained, which have improved both our knowledge and understanding considerably. The work was stimulated by the extraordinary properties of hydrogen-storing intermetallic compounds and by the energy crisis. The recent progress, which has more often concerned *intermetallic compounds* and alloys than elemental metals, has been reviewed in many good articles either at a technical, purely scientific, or at a more popular level. Until now, however, the details of this important field have never been drawn together and presented in the form of a book.

Thus, the aim of the publication of the two volumes "Hydrogen in Intermetallic Compounds" is to give a thorough description of the various aspects of the topic in a series of chapters written by specialists in the field and to review major progress on hydrogen in and on elemental metals. This volume begins with a description of the preparation of intermetallics and their hydrides. It contains further chapters on crystallographic, thermodynamic, electronic and magnetic properties and on heat of formation models. Volume II will be devoted to aspects of the kinetics and dynamics of hydrogen, to surface phenomena and to applications and experimental techniques.

As the interaction of hydrogen with metals and alloys is, and will be, of significant importance for basic research as well as for hydrogen energy technology, fusion, catalysis, getters, electrochemical cells and for many more applications, I hope that these two volumes will help many scientists to find the information they are looking for, to spread the fascination which we the authors already share, and to stimulate further work.

I should like to express my thanks to all the authors for their individual contributions and for their willing and fruitful cooperation and to Angela Lahee and to my wife Christine for careful reading of the manuscripts and proofs.

Zurich, January 1988 *Louis Schlapbach*

Contents

1. Introduction
By L. Schlapbach (With 2 Figures) 1
1.1 Metal–Hydrogen Systems and Related Phenomena 1
1.2 Scope of These Two Volumes 3
1.3 Recent Highlights and Outlook 6
References 8

2. Preparation of Intermetallics and Hydrides
By A. Percheron-Guégan and J.-M. Welter
(With 30 Figures and 1 Table) 11
2.1 Introduction 11
2.2 Thermodynamic Aspects of Intermetallic Phases and Their Synthesis 12
2.3 Preparation of Intermetallic Phases: Laboratory Scale 17
2.3.1 The Starting Materials 17
2.3.2 Melting Techniques 19
2.3.3 Annealing Treatments 24
2.3.4 Single Crystal Growth Techniques 27
2.3.5 Noncrystalline Materials: Rapid Solidification and Solid-State Reactions 28
2.3.6 Synthesis from the Gas Phase and Thin Films 28
2.4 Characterization of Intermetallic Phases 29
2.5 Preparation of Intermetallic Phases: Industrial Scale 32
2.6 Hydrogenation of Intermetallic Phases 37
2.7 Concluding Remarks 40
References 44

3. Thermodynamics of Intermetallic Compound–Hydrogen Systems
By T. B. Flanagan and W. A. Oates (With 14 Figures and 3 Tables) . 49
3.1 Introduction 49
3.2 Macroscopic Thermodynamics 50
3.2.1 Single and Two Condensed Phase Equilibria 50
a) Integral/Formation Properties 50
b) Partial Molar Properties 52
c) Plateau Properties 53
3.2.2 Three Condensed Phase Equilibria 56

3.3 Experimental Aspects 57
3.3.1 p-r-T Measurements 57
3.3.2 Calorimetric Measurements 58
a) Reaction Calorimetry 58
b) Thermal Analysis 58
c) Heat Capacity 59
3.3.3 Problems in Obtaining Accurate Thermodynamic Data . 59
a) Hysteresis 59
b) Characterization of the Starting IMC – Activation and Annealing 60
c) Sloping Plateaux 61
d) Decomposition 62
e) Amorphization 62
3.4 Results for Representative Systems 63
3.4.1 $LaNi_5$ and Some $LaNi_5$-Based Systems 63
3.4.2 TiFe 64
3.4.3 ZrV_2 and Some $ZrMn_2$-Based Systems 66
3.4.4 $ErFe_2$ 67
3.4.5 Mg-Based IMCs 68
3.5 Microscopic Thermodynamics 69
3.5.1 Configurational Models 69
a) Site and Interaction Energies 69
b) Hard Core, Constant $V_0(j)$ Models 71
c) Multiple Plateaux Models 73
3.5.2 Non-Configurational Contributions 76
3.5.3 Phase Transitions 78
a) The $\alpha-\alpha'$ Transition 79
b) Order/Disorder Transitions 79
3.5.4 Complete and Partial Equilibrium 80
3.5.5 Hysteresis 81
3.6 Conclusions 82
References 83

4. Crystal and Magnetic Structures of Ternary Metal Hydrides: A Comprehensive Review
By K. Yvon and P. Fischer (With 4 Figures and 3 Tables) 87
4.1 Introduction 87
4.2 Experimental Method 89
4.2.1 Sample Preparation 89
4.2.2 Neutron Diffraction 90
4.2.3 Structure Refinement 91
4.3 Structural Results 93
4.3.1 Metal Atom Substructure 93
4.3.2 Distribution of Hydrogen Atoms 106
4.3.3 Interatomic Distances 109

4.4 Stability Aspects 112
4.4.1 Interstitial Hole Size 113
4.4.2 Electrostatic Repulsion 114
4.4.3 Electronic Factors 115
4.4.4 Attractive Nearest-Neighbor Interactions 116
4.5 Magnetic Ordering 117
4.6 Conclusions 118
4.A Appendix. Table: Structural Data for Ternary Hydrides Studied by Neutron Diffraction 119
4.B Appendix. Table: Magnetic Structures 130
References 132

5. Electronic Properties
By M. Gupta and L. Schlapbach (With 25 Figures and 2 Tables) . . 139
5.1 Introduction 139
5.2 Theoretical and Experimental Methods: General Features . . . 140
5.2.1 Theoretical Methods 140
a) Stoichiometric Hydrides 142
b) Nonstoichiometric Hydrides 146
c) H Impurity in Metals 146
d) Calculation of Other Observables 147
5.2.2 Experimental Methods 148
5.3 Results for Binary Hydrides 150
5.3.1 Hydrides of Pd and Ni 150
a) Energy Bands and Densities of States 150
b) Fermi Surfaces (FS) 160
c) Study of Other Observables 160
5.3.2 Hydrides of Ti, Zr, and Hf 162
5.3.3 Hydrides of Nb, V, and Ta 166
5.3.4 Hydrides of Rare Earth Metals Including Sc, Y, and La 168
5.3.5 Hydrides of Ca and Mg 176
5.3.6 Hydrides of Actinide Metals 176
5.4 Results on Ternary Hydrides 177
5.4.1 Hydrides of AB_5 Compounds (Haucke Compounds) . . 177
5.4.2 Hydrides of FeTi 185
5.4.3 Mg_2Ni and Isoelectronic Compounds and Their Hydrides . 193
5.4.4 The AB_2 Laves Phase, and Related Compounds and Their Hydrides 201
5.4.5 Amorphous Alloys and Their Hydrides 205
5.5 Conclusions 207
References 209

6. Heat of Formation Models
By R. Griessen and T. Riesterer (With 23 Figures and 3 Tables) . . 219
6.1 Introduction 219

6.2 Thermodynamics of Hydrogen Absorption in Metals 222
6.3 Empirical Correlations 231
6.4 Semi-empirical Models 234
6.4.1 General Properties of Semi-empirical Models 235
a) Binary Alloys 235
b) Binary Metal Hydrides 237
c) Ternary Metal Hydrides 239
d) Local or Site-Dependent Enthalpies 241
6.4.2 The Atomic-Cell Model 244
a) Intermetallic Compounds and Alloys 244
b) Binary Metal Hydrides 245
c) Ternary Metal Hydrides 247
d) Quaternary Metal Hydrides 248
6.4.3 The Band-Structure Model 249
a) Binary Metal Hydrides 249
b) Ternary Metal Hydrides 255
c) Volume Effects 260
6.4.4 The Local Band-Structure Model 262
6.5 Conclusions 265
6.6 Tables 266
References 279
Additional References 284

7. Magnetic Properties, Mössbauer Effect and Superconductivity
By G. Wiesinger and G. Hilscher (With 20 Figures) 285
7.1 Introduction and Scope 285
7.2 Theoretical Background 286
7.2.1 Theoretical Concepts in Magnetism 286
7.2.2 Physical Concepts of Mössbauer Spectroscopy 291
a) Introductory Remarks 291
b) Experimental Techniques 292
c) Hyperfine Interactions 293
7.3 Magnetism and Mössbauer Spectroscopy – Experimental Results and Discussion 295
7.3.1 General Remarks 295
7.3.2 Transition-Metal-Rich *R*-TM Compounds 296
a) R_2TM_{17} 296
b) RTM_5 296
c) R_2TM_7 (Space Group $P6_3/mmc$ No. 194) 301
d) $R_2Fe_{14}B$-Type Compounds 302
e) R_6TM_{23} (TM = Mn, Fe) 303
f) RTM_3 (TM = Fe, Co, Ni) 307
7.3.3 Laves-Phase-Type Compounds 309
a) Mn Compounds 309
b) Fe Compounds 312

c) Fe-Containing Pseudobinaries 315
d) Compounds of Co, Ni and Others 317
7.3.4 TiFe and Related CsCl-Type Compounds (Space Group Pm3m, No. 221) 318
7.3.5 Compounds of Low Transition Metal Content 323
7.3.6 Amorphous Alloys 325
7.4 Superconductivity 329
References 332

List of Tables 343

Subject Index 345

Contributors

Fischer, Peter

Labor für Neutronenstreuung, ETH Zürich,
CH-5303 Würenlingen, Switzerland

Flanagan, Ted B.

Chemistry Department, University of Vermont,
Burlington, VT 05405, USA

Griessen, Ronald

Natuurkundig Laboratorium, Vrije Universiteit, De Boelelaan 1081,
NL-1081 HV Amsterdam, The Netherlands

Gupta, Michèle

Institut des Sciences des Materiaux, U.A. 446 du C.N.R.S.,
Bât. 415, Université Paris-Sud, F-91405 Orsay, France

Hilscher, Gerfried

Institut für Experimentalphysik, TU Wien,
A-1040 Wien, Austria

Oates, W. Alan

Department of Chemical and Materials Engineering,
The University of Newcastle, Shortland, N.S.W. 2308, Australia

Percheron-Guégan, Annick

Laboratoire de Chimie Métallurgie des Terres Rares, C.N.R.S.,
1, Place A. Briand, F-92190 Meudon, France

Riesterer, Thomas

Laboratorium für Festkörperphysik, ETH Zürich,
CH-8093 Zürich, Switzerland

Schlapbach, Louis

Laboratorium für Festkörperphysik, ETH Zürich,
CH-8093 Zürich, Switzerland and
Institut de Physique, Université de Fribourg,
CH-1700 Fribourg, Switzerland

Welter, Jean-Marie

Institut für Festkörperforschung der Kernforschungsanlage Jülich GmbH,
Postfach 1913, D-5170 Jülich, Fed. Rep. of Germany
Permanent address: Tréfimetaux, Centre de Recherche, B.P. 11,
F-60590 Serifontaine, France

Wiesinger, Günter

Institut für Experimentalphysik, TU Wien,
A-1040 Wien, Austria

Yvon, Klaus

Laboratoire de Cristallographie aux Rayons X, Université de Genève,
24, Quai E. Ansermet, CH-1211 Genève 4, Switzerland

1. Introduction

Louis Schlapbach

With 2 Figures

This short introduction to the more general topic of hydrogen in and on metals starts with a simple description of metal-hydrogen systems and of related phenomena. Then the relevance of research on metal–hydrogen systems for solid state science and for applications will be emphasized, the discovery of hydrides of intermetallics and their differences to binary hydrides will be sketched and references to standard literature will be given. Finally, in the third part, Sect. 1.3, some highlights of the topics treated in this volume will be outlined.

1.1 Metal–Hydrogen Systems and Related Phenomena

A metal–hydrogen system consists of a metal (possibly containing dissolved hydrogen), hydrogen in the gaseous or in a condensed phase, and of an interface in between. The metal can be a high purity single crystal, a crystalline ordered or disordered or amorphous single-phase alloy or a multiphase alloy of technical purity. Hydrogen, and also its isotopes deuterium and tritium, can be present as a clean molecular gas, a gas mixture, dissolved in an organic or aequous liquid, as hydrogen atoms, or as ionized atoms in a hydrogen plasma or for hydrogen implantation. The interface may also be a clean crystalline, ordered or disordered or amorphous metallic surface, it may be oxidized, contaminated or really dirty. Some examples are

- H_2 gas and a Pd single crystal with the (H-induced) reconstructed (110) surface as interface,
- a $LaNi_5$ powder electrode in an aequous KOH solution as electrolyte,
- H_2-containing coal gas in a carbon steel pipeline.

A simplified and frequently used (one-dimensional) model to describe H_2 gas and a hydrogen dissolving metal is shown in Fig. 1.1. A hydrogen molecule approaching the metal can be dissociated at the interface, adsorbed at appropriate surface and near surface sites and dissolved at interstitial sites of the host metal. If the local hydrogen concentration exceeds a certain limit, a hydride phase precipitates.

The thermodynamic aspects of hydride formation (gaseous hydrogen) can be described by pressure–composition isotherms (pcT curves, Fig. 1.2). The host metal dissolves some hydrogen as a solid solution (α-phase). As the H_2 pressure,

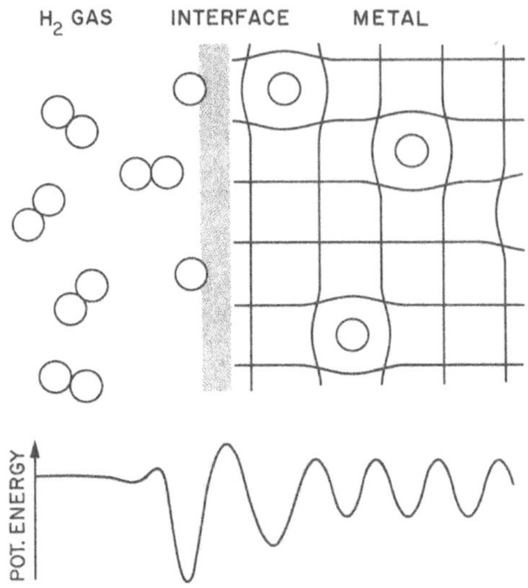

Fig. 1.1. Simplified model of the dissociation of molecular hydrogen at an interface and of the solution of hydrogen atoms in the bulk. The potential energy indicates from left to right a shallow minimum for physisorbed H_2, a deep minimum for chemisorbed dissociated H, a rather deep minimum for near surface hydrogen and periodic minima for H dissolved on interstitial sites of the host metal, separated from each other by diffusion barriers

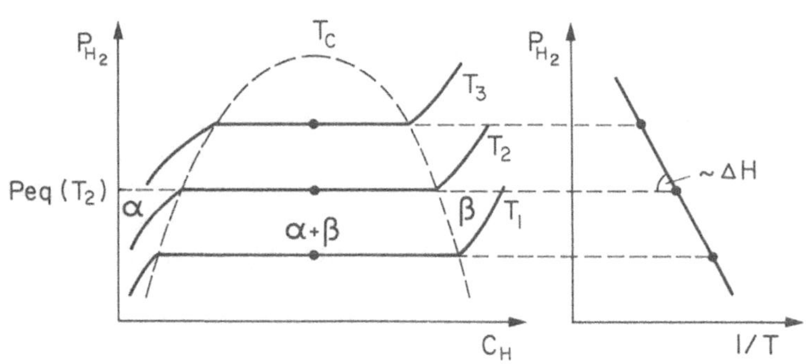

Fig. 1.2. Pressure–composition isotherms for the solid solution of hydrogen (α-phase) and hydride formation (β-phase). The region of coexistence of the two phases is characterized by the flat plateau at the equilibrium pressure $p_{eq}(T)$ and ends at the critical temperature T_c. The enthalpy of hydride formation ΔH is obtained from the variation of the equilibrium pressure (on a logarithmic scale!) with temperature in a van't Hoft plot

and herewith the concentration c_H of dissolved H, is increased, the H–H interaction becomes locally important and nucleation and growth of the hydride phase (β) start. Whilst the two phases α and β coexist, the isotherms show a plateau, the length of which determines how much H_2 can be stored reversibly with small pressure variations. In the pure β-phase the H_2 pressure rises steeply with the concentration. At higher H_2 pressure further plateaux and further hydride phases may be formed. The two-phase region ends in a critical point T_c. The plateau pressure $p_{eq}(T)$, strongly depends on temperature. From the slope of a so-called van't Hoff plot of the plateau pressure (on a logarithmic scale) versus T^{-1}, the heat of hydride formation ΔH can be evaluated (see Chap. 3).

The solution of hydrogen in the bulk at interstitial sites of the host metal and the formation of metal hydride phases affect the host metal properties and induce a variety of phenomena:

Crystal Structure. The lattice expands, the crystal structure changes mostly with a reduction of symmetry, a H-sublattice is formed, order-disorder transitions of the H-sublattice occur, lattice defects are formed, they interact and diffuse, strain fields grow, and even non crystalline and probably quasi-crystalline materials can be formed.

Phase Diagram. The hydrogen–metal and hydrogen–hydrogen interaction (nearest neighbours, next nearest neighbours, ...) are the origin of the occurrence of a variety of phases as a function of temperature and hydrogen concentration and phase transitions. Hydrogen in metals can often be treated as a lattice gas.

Electronic and Magnetic Properties. The lattice expansion, the additional electron and proton strongly perturb electrons and phonons. Metal–semiconductor transitions, loss and appearance of magnetic moments, of magnetic order, and also of superconductivity can be observed.

Cohesion. The change of the phonon spectra and of the electronic properties deteriorates the mechanical, particularly elastic properties and leads to decohesion, embrittlement, and even to disintegration into powder.

Dynamics of Hydrogen. The hydrogen atoms and their isotopes (D, T, μ^+) vibrate about their equilibrium positions, perform local motions and long-range diffusion. The high mobility allows a redistribution of hydrogen atoms and the formation of concentration gradients under the influence of mechanical deformation (strain field), temperature gradient (thermotransport), electric and magnetic fields (electromigration, proton Hall effect). Quantum diffusion occurs down to the 10 K range.

Surface. The sorption of hydrogen covers adsorption and desorption, surface migration, surface reconstruction and surface segregation, chemical reduction, surface hydride formation and embrittlement.

It is evident that these effects and phenomena are at the same time exciting for basic research and important for technical applications and all steps in between.

1.2 Scope of These Two Volumes

Since *T. Graham's* observation of the ability of Pd to absorb large amounts of hydrogen in 1866, i.e. almost a hundred years after the discovery of hydrogen by Lavoisier in 1783, who had followed Cavendish's work on "factitious air", research on hydrogen in and on metals has seen various motivations and many ups and downs. In addition to the continuous interest of physicists and chemists in metal hydrides as an interesting class of materials as well as in the hydrogen adsorption as a prototype step of a catalyzed reaction, and of metallurgists fighting against hydrogen embrittlement, metal–hydrogen systems frequently became "runners". This was the case e.g. after the recognition of their potential application as moderators in nuclear reactors or after the discovery of the astonishing hydrogen sorption properties of intermetallic compounds and their use for the reversible storage of hydrogen as an energy carrier.

A no less important part of the motivation is, however, related to basic research and to the fact that many key phenomena of solid state science can be studied in metal–hydrogen systems: magnetism and superconductivity, heavy electron behaviour, alloy formation, superlattices, fractals, metal–semiconductor transitions in the bulk and at the surface, surface reconstruction and surface segregation, phase diagrams, order–disorder transitions, quantum diffusion, formation of disordered and non-crystalline materials and many more.

The results of the world wide research efforts on metal–hydrogen systems are published in a variety of journals covering physics, chemistry, metallurgy, and engineering. A large part of the results concerning metal hydrides are regularly presented at the "International Symposium on the Properties and Applications of Metal Hydrides" and at the "International Conference on Hydrogen in Metals", whose proceedings are published as special issues of the *Journal of Less-Common Metals* [1.1] and *Zeitschrift für physikalische Chemie* [1.2]. The two conferences will merge in 1988. Hydrogen–energy related papers are often presented at the World Hydrogen Energy Conference [1.3] or published in the *International Journal of Hydrogen Energy.*

Two NATO Conferences were dedicated to the electronic structure of hydrogen metals [1.4] and to the effect of disorder and amorphicity on metal–hydrogen systems [1.5]. Energy storage in metal hydrides was reviewed with emphasis on automotive applications [1.6] and on energy technology in general [1.7].

Among the further review papers written within the last few years some introductory papers at a popular level should be mentioned [1.8], as well as reviews which focus on particular materials: Zr based alloys and AB_2-Laves phase alloys [1.9]; the physicochemistry of intermetallic compounds and their hydrides [1.10]; atomistic and electronic approaches to hydrogen in *bcc* metals [1.11]; and finally two rather extended reviews focussed on thermodynamic and electronic properties of hydrides formed from intermetallic compounds of rare earth and transition metals [1.12], which were finished at the end of 1981 and 1983, respectively.

As is evidenced by the title these two books on *Hydrogen in Intermetallic Compounds*, Vols. I and II, give a rather complete description of our knowledge and understanding of hydrogen in and on intermetallic compounds of the major families. Furthermore, some important results on hydrogen in and on elemental metals and in some amorphous alloys are given in order that the description of the state of the art at the end of 1986 is as complete as possible.

Intermetallic compounds are, as will be described in detail in Chap. 2 on preparation, a special case of ordered single-phase alloys, characterized by stoichiometric (or near stoichiometric) concentration ratios of the two metallic components. The major differences between hydrides of intermetallic compounds and those of elemental metals are:

- larger variety of interstitial sites
- wide range of stability, which can be adjusted by substitutions
- tendency to decompose into a binary hydride and another intermetallic compound (metastable)
- very reactive to hydrogen due to surface segregation, which prevents passivation
- formation of hydrides of intermetallic compounds whose components do not form binary hydrides and vice versa.

The first extended reports on the reaction of hydrogen with alloys and intermetallic compounds were written by *Beck* [1.13] and by *Pebler* and *Gulbranson* [1.14], who studied very stable hydrides and the detrimental effect of hydrogen on the use of Zr alloys in nuclear reactions.

The discovery and description of hydride formation of Mg_2Ni [1.15] and FeTi [1.16] by *Reilly* and *Wiswall* and particularly of the easy hydride formation of $LaNi_5$ by *van Vucht* et al. [1.17] initiated a worldwide series of studies of hydrogen storage in intermetallic compounds. The solution of hydrogen and deuterium in $LaNi_5$ had already been studied a few years earlier by *Neumann* [1.18]. Although he observed the formation of a new phase with expanded lattice constants (those of $LaNi_5H_6$) and the onset of disintegration of the sample into small pieces which desorbed hydrogen gas, he did not realize the consequences of this observation. Several years later hydrides of AB_2 type Laves phase compounds were independently discovered by *Shaltiel* et al. [1.19] and *Ishido* et al. [1.20]. Then hydrogen absorption was reported for oxygen-stabilized compounds like $Ho_3Fe_9O_x$ [1.12] and $Ti_4Fe_2O_x$ [1.22] and already much earlier by Beck [1.13]. Investigations of hydrogen absorption by amorphous alloys were initiated by *Maeland* [1.23]. Finally the series of new hydrides came to a – hopefully only temporary – stop with the successful synthesis of Mg_2FeH_6 by *Didisheim* et al. [1.24].

Volume I of *Hydrogen in Intermetallic Compounds*, "Electronic, Thermodynamic and Crystallographic Properties", contains the chapters Preparation of Intermetallics and their Hydrides, Thermodynamic Properties, Crystal and Magnetic Structure, Electronic Properties, Heat of Formation Models, and Magnetism and Superconductivity. Volume II, "Surfaces, Dynamics, and Applications", contains the chapters Surface Properties, Dynamics, Kinetics, Applications, and Experimental Techniques.

Most chapters of these two volumes are structured as follows:

- Introduction
- Methods
- Results on binary hydrides
- Results on $LaNi_5$ type hydrides
- Results on FeTi type hydrides
- Results on Mg alloy hydrides
- Results on AB_2 Laves phase hydrides
- Results on other hydrides (amorphous).

1.3 Recent Highlights and Outlook

To round up the introductory chapter we would like to mention some highlights in this field which should give the reader a taste of the most exciting current research activities. Some possible new developments in the near future are also indicated.

Phase Diagrams, Elastic Interactions, Hydrogen Ordering: Considerable efforts are made to study hydrogen–metal phase diagrams and phase transitions, and to understand the underlying hydrogen–metal and hydrogen–hydrogen interactions. The fundamental mechanism is the elastic expansion of the metal lattice around the dissolved hydrogen atom and the resulting elastic interaction between neighbouring hydrogen atoms. The theory concentrates on experimentally well-known phase diagrams such as Nb–H, for which transitions from disordered to disordered phases were successfully described by the lattice gas model by calculating pairwise interactions. Now the models are extended to include the effects of long-range electronic and *many-body elastic interactions* on the occurrence of ordered phases on the one hand, and the effect of *disorder* of the metal host lattice (disordered alloy, amorphous metal, disordered impurities) on the phase diagram, on the other hand [1.25–27]. Detailed experimental studies of the phase diagrams of rare earth (RE) hydrides have been initiated particularly at low temperatures and in the concentration ranges of solid solubility and from REH_2 to REH_3 [1.28, 29]. The occurrence of miscibility gaps has been discussed [1.29]. In addition to the hydrogen ordering, which is known to occur in REH_{2+x} around 240 K, *linear ordering* and hydrogen pairing were found in the solid solution range of LuH_x and YH_x at low temperature, caused by an anisotropic hydrogen potential [1.30, 31].

Materials, Preparation Methods. Only very few new hydrides have been found recently, but the successful synthesis of Mg_2FeH_6 [1.24] opened further paths for the discovery of ternary hydrides whose corresponding intermetallic compounds do not exist. Implantation techniques [1.32] allow the preparation of hydrides with hydrogen contents higher than any known to date and with as yet unknown properties. The development of a catalyzed reaction to precipitate MgH_2 from an organometallic solution opened an extremely promising field for producing highly reactive high surface area hydrides [1.33]. The development of high pressure technology up to the 100 GPa range in diamond anvil pressure cells is soon likely to lead to the synthesis of new hydrides [1.34].

Hydride Formation, Formation of Disordered, Amorphous of Quasi-Crystalline Materials, Superlattices, Hydrogen as a Probe. The solution and rapid diffusion of H in metals and the subsequent hydride formation together with the rise of temperature due to the evolution of the heat of formation affect the local structure and order of the host metal in many different ways:

The implantation of a few percent of H into Al grains (in an Al–Ge mixture) induces an Anderson type *metal-insulator transition* by increasing the *local disorder* [1.35].

Nb–Ta superlattices (multilayered structures) not only show greatly increased H solubility (as compared to bulk Ta and Nb), but also H-induced strain modulation [1.36].

The short-range order of intermetallic compounds can be destroyed completely by hydride formation so that H absorption may well develop into an important solid state reaction to produce *amorphous metals* [1.37, 38]. As other solid state reactions have already been successfully applied to form quasi-crystalline materials [1.39], and since the rate of hydride formation is easily controlled by parameters such as H_2 pressure, sample size and heat transfer, hydrogen absorption might soon be used to form *quasicrystals*. Icosahedral ordering was recently suggested in a hydrogen site occupancy model for amorphous transition metal hydrides [1.50].

Some hydrides of binary alloys are thermodynamically unstable and tend to decompose into a stable binary hydride and a binary alloy of different composition. Early stages of this decomposition can be viewed as local disorder. The formation of "antistructure" Fe atoms (i.e. Fe atoms on Ti sites) in the bulk of FeTi-like ordered alloys has been discussed [1.40]. TiH_2 precipitates out of a hydrogenated amorphous CuTi matrix. A small angle neutron scattering study of the aggregation of these TiH_2 precipitates demonstrated their *fractal* structure [1.41].

Disordered and amorphous systems have attracted much attention, and considerable progress is being made particularly in the theory. Again, H-metal systems provide challenging examples [1.5]. Amorphous metals absorb hydrogen in amounts which are – apart from some interesting exceptions – comparable to those of their crystalline counterparts at comparable pressure and temperature. Many properties, e.g. pressure–composition isotherms or diffusion, depend on the configurational disorder in the host lattice, and particularly on the spread in energy of the hydrogen sites [1.42, 1.43]. Recently it was shown the other way round that the slope of pressure-composition isotherms of hydrogen in amorphous metals can be used to probe the hydrogen sites [1.44, 50].

Electronic and Magnetic Properties. Theoretical approaches to the electronic properties of stoichiometric hydrides of elemental metals by total energy calculations look promising [1.45]. Band structure calculations are being tentatively extended to nonstoichiometric and disordered hydrides and very successfully to ordered hydrides of intermetallic compounds (Chap. 5). Purely phenomenological correlations, for example, between electronic and thermodynamic properties or between embrittlement and decohesion are being replaced by semiempirical models based on the electronic band structure [1.46] (Chap. 6). Hydrides of rare earth metals illustrate how advances in the understanding can sometimes lead to new contradictions: Band structure calculations for stoichiometric REH_2 and REH_3 support a metal–semiconductor transition. However, explanations on the origin and nature of the gap and on the opening of that gap at nonstoichiometric concentrations are very controversial [1.47] (Chap. 5).

Hydrides of Ce are heavy-electron-like materials. A weak $f-d$ interaction seems, at low temperatures, to cause a narrow band at the Fermi level which enhances the electronic specific heat by a factor of more than 2000 relative to that of La hydrides [1.48]. Core level spectroscopy revealed that $YbH_{2.6}$ is a mixed valence compound [1.49]. Several hydrides of rare earth metals and rare earth intermetallic compounds show narrow band phenomena.

Some analogies between hydrides and oxides allow for speculation about the existence of high T_c hydrides: Hydrogen absorption can affect the T_c of superconducting metals; some high T_c oxides absorb hydrogen and thereupon change their T_c; hydrides of alloys with inhomogeneous valence of the constituents are known and their phonon frequencies as well as electron-phonon coupling can be modified over a wide range. Let's begin the search for these high T_c hydrides!

References

1.1 Intl. Symp. on the Properties and Applications of Metal Hydrides V, Maubuisson, France, 1986, ed. by A. Percheron-Guegan, M. Gupta, J. Less-Common Met. **129, 130, 131** (1987), with references to earlier conferences of this series

1.2 Hydrogen in Metals, Proceedings of the Intl. Symp., Belfast 1985, ed. by F.A. Lewis, E. Wicke, Z. Phys. Chemie N.F. **143–147**, (1986), with references to earlier conferences of this series

1.3 *Hydrogen Energy Progress VI*, ed. by T.N. Veziroglu, N. Getoff, P. Weinzierl, Vol. 2 and 3, (Pergamon, New York 1986) with references to earlier conferences of this series

1.4 P. Jena, C.B. Satterthwaite (eds.): *Electronic Structure and Properties of Hydrogen in Metals* (Plenum Press, New York 1983)

1.5 *Hydrogen in Disordered and Amorphous Solids*, ed. by G. Bambakidis, R.C. Bowman, NATO ASI Series, Series B: Physics Vol. **136** (Plenum Press, New York 1986)

1.6 H. Buchner: *Energiespeicherung in Metallhydriden* (Springer, Wien 1982)

1.7 C.J. Winter, J. Nitsch: *Wasserstoff als Energieträger* (Springer, Berlin 1986); *DECHEMA*, *Wasserstofftechnologie*, ed. by D. Behrens, Frankfurt (1986)

1.8 D.G. Westlake, C.B. Satterthwaite, J.H. Weaver: Physics Today **31**, 32 (Nov. 1978) J.J. Reilly, G.D. Sandrock: Sci. Am. **242**, No. 2, 118 (1980)

1.9 D.O. Northwood, D.G. Ivey: Proc. Second. Int. Symp. on *Hydrogen Produced from Renewable Energy*, 1985, ed. by O.G. Hancock, K.G. Sheinkopf, p. 199, Florida Solar Energy Center;
D.G. Ivey, D.O. Northwood: Z. Physik. Chemie N.F. **147**, 191 (1986)
1.10 H. Oesterreicher: Appl. Phys. **24**, 169 (1981);
K.N. Semenenko, V.V. Burnasheva: J. Less-Common Met. **105**, 1 (1985)
1.11 Y. Fukai: Cryst. Latt. Def. and Amorph. Mater. **11**, 85 (1985)
1.12 K.H.J. Buschow, P.C.P. Bouten, A.R. Miedema: Rep. Prog. Phys. **45**, 937 (1982)
K.H.J. Buschow: "Hydrogen Absorption in Intermetallic Compounds" in *Handbook on the Phys. and Chem. of Rare Earths*, ed. by K.A. Gschneidner, Jr., L. Eyring (North Holland, Amsterdam 1984) Vol. 6, Chap. 47, 1–111
1.13 R.L. Beck: Investigation of Hydriding Characteristics of Intermetallic Compounds, DRI 2059 (University of Denver, 1962)
1.14 A. Pebler, E.A. Gulbranson: Electrochemical Technology **4**, 2011 (1966)
1.15 J.J. Reilly, R.H. Wiswall: Inorg. Chem. **7**, 2254 (1968)
1.16 J.J. Reilly, R.H. Wiswall: Inorg. Chem. **13**, 218 (1974), where reference to earlier work on FeTi–H is given
1.17 J.H.N. van Vucht, F.A. Kuijpers, H.C.A.M. Bruning: Philips Res. Rep. **25**, 133 (1970)
1.18 H.-H. Neumann: "Löslichkeit von Wasserstoff und Deuterium in $LaNi_5$"; Ph. D. thesis, Faculty of Chemistry, Technische Hochschule Darmstadt, 1969
1.19 D. Shaltiel, I. Jacob, D. Davidov: J. Less-Common Met. **53**, 117 (1977)
1.20 Y. Ishido, N. Nishimiya, A. Suzuki: Denki Kagaku **45**, 52 (1977) and Energy Developments in Japan **1**, 207 (1978)
1.21 M.P. Dariel, M.H. Mintz, Z. Hadari: J. de Phys. **40**, Suppl., C5-213 (1979)
1.22 K. Hiebl, E. Tuscher, H. Bittner: Monatshefte für Chemie **110**, 9 (1979)
1.23 A.J. Maeland: "Comparison of Hydrogen Absorption in Glassy and Crystalline Structures" in *Hydrides for Energy Storage*, ed. by A.F. Andresen, A.J. Maeland (Pergamon, Oxford 1978)
1.24 J.-J. Didisheim, K. Yvon, P. Fischer, J. Schefer, M. Gubelmann, A.F. Williams: Z. Kristallogr. **162**, 61 (1983)
1.25 J. Peisl: *Lattice Distortion, Elastic Interaction and Phase Transitions of Hydrogen in Metals*, Festkörperprobleme XXIV, 45, ed. by P. Grosse (Vieweg, Braunschweig 1984)
1.26 C.K. Hall, A.I. Shirley, P.S. Sahni: Phys. Rev. Lett. **53**, 1236 (1987);
C.K. Hall, C. Soteros, I. Macgillivray, A.I. Shirley: J. Less-Common Met. **130**, 319 (1987)
1.27 W. Frenzl, J. Peisl: Phys. Rev. Lett. **54**, 2064 (1985)
1.28 T. Ito, B.J. Beaudry, K.A. Gschneidner, T. Takeshita: Phys. Rev. B **27**, 2830 (1983)
1.29 E. Kaldis, M. Tellefsen, R. Bischof: J. Less-Common Met. **129**, 57 (1987)
1.30 O. Blaschko, G. Krexner, J.N. Daou, P. Vajda: Phys. Rev. Lett. **55**, 2876 (1985)
1.31 I.S. Anderson, J.J. Rush, T. Udovic, J.M. Rowe: Phys. Rev. Lett. **57**, 2822 (1986)
1.32 X.W. Lin, M.O. Ruault, A. Traverse, J. Chaumont, M. Salomé, H. Bernas: Phys. Rev. Lett. **56**, 1835 (1986)
1.33 B. Bogdanovic, K.-H. Claus, S. Gürtzgen, B. Spliethoff, U. Wilczok: J. Less-Common Met. **131**, 163 (1987)
1.34 H. Hemmes, A. Driessen, R. Griessen: J. Phys. C **19**, 3571 (1986)
1.35 P. Nédellec, A. Traverse, L. Dumoulin, H. Bernas, L. Amaral, G. Deutscher: Europhys. Lett. **2**, 465 (1986)
1.36 P.F. Miceli, H. Zabel, J.E. Cunningham: Phys. Rev. Lett. **54**, 917 (1985)
1.37 K. Samwer, in [1.5] p. 173
1.38 K. Chattopadyhay, K. Aoki, T. Masumoto: Scripta Met. **21**, 365 (1987)
1.39 W.A. Cassada, G.J. Shiflet, S.J. Poon: Phys. Rev. Lett. **56**, 2276 (1986)
1.40 G. Hilscher, G. Wiesinger, R. Hempelmann: J. Phys. F **11**, 2161 (1981)
1.41 Ph. Mangin, B. Rodmacq, A. Chamberod: Phys. Rev. Lett. **55**, 2899 (1985)
1.42 R. Kirchheim, F. Sommer, G. Schluckebier: Acta Metall. **30**, 1059 (1982)
R. Kirchheim: Acta Metall. **30**, 1069 (1982)
1.43 R. Griessen: Phys. Rev. B **27**, 7575 (1983)

1.44 R. Griessen et al.: to be published
1.45 A.C. Switendick: J. Less-Commun Met. **130**, 249 (1987)
1.46 M.S. Daw, M.I. Baskes: Phys. Rev. Lett. **50**, 1285 (1983)
1.47 J. Shinar, B. Dehner, B.J. Beaudry, D.T. Peterson: Phys. Rev. B (1987) in press
1.48 L. Schlapbach, H.R. Ott, E. Felder, H. Rudigier, P. Thiry, J. Bonnet, Y. Petroff, J.P. Burger: J. Less-Common Met. **130**, 239 (1986)
1.49 St. Büchler, L. Schlapbach, R. Monnier, L. Degiorgi: Proc. 14th Conf. on X-ray and Inner Shell Physics, 1987, Paris (to be published)
1.50 J.H. Harris, W.A. Curtin, M.A. Tenhover: Phys. Rev. B **36**, 5784 (1987)

2. Preparation of Intermetallics and Hydrides

Annick Percheron-Guégan and Jean-Marie Welter

With 30 Figures and 1 Table

This chapter describes the preparation and the characterization of the intermetallic compounds and their hydrides. It is not an exhaustive review, but by means of some examples, emphasis is put on the correlation between the quality of base intermetallic compounds and the properties of their hydrides.

After a recapitulation on the thermodynamic properties of the most representative families, the different preparation techniques on laboratory and industrial scale are discussed.

2.1 Introduction

The aim of this chapter is first of all to draw the attention of scientists and engineers who work in the field of hydrides to the importance of the synthesis and characterization of the intermetallic compounds in determining the properties of their hydrides. Indeed, the standard route to obtain the hydrides consists in preparing first the intermetallic compound and then in hydrogenating it. This is usually acceptable because in most intermetallic phases the metal atoms do not diffuse when hydrogen is dissolved in the lattice. Secondly, we will give a short description of the different ways of preparing hydrides.

We do not give here an exhaustive review of the preparation of all the different intermetallic compounds which are of interest since in most work dealing with the hydrides of intermetallics little is said about the preparation of the compounds and almost nothing about their characterization. So we want to focus on general ideas needed to prepare well-defined intermetallic compounds of high quality, and on the intrinsic properties of the hydrides. We will report mostly on methods used in our own laboratories. General formation methods for intermetallic synthesis have been well described by *Brown* and *Westbrook* [2.1].

As various hydrides of intermetallic compounds are also used for industrial applications, it is necessary to cover the preparation of the intermetallics both for basic research and for large-scale applications. The prevailing points of view in the two fields are different: in the first case it is essential to have as simple and well-defined materials as possible (low impurity level, usually only two elements...) whereas on an industrial level many elements have to be added to the base material to tailor it for a given application. Furthermore, batch sizes and economic considerations are drastically different.

The chapter is therefore structured as follows: First, we will review the thermodynamic properties of the most representative intermetallic compound families which form hydrides. A knowledge of these properties is needed to optimize the preparation of the compounds. For the other intermetallics a review of phase diagrams will be given in the Sect. 2.2.

Second, we will discuss the preparation techniques of intermetallics which are mainly solidification methods from a melt, the importance of several parameters (purity of the starting materials, annealing treatments...) and the characterization of the resulting materials, both on a laboratory and on industrial scale.

The hydrogenation techniques will then be treated rather briefly since they are almost identical to those used for binary hydrides [2.2]. Finally, we will illustrate with some examples the variation of the properties of the hydrides induced by the impurity content, the homogeneity of the compounds and the partial substitution of main components. We will concentrate on crystalline materials and restrict ourselves to the following four major families of intermetallic compounds and their hydrides:

Family	Representative compound	
AB_5	$LaNi_5$	
AB_2	$TiCr_2$	
AB	TiFe	(FeTi)
$AB_{0.5}$	$MgNi_{0.5}$	(Mg_2Ni)

In the short hand formula B represents in general a late transition metal of the iron group whereas A may be an alkaline earth, an early transition metal, a rare earth or an actinide element.

The formulae of minor families can be found in *Buschow* et al. [2.3] and *Ivey* et al. [2.4].

2.2 Thermodynamic Aspects of Intermetallic Phases and Their Synthesis

The question which arises first is: what is an intermetallic compound AB? It is an intermediate phase between A and B with a different structure to the parent elements and to their terminal solid solutions which is based either on symmetry or order properties. In all cases the intermediate phase exhibits a domain of homogeneity which may be of variable extent according to the relative stability of the neighbouring phases. Having two constituents increases the degrees of freedom of the material: besides the intrinsic variation in composition, within the domain of homogeneity, various structural defects such as vacancies and antiphase boundaries may be present. In comparison, elemental materials may be considered as well-defined materials. As this is no longer true for intermetallic compounds, experimental reproducibility is more difficult to achieve. Of course both kinds of systems have in common the problem of purity, but here again intermetallics are in a less favorable situation because impurities may either stabilize or destabilize the phase in a much more dramatic way. Furthermore, whereas in pure elements the surface layers are chemically identical to the bulk, a shift of composition exists between the bulk and the surface of intermetallics [2.5]. Here too, impurities may play a more important role in enhancing the segregation of one of the constituents.

As a consequence the preparation of the intermetallic compound, its chemical and structural characterization and the study of its interaction with hydrogen are very strongly correlated. For instance, a spatial fluctuation of the composition can affect the solubility of hydrogen. Indeed, in a chemically inhomogeneous binary system the chemical potential μ of e.g. element A not only varies with the local composition, but depends also on the variation of the composition [2.6]

$$\mu(c)=\mu_0(c)+K^2\nabla^2 c\,, \tag{2.1}$$

where $\mu_0(c)$ denotes the chemical potential of a uniform system with composition c, K is a constant and ∇^2 the Laplace operator. Equation (2.1) shows that fluctuations with short wavelengths affect $\mu(c)$ most. If the highly mobile hydrogen is dissolved in such a material, it will try to have the same chemical potential everywhere, as imposed e.g. by the gas phase. Because of the coupling of the chemical potentials of the hydrogen and of the metallic components, this situation can only be achieved through a local adjustment of the hydrogen concentration: the total content of hydrogen may now differ considerably from the content in a uniform system with a mean composition c.

To prepare these intermetallic compounds it is first necessary to study the phase diagram: this gives us the basic information on the homogeneity range, on the solidification reaction and on possible solid state transformations. These data are of primary importance in determining the technique of preparation, the heat treatment and the allowed impurity levels and composition shifts. For binary systems the phase boundaries are satisfactorily well known [2.7–13]. Here also, a careless preparation technique of the samples used in phase diagram investigations is often the source of controversy over topographic details [2.14].

The phase diagrams of the systems La–Ni, Mg–Ni, Ti–Cr, Ti–Fe are shown for illustrative purposes in Figs. 2.1–4. These diagrams, and more specifically the microregions around the compounds of interest, should only be considered as a general guideline because many uncertainties remain. These involve first the nature of the solidification reaction. Whereas the situation appears to be quite clear for $LaNi_5$, which solidifies congruently and for Mg_2Ni, which crystallizes through a peritectic decomposition, it was long a matter of uncertainty in the case of FeTi. For this compound both kinds of solidification reactions were quoted in the literature and only recently was the peritectic reaction clearly established [2.17]. For $TiCr_2$ one should first note the large

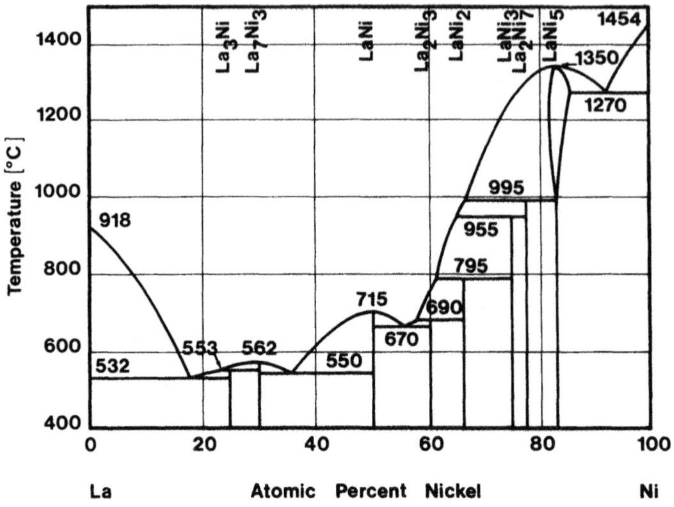

Fig. 2.1. Phase diagram of the system La–Ni [2.15, 16]

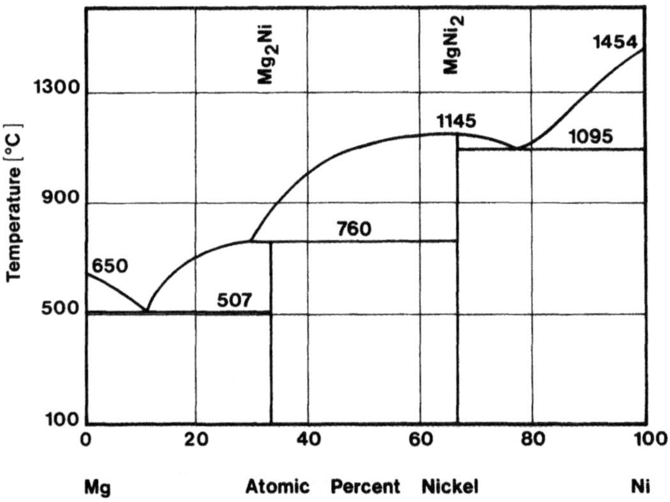

Fig. 2.2. Phase diagram of the system Mg–Ni [2.7]

solidification interval ($\Delta T \simeq 95$ K) which leads to the solid solution. The compound forms through solid state reactions which involve three polymorphic phases [2.18]. Until now no agreement exists on the exact nature of these transformations.

The second aspect is the domain of homogeneity of the compounds: $TiCr_2$ and FeTi have an extended homogeneity range [2.17, 18], whereas $LaNi_5$ and Mg_2Ni were thought to belong to the line-compound type. However, recent studies have revealed that $LaNi_5$ has a large domain of solubility at high temperatures, but the exact phase lines are still unclear [2.15].

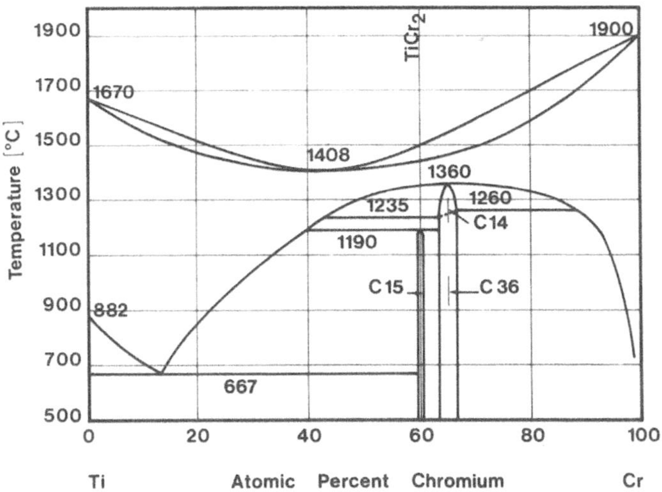

Fig. 2.3. Phase diagram of the system Ti–Cr [2.19]

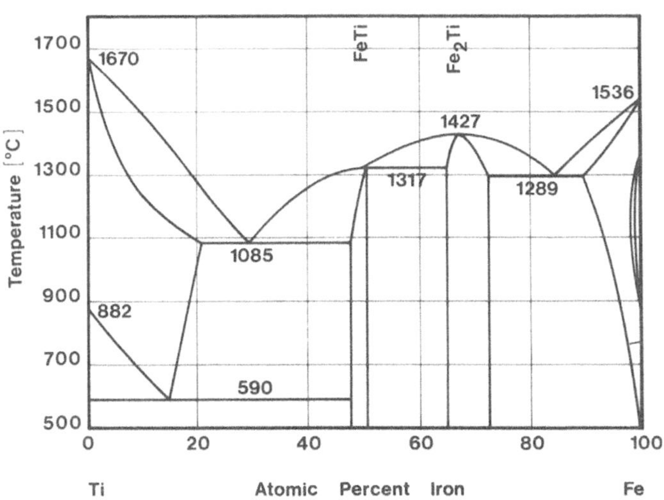

Fig. 2.4. Phase diagram of the system Ti–Fe [2.17]

A third point concerns the arrangement of the atoms in the compounds. The data on the crystal structures and lattice parameters of hydrides and some intermetallic compounds are reviewed in Chap. 4. A very complete collection of crystallographic data of intermetallic compounds was published by *Villars* and *Calvert* [2.19]. But little information exists to indicate how off-stoichiometry and thermal disordering affect the occupancy of the various sublattices of the metallic components. The degree of ordering at the stoichiometric composition has been analysed both experimentally and theoretically only in FeTi [2.20, 21]. It was found that the order parameter is one up to the melting point. In $LaNi_5$ and FeTi deviations from the stoichiometric composition seem not to create vacancies on one of the sublattices. In both cases the variation of the lattice parameter across the range of homogeneity has been determined quite accurately, but a good determination of the density is still missing. Thus the conclusions have been drawn on a comparative basis. In analogy to YCo_5 [2.22] and $SmCo_5$ [2.23] an excess of nickel atoms, is accommodated in $LaNi_5$ by the replacement of lanthanum atoms by nickel atoms which associate with a neighbouring nickel atom in the form of a dumbell. A 2:17 compound is formed locally. In FeTi the excess atom is just substituted on the deficient sublattice [2.21]. The question of vacancies and other structural defects is also a critical aspect for mass transport in the compound. A knowledge of the transport coefficient is needed to optimize the conditions for sample homogenization beyond an empirical approach.

Phase diagrams of ternary or higher intermetallic compounds are almost unknown. Therefore in such cases the adjustment of the composition to obtain pure phases is mostly a question of trial and error. A guideline was in many cases the search for pseudo-binary compounds [2.36, 37, 47]. Another problem introduced by the lack of knowledge on ternary compounds is the question of how many impurity atoms a phase can support before being destabilized, especially when the hydride is to be formed. This is especially true for the light

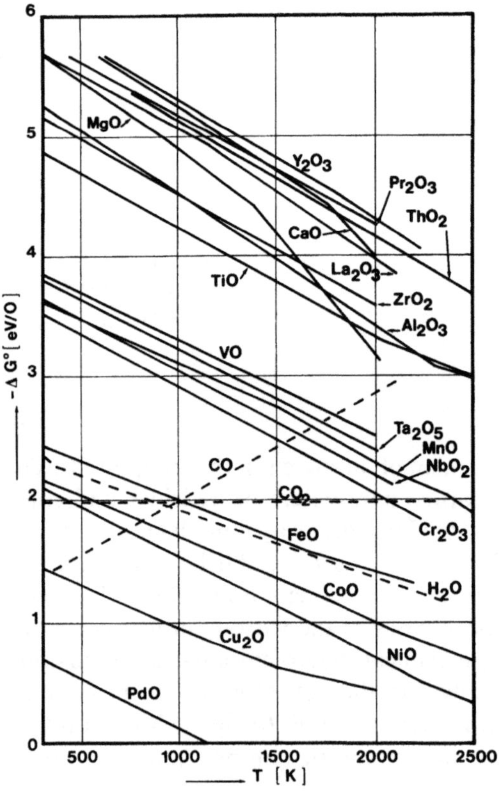

Fig. 2.5. The standard free energies of formation of the oxides (Ellingham diagram) [2.27, 28]. Conversion factors: 1 eV/O atom = 24 kcal/mole of O atoms = 100.34 kJ/mole of O atoms

elements like oxygen, nitrogen and carbon. Here most results exist in connection with the system Ti–Fe [2.24–26]; see Sect. 2.6.

The light elements C, N, O... are also the major source of contamination during the synthesis of the compounds. The standard technique consists of preparing a melt of the components which is then solidified. This is the only technique used on a large industrial scale whereas on a laboratory scale other techniques like evaporation/condensation or solid state reaction are also used. One characteristic of the AB_n compounds is that the elements A form very stable oxides, whereas the B elements have a much lower affinity for oxygen. This is clearly revealed by the Ellingham diagram shown in Fig. 2.5 [2.27, 28]. The analogy with the stability of the corresponding hydrides is striking. It is the coupling of dissimilar elements which create the major metallurgical difficulties. Whereas the metallurgy and shaping of the iron group elements are well established techniques which are not too sensitive to crucible materials and gaseous environment, the presence of the reactive element drastically complicates the situation and new melting and shaping techniques have to be developed. Eventually, one must also account for the differences in volatility of the elements involved [2.29, 30].

2.3 Preparation of Intermetallic Phases: Laboratory Scale

In this chapter we describe and discuss how to make well-defined samples which will be used for investigating the intrinsic physical and chemical properties of the intermetallic compound–hydrogen systems. Well-defined means two things here: first, a material with a low level of impurities and structural defects and with a high degree of homogeneity and stability; secondly, an indication how well this goal has been achieved.

2.3.1 The Starting Materials

As the chemical purity of the final material depends to a large extent on the impurity level of the constituents, it is worth selecting the starting materials carefully, keeping in mind the costs of the substances, the requirements of the experiment and possible alterations of the purity during further processing [2.31]. As a guideline, Fig. 2.6 gives a survey of the extent to which the elements can be purified nowadays. The aim of this chart is not to list world records in purity, but to give the order of magnitude of foreign atoms present in high-purity brands, which are commercially available in sufficient quantities at reasonable prices [2.32]. Here all the impurities have been included, the main group being the interstitial ones followed by the group of neighbouring elements. Interstitials usually have a very harmful influence, whereas the neighbouring elements affect the hydrogen related properties very little. One should be aware that many commercial data sheets only quote the detected

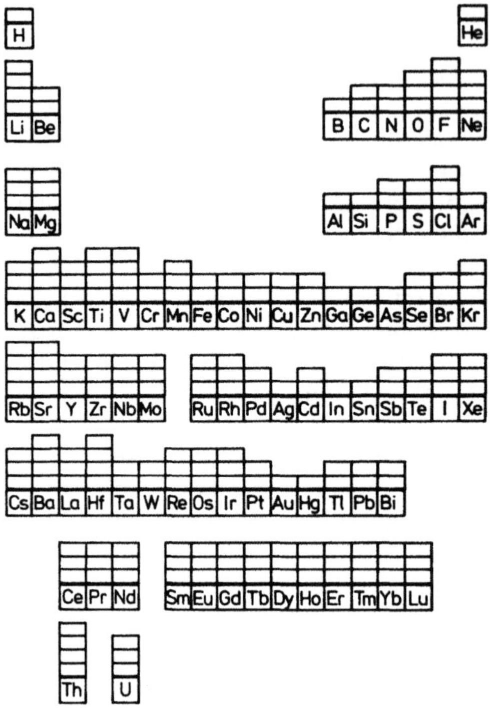

Fig. 2.6. Impurities contents in commercially available high purity brands of the elements. The scale is logarithmic: one case = 1 atomic ppm, two cases = 10 atomic ppm... [2.32]

metallic impurities. These are usually detected by optical emission spectroscopy and in a few cases by mass spectrometry. It is clear that an accurate determination of e.g. the oxygen content in a very reactive material such as a rare earth is a very arduous and expensive task because of the interference of the surface oxide layer. Here nuclear activation methods can lead to more reliable data. In the following we comment on the purity of a few selected materials. They are listed according to increasing prices, which range from a few $/kg for magnesium to more than thousand $/kg for lanthanum and chromium of very high purity.

Magnesium: the purity of selected electrolytic magnesium batches can reach 99.9%, for distilled magnesium it exceeds 99.99%; here the main impurities are Fe 30 w.ppm, Ni 5 w.ppm, Cr 5 w.ppm.

Nickel: electrolytic nickel, which is prepared in large quantities for superalloys, can be obtained with a purity better than 99.995%; in comparison to nickel pellets it has a very low carbon content – typically less than 20 w.ppm.

Iron: electrolytic material has the highest purity at better than 99.995%; whereas the level of metallic impurities is below 20 w.ppm, larger amounts of light elements like O, C, Cl are present up to 200 w.ppm.

Titanium: electrorefined titanium crystals have a purity of 99.9%; the main metallic impurity is usually Al at 100 w.ppm, but O and N can easily exceed 500 w.ppm.

Manganese: electrolytic flakes can be obtained with purity 99.99%; the main impurity is magnesium with 50–100 w.ppm; sometimes calcium and silicon may be present up to 10 w.ppm; the typical oxygen level is around 50 w.ppm.
Zirconium: crystal bars obtained through the decomposition of the iodide have a purity of 99.95%; the main metallic impurities are Hf, 100 w.ppm and Fe, 100 w.ppm and the light elements are usually present at a level below 100 w.ppm.
Lanthanum: calciothermic reduced lanthanum fluoride can be obtained with less than 50 w.ppm of rare earth and common metal impurities; the oxygen level usually exceeds a few 100 w.ppm.
Mischmetal: is an unrefined alloy of a number of rare earths elements, typically: 50% Ce, 32–34% La, 13–14% Nd, 4–5% Pr and about 1.5% other rare earths; this composition changes slightly according to the origin of the ones [2.33].
Chromium: electrolytic chromium has less than 50 w.ppm of metallic impurities, but an oxygen level as high as 3000 w.ppm; it can be reduced by distillation or through an iodide decomposition process, but even here 100 w.ppm of light elements are still present.

Batches of less pure materials (with the possible exception of lanthanum and chromium) are only slightly cheaper. On the other hand, an increase of the purity can only be achieved at very high costs. Therefore a post-purification treatment to remove primarily the interstitial elements should only be undertaken if one makes sure that all sources of contamination have been removed from the following preparation steps. The most effective technique to remove is a heat treatment under ultra-high vacuum [2.34] (which is usually not available in industrial companies). Either oxygen leaves the metal as a volatile oxide like in the IV B, V B elements and non-volatile rare earths [2.35], or the metal is distilled and leaves the oxide as a residue as in Mg, Mn, and the volatile rare earths [2.35]. The method of solid state electrotransport has produced rare earth metals of higher purity than any other methods [2.35]. Carbon and oxygen can be removed from the iron group elements by a fusion treatment in wet and dry hydrogen.

Once the interstitials have been removed the reactive elements must be handled carefully: this requires minimal exposure to air and no contact with humidity and fingers. If extreme demands exist, the use of glove-boxes with high purity inert gases is necessary.

2.3.2 Melting Techniques

Having obtained the pure components, the question arises of how to mix them thoroughly to get a homogeneous material. Due to the kinetic limitations of solid state reactions, the components first have to be dissolved in a liquid phase. In this section we will concentrate mainly on melts, because it is the usual procedure and has the advantage of giving a high yield of material. The subsequent solidification process is at least as important as appropriate melting, so we will also make a few remarks about this problem.

The melt is prepared by heating the constituent elements in a suitable container at least up to the melting temperature given by the phase diagram.

The following melting techniques are frequently used to prepare the alloys:

a) RF levitation melting in a cold crucible: best and very versatile method to alloy reactive metals in high vacuum or inert gas atmosphere in quantities of 1–100 g. No temperature gradient, rapid cooling possible.

b) Arc melting in a water cooled copper hearth in an argon atmosphere in quantities of 1–100 g. Strong temperature gradients across the sample which may induce inhomogeneities; inert gas concentration of the solidified sample might be high.

c) Resistance or RF induction heating in crucibles in vacuum, inert gas atmosphere or in air under a protective layer to prevent strong oxidation (quantities 10 g–100 kg). Refractory metals as well as oxide and non-oxide ceramics are the most suitable materials for crucibles. Unfortunately the presence of the reactive element in the liquid state usually leads to the corrosion of crucibles made from ceramics and hence to a contamination of the melt. The inertness of refractory metals like tantalum, molybdenum, and tungsten for molten rare earth elements no longer holds when the iron group elements are introduced in the melt.

Magnesium however is an exception. Because of the low melting temperature (923 K) and protective oxide layer, this element can be melted in almost any crucible and the final choice will depend mainly on the nature of the second element. In many cases a very convenient crucible material appears to be iron or stainless steel. One advantage is that the lid can be welded on the crucible either in vacuum or in an inert gas atmosphere. The reason for this precaution is that magnesium has a high vapour pressure: it is approximately 0.4 kPa at the melting point and the boiling point is reached at 1391 K. Many heating techniques have been considered for this kind of crucible: they include the use of standard tubular furnaces, RF-generators or by passing the current directly through the crucible. In all cases the crucible should be rocked to ensure a good homogeneity. If extreme metallic purity is not essential, iron crucibles can also be used for Mg_2Ni, Mg_2Cu, and $MgLa_{0.12}$.

In the next section we show a metallographic view of a $Mg_{0.68}Ni_{0.31}La_{0.01}$ sample prepared by such a technique (Fig. 2.14). A very complete study of the effect of crucible on the melt contamination during vacuum induction melting of FeTi has been made by *Sandrock* et al. [2.24] showing the oxygen and metal contamination for three common materials. The results are summarized in Table 2.1.

Upon the melting of "Mischmetal-calcium-nickel" alloys of the $LaNi_5$ type in crucibles of Al_2O_3, *Sandrock* [2.37] noticed a rather low oxygen solubility of the melt as long as the temperature did not exceed the melting temperature by more

Table 2.1. Effect of crucible on melt contamination during vacuum induction melting of FeTi [2.24]

Crucible material	O Contamination wt. %	*M* Contamination wt. %
MgO	0.072	<0.001 Mg
Al_2O_3	1.14	1.23 Al
ZrO_2	0.30	0.45 Zr

than about 100 °C. Best results were obtained by adding solid Mischmetal and calcium to the Ni melt immediately prior to pouring. For the rare earths and titanium-based compounds cold crucible techniques have proved to be most convenient, also here the size of the batch is limited to a few cm^3. The cheapest technique is arc melting. It has furthermore the advantage that the melt and hence the solidified ingot can be somewhat shaped. Its disadvantage is that inhomogeneities due to segregation and foreign phases frequently occur. A better – but also much more expensive technique – includes RF-heating because here levitation effects can be exploited. Full levitation crucibles of the basket type are ideally suited to prepare rapidly solidified samples e.g. in a piston and anvil type system. Figure 2.7 shows a picture of a machined levitation crucible made from

Fig. 2.7. Water cooled copper cold crucible for RF induced heating and levitation with titanium disc getter [2.38]

copper [2.38]. The stems are watercooled. The copper contamination is typically less than 50 ppm even for high melting temperature metals like niobium. Furthermore the melt can be easily stirred either through mechanical agitation of the crucible or through electromagnetic forces. Finally a very large flexibility exists with regard to the gaseous atmosphere. This flexibility is lost again when the melting is performed by bombardment with high energy electrons. When inert gas is used – e.g. for arc melting or when volatile elements are heated – high purities are imperative. Therefore it is useful to introduce into the reaction chamber a getter material like titanium or zirconium which first has to be activated. Figure 2.7 clearly shows the bright shining titanium disc above the melt.

An important point is to ensure that the melt becomes homogeneous. Difficulties arise when a solid crust remains in the cold part of the crucible or when one component is trapped within its surface oxide layer. This occurs frequently when rare earths are used in a poor atmosphere. Eventually the rate of dissolution may be slow with small gains in mixing free energy [2.11]. An excessive increase of the temperature is not necessarily the solution, because the rate of parasitic reactions such as contamination and evaporation also increases. Some help can be obtained from mechanically and electromagnetically induced forced convection. On the other hand, an excessive gain in mixing enthalpy can lead to an explosive reaction where some material is spilled out of the crucible.

Once a homogeneous melt has been obtained, it must be solidified without segregation. This problem has been discussed in various recent reviews [2.39, 40]. Therefore we will limit ourselves here to a few remarks. Segregation-free freezing can only occur if the liquidus curve of the melt and solidus curve of the intermetallic compound have a point of congruency. At this particular concentration the equilibrium partition coefficient k, which expresses the ratio of the solubilities c_s and c_l of e.g. component B at a given temperature in the solid and liquid phases, respectively, is unity. In reality, even for congruent compounds, this ideal condition is never met. Minute shifts in composition or the addition of dopants and impurities lead to a spatially varying redistribution of the components during solidification. External parameters which influence this redistribution are the heat flow on both sides of the liquid-solid interface and the mass flow in the melt. The heat flow determines the rate at which the latent heat of solidification is removed and hence the speed of freezing v. Furthermore it governs to a large extent the stability of the interface. If the latent heat is evacuated through the undercooled melt the growth front of a solid nucleus becomes unstable and radially propagating dendrites develop. The resulting solid morphology consists of strongly inhomogeneous equiaxed grains. If the melt is overheated and the latent heat is extracted through the solid, the interface remains stable and microsegregation is avoided. This holds as long as the build up of the rejected component at the solidification front remains small. Otherwise a thermodynamically unstable layer develops ahead of the interface which decomposes tangentially into microregions of different

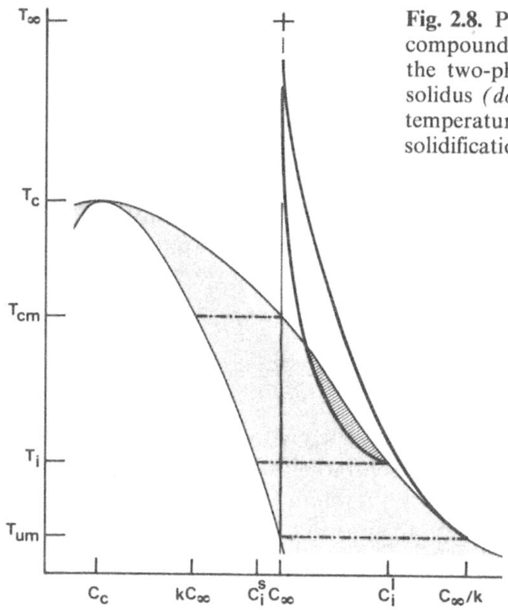

Fig. 2.8. Partial phase diagram of an intermetallic compound solidified congruently at (T_c, C_c) showing the two-phase gap between the liquidus and the solidus *(dotted region)*. The heavy lines represent temperature concentration trajectories ahead of the solidification front

composition. A criterion which describes roughly the onset of this instability can be derived with the help of Fig. 2.8 for a binary system. The evolution of the temperature and of the solute concentration along the ordinate z ahead of the solidification front can be described by the two functions $T(z)$ and $c(z)$, respectively. Combined they give the trajectory $T(c)$ in the phase diagram: it starts at (T_∞, c_∞) far ahead of the solidification front and ends at (T_i, c_i^l) on the melt side of the interface. If the trajectory crosses the two-phase field an instability occurs. To avoid this, the following condition must be fulfilled:

$$\left.\frac{\operatorname{grad} T}{\operatorname{grad} c^l}\right|_{c_i^l} > m_i\,, \tag{2.2}$$

where m_i is the slope of the liquidus at c_i^l. The location of c_i^l and the strength of $\operatorname{grad} c_i^l$ depend on the importance of the buildup of the solute. Two extreme cases can be considered. If the mass flow in the melt is very rapid because of convective transport, almost no solute can accumulate in front of the moving interface. Hence $c_i^l \simeq c_\infty$ and $\operatorname{grad} c_i^l \simeq 0$: The condition expressed by (2.2) is always fulfilled. On the other hand, maximal pile-up occurs for a regime of pure diffusive mass transport. Under stationary conditions $c_i^l \simeq c_\infty/k$ and $\operatorname{grad} c_i^l \simeq -v(c_i^l - c_\infty)/D = -vc_\infty(1-k)/kD$. D is the diffusion coefficient and is usually of the order of 10^{-4} cm^2/s. As an example we can ask at which maximum speed v, a $TiCr_2$ melt can crystallize in order to give a microsegregation-free solid assuming a temperature gradient of 100 K/cm. With $m_i = 750$ K/unit concentration and $c_\infty - c_i^l \simeq 0.1$, one obtains $v < 1.4 \times 10^{-4}$ cm/s $\simeq 0.5$ cm/hour. Although this condition would be strongly relaxed in reality because pure diffusive mass transport is almost never achieved, the example shows the advantage of low freezing speeds to avoid microsegregation. Of course v should not be so low as to enhance macrosegregation. This takes place in the melt layers which solidify first and last and have width of the order of D/vk. At very high freezing speeds ($v > 10$ cm/s) the melt is quenched without spatial concentration shifts of the components, but with these ultrarapid solidification techniques only minute amounts of material can be prepared.

2.3.3 Annealing Treatments

Although the compounds which melt congruently can be obtained directly from the liquid, an annealing treatment is useful to eliminate microsegregation by solid state diffusion or to allow the system to establish a strictly homogeneous composition and full structural order.

For a peritectic formation compound, annealing is in all cases required. The annealing temperature should be below the peritectic temperature to dissolve precipitates of neighbouring phases. This aspect was discussed by *Buschow* [2.16] for the La–Ni system. Since the reaction rates are diffusion controlled, high temperatures are preferred to shorten the annealing time.

The problem during annealing is to avoid excessive oxidation of the material and evaporation of the more volatile compound (see Fig. 2.29). Depending on the nature of the compound it is useful to anneal the sample either in vacuum or in a high-purity inert gas atmosphere. In both cases a good technique is to wrap the sample in an envelope of a getter material with a low vapour pressure like zirconium or tantalum. The effect of the annealing on the hydrogen absorption properties will be discussed in the Sect. 2.6. An example of the influence of the conditions of annealing, time and temperature, is given in Fig. 2.9 for $LaNi_{5-x}Mn_x$. The composition of the samples is analyzed by electron microprobe. The 1150 °C annealing leads to a large dispersion of the composition with some microprecipitation of secondary phase, whereas an annealing at 900 °C for 36 to 72 hours allows a satisfactory homogeneity of the strictly single phase to be reached. The effect of homogenization on the shape of the isotherms can be seen on Fig. 2.10: the inhomogeneous compound annealed for 3 hours at only 1150 °C exhibits very sloped plateaus with large hysteresis. Another example is given by *Sandrock* [2.42] for the Mm $Ni_{4.5}Al_{0.5}$.

In the case of pseudobinary compounds such as the $LaNi_{5-x}M_x$ series, the limit of substitution varies according to the substitutional element: X(Ga, In, Sn) = 0.4 [2.43]; X(Si) = 0.6 [2.44]; X(Fe) = 1.2 [2.45, 46]; X(Al) = 1.3 [2.47]; X(Al) = 1.5 [2.48]; X(Mn) = 2.2 [2.49]; X(Mn) = 0.99 [2.50]; X(Cu) = 5 [2.51, 52]; X(Co) = 5 [2.53]; X(Pt) = 5 [2.54]. These limits depend strongly on the heat treatment (number of fusions temperature and duration of the annealing of the samples) which has to be fitted to each family of compounds. For example this

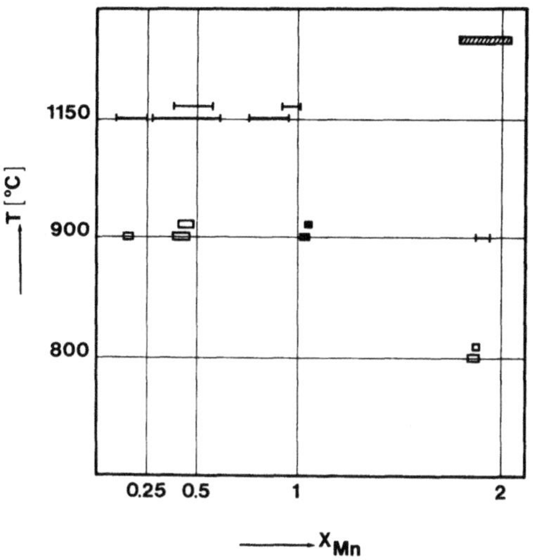

Fig. 2.9. Chemical homogeneity of $LaNi_{5-x}Mn_x$ alloys after casting [▨] and annealing treatments at three temperatures for 3 (▭) and 72 [■] hours. The bars represent the compositional fluctuations as observed by electron microprobe. For samples of very similar compositions the bars on the temperature scale are slightly shifted for clarity [2.41]

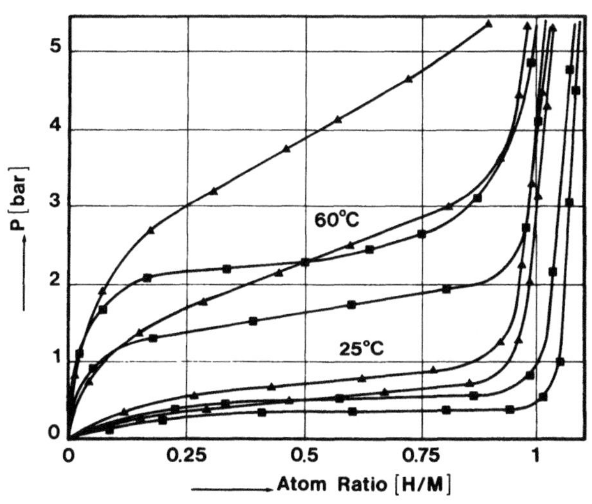

Fig. 2.10. Isotherms of $La_{1.01} \cdot Ni_{4.83}Mn_{0.17}$ (▲) annealed for 3 hours at 1150 °C and of $La_{0.99}Ni_{4.74}Mn_{0.27}$ (■) annealed for 84 hours at 900°C [2.41]

limit can be increased either by lowering the annealing temperature as for the $LaNi_{5-x}Mn_x$ system. In this case annealing at 800 °C leads to X(Mn) = 2 as the limit of substitution, whereas for X(Mn) = 0.99 with 1175 °C annealing a two phase compound was observed [2.50].

For $LaNi_{5-x}Fe_x$ only a high temperature annealing (1090 °C) for three days enables a single phase to be obtained for X(Fe) = 1. An example of the disappearance of microprecipitations is given in Fig. 2.11. This effect can explain the different hydrogen contents given in the literature: 5.7 H/mole [2.56, 45], 4.5 H/mole [2.57] at 40 °C under 10 atm; and 4 H/mole [2.42] at 25 °C under 10 atm for an as-cast compound.

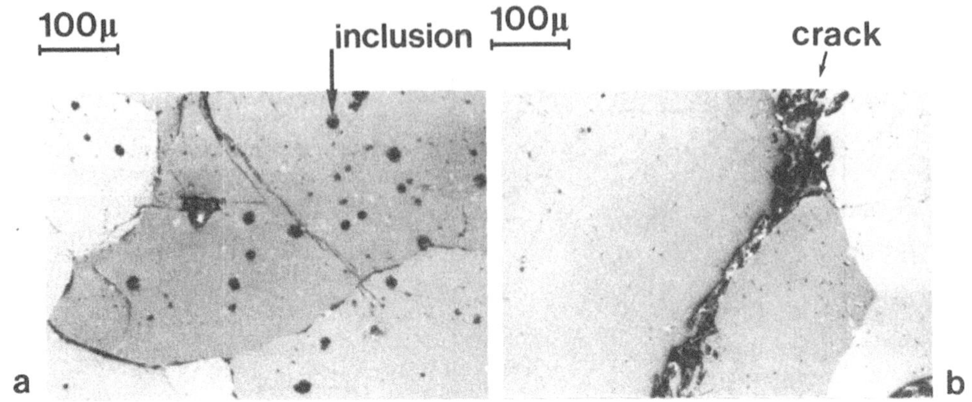

Fig. 2.11a, b. Micrographs showing the disappearance of $Fe_{1.5}Ni$ inclusions present in as-cast $LaNi_4Fe$ compound (**a**) and after 3 days annealing at 1070 °C (**b**) [2.55]

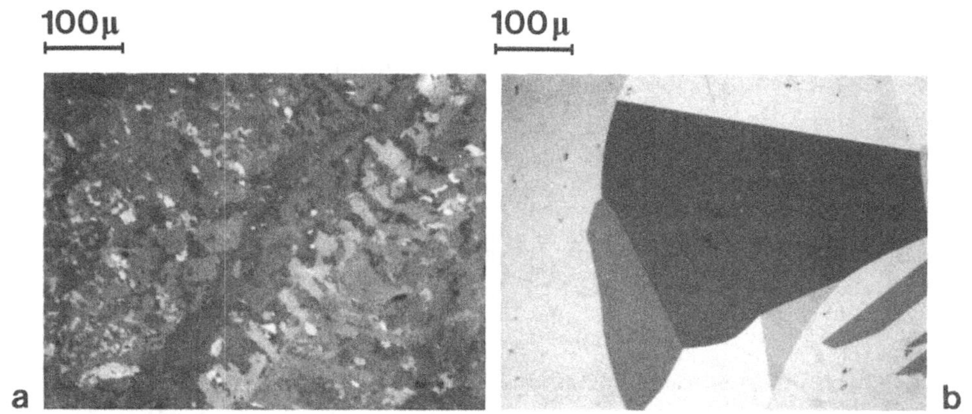

Fig. 2.12a, b. Comparison of multiphase annealed (2 months at 900 °C) $LaCu_5$ compounds (**a**) and monophase $LaCu_5$ compound obtained by partial unidirectional solidification (**b**) [2.55]

In some cases a long annealing is not sufficient to get a single phase sample. For example $LaCu_5$, a compound which decomposes peritectically at 805 °C [2.58] did not become a single phase even after 2 months' annealing at 900 °C as we show in Fig. 2.12a. The electron microprobe analysis reveals three phases: $LaCu_{5.9}$, $LaCu_{4.9}$ and some inclusion of almost pure copper ($La_{0.05}Cu_{0.95}$). Only a partial unidirectional solidification allows one to obtain a strictly single-phase compound (Fig. 2.12b) [2.55].

Such annealing treatments are also useful to remove the structural defects introduced during crushing and grinding of the materials. Indeed, it has been shown that cold worked and annealed material react differently with hydrogen [2.59].

2.3.4 Single Crystal Growth Techniques

For fundamental research the best defined materials are single crystals. In the case of many binary hydrides, like those of vanadium, niobium and tantalum, single crystals have led to many interesting results, especially when the dynamics of the base metal and of the hydrogen were studied. Unfortunately such samples cannot be prepared in general for the ternary hydrides. The large increase of the cell volume upon hydriding and the brittleness of the compound create a decrepitation of the sample. Nevertheless single crystals are useful to study the properties of the solid solution with hydrogen or just to make a starting material of high quality.

Single crystals are usually prepared from the melt. Therefore many aspects discussed in the previous section must also be considered here, e.g. contamination due to crucible corrosion or interaction with the reactive impurities of the atmosphere will also play an important role. This is an even more serious problem during crystal growth because of the long time that the material is kept in the molten state or at high temperature. Indeed, to avoid segregation, the solidification speeds are of the order of a few mm/h. The optimal situation is to give up crucibles completely and to work in an atmosphere where the partial pressures of the reactive gas molecules are less than 10^{-6} Pa. This means that floating zone techniques in ultra high vacuum should be used. A difficulty is the preparation of the cylindrical feed rods. Because intermetallic compounds are very brittle, it is not possible to swage them. A good method consists in casting the rods into a sucking mold attached to an arc melting unit. These techniques were successfully applied to prepare single crystals of $LaNi_5$ and FeTi.

Because of the peritectic solidification reaction of FeTi, a slight off-stoichiometry with an excess of titanium had to be accepted. The components were first melted by RF-induction heating in a water-cooled horizontal boat crucible. Homogenization was achieved by levitation melting. The further treatment of the homogenized knob depended on the specimen requirements. If powder was needed, the knob was crushed in a mortar. This procedure improved, if the material was, at least partially, hydrided. Because of the hardness of iron-titanium, the best technique to obtain slabs and discs was spark cutting. The minimal thickness which could be obtained was approximately 0.5 mm. The damaged and contaminated surfaces were usually cleaned with a mixture of $HF + HNO_3 + H_2O$. Single crystals were grown in a vacuum which had to be better than 10^{-5} Pa. Otherwise a bluish layer immediately covered the crystal.

$LaNi_5$ single crystals can be obtained by pulling (5 mm/hour) in ultra high vacuum (10^{-7} Pa) by the Czochralsky technique (MgO–Al_2O_3 crucible) from the melt of polycrystalline sample prepared by induction melting and subsequent quenching in a water cooled boat at 2×10^{-4} Pa. The starting stoichiometry is $LaNi_{4.97}$ to avoid Ni precipitations [2.61].

Single crystals are usually shaped by acid sawing or by spark erosion. Here also, the high reactivity of the compounds with hydrogen makes it mandatory to minimize the presence of hydrogen. Therefore the best technique is spark erosion cutting at low speed.

2.3.5 Noncrystalline Materials: Rapid Solidification and Solid-State Reactions

Maeland [2.62] has shown that under similar conditions amorphous or glassy alloys can have larger hydrogen absorption capacities than their corresponding crystalline alloys, e.g. in the case of TiCu and Ti_2Cu. Furthermore often they do not disintegrate into powder. Hydrides of amorphous alloys are also amorphous if upon hydride formation the recrystallization temperature is not exceeded. Since these first results the formation and properties of amorphous metal hydrides have become a quite active research area. Among the different alloy systems studied we can quote Zr–Cu, Zr–Ni, Zr–Fe, Zr–Pd, and Zr–Rh. For reviews we refer to [2.63–67].

Amorphous alloys are usually prepared by rapid quenching of the melt by the melt spinning method or by the splat cooling technique [2.68]. These ultrafast cooling methods allow the preparation of many amorphous alloys which are normally crystalline, but restrict their form to thin ribbons or sheets. On the other hand, the formation of amorphous alloys by solid-state reactions [2.67], e.g. diffusion in layered structures, does not in principle limit the shape or size of the samples. It is particularly interesting to note that hydrogen absorption is also a way to form amorphous alloys and hydrides, e.g. in the Zr–Rh system [2.67].

2.3.6 Synthesis from the Gas Phase and Thin Films

To avoid the problems due to the peritectic decomposition of the melt it may sometimes be useful to use an evaporation-condensation technique in a closed crucible. This is only possible if the components have a sufficiently high vapour pressure at temperatures below the peritectic temperature and if no reaction occurs with the crucible material.

This case can be illustrated by the preparation of Mg_2Ni: in a recent publication *Post* et al. [2.69] described the preparation of single phase Mg_2Ni, by vapour exchange and diffusion reaction of elements in the absence of a liquid phase, in a sealed stainless steel container under high purity argon. In previous studies samples of Mg_2Ni prepared by induction melting of the elements under argon have been reported for multiphases including usually Mg, MgO, and Ni or $MgNi_2$, often in substantial amounts due to peritectic decomposition and atmosphere contamination.

Thin films are extremely well suited for electron microscope and resistivity studies. Good results have been obtained in the case of binary hydrides. They would even be more suited in the case of ternary hydrides because on one hand it is very difficult to prepare thin samples from a massive ingot and on the other hand decrepitation problems could in principle be avoided. Little success has been obtained until now, however, *Adachi* et al. [2.70] succeeded in preparing thin films of $LaNi_5$ by conventional vacuum deposition: by heating $LaNi_5$ powder at 2000 °C on a tungsten filament at 10^{-2} Pa. Under these conditions both the two components evaporated immediately and were deposited on quartz plate. The thickness was about 3000 Å, the characterization confirmed the composition $LaNi_5$ with presence of lanthanum oxide due to superficial oxygen contamination. *Livesay* et al. [2.71] also prepared AB_5 alloys in thin films with thickness less than about one micron, which remained coherent even with repeated hydrogenation cycles.

The RF sputtering technique has been widely used to produce amorphous materials; some examples are given for FeTi compounds [2.72, 73].

2.4 Characterization of Intermetallic Phases

It is obvious that for a good chemical and structural characterization of the material all the classical tools can be used. They include characterization by analytical chemistry, diffraction techniques, the measurements of physical properties and of the pressure-composition-temperature (pcT) curves in this peculiar case.

Usually the nature of the main phase is obtained by x-ray powder diffraction, which also enables one to check for the presence of minor phases, but only at a level greater than 5%. More detailed information is obtained by metallographic examination either by optical or electron microscopy. Here, not only minor phases at a level below 1% are seen, but also compositional shifts due to segregation during solidification are revealed. Furthermore the electron microscope can be equipped with an x-ray fluorescence analyser to give a quantitative chemical analysis of the phases present. Electron microprobe analysis can achieve the same results. Nevertheless one should be aware that the level of detection depends on the element, but it is usually around 1% and that the analytical information is averaged over a volume a few μm^3.

For the metallographic preparation of the samples it is recommended to use organic solvents. Otherwise, depending on the reactivity of the material, artefacts may be induced by hydrogenation or oxidation due to the decomposition of water. In some cases it is better to completely avoid chemical etching and use only mechanical polishing (diamond paste under organic lubricant) to prepare the surface for further examination.

If the material is a non-cubic structure, the use of polarized light for optical microscopy examination is essential. An example is given in Fig. 2.13a which shows the grain morphology of a $LaNi_5$ sample prepared on an industrial scale by Japan Metals and Chemicals. In contrast to this, the grain structure is not revealed when the same sample is examined with unpolarized light (Fig. 2.13b).

A second example is a Mg_2Ni type compound ($Mg_{0.68}Ni_{0.31}La_{0.01}$). The sample shown in Fig. 2.14 is representative of a 0.5 kg batch melted in an evacuated steel crucible and cooled down slowly (50 K/hour). The very different nature of magnesium and lanthanum did not allow us to contrast the phases by

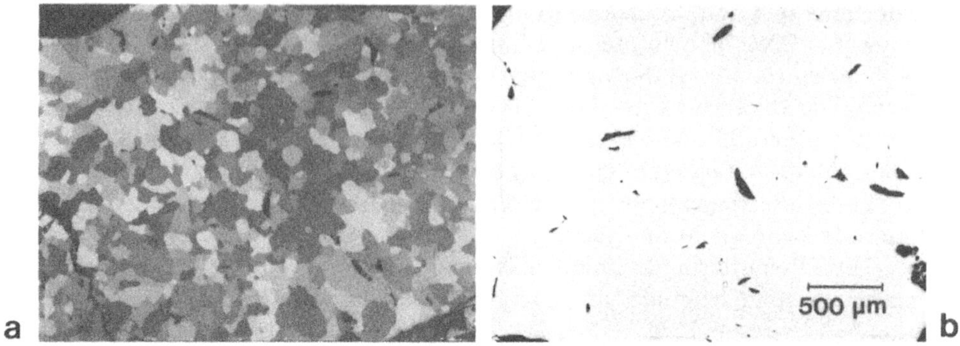

Fig. 2.13a, b. Micrographs of $LaNi_5$ examined by using polarized light (100 ×) (**a**) and unpolarized light (**b**) [2.74]

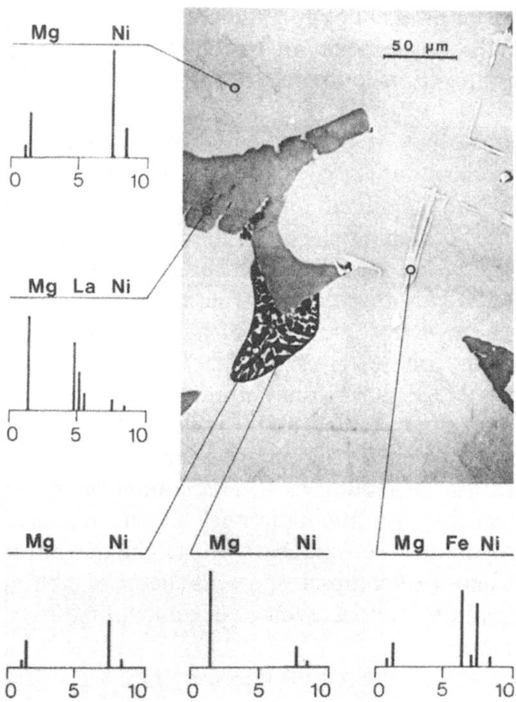

Fig. 2.14. Scanning electron micrograph of a slowly solidified partially substituted Mg_2Ni ($Mg_{0.68} \cdot Ni_{0.31}La_{0.01}$) with energy dispersive x-ray spectra of the different phases [2.75]

chemical attack. Therefore the polished surface was examined with secondary electrons in a scanning electron microscope. This also had the advantage that the composition of the different phases could be determined at least on a relative scale by x-ray analysis. The micrograph of Fig. 2.14 reveals that besides the major phase Mg_2Ni and the minor phase $Mg_{17}La_2$ there is also some magnesium forming an eutectic with Mg_2Ni and various (Fe, Ni) inclusions resulting from a reaction with the crucible material. Neither Mg nor the inclusions were revealed by x-ray powder diffraction. Special care was taken to eliminate the fluorescence radiation of nickel and to improve the peak to background ratio by using a graphite monochromator in front of the detector. Therefore the limit of detection can be set as low as 5%. It is good to point out in the present context that many JCPDS-cards need revision. For example, the card 1-1268 for Mg_2Ni (up to 1982) still contains one magnesium line but many important reflections of the compound are missing. This, of course, is again the result of poor sample preparation. This point has also been noted by *Post* et al. [2.69] (see Sect. 2.3.6) who have recently presented a refined set of x-ray data.

For the compounds which exhibit a fairly wide homogeneity range such as $LaNi_{5\pm0.3}$ metallographic examination and microprobe analysis are not able to give accurate information about the stoichiometry. Chemical analysis and measurements of the lattice parameters and of the magnetic susceptibility can give good results [2.15, 76, 79].

Since the pcT curves are also very sensitive to the variation of the composition [2.15] and homogeneity of the intermetallic compounds, it is a good way to assess the quality of the materials. This is particularly true to check the stoichiometry deviation of $LaNi_5$ [2.15] or the rate of substitution of one component [2.55]. This was already mentioned in Sect. 2.3.3 and will be emphasized again in Sect. 2.6 where we will report the correlations between the properties of hydrides and the microstructure of intermetallic compounds.

The structural quality of single crystals can be derived from the widths of the rocking curves measured with various types of monochromatic radiation like neutrons, x- and γ-rays. The curves are obtained by rocking the crystal through

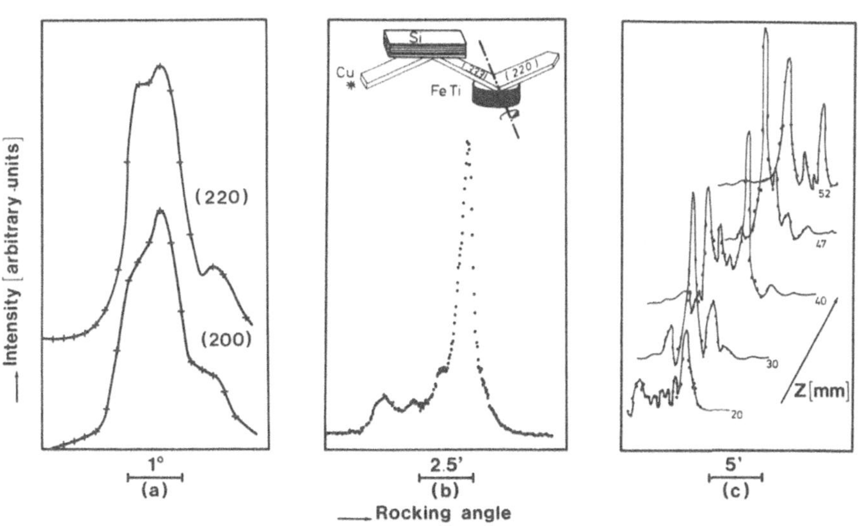

Fig. 2.15a–c. Rocking curves of a FeTi single crystal obtained with neutrons (**a**), x-rays (**b**), and γ-rays (**c**) [2.60]

the beam in such a way that the different subgrains successively fulfill the Bragg condition. Figure 2.15 shows the rocking curves obtained with the three probes (neutrons: $\lambda = 2.4$ Å; x-ray: $\lambda = 1.54$ Å; γ-ray: $\lambda = 0.03$ Å) from the same FeTi crystal. It is clear that each radiation gives a different information depending on the beam dimension, angular spread and on the absorption and extinction lengths.

For the analysis of thin films and their surfaces high and low energy electron beams are very useful. The structure and the composition of thin films are best checked in an analytical transmission electron microscope [2.77, 78] whereas Auger spectroscopy is very sensitive to surface contamination and segregation. For this last case magnetic measurement has emerged as a very sensitive physical technique when a ferromagnetic component is involved in a weak paramagnetic compound [2.79].

2.5 Preparation of Intermetallic Phases: Industrial Scale

Although industrial applications of reversible metallic hydrides have not blossomed as was expected in the seventies, there exists nevertheless a limited market for which approximately 50 tons of material were prepared during the last few years. Such amounts could no longer be produced on a laboratory scale. Therefore various metallurgical companies started the fabrication of intermetallic compounds with batch sizes of approximately 100 kg.

It is very difficult to get detailed information on the state of the art. It is obvious that industry is less and less willing to describe how the fine tuning of the melting, casting and shaping techniques is performed. Nevertheless some general information could be obtained from various companies.

Historically the FeTi-based hydrides were the first ones to be prepared on a large scale. Although they no longer represent the prototype system for automotive applications, they are still of interest for large-scale, stationary hydrogen storage units. Furthermore it is interesting for comparison with the laboratory techniques described previously to show how industrial size batches are made.

For example batches up to 50 kg of manganese-alloyed FeTi are prepared at the Nürnberg plant (FRG) of the Gesellschaft für Elektrometallurgie [2.80]. As starting materials, vacuum remelted titanium sponge (99.5%) and electrolytic manganese (99%) are used. For iron a source with low carbon and oxygen levels is selected: it is either sponge which has been treated by hydrogen (e.g. by the Höganäs process), or cathodes. The compound is prepared by a standard vacuum (medium frequency) induction melting (VIM) process and is obtained within one day. To minimize manganese losses during melting and casting the vacuum unit is backfilled with argon up to approximately 0.5 bar. Ceramic crucibles and steel molds are used. The ingot usually breaks apart and the pieces are crushed and ground to granules with a diameter less than 3 mm in an inert gas atmosphere to avoid excessive oxidation.

The granules are packed in an argon atmosphere into plastic bags which are hermetically sealed. All these precautions are necessary to obtain a material with a low oxygen content. Because of the careful selection of the starting materials and of the VIM procedure no deoxidation substances are used. A tight control of the melting sequence and duration leads to a high yield of usable batches of 80%. Failures are mainly due to an oxygen contamination during the process and in most cases its source remains unclear. The chemical composition and purity are checked for each batch: metallic elements are analyzed by x-ray fluorescence, gaseous impurities by inert gas extraction. Occasionally a metallographic examination is made. The hydrogen absorption properties are determined for each batch by measuring the pressure-composition isotherm at room temperature. If titanium with a higher oxygen content is used, it is necessary to deoxidize the melt by the addition of mischmetal. This technique was developed at Ergenics (USA) [2.81] and has the advantage of giving a compound with a good activation behaviour although no manganese is added to the material. FeTi with niobium and zirconium additives ($Fe_{0.94}Ti_{0.96}Zr_{0.04}Nb_{0.04}$) is also prepared by Japan Metals & Chemicals on a medium size batch scale of approximately 10 kg [2.82]. Small blocks of iron, titanium, zirconium, and niobium are cleaned with compressed air and melted together in an arc or induction furnace. The working atmosphere is argon. Crucibles are made from carbon. Trouble with the crucible is the major reason for failures. After keeping the alloy for 20 minutes in the molten state it is cast to obtain cylindrical blocks. Heat treatments, crushing and grinding are performed as required by the customer and the final material is sealed in plastic bags. The quality control of the starting materials is based on the certificates of the producers, but for each batch the chemical purity and composition are controlled and the pcT curve is measured.

For the more recently developed multicomponent $TiMn_2$-based compounds, low oxygen levels are difficult to achieve in a one stage VIM process, unless excessive amounts of deoxidation substances are used. Therefore a two stage process was developed by the Gesellschaft für Elektrometallurgie (FRG) [2.83]. To produce a compound e.g. of the type $(TiZr)(MnVFeCr)_2$ a fraction of the manganese is first melted with ferrovanadium to make an oxygen poor MnVFe alloy. The ingot is crushed and then remelted together with the rest of the manganese and with titanium and zirconium. Small amounts of cerium-mischmetal are also added for deoxidation purposes. With this technique a yield of 90% is obtained even for batches with sizes up to 250 kg. When a single stage procedure is used, the yield does not exceed 50%. Absorption isotherms and a micrograph of a grain are shown in Figs. 2.16 and 2.17.

Rare earth (R) compounds of the type $RNi_{5-x}M_x$ with M being aluminium, iron or manganese, are prepared e.g. by Santoku Metal Industries (Japan) [2.85] with heat sizes of 100 kg. The purity of the rare earths is typically 99%. Melting is performed in an induction furnace using alumina crucibles. An

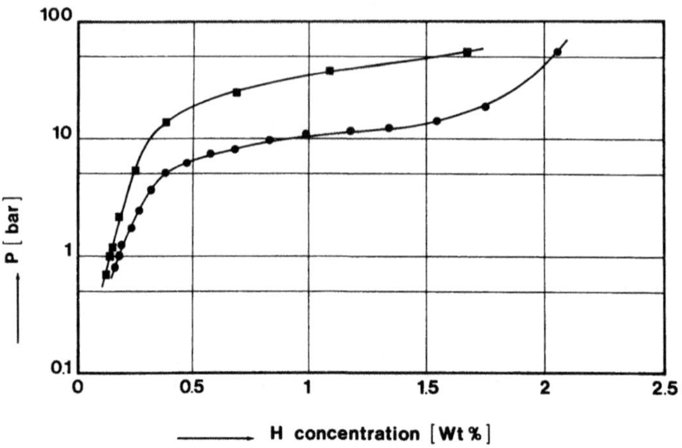

Fig. 2.16. Desorption isotherms at 20 °C (●) and 60 °C (■) of the partially substituted $TiMn_2$ type compound $Ti_{0.98}Zr_{0.016}Mn_{1.508}V_{0.431}Fe_{0.084}Cr_{0.049}$ prepared on industrial scale [2.84]

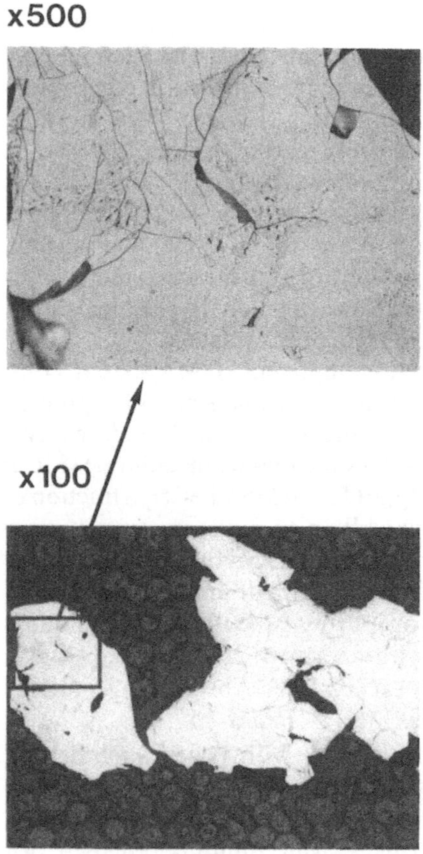

Fig. 2.17. Micrograph for a crunched grain of a batch of the compound described in Fig. 2.16 [2.74]

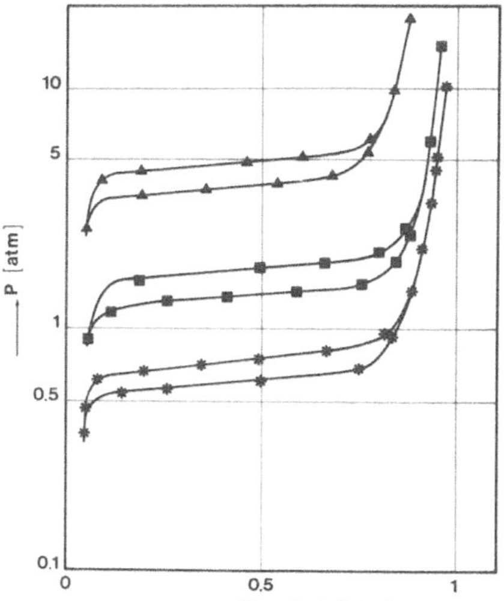

Fig. 2.18

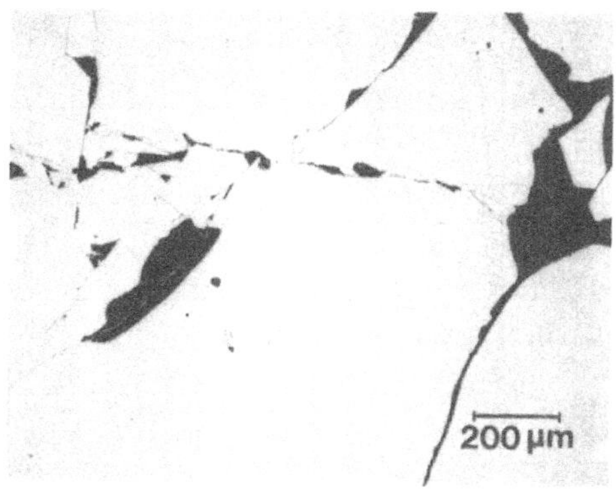

Fig. 2.19

Fig. 2.18. Isotherms at 30 °C (✱), 50 °C (■), and 80 °C (▲) of partially substituted $LaNi_5$ compound ($LaNi_{4.7}Al_{0.3}$) prepared on industrial scale [2.86]

Fig. 2.19. Micrograph for a crunched grain of a batch of the compound described in Fig. 2.18 [2.84]

important reason for failure is that the melt adheres to the crucible. The melting atmosphere is argon. Deoxidation is made in vacuum obtained by diffusion pumps. The temperature and the duration of the melt are tightly controlled. The melt is cast into a copper mold. After shaping it is packed in vacuum. The chemical purity and composition are checked by x-ray fluorescence, the phase purity by x-ray diffraction and pcT curves are usually measured. The yield of usuable batches is approximately 97%. Similar techniques are used by Ergenics (USA) [2.81] and by Japan Metals & Chemistry (Japan) [2.82]. A characterization of $LaNi_{4.7}Al_{0.3}$ material produced by this company has given the isotherms shown in Fig. 2.18 and the micrograph given in Fig. 2.19.

It is interesting to note that with some care a material prepared on a large industrial scale can achieve almost the same performance as a material prepared on a laboratory scale. This is revealed by the sorption-rate curves obtained with JMC material (Fig. 2.20a) and with a reference material prepared in one of our laboratories (Fig. 2.20b).

A large variety of binary and pseudobinary compounds are made by Chuo Denko Kogyo (Japan) [2.88] including Mg_2Ni. The purity and shape of their raw materials are: Ti (99.6%, sponge); Fe (99.98%, flakes); Mn (99%, rods); Mn (99.9%, flakes); Ni (99.97%, granules); Mg (99.9%, ingots); Ca (99.8%, granules) and Al (99.7%, granules). These materials are – according to the compound – loaded in alternate layers up to 100 kg in crucibles made from

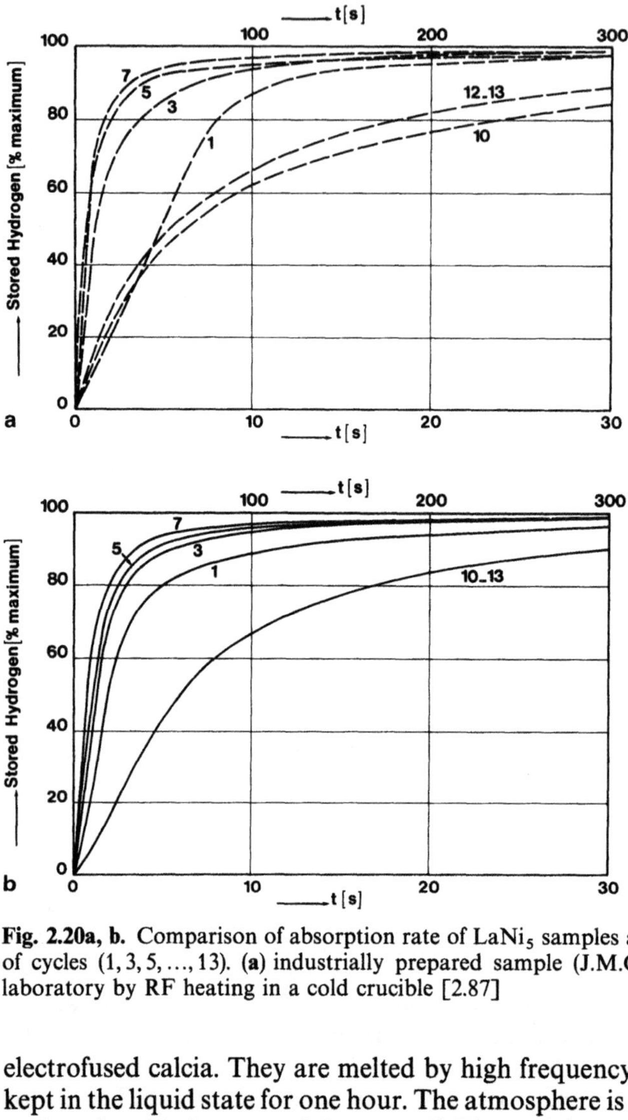

Fig. 2.20a, b. Comparison of absorption rate of $LaNi_5$ samples as a function of the number of cycles (1, 3, 5, ..., 13). (**a**) industrially prepared sample (J.M.C.); (**b**) sample prepared in laboratory by RF heating in a cold crucible [2.87]

electrofused calcia. They are melted by high frequency induction heating and kept in the liquid state for one hour. The atmosphere is either vacuum or argon and no deoxidation materials are used. The melt is cast in a mold and a plate-like block is obtained. The material is crushed to medium granules of approximately 10 mm which are packed under argon and sealed in vinyl bags. During the melting procedure the pumping time, the residual pressure, the melting time, and the tapping temperature are controlled. Each batch is analyzed and the pcT curve is determined. The yield of usuable batches is approximately 95%. A batch is rejected when the hydrogen plateau pressure deviates from the reference value by more than 0.5 bar.

An economically promising method of preparing FeTi compound was studied by *Shen* et al. [2.89]. A chemical reduction was used to form FeTi from a

Fe–O–Ti–O ore (illmenite), thus avoiding the process of chemical separation of Fe and Ti. Other chemical methods to prepare FeTi, $LaNi_5$, and Mg_2Ni from chlorides or mixed oxides were also developed in China [2.90, 91].

2.6 Hydrogenation of Intermetallic Phases

The hydrogenation techniques will be treated rather briefly because they are almost identical to those used for binary hydrides [2.92].

The usual way to hydrogenate the intermetallic compounds consists of exposing the sample to hydrogen gas at a convenient pressure and temperature.

This is usually acceptable because in most intermetallic phases the arrangement of the metallic components remains unchanged when hydrogen is dissolved in the lattice. Nevertheless, a rapid decay of the metastable state occurs sometimes even at quite low temperature like in Mg_2CuH_4, which disproportionates to MgH_2 and $MgCu_2$ [2.93]. On the other hand, some compound-like structures may only exist together with hydrogen. An example is given by magnesium and iron. In the binary system they are almost immiscible, but they form $MgFe_2H_6$ in presence of hydrogen [2.94].

Intermetallic compounds are generally much more reactive towards hydrogen than elemental metals and less sensitive to impurities in hydrogen gas. Nevertheless the gaseous contaminents (mainly H_2O and O_2) have an inauspicious influence on the hydrogenation [2.95], thus a tight control of the purity of the hydrogen is important. The apparatus should have a low degassing rate and low leak rate [2.96]. In some cases it is necessary to use ultra high vacuum equipment [2.34]. This is especially true when hydrogenation takes place at high temperatures (>200 °C) where oxidation is a very severe problem. Secondly, high-purity hydrogen gas should be used, obtained e.g. by the decomposition of a hydride. Nowadays containers filled with FeTi hydrides are commercially available. They deliver suprapure hydrogen with an impurity level well below 1 ppm [2.97, 98].

A guide-line for the choice of the pressure and temperature considerations are the pcT diagrams. Isotherms with plateau pressures in the range 1–10 atm are shown in Fig. 2.21 for the representative compounds mentioned in Sect. 2.2. One should especially note the wide temperature range between -90 and 300 °C according to the compound. Further isotherms are given in Chaps. 3 and 6 of this volume. Very thorough surveys of experimental results can be found in [2.4, 16, 99–102].

The pressure-temperature-time (p-T-t) program is mainly dictated by the passivity of the surface and by the resistance to decrepitation. The first property governs the dissociative chemisorption of the hydrogen molecules and hence the incorporation of the atoms in the lattice. The fundamental aspects for these reactions are discussed in Chap. 2 of volume II.

Hydrogen transport in intermetallic compounds is usually a slow process. This can be seen from the values of the diffusion coefficients given in Chap. 3 of

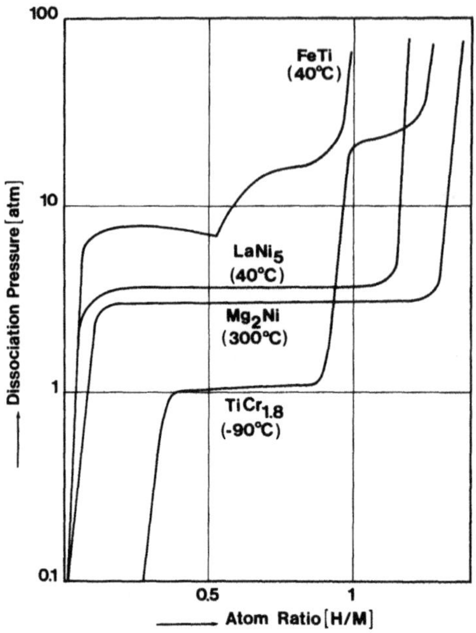

Fig. 2.21. Comparison of desorption isotherms of representative compounds at different temperatures

volume II. Nevertheless the sorption time is drastically reduced if decrepitation occurs [2.96]. This happens in compounds where the lattice expansion due to hydride formation is so large that it cannot be accommodated by plastic deformation. As a consequence the effective diffusion lengths are shortened. Furthermore, new clean surfaces for hydrogen adsorption are created spontaneously. A drawback of this phenomena is the impossibility of preparing large massive samples. To avoid an excessive dispersion of the freshly formed grains an embedding technique with aluminium or copper can be very useful [2.103, 104]. This technique also has the advantage of improving the thermal conductivity of the hydride bed.

Rare earth based compounds like $LaNi_5$ and its substitutes and some titanium and zirconium Laves phases have the advantage of an active surface which is rather insensitive to the presence of oxygen and of a strong tendency to decrepitation. Figure 2.22a shows a decrepitated grain of $LaNi_5$. With these compounds one could basically start from the ingot and simply by hydriding and dehydriding obtain a fine powder on the μm scale. Nevertheless, to accelerate the kinetics the ingot should first be ground to grains with an intermediate size of approximately 1 mm^3. Because grinding heats up the particles, it is recommended to work in an inert gas atmosphere. With a compound like TiFe a passive surface layer forms quite rapidly. Here it is necessary first to activate the surface by running heating/cooling cycles up to a few hundred degrees. Sometimes, judicious additions of different alloying elements such as manganese lead to a material which is sufficiently active at room temperature, and which does not necessitate any preliminary activation treatment. Once hydrogenation has started, the material also decrepitates:

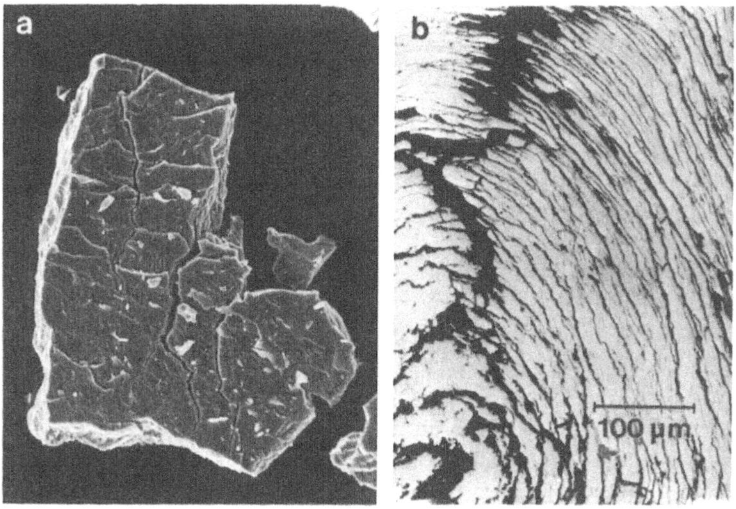

Fig. 2.22. (**a**) Decrepitated grain of $LaNi_5$ (3000 ×). (**b**) Decrepitated grain of FeTi

Fig. 2.22b reveals very nicely the peeling behaviour of decrepitation. The specific surface areas of most intermetallic compounds decrepitated by some hydrogen sorption cycles (about 10) amount to 0.1–1 m^2/g [2.24, 79, 96, 105].

Once the hydrogenation has started, it is sufficient to keep the pressure above the equilibrium pressure to have a high driving force. The amount of hydrogen which is taken up by the sample can be followed by measuring the weight increase of the sample or the pressure drop and the gas flow when working at constant volume or pressure, respectively.

A modification of the gas-loading technique consists of dissolving the hydrogen molecules in an organic solvent in which the sample is suspended [2.106].

Because hydrogenation may be quite inhomogeneous, especially when different hydride phases can be formed, it is worthwhile checking the final product by x-ray, and if possible by metallographic analysis. These two techniques reveal mainly large inhomogeneities. At the present time no good simple analytical method exists with a high spatial and/or compositional resolution.

For these examinations, and in a more general way for many experiments, it is necessary to expose the hydrided material to air. Precautions must be taken to ensure that the fine powder does not start to burn or the hydrogen does not escape. In many cases the hydride can be sealed quite effectively by the adsorption of a layer of SO_2 or CO [2.95, 103, 107]. Air should be introduced into the reaction vessel progressively in small amounts using an inert gas as a ballast. In all cases the hydrogen content should be controlled after the experiments, e.g. by desorption at high temperature. Furthermore, it is also

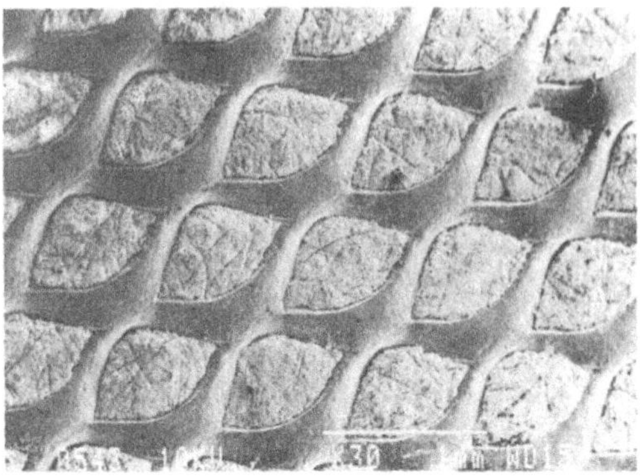

Fig. 2.23. Scanning electron micrograph of a $LaNi_5$-Teflon composite electrode pressed on a nickel grid (1000 ×) [2.111]

necessary to check whether disproportionation processes have taken place as a consequence of internal hydriding.

A very new method to prepare highly reactive magnesium and Mg_2Ni with large specific surface areas ($\sim 100\ m^2/g$) was described by *Bogdanovic* et al. [2.108]. An organic solvent in the presence of soluble organo–transition-metal catalysts is used to hydrogenate magnesium powder. By doping the catalytically prepared magnesium hydride with small amounts of a nickel organo–metallic compound, a well defined Mg_2Ni or hydride in amorphous form is obtained.

Another way to charge the intermetallic compound is to use hydrogen ions in an electrochemical medium. The electrolyte may be either an acid or a KOH aqueous solution. The experimental set-up is quite inexpensive, but it is limited to rather small samples [2.64, 78]. Furthermore, the technique is suited for hydrides with a low equilibrium pressure [2.109]. Sometimes, to avoid hydrogen losses from materials which form hydrides of low stability, it is necessary to pressurize the electrochemical cell with hydrogen gas at the equilibrium pressure of the hydride [2.110]. The limitations of the technique are the difficulties in controlling the amount and spatial distribution of the dissolved hydrogen. The problem associated with the decrepitation of the cathode can be eliminated by mixing the active particles with metallic or polymeric powder. Furthermore, the cathode must be pressed into an envelope of porous metal to ensure a good electrical contact. An example of such a mounting is shown in Fig. 2.23 where $LaNi_5$ was pressed into a nickel grid [2.111].

2.7 Concluding Remarks

To conclude this chapter we will give a few examples of how the preparation of the intermetallic compounds affects their aptitude to form hydrides. This aptitude will be characterized by the plateau pressure, the hydrogen loading capacity, the kinetics and the cyclic life-time.

An important factor which influences these properties is the level of impurities contained in the compound. We do not know of any work where the influence of metallic impurities at less than 100 at. ppm level has been studied. To get a feeling for the importance of such impurities, it is necessary to use the data obtained with metallic substitutions at a 1–10 at. % level. Using data available in the literature [2.42, 112], the temperature dependence of the plateau pressures at a H/M ratio of 0.5 is shown in Fig. 2.24 for various substitutes in FeTi. Similar variations are found for non-metallic impurities like C, O, and S as shown in Fig. 2.25. Typically 1 at. % of impurities induces a variation of the order of 1 atm in the plateau pressure. A closer examination of the data shows that at least for the metallic substitutions the plateau pressure depends exponentially on the volume expansion of the compound due to the

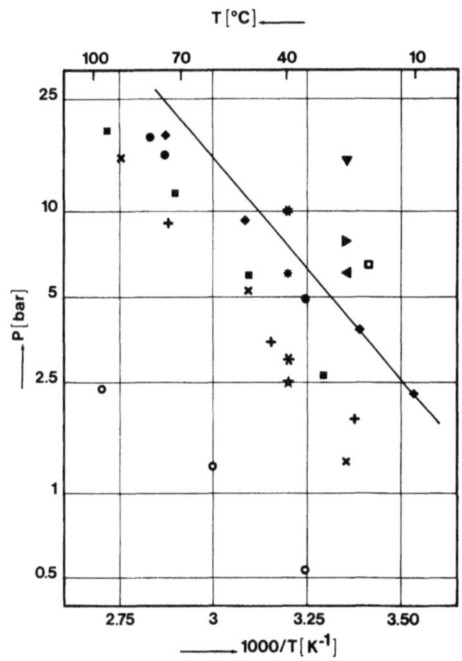

Fig. 2.24. Stability of $FeTiH_{0.5}$ type hydrides for various partial substitutions of iron or titanium by other metallic elements. Data compiled from [2.112] and [2.113]. The solid line is representative for pure FeTi and has been evaluated in [2.59]. (▸) $FeTi_{1-x}Al_x$; $x<0.1$; (□) $FeTi_{0.95}V_{0.05}$; (◆) $FeTi_{0.95}Cr_{0.05}$; (○) $FeTi_{0.80}Zr_{0.20}$; (+) $FeTi_{0.85}Si_{0.15}$; (×) $FeTiMn_{0.2}$

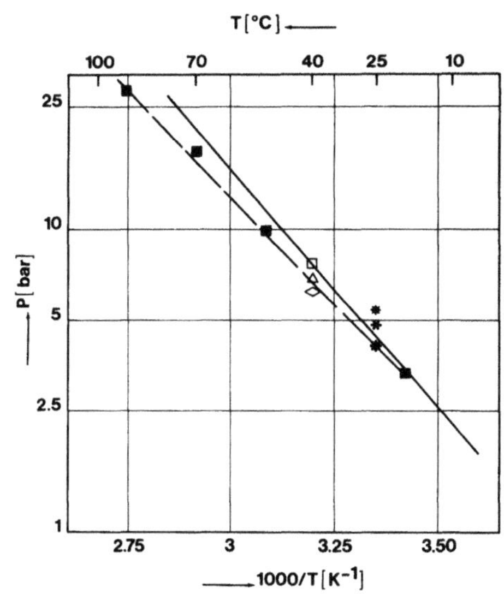

Fig. 2.25. Influence of various oxygen (□ △ ◇) contents (wt. % = 0.01 – 0.87) at 40 °C [2.24], of various sulphur (✳ * ✸) contents (0.004 < S/Ti at. ratio < 0.08) at 25 °C [2.25] and of 2 at. % carbon (■) (20 °C < T < 91 °C) [2.26] dissolved in the base compound on the stability of $FeTiH_{0.5}$

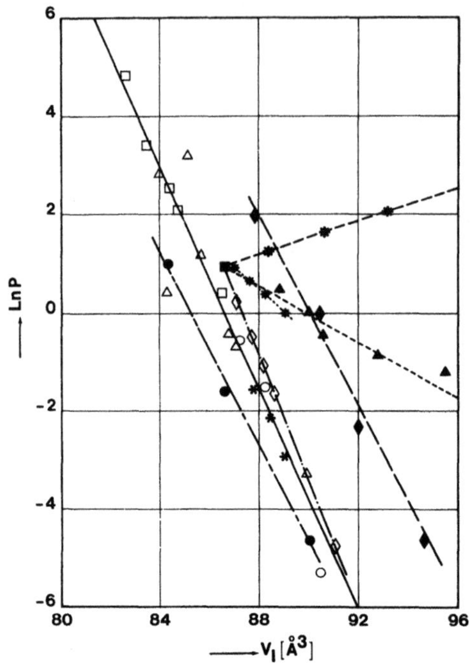

Fig. 2.26. Stability of $LaNi_5$ type hydrides for various partial substitutions of nickel or lanthanum by other metallic elements. Data compiled from [2.56]. (✳) $La(Ni, Pt)_5$; (▲) $La(Ni, Cu)_5$; (◆) $Th(Ni, Al)_5$; (◇) $La(Ni, Mn)_5$; (✸) $La(Ni, Fe)_5$; (○) $La(Ni, Al)_5$; (□) RNi_5. Plateau pressure [bar] is plotted versus unit cell volume V_i of the intermetallic compound

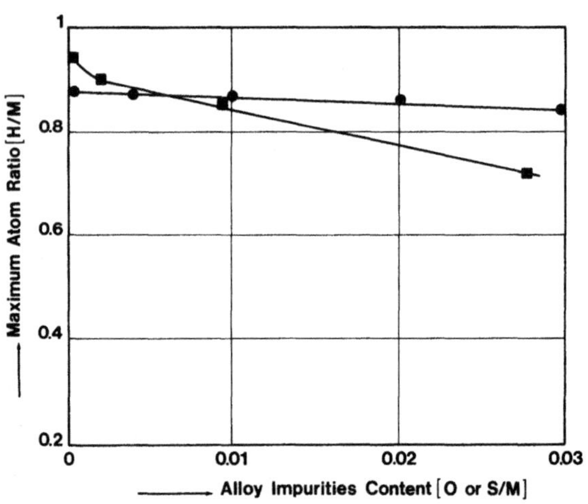

Fig. 2.27. Influence of oxygen (■) [2.24] and sulphur (●) [2.25] content dissolved in FeTi on the maximum loading capacities

foreign elements. This is clearly demonstrated in Fig. 2.26 for $LaNi_5$ [2.47, 113, 114]. Such an argument does not hold for the non-metallic elements: here a second phase is precipitated. In the case of FeTi one has observed the formation of TiC [2.26, 42], $Fe_7Ti_{10}O_3$ [2.42], and Ti_2S [2.25]. This leads, of course, to a shift of stoichiometry in the bulk material. All these impurities also influence the shape of the pcT curves and, in particular, the loading capacity of the compound. As an example we show in Fig. 2.27 how the capacity is reduced by the presence of oxygen and sulphur in FeTi.

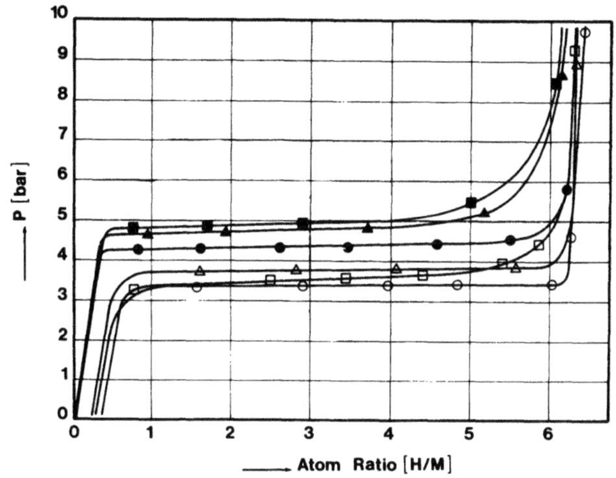

Fig. 2.28. Comparison of 40 °C isotherms of $LaNi_5$ prepared by RF melting in cold crucible under vacuum (○), arc melting (□) and arc melting followed by an annealing treatment at 1100 °C for 1½ hours (△). The lattice parameters *a* and *c* in Å of the compounds are respectively: (○) 5.014, 3.977; (□) 5.005, 3.974; (△) 5.007, 3.973 [2.86]

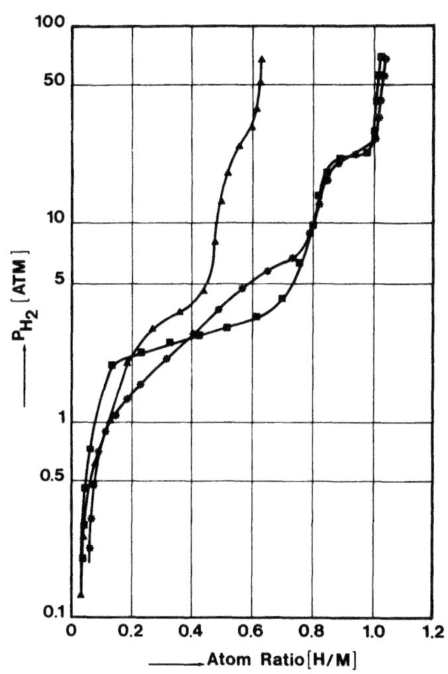

Fig. 2.29. Desorption isotherms at 25 °C of Mischmetal substituted calcium-nickel compound ($Ca_{0.7}M_{0.3}Ni_5$): As-cast (✱) and annealed for 24 hours at 1000 °C (■) and at 1100 °C (▲) [2.42]

Even if all impurities influence the plateau pressure and the loading capacity in a similar way, the problem is more severe for the non-metallic impurities. As we mentioned in Sect. 2.3.1 the level of foreign metallic elements can be kept below 100 at. ppm. This is not true for non-metallic elements, either because the starting materials are strongly doped with them or because they are introduced during the process. Levels of 1 at. % frequently occur.

Another important factor which governs the hydrogenation behaviour of the compounds is the solidification structure. This aspect is revealed in Fig. 2.28

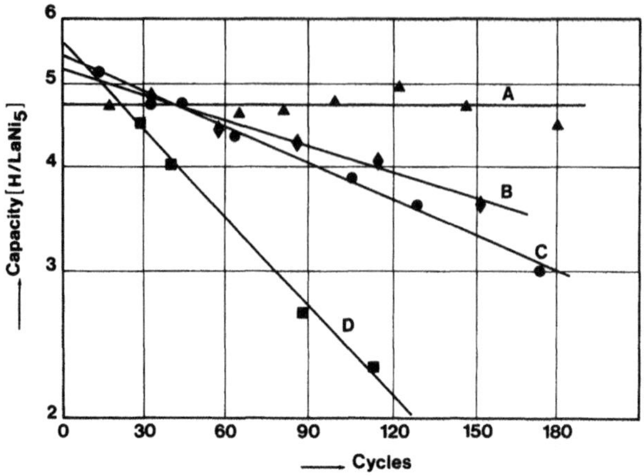

Fig. 2.30. Hydrogen storage capacities of $LaNi_5$ samples from various suppliers during absorption cycling: (*A*) Research Chemicals sample, dry H_2; (*B*) Ergenics (*C*) Research Chemicals; (*D*) Molycorp samples are cycled in 12 atm. hydrogen with about 10 Torr H_2O [2.115]

which shows the different behaviour of $LaNi_5$ obtained by melting together lanthanum and nickel from the same batches in an arc melter and in a cold crucible (see Sect. 2.3.2). Probably for homogeneity reasons the cold crucible sample exhibits flatter plateaus at lower pressures, a smaller hysteresis loop and an increased absorption capacity. By a heat treatment the properties of the arc melt material are slightly improved. The same effect was previously mentioned in Sect. 2.3.3. and illustrated in Fig. 2.10. Nevertheless, some precautions must be taken when a material containing a highly volatile component is heated. This is clearly seen in Fig. 2.29 where isotherms of three $Ca_{0.7}M_{0.3}Ni_5$ (M= mischmetal) samples, which have experienced different heat treatments [2.42], are shown.

As a conclusion, we hope that we have illustrated the importance of the metallurgical state of the base intermetallic compounds which will be used for research and application purposes. According to our experience, a tight control of the whole fabrication process needs a high degree of care. As a final example a comparative study of the cyclic life-time of different commercially available $LaNi_5$ samples should be quoted (Fig. 2.30) [2.115]. Furthermore, we can only repeat that a painstaking characterization of the material and of the process is mandatory for the use of hydrides in basic research as well as in their applications.

References

2.1 A. Brown, J.H. Westbrook: "Formations Techniques", in *Intermetallic Compounds*, ed. by J.H. Westbrook (Wiley, New York 1967) Chap. 17, pp. 303–339

2.2 W.M. Mueller, J.P. Blackledge, G.G. Libowitz: *Metal Hydrides* (Academic, New York 1968)

2.3 K.H.J. Buschow, P.C.P. Bouten, A.R. Miedema: Rep. Prog. Phys. **45**, 937–1039 (1982)

2.4 D.G. Ivey, D.O. Northwood: J. Mater. Sci. **18**, 321–347 (1983)

2.5 A.R. Miedema: Z. Metallkde. **69**, 445–461 (1978)

2.6 J.W. Cahn, J.E. Hilliard: J. Chem. Phys. **28**, 258–267 (1958)

2.7 M. Hansen, K. Anderko: *Constitution of binary alloys*, 2nd ed. (McGraw-Hill, New York 1958)
2.8 R.P. Elliott: *Constitution of binary alloys*. 1st Suppl. (McGraw-Hill, New York 1965)
2.9 F.A. Shunk: *Constitution of binary alloys*, 2nd Suppl. (McGraw-Hill, New York 1969)
2.10 N.V. Ageev: *Diagrammy Sostoyaniya Metallicheskik Sistem* (Metallurgiya, Nowsow 1969)
2.11 R. Hultgreen, R.L. Orr, P.D. Anderson, K.K. Kelley: *Selected Values of Thermodynamic Properties of Metals and Alloys* (Wiley, New York 1963, ASM; Metals Park/04 1973)
2.12 W.G. Moffatt: *The Handbook of Binary Phase Diagrams* (General Electric Co. Shenecdaty NY 1977)
2.13 Bulletin of alloy phase diagrams since (1980)
2.14 R.D. Shull: Bulletin of Alloy Phase Diagrams **4**, 5–15 (1983)
2.15 K.H.J. Buschow, H.H. Van Mal: J. Less-Common Met. **29**, 203–210 (1972)
2.16 K.H.J. Buschow: "Hydrogen Absorption in Intermetallic Compounds" in *Handbook on the Phys. and Chem. of Rare Earths*, ed. by K.A. Gschneidner, Jr., L. Eyring (North Holland, Amsterdam 1984) Vol. 6, Chap. 47, 1–111
2.17 J.L. Murray: Bulletin of Alloy Phase Diagrams **2**, 320–334 (1981)
2.18 J.L. Murray: Bulletin of Alloy Phase Diagrams **2**, 174–181 (1981)
2.19 P. Villars, L.D. Calvert: *Pearsons Handbook of Crystallographic Data for Intermetallic Phases*, Vols. 1–3 (American Soc. for Metals, Metals Park, Oh. 1985)
2.20 H. Huthman, G. Inden: Phys. Status Solidi A **28**, K129–130 (1975)
2.21 G. Inden: Z. Metallk. **66**, 648–653 (1975)
2.22 J. Schweizer, F. Tasset: J. Phys. F **10**, 2799–2817 (1980)
2.23 D. Givord, J. Laforest, J. Schweizer, F. Tasset: J. Appl. Phys. **50**, 2008–2010 (1979)
2.24 G.D. Sandrock, J.J. Reilly, J.R. Johnson: Proc. 11th Intersociety Energy Conversion Engineering Conference, Vol. I, 965–971 (1976)
2.25 R. Suzuki, J. Ohno, H. Gondoh: J. Less-Common Metals, **104**, 199–206 (1984)
2.26 D. Khatamian, N.S. Kazama, F.D. Manchester, G.C. Weatherly, C.B. Alcock: J. Less-Common Met., **91**, 267–273 (1983)
2.27 Ellingham: Proceedings of the Electric Steel Conference Amer. Inst. Min. Met. Eng. 1949 in: "*Aspects Physicochimiques de l'Elaboration des Métaux*" ed. by D.W. Hopkins (Dunod, Paris 1958)
2.28 K.A. Gschneidner, N. Kippenhan, O.D. Mc. Masters: *Thermochemistry of the Rare Earths*, Rare Earth Information Center Report IS.RIC.6, 1973
2.29 A.N. Nesmeyanov, R. Gary: *Vapor Pressure of the Chemical Elements* (Elsevier, Amsterdam 1963)
2.30 R.E. Honig, D.A. Kramer: in *Techniques of Metals Research* ed. by R.A. Rapp (Interscience Publishers New York 1970) Vol. IV, Part 1
2.31 J.M. Welter: Metall. **28**, 731–735 (1974)
2.32 J.M. Welter, H. Wenzl: Metall. **39**, 116–125 (1985)
2.33 Y. Osumi, H. Suzuki, A. Kato, K. Oguro, M. Nakane: J. Less-Common Met. **74**, 271–277 (1980)
2.34 H. Wenzl, J.M. Welter: "Properties and Synthesis of Nb–H Interstitial Alloys. Current Topics" in *Mat. Science*, Vol. 1, ed. by E. Kaldis (North Holland, Amsterdam 1978)
2.35 B.J. Beaudry, K.A. Gschneidner, Jr.: in *Handbook on the Physics and Chemistry of Rare Earth*, ed. by K.A. Gschneidner, Jr., L. Eyring (North Holland, Amsterdam 1978) Vol. 1, Chap. 2, p. 173–232
2.36 Z. Zhou, H. Zhang, D. Tang, D. Zhou, X. Ping, J. Ni: Proc. 6th World Hydrogen Energy Conf., Vienna 1986, ed. by T.N. Veziroglu, N. Getoff, P. Weinzierl. (Pergamon, Oxford 1986) Vol. 2, p. 898
2.37 G.D. Sandrock: Proceedings of the 12th Intersociety Energy Conversion Engineering Conference, Am. Nuclear Soc., **1**, 951–958 (1977)
2.38 M. Beyss, J.-M. Welter, T. Kaiser: J. Crystal Growth **50**, 419–428 (1980)
2.39 H. Biloni: "Solidification", in *Physical Metallurgy*, ed. by R.W. Cahn, P. Haasen (North Holland, Amsterdam 1983) pp. 477–579

2.40 W. Kurz, D.J. Fisher: *Fundamentals of Solidification* (Trans. Tech. Publications, 1984)
2.41 C. Lartigue: "Etude Structural et Thermodynamique du Système $LaNi_{5-x}Mn_x$-Hydrogène" Thèse 3ème cycle, Université Paris VI, 1979
2.42 G.D. Sandrock: Proceedings of the 2nd World Hydrogen Energy Conference **3**, 1625–1656, 1978, ed. by International Association for Hydrogen Energy
2.43 M.H. Mendelsohn, D.M. Gruen, A.E. Dwight: Mat. Res. Bull. **13**, 1221–1224 (1978)
2.44 M.H. Mendelsohn, D.M. Gruen: in *Rare Earths in Modern Science and Technology*, ed. by G.J. Carthy, H.B. Silber, J.J. Rhyne (Plenum, New York 1980) **2**, 593–598
2.45 J. Lamloumi, C. Lartigue, A. Percheron-Guegan, J.C. Achard: in *The Rare Earths in Modern Science and Technology*, ed. by G.J. Mc Carthy, H.B. Silber, J.J. Rhyne (Plenum, New York 1982) **3**, 487–488
2.46 J. Lamloumi, A. Percheron-Guégan, J.C. Achard, G. Jehanno, D. Givord: J. de Phys. **45**, 1643–1652 (1984)
2.47 J.C. Achard, A. Percheron-Guégan, H. Diaz: Proceedings Second International Congress on Hydrogen in Metals, Paper 1E12, Paris, 1977
2.48 M.H. Mendelsohn, D.M. Gruen, A.E. Dwight: Nature **269**, 45–47 (1977)
2.49 C. Lartigue, A. Percheron-Guégan, J.C. Achard, F. Tasset: J. Less-Common Met. **75**, 23–29 (1980)
2.50 C.E. Lundin, F.E. Lynch: Proc. International Conf. on Alternative Energy Sources, Miami Beach, FL, pp. 3803–3820, 1977
2.51 J.J. Reilly, R.H. Wiswall: Summary Rep. BNL 17136 Upton, NY: Brookhaven Natl. Lab., p. 28, 1972
2.52 J. Shinar, I. Jacob, D. Davidod, D. Shaltiel: *Hydrides for Energy Storage*, ed. by A.F. Andresen, A.J. Maeland (Pergamon, Oxford 1978) pp. 337–352
2.53 H.H. Van Mal, K.H.J. Buschow, F.A. Kuijpers: J. Less-Common Met. **32**, 289–296 (1973)
2.54 I.D. Weisman, L.H. Bennet, A.J. Mc Alister, R.E. Watson: Physical Review B **11**, 82–91 (1975)
2.55 A. Percheron-Guégan, C. Lartigue, J.C. Achard: J. Less-Common Met. **109**, 287–309 (1985)
2.56 J. Lamloumi: "Etude Structurale, Thermodynamique et Magnétique du Système $LaNi_{5-x}Fe_x$-hydrogène", Thèse 3ème cycle, Université Paris VI, 1982
2.57 H.H. Van Mal, K.H.J. Buschow, A.R. Miedema: J. Less-Common Met. **35**, 65–76 (1974)
2.58 S. Cirafici, A. Palenzona: J. Less-Common Met. **53**, 199–203 (1977)
2.59 J.-M. Welter, G. Arnold, H. Wenzl: J. Phys. F **13**, 1773 (1983)
2.60 M. Beyss, T. Kaiser, A.J. Singh, J.-M. Welter: Proc. 2nd European Conference on Crystal Growth, Lancaster (1979)
2.61 L. Schlapbach, A. Seiler, H.C. Siegman, T.V. Waldkirch, P. Zürcher: Int. J. Hydrogen Energy **4**, 21–28 (1979)
2.62 A.J. Maeland: In *Hydrides for Energy Storage*, ed. by A.F. Andresen, A.J. Maeland (Pergamon, Oxford 1978) p. 427–462
2.63 A.J. Maeland: In *Hydrogen in Disordered and Amorphous Solids*, ed. by G. Bambakidis, R.C. Bowman, NATO ASI Series B **136**, p. 127 (Plenum, New York 1986);
G.G. Libowitz, A. Maeland: J. Less-Common Met. **101**, 131 (1984);
A.J. Maeland, L.E. Tanner, G.G. Libowitz: J. Less-Common. Met. **89**, 183 (1983)
2.64 K. Suzuki: J. Less-Common. Met. **89**, 183–195 (1983)
2.65 R.C. Bowman, J.S. Cantrell, A. Attala, D.E. Etter, B.D. Craft, J.E. Wagner, W.L. Johnson: J. Non-Crystalline Solids **6162**, 649–654 (1984);
R.C. Bowman, J.S. Cantrell, E.L. Venturini, R. Schulz, J.E. Wagner, A. Attala, B.D. Craft: In *Rapidly Quenched Metals*, ed. by S. Steeb, H. Warlimont (Elsevier, Amsterdam 1985) p. 1541–1544
2.66 H. Bernas, A. Traverse, C. Janot: In *Amorphous Metals and Semiconductors*, ed. by P. Haasen, R.I. Jaffee, (Pergamon, Oxford 1985) p. 435;
A. Traverse, H. Bernas: J. Less-Common. Met. **129**, 1–12 (1987)

2.67 K. Samwer: J. Less-Common Met. **131**, 210 (1987) and NATO ASI Series B **136** on Hydrogen in Disordered and Amorphous Solids, ed. by G. Bambakidis, R.C. Bowman (Plenum, Oxford 1986);
H. Schröder, K. Samwer, U. Köster: Phys. Rev. Lett. **54**, 197 (1985)
2.68 H.J. Güntherodt, H. Beck (eds.): *Glassy Metals*, I, II, Topics Appl. Phys. Vols. 46, 53 (Springer, Berlin, Heidelberg 1981, 1983)
2.69 M.L. Post, J.J. Murray, G.J. Despault, J.B. Taylor: Mat. Res. Bull. **20**, 337–342 (1985)
2.70 G.Y. Adachi, K.I. Niki, J. Shiokawa: J. Less-Common Met. **81**, 345–348 (1981)
2.71 B.R. Livesay, J.W. Larsen: Proceedings of JIMIS-2, *Hydrogen in Metals*, November 26–29, 1979, Minakami, Japan, Suppl. Transact. Jap. Inst. of Metals **21**, 345 (1980)
2.72 K. Nakamura: Scripta Met. **18**, 793 (1984)
2.73 K. Sumiyama, Y. Hashimoto, Y. Nakamura: Transact. Jap. Inst. of Metals **24**, 61 (1983)
2.74 E.M. Würtz: private communication (1985)
2.75 H. Baars: Diplomarbeit, Fachhochschule Aachen, Abt. Jülich (1985)
2.76 J.F. Lynch, J.J. Reilly: J. Less-Common Met. **87**, 225–236 (1982)
2.77 T. Schober, G. Arnold, A.W. Nellen: Z. Praktische Metallographie **16**, H.7, 309 (1979)
2.78 T. Schober: Scri. Metall. **13**, 107 (1979)
2.79 L. Schlapbach: J. Phys. F **10**, 2477–2490 (1980)
2.80 Gesellschaft für Elektrometallurgie: private communication (1984)
2.81 Ergenics: Hy-Stor Marketing Leaflet (1984)
2.82 Japan Metals & Chemicals: private communication (1984)
2.83 Gesellschaft für Elektrometallurgie: Patent Application (1984)
2.84 Gesellschaft für Elektrometallurgie: private communication (1985)
2.85 Santocku Metal Industry: private communication (1984)
2.86 F. Briaucourt, F. Demany, A. Percheron-Guégan: private communication (1985)
2.87 L. Touron, A. Percheron-Guégan: private communication (1985)
2.88 Chuo Denki Kogyo: private communication (1984)
2.89 P. Shen, G. Wang, W. Yin, D. Song, D. Zhan: *Rare Metals*, The Chinese Society of Metals, Vol. **2**, 38 (1983)
2.90 P. Shen, G. Wang, D. Song, Z. Zhou, D. Zhang, H. Yuan: Proc. 5th World Hydrogen Energy Conf., Toronto 1984, ed. by T.N. Veziroglu, J.B. Taylor, Vol. 3, p. 1303, Int. Assoc. Hydrogen Energy;
P. Shen, Y. Zhang, S. Zheng, X. Feng, H. Yuan, S. Chen: Proc. 6th World Hydrogen Energy Conf., Vienna 1986, ed. by T.N. Veziroglu, N. Getoff, P. Weinzierl (Pergamon, Oxford 1986) Vol. 2, p. 831
2.91 D. Song, G. Wang, Z. Zhou, F. Wang, S. Li, S. Lu, J. Wu: Proc. 6th World Hydrogen Energy Conf., Vienna 1986, ed. by T.N. Veziroglu, N. Getoff, P. Weinzierl (Pergamon, Oxford 1986) Vol. 2, p. 845
2.92 Coy L. Huffine: In *Metal Hydrides*, ed. by W.M. Mueller, J.P. Blackledge, G.G. Libowitz (Academic Press, New York 1968) Chap. 13, pp. 675–747
2.93 J.J. Reilly, R.H. Wiswall: Inorg. Chem. **6**, 2220–2223 (1967)
2.94 J.J. Didisheim, P. Zolliker, K. Yvon, P. Fischer, J. Schefer, M. Gubelmann, A.F. Williams: Inorg. Chem. **23**, 1953–1957 (1984)
2.95 G.D. Sandrock, P.D. Goodell: J. Less-Common Met. **73**, 161–168 (1980)
2.96 H.H. Van Mal: Philips Res. Repts. Suppl. **1**, 1–85 (1976)
2.97 K.-H. Klatt, S. Pietz, H. Wenzl: Z. Metallkde. **69**, 170–173 (1978)
2.98 HWT Gesellschaft für Hydrid- und Wasserstofftechnik mbH, D-4330 Mühlheim
2.99 R. Wiswall: "Hydrogen storage in Metals" in *Hydrogen in Metals II*, ed. by G. Alefeld, J. Völkl, Topics Appl. Phys., Vol. 29 (Springer-Verlag, Berlin, Heidelberg 1978) Chap. 5, pp. 201–242
2.100 W.E. Wallace, R.S. Craig, V.U.S. Rao: Adv. Chem. **186**, 207–240 (1980)
2.101 H. Oesterreicher: Appl. Phys. **24**, 169–186 (1981)
2.102 P.S. Rudman, G. Sandrock: Ann. Rev. Mat. Sci. **12**, 271–294 (1982)
2.103 M. Ron, D. Gruen, M. Mendelsohn, I. Sheft: J. Less-Common Met. **74**, 445–448 (1980)
2.104 J. Toepler, O. Bernauer, H. Buchner: J. Less-Common Met. **74**, 385–399 (1980)

2.105 L. Belkbir, E. Joly, N. Gerard, J.C. Achard, A. Percheron-Guégan: J. Less-Common Met. **73**, 69–77 (1980)
2.106 J.J. Reilly, J.R. Johnson: J. Less-Common Met. **104**, 175–190 (1984)
2.107 D.M. Gualtieri, K.S. Narasimhan, T. Takeshita: J. Appl. Phys. **47**, 3432 (1976)
2.108 B. Bogdanovic: Int. J. Hydrogen Energy **9**, 937–941 (1984); **12**, 863–873 (1987)
B. Bogdanovic, K.H. Claus, S. Gürtzgen, B. Spliethoff, U. Wilczok: J. Less-Common Met. **131**, 163–172 (1987)
2.109 J.C. Achard, A. Percheron-Guégan, J. Sarradin, G. Bronoel: In: A.F. Andresen, A.J. Maeland (eds.) Proc. Int. Symp. on Hydrides for Energy Storage, Geilo, 1977 (Pergamon, Oxford 1978) pp. 485–490
2.110 M.H.J. Van Rijswick: In A.F. Andresen, A.J. Maeland (eds.) Proc. Int. Symp. on Hydrides for Energy Storage, Geilo 1977 (Pergamon, Oxford 1978) pp. 261–271
2.111 A. Percheron-Guégan: L'Actualité Chimique **2**, 11–18 (1982)
2.112 H. Buchner: Energiespeicherung in Metallhydriden (Springer, Wien, New York 1982)
2.113 K.A. Gschneidner Jr., T. Takeshita, Y. Chung, O.D. McMasters: J. Phys. F **12**, L1–L6 (1982)
2.114 M.H. Mendelsohn, D.M. Gruen, A.E. Dwight: J. Less-Common Met. **63**, 193–207 (1979)
2.115 G.L. Holleck, J.R. Driscoll, B.E. Paul: J. Less-Common Met. **74**, 379, 384 (1980)

3. Thermodynamics of Intermetallic Compound–Hydrogen Systems

Ted B. Flanagan and W. Alan Oates

With 14 Figures and 3 Tables

This chapter concerns the thermodynamics of intermetallic compound–hydrogen systems. The single phase and two condensed phase situations are considered. Integral and partial thermodynamic quantities are defined with reference to these systems. Experimental approaches are described and compared. Special difficulties encountered with intermetallic compound hydrides are discussed, e.g., effects of activation and annealing, sloping plateaux, decomposition. Results for some representative systems are described. Phase transitions in these systems are considered, as is the relationship between partial and complete equilibrium.

3.1 Introduction

The suitability of intermetallic compounds for energy storage and other technological purposes rests principally on their thermodynamic properties and thus a knowledge and understanding of these properties is of paramount importance.

The need for this review arises because the properties of intermetallic compounds introduce features of behavior towards hydrogen not found in the pure metal-hydrogen systems. For example, the range of stoichiometry over which most intermetallic compounds can exist introduces trapping sites which influence the dilute phase hydrogen solubility. The variety of interstices available for hydrogen occupation in most intermetallic compounds introduces complicating features in the statistical thermodynamic modeling of these systems. The well-known instability of intermetallic compound hydrides towards decomposition into a component metal hydride also leads to complications in the thermodynamic description of these metastable systems.

The present review will not attempt to survey all aspects of the thermodynamics of intermetallic compound–hydrogen systems but will concentrate on the more fundamental aspects. Results for several prototype systems will be described in detail. Other recent reviews, which overlap with the present one to a certain extent, have been given by *Buschow* et al. [3.1], *Rudman* and *Sandrock* [3.2], *Ivey* and *Northwood* [3.3] and *Oesterreicher* [3.4]. A comprehensive tabulation of the plateau pressures and hydrogen capacities of intermetallic hydrides is given in [3.1] and Chap. 6 of this volume should also be consulted.

The words “intermetallic compound”, “intermetallic hydride” and “intermetallic compound–hydrogen system” occur so frequently in this review that we will use the abbreviations IMC (plural IMCs), IMH (plural IMHs) and IMCH (plural IMCHs), respectively.

3.2 Macroscopic Thermodynamics

At low temperatures, IMCHs behave in a pseudo-binary, partial equilibrium or para-equilibrium (PE) [3.5] manner. This type of behavior, which gives rise to a maximum of two solid phases in equilibrium with hydrogen gas, is discussed in Sect. 3.2.1. However, an IMCH may, under the right conditions, behave as a ternary system in complete equilibrium (CE), when a maximum of three solid phases may be in equilibrium with hydrogen gas. CE, which is most likely to occur at higher temperatures, is discussed in Sect. 3.2.2. We shall stress later (in Sect. 3.5.4) that the more commonly observed PE behavior is not independent of CE, so that an understanding of the latter can provide insight into the more commonly met PE behavior.

The following notation will be employed throughout:

The IMC is composed of the metals A and B and, in substituted compounds, of metal C also. Metal A is assumed to have a stronger affinity for hydrogen than B.

The composition of the host IMC is expressed as $c=n_B/(n_A+n_B)$ so that the IMC may be written $A_{1-c}B_c$. In substituted IMCs we use $A_{1-c}(B_{1-x}C_x)_c$, where $x=n_C/(n_B+n_C)$ and $c=(n_B+n_C)/(n_A+n_B+n_C)$. The H concentration is expressed as $r=n_H/(n_A+n_B)$ so that the IMH composition is $A_{1-c}B_cH_r$. Alternatively, the H concentration is sometimes expressed in terms of θ, the fraction of interstitial sites occupied. Sometimes the IMC will be denoted by M and the IMH by MH_r. Thus the compound $LaNi_5H_{6.4}$, for example, may also be written as $La_{0.17}Ni_{0.83}H_{1.07}$ or as $MH_{1.07}$. The composition “a” refers to the value of r in the metal-rich phase which is in equilibrium with the hydride, and “b” refers to the value of r in the hydride phase which is in equilibrium with the metal-rich phase, i.e., MH_a is in equilibrium with MH_b. Saturation occurs at the composition $r=y$, i.e., MH_y is a stoichiometric compound. In some systems several such saturations occur at increasingly higher H concentrations as the hydrogen chemical potential is increased.

3.2.1 Single and Two Condensed Phase Equilibria

a) Integral/Formation Properties

Figures 3.1–3 illustrate, schematically, the influence of H concentration on the integral and partial thermodynamic properties G, S, and H. The integral quantities are expressed per unit amount of M, e.g., $\Delta G/n_M$, which we denote by ΔG_m. Pure M and $H_2(g)$ at 1 atm are chosen as the standard states, i.e., ΔG_m

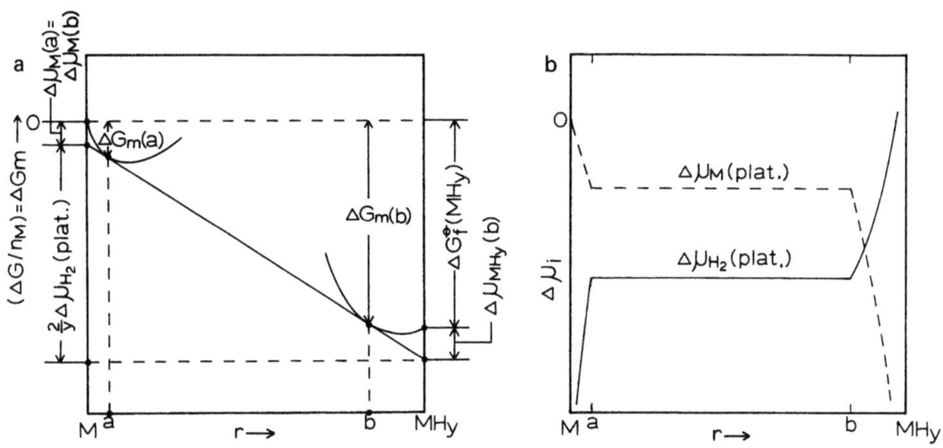

Fig. 3.1a, b

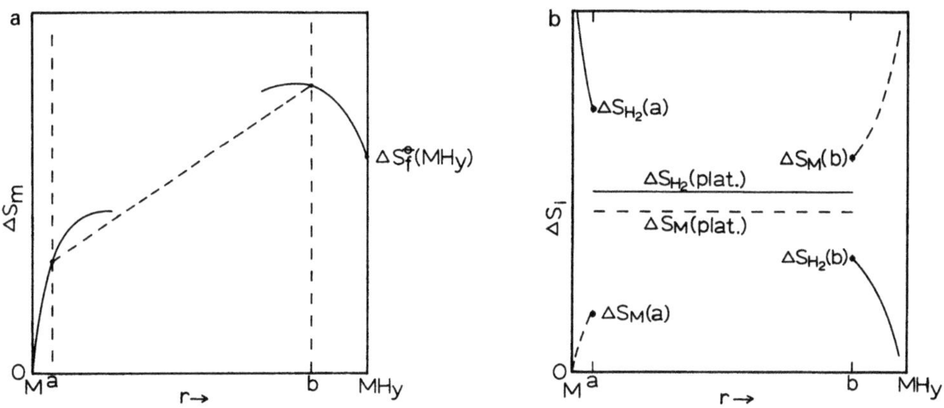

Fig. 3.2a, b

Fig. 3.1. **(a)** Schematic plot of ΔG_m against hydrogen content. The phase boundaries are shown as the values of r at the points of common tangent. **(b)** Schematic plot of $\Delta\mu_i$ against hydrogen content where $\Delta\mu_i$ represents $\Delta\mu_{H_2}$ or $\Delta\mu_M$. (——), $\Delta\mu_{H_2}$, and (---), $\Delta\mu_M$

Fig. 3.2. **(a)** Schematic plot of ΔS_m against hydrogen content. ΔS_{H_2}(plat) is given by $2(b-a)^{-1}[\Delta S_m(b)-\Delta S_m(a)]$ where $\Delta S_m(b)$ and $\Delta S_m(a)$ are shown by the small filled circles. **(b)** Schematic plot of ΔS_i against hydrogen content where ΔS_i represents ΔS_{H_2} or ΔS_m. (——), ΔS_{H_2} and (---), ΔS_M. The values of ΔS_{H_2} and ΔS_M at the co-existing phase boundaries are indicated by the small filled circles

refers to the free energy change for the mixing or formation reaction:

$$M(\mathrm{s})+r/2\,\mathrm{H_2(g)}=M\mathrm{H}_r(\mathrm{s})\,, \qquad (3.1)$$

where s and g denote solid and gaseous states, respectively. Using these standard states, the free energy of mixing at $r=y$ is the same as the standard free energy of formation of $M\mathrm{H}_y$, $\Delta G^{\circ}(M\mathrm{H}_y)$, i.e.,

$$\Delta G^{\circ}_{\mathrm{f}}(M\mathrm{H}_y)=\Delta G_m(y)\,. \qquad (3.2)$$

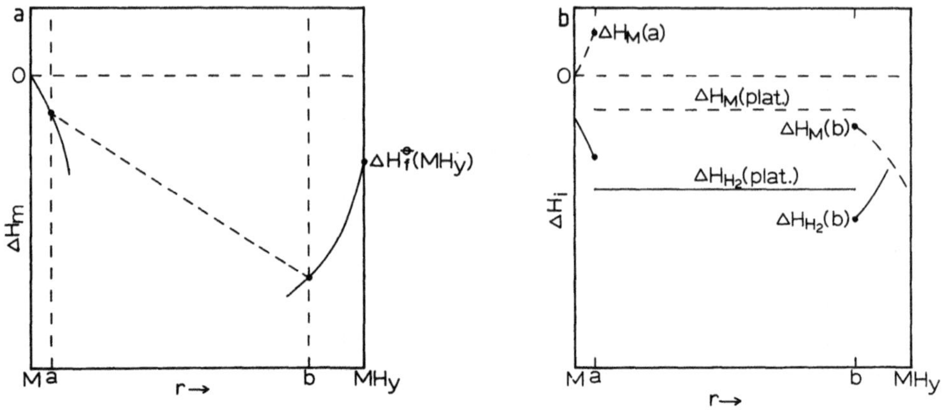

Fig. 3.3. **(a)** Schematic plot of ΔH_m against hydrogen content. The values of ΔH_m at the phase boundaries are indicated by the small filled circles. **(b)** Schematic plot of ΔH_i against hydrogen content where ΔH_i represents ΔH_{H_2} or ΔH_M. (——), ΔH_{H_2} and (- - -), ΔH_M. The values of ΔH_{H_2} and ΔH_M at the co-existing phase boundaries are indicated by the small filled circles

b) Partial Molar Properties

When ΔG_m and r are used as variables, the partial quantities in a single-phase region may be obtained from the tangent slope and tangent intercept to the integral quantity [3.6]. The integral and partial quantities are related to one another at constant temperature and total pressure by the following equations:

$$\Delta G_m(r) = (r/2)\Delta\mu_{H_2}(r) + \Delta\mu_M(r) \tag{3.3}$$

$$\Delta\mu_{H_2}(r) = 2\frac{d[\Delta G_m(r)]}{dr} \tag{3.4}$$

$$\Delta\mu_M(r) = \Delta G_m(r) - r\frac{d[\Delta G_m(r)]}{dr} \quad \text{where} \tag{3.5}$$

$$\Delta\mu_{H_2} = 2\mu_H(r) - \mu^{\ominus}_{H_2} = RT\ln(p_{H_2}/\text{atm}) \quad \text{and}, \tag{3.6}$$

$$\Delta\mu_M = \mu_M(r) - \mu^{\ominus}_M, \tag{3.7}$$

where $\mu^{\ominus}_M$ refers to the structural form existing in the pure metal. $\Delta\mu_{H_2}$ is the property most accessible to experimental determination. The other partial and the integral quantity are readily obtained from it as follows:

$$\Delta G_m(r) = \tfrac{1}{2}\int_0^r \Delta\mu_{H_2}\, dr \tag{3.8}$$

$$\Delta\mu_M(r) = \Delta G_m(r) - \tfrac{1}{2}r\Delta\mu_{H_2}(r). \tag{3.9}$$

The chemical potentials of the individual metallic components may be obtained from the equations:

$$\Delta\mu_A(r) = \Delta G_m(r) - \tfrac{1}{2} r \Delta\mu_{H_2} - c\left(\frac{\partial \Delta G_m}{\partial c}\right)_r, \tag{3.10}$$

$$\Delta\mu_B(r) = \Delta G_m(r) - \tfrac{1}{2} r \Delta\mu_{H_2} + (1-c)\left(\frac{\partial \Delta G_m}{\partial c}\right)_r. \tag{3.11}$$

Since $\Delta\mu_M = (1-c)\Delta\mu_A + c\Delta\mu_B$ it can be seen that (3.10) and (3.11) are consistent with (3.9).

When using models for the solid phases it is desirable to choose some standard state for the hydrogen potential, μ_H°, in the particular phase of interest. For pure metal-hydrogen systems the most convenient choice is the hypothetical infinitely dilute solution standard state [3.7]. With this choice the hydrogen potential standard state difference, $\Delta\mu_H^\circ = \mu_H^\circ - \frac{1}{2}\mu_{H_2}^\circ$, can be obtained from

$$\Delta\mu_H^\circ = \lim_{r\to 0}\left[RT\ln\left(p_{H_2}^{1/2}\frac{\beta - r}{r}\right)\right], \tag{3.12}$$

where β is the number of interstitial sites per metal atom. An excess H potential may then be defined by

$$\mu_H^E = \tfrac{1}{2}\Delta\mu_{H_2} - \left[\Delta\mu_H^\circ + RT\ln\frac{r}{\beta - r}\right], \tag{3.13}$$

where the term in square brackets is the H potential of the ideal solution.

Because of the multiplicity of the different types of sites for H occupation in IMCs a suitable definition of an ideal solution and hence a reference state for H in the solid is not as straightforward as for pure metals. The best solution is to use (3.12) and (3.13) with β being taken as the value of r equal to the experimentally observed saturation composition, y.

c) Plateau Properties

When *PE* behavior by the IMCH is being considered the IMC behaves as one component and the coexistence of the dilute and the pseudo-binary IMH phase leads to one degree of freedom or a constant pressure of hydrogen at constant temperature, the plateau pressure.

This co-existence of the two condensed phases may arise from either the presence of H–H attractive interactions, which lead to a phase separation at low temperatures, or from a structural difference between the IMC and the IMH, which can occur in the presence or absence of H–H interactions. In practice, the phase transformations are accompanied by hysteresis, which is discussed in Sects. 3.3.3, 3.5.5 but in this section we will assume that the plateau pressures represent equilibrium states.

The most important thermodynamic property from a technological viewpoint is the relative hydrogen potential when MH_a is in equilibrium with MH_b, since it is this which is related to the plateau pressure:

$$\Delta\mu_{H_2}(\text{plat}) = RT\ln[p_{H_2}(\text{plat})/\text{atm}]. \tag{3.14}$$

$\Delta\mu_{H_2}(\text{plat})$ is related to the integral quantities by:

$$\Delta\mu_{H_2}(\text{plat}) = \frac{2}{b-a}[\Delta G_m(b) - \Delta G_m(a)] \tag{3.15}$$

and, for the related partial quantity for the metal:

$$\Delta\mu_M(\text{plat}) = \frac{1}{b-a}[b\Delta G_m(a) - a\Delta G_m(b)] \tag{3.16}$$

whilst the integral plateau property at composition r is:

$$\Delta G_m(\text{plat}, r) = \frac{1}{(b-a)}[(r-a)\Delta G_m(b) + (b-r)\Delta G_m(a)]. \tag{3.17}$$

As may be seen from Fig. 3.1b, in the case of the chemical potential, the plateau property is the same as the partial molar hydrogen quantity in the single phases which are in equilibrium, i.e.,

$$\Delta\mu_{H_2}(\text{plat}) = \Delta\mu_{H_2}(a) = \Delta\mu_{H_2}(b). \tag{3.18}$$

However, this is *not* the case for the plateau entropy and enthalpy properties (cf. [3.8, 9]), i.e.,

$$\Delta S_{H_2}(\text{plat}) \neq \Delta S_{H_2}(a) \neq \Delta S_{H_2}(b) \tag{3.19}$$

$$\Delta H_{H_2}(\text{plat}) \neq \Delta H_{H_2}(a) \neq \Delta H_{H_2}(b). \tag{3.20}$$

This distinction may be seen by comparing Fig. 3.1b with Figs. 3.2b and 3.3b.

Frequently, the experimentally determined values of $\Delta S_{H_2}(\text{plat})$ and $\Delta H_{H_2}(\text{plat})$ remain approximately constant over fairly wide ranges of temperature, even though substantial changes in a and b may occur over this temperature range. *Rudman* and *Sandrock* [3.2] suggested that $S_{H_2}(\text{plat})$ is small compared with $S^{\ominus}_{H_2}$ and, since the latter varies strongly with temperature, $\Delta S_{H_2}(\text{plat})$ should also vary with temperature. They conclude that any experimental observation to the contrary reflects inaccuracies in the measurements. This is not necessarily so, however, as can be illustrated for the case of a regular interstitial solution [3.6] where a and b vary markedly with temperature. In this case it is readily shown that $\Delta S_{H_2}(\text{plat})$ is given by:

$$\Delta S_{H_2}(\text{plat}) = 2\Delta S^{\ominus}_{H} = 2S^{\ominus}_{H} - S^{\ominus}_{H_2}. \tag{3.21}$$

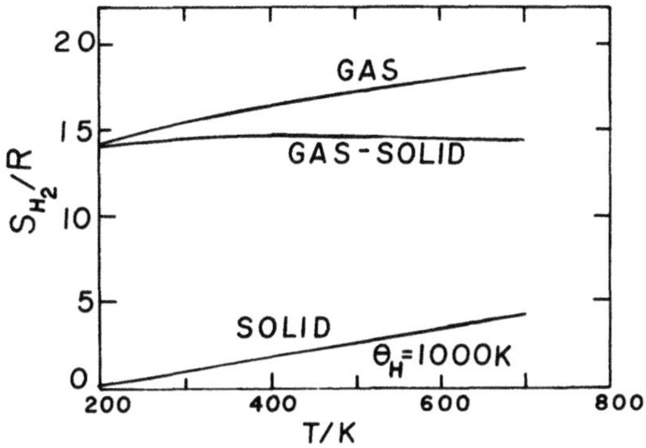

Fig. 3.4. Plot of S_{H_2}/R against temperature for the gas phase ($S^{\circ}_{H_2}/R$), the solid phase ($2S^{\circ}_{H}/R$) and the difference between these two quantities

If the principal contribution to S°_{H} is assumed to be vibrational and this is approximated by the entropy for an Einstein oscillator then, as shown in Fig. 3.4, we can understand how ΔS_{H_2}(plat) can remain approximately constant over a quite wide temperature range. Similar arguments apply to the approximate constancy of ΔH_{H_2}(plat) with temperature.

An alternative way to approach the plateau equilibrium is to consider the reaction

$$\frac{2}{b-a}MH_a + H_2 = \frac{2}{b-a}MH_b \tag{3.22}$$

and use the equilibrium condition, $\sum_i n_i\mu_i = 0$ so that

$$\frac{2}{b-a}(\mu_{MH_b} - \mu_{MH_a}) - \mu_{H_2} = 0\,. \tag{3.23}$$

For the same reaction under standard conditions we have

$$\frac{2}{b-a}(\mu^{\circ}_{MH_b} - \mu^{\circ}_{MH_a}) - \mu^{\circ}_{H_2} = \Delta G^{\circ}(a \to b) \tag{3.24}$$

so that, by subtraction

$$\Delta G^{\circ}(a \to b) = \Delta\mu_{H_2}(\text{plat}) = RT\ln[p_{H_2}(\text{plat})/\text{atm}] \tag{3.25}$$

since MH_a and MH_b are in their standard states.

It should be noted that in (3.25) we are dealing with the standard free energy change for (3.22) and not free energy change, which is, of course, zero.

The relations between the plateau properties and the formation properties are important and, for a system with a single plateau, are readily obtained by

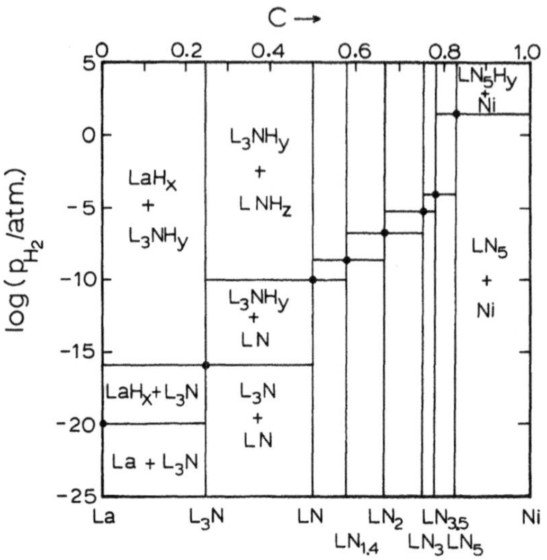

Fig. 3.5. Log$[p_{H_2}$/atm.] against c where $c = n_{Ni}/(n_{La} + n_{Ni})$ for the La–Ni–H system under conditions of para-equilibrium at 373 K from *Oates* and *Flanagan* [3.10]. For clarity of presentation not all of the phase fields have been indicated. The small filled circles indicate the hydrogen pressures where the various hydride phases form under PE conditions. L = La and N = Ni in the formulas for the various intermetallic compounds

considering the reaction

$$M(\mathrm{s}) + y/2\,\mathrm{H_2(g)} = M\mathrm{H}_y(\mathrm{s}) \tag{3.26}$$

for which it can be shown using (3.6) and (3.7) that

$$\Delta G_f^{\circ}(M\mathrm{H}_y) = y/2\,\Delta\mu_{\mathrm{H_2}}(\mathrm{plat}) + \Delta\mu_M(a) - \Delta\mu_{M\mathrm{H}_y}(b)\,. \tag{3.27}$$

As can be seen from Fig. 3.1a both the second terms are fairly small when $a \to 0$ and $b \to y$, and they also offset one another because of their different signs in (3.27). Consequently, a good approximation, especially at low temperatures, is:

$$\Delta G_f^{\circ}(M\mathrm{H}_y) \approx \frac{y}{2}\,\Delta\mu_{\mathrm{H_2}}(\mathrm{plat})\,, \tag{3.28}$$

so that $\Delta\mu_{\mathrm{H_2}}$(plat) may usually be identified with the stability of the hydride. In some IMCHs multiple plateaux are found. This gives rise to no significant differences in the thermodynamic treatment from that discussed above.

Another feature of interest is the hydriding behavior of different IMCs belonging to the same binary alloy system. The type of isothermal representation shown in Fig. 3.5 for PE in the La–Ni system is useful in that it permits the isothermal behavior of the whole system to be taken in at a glance [3.10].

3.2.2 Three Condensed Phase Equilibria

Three condensed phases may co-exist in an IMCH if the system is in CE. The IMC $A_{1-c}B_c$ dissolves H to the composition $A_{1-c}B_c\mathrm{H}_a$ which then breaks down into AH_x and another IMC with H in solution, $A_{1-c'}B_{c'}H_{a'}$. The standard state reaction equation for CE is

$$\frac{2c'}{y\Delta c}A_{1-c}B_c(s)+H_2(g)=\frac{2}{y}AH_y(s)+\frac{2c}{y\Delta c}A_{1-c'}B_{c'}(s), \tag{3.29}$$

where $\Delta c=c'-c$ and $c'>c$.

The standard free energy for this reaction, ΔG°, can be obtained from the standard free energies of formation of the the three solids in (3.29). However, in order to calculate $\Delta\mu_{H_2}$ for the three condensed phase equilibrium it is necessary to know $\Delta\mu_{AH_y}$, $\Delta\mu_{A_{1-c}B_c}$, and $\Delta\mu_{A_{1-c'}B_{c'}}$ as a function of H concentration, so that they can be obtained at AH_r, $A_{1-c}B_cH_a$, and $A_{1-c'}B_{c'}H_{a'}$, respectively. Since, in general, this information is not known, it is not possible to calculate $\Delta\mu_{H_2}$ accurately. Nevertheless, as is discussed in Sect. 3.5.4, calculations based on taking $\Delta\mu_{H_2}=\Delta G^{\circ}$ can be useful in predicting the PE behavior to be expected in the same system.

3.3 Experimental Aspects

Some experimental difficulties are encountered in the determination of the thermodynamics of IMCHs which are not found for pure metal–H systems. For example, deviations in the stoichiometry of the starting material can cause differences in the dilute phase solubilities and the observed plateau pressures. The preparation of IMCs and their hydrides is described in Chap. 2 and should be referred to. Only the special problems related to the determination of the thermodynamic properties of IMCHs will be stressed in this section.

3.3.1 p-r-T Measurements

In single phase regions $\Delta\mu_{H_2}$ in the solid can be determined from the equilibrium hydrogen pressures as a function of r through the use of (3.6). ΔH_{H_2} and ΔS_{H_2} can be obtained from the temperature dependence of $\Delta\mu_{H_2}$ and r can be measured volumetrically, gravimetrically or by monitoring the change in a physical property whose variation with r is known. Errors in r can arise if the starting IMC contains residual hydrogen following the attempted removal of trapped hydrogen at the temperatures where the isotherms are being determined.

Results from p-r-T measurements are usually presented as a series of isotherms. Phase diagrams can be deduced from these p-r-T results, but the precise locations of the phase boundaries are often difficult to determine because of the hysteresis problem (see Sect. 3.3.3a) and because of the tendency for the isotherms to change very gradually in the regions of these phase transitions.

3.3.2 Calorimetric Measurements

a) Reaction Calorimetry

The heat evolved or absorbed on the addition or removal of a given amount of hydrogen, n_{H_2}, to an IMC can be measured calorimetrically. An advantage of IMCHs for such measurements is that the kinetics are generally fast, thereby increasing the accuracy of the calorimetry. If n_{H_2} is small enough, the measured heats approach ΔH_{H_2} in the single phase regions, so that either ΔH_{H_2} or ΔH_m can be measured. It should be pointed out that, with this technique, the largest errors often arise from the determination of the Δn_{H_2} which causes the heat absorption or evolution. This is particularly the case for measurements of ΔH_{H_2} in those regions of an isotherm where the pressures are relatively large and change rapidly with r.

Different types of calorimeters have been used for studies on IMCHs. *Lebsanft* and *Wenzl* [3.11] employed a heat flow calorimeter of the Tian-Calvet type in their investigations of the Fe–Ti–H system, whilst *Murray* et al. [3.12] used a twin cell calorimeter of similar type. *Bowerman* et al. [3.13] used a single cell, quasi-isothermal calorimeter manufactured by the LKB Instrument Corporation for high pressure applications. *Hubbard* et al. [3.14] used a single cell Rossini-type iso-peribolic flow calorimeter. *Murray* et al. [3.12], *Bowerman* et al. [3.13] and *Hubbard* et al. [3.14] all investigated the $LaNi_5$–H system and in addition *Murray* and his group have examined the $CaNi_5$–H and $MgNi_2$–H systems [3.15, 16].

The heat correction due to work, (pV), in these calorimeters has been discussed by *Nace* and *Aston* [3.17] and more recently by *Boureau* and *Kleppa* [3.18]. This correction, $V_{cal}\Delta p$, where $\Delta p = p_f - p_i$, has been employed by most workers in this field except for *Murray* et al. [3.12], who argue that no correction is needed for the twin-cell calorimeter. The correction is negligible in the plateau regions where there is zero pressure change but it is important in single phase regions. The need for the correction should be re-examined in the light of the work of *Hill* [3.19], who considered the analogous problem for adsorption and concluded that it is not possible to make a correction for the work term except in the two extreme cases where the gas is added reversibly or completely irreversibly.

b) Thermal Analysis

Differential thermal analysis has been employed by *Shilov* et al. [3.20] to determine ΔH_{H_2}(plat) from the temperature at which the decomposition of the hydride occurs. Alternatively, ΔH_{H_2}(plat) values can be determined from the area under the DTA curves [3.21]. These methods are possibly useful for rapid surveys of systems but in view of the relative ease with which values of ΔH_{H_2}(plat) can be determined from van't Hoff plots, these methods seem to be of limited utility.

c) Heat Capacity

Only a few heat capacity measurements have been made on IMHs. *Ohlendorf* and *Flotow* [3.22] determined the heat capacities of $LaNi_5$, $LaNi_5H_{0.36}$, and $LaNi_5H_{6.39}$ from 5 to 300 K. Activated $LaNi_5$ was used in the preparation of the hydrides and it is not known if this activation affected the results due, for example, to strain removal during heating. Standard entropies at 298 K were evaluated from the results and the vibrational and electronic contributions were also evaluated from the low-temperature heat capacities.

Wenzl and *Pietz* [3.23] measured the heat capacities of $Fe_{0.5}Ti_{0.5}H_r$ samples with $r=0.31$, 0.47, and 0.66 at 280 K. From their measured heat capacities they calculated the heat capacities due to H and, from these, the values of the Einstein temperatures. It was observed that the heat capacities due to H increased (Einstein temperatures decreased) rather markedly with r.

3.3.3 Problems in Obtaining Accurate Thermodynamic Data

a) Hysteresis

Hysteresis, which is almost always observed in metal-hydrogen systems, leads to the results that (i) $p_f > p_d$, and (ii) $a_f > a_d$, $b_f > b_d$, where the subscripts f and d refer to formation and decomposition, respectively.

There are thus two sets of "thermodynamic" parameters which correspond to the reactions:

$$\frac{2\,MH_{a_f}}{b_f - a_f} + H_2 = \frac{2\,MH_{b_f}}{b_f - a_f}, \tag{3.30}$$

$$\frac{2\,MH_{a_d}}{b_d - a_d} + H_2 = \frac{2\,MH_{b_d}}{b_d - a_d}, \tag{3.31}$$

where $a_f > a_d$ and $b_f > b_d$.

For example, two ΔH_{H_2}(plat) are obtainable from:

$$\frac{-R\partial \ln p_f}{\partial(1/T)} = \Delta H_{H_2}(\text{plat, f}) \quad \text{and} \tag{3.32}$$

$$\frac{-R\partial \ln p_d}{\partial(1/T)} = \Delta H_{H_2}(\text{plat, d}). \tag{3.33}$$

Hysteresis can be expressed quantitatively by the free energy difference, $\Delta G_{H_2}(\text{hyst}) = RT\ln(p_f/p_d)$. This is small relative to the enthalpy changes for the above reactions so that only small differences are found between ΔH_{H_2}(plat) determined from van't Hoff plots of p_f and p_d. It has been argued [3.24, 25] that calorimetrically measured enthalpies will not be influenced by (i) but will be affected by (ii) and that entropies determined from the temperature dependence of p_f or p_d will not be affected by (i) but will also be affected by (ii).

It has also been argued by some [3.26, 27] that p_d closely represents the equilibrium plateau pressure, whereas *Flanagan* and co-workers [3.28, 29] have presented evidence that the equilibrium pressure lies somewhere between p_f and p_d (see Sect. 3.5.4 for more details). In the case of the Pd–H system, where the data are probably more accurate than for any other metal-hydrogen system, the equilibrium plateau pressures, calculated from the single phase thermodynamic data, seem to lie nearly half-way between p_f and p_d [3.30], thus favoring the Flanagan model.

b) Characterization of the Starting IMC – Activation and Annealing

The IMCs employed in thermodynamic investigations of the IMCHs are generally not those resulting from the initial preparation and annealing treatment. They are usually first activated by subjecting the IMC to a series of hydriding-dehydriding cycles and, although it is still not entirely clear from a fundamental viewpoint what occurs during this treatment, from a practical viewpoint the material is mechanically disintegrated into a finely divided material which reacts very rapidly with hydrogen. Different IMCs are affected differently by activation. In the case of $LaNi_5$, x-ray examination of the dehydrided, unannealed IMC [3.31] reveals that there is considerable line broadening due to strain which does not appear in the unactivated powder. This broadening is greatest in the (*hh*0) and (0*k*0) directions and is zero in the (00*l*) direction. It has been shown [3.32] that energy is released upon heating activated, dehydrided $LaNi_5$. This is presumably due to the release of the strain energy. The strain energy introduced during activation may also be responsible for the effect of activation on the plateau pressure p_f (see below).

Hempelmann et al. [3.33] have recently shown from inelastic neutron scattering spectra that activated, annealed $LaNi_5$ exhibits well-defined discrete localized modes in the dilute phase whereas the unannealed, activated material does not. They supposed that the defects in the activated, unannealed material lead to a nearly continuous distribution of frequencies. The tendency for $LaNi_5$ hydride to decompose upon repeated activation cycles has recently been discussed by *Goodell* [3.34], who suggests that repeated cycling leads to disordering of the $LaNi_5$ structure. This was also suggested earlier by *Flanagan* et al. [3.35] on the basis of the non-ideal solubility behavior in the dilute phase of activated $LaNi_5$. With reference to decomposition of $LaNi_5$ during cycling, Sect. 3.5.4 should also be consulted and Chap. 6 of volume II of this work and a recent contribution of *Buschow* [3.36].

It is the behavior of the defect-free material which is important for a detailed theoretical understanding of the system, whilst the behavior of the activated material is important for an understanding of processes occurring under technological conditions. Other than for the FeTi–H system [3.37], there appear to be no unequivocal data available for the dilute metal-rich region for systems which show large differences in thermodynamic properties between annealed and activated material. Results for the $LaNi_5$–H system [3.38, 39] are conflicting.

It is also possible to study the thermodynamic properties of a material which has been annealed after activation, i.e., when the IMC is in a finely divided state. The annealing of activated, H-free $LaNi_5$ markedly reduces the dilute phase solubility in $LaNi_5$ and leads to a transition from non-ideal behavior, characteristic of trapping and other factors introduced by activation, to nearly ideal dilute-phase behavior [3.38, 39]. *Flanagan* and *Biehl* [3.40] also found that the plateau pressure p_f was markedly increased by this annealing treatment which commences at relatively low temperatures (<400 K) suggesting that the effect on p_f may arise from micro-strain. There was only a small effect of annealing on p_d which is not an unreasonable observation because the formation of the hydride phase results in activation and decomposition and therefore always reflects the behavior of the activated IMC.

It is clear that this annealing effect is important because it can introduce variations in plateau pressures p_f, even for IMCs which are otherwise identical.

Rapid hydriding and the accompanying self-heating might introduce localized annealing which could lead to anomalous plateau behavior.

The generality of the annealing-after-activation effect has not been established. It has been found to be small in $LaCo_5$–H [3.41] and non-existent in $ErFe_2$–H [3.39]. The effect of annealing on activated TiFe appears rather different from that of the ANi_5 compounds (see below). It would seem that it will be a factor in those IMCHs where the effects of activation appear to be profound, e.g., when it leads to mechanical disintegration as in the ANi_5–H IMCHs, and not so important in those IMCs where activation does not so markedly disrupt the sample, e.g., $ErFe_2$–H [3.39].

Besides the influence of activation on the measured thermodynamic properties, structural disorder, small variations from the desired IMC stoichiometry or second phases can also be unwanted sources of complicating behavior in studies of IMCH thermodynamics.

c) Sloping Plateaux

Sloping two-solid-phase tie lines or "plateaux" are frequently observed in IMCHs. The phenomenon seems to be most pronounced in the case of ternary IMCs and in non-stochiometric binary IMCs. Sloping plateaux can result from inhomogeneities in the IMC [3.42] but may have a more fundamental origin [3.43]. *Ivey* and *Northrup* [3.44] have shown from x-ray diffraction that the hydrogen content of a homogeneous hydride phase increases with hydrogen content along the sloping plateau of the $Zr(Fe_xCr_{1-x})_2$–H system. This observation seems to rule out inhomogeneities in the starting material as the source of the sloping plateaux because otherwise the x-ray reflections of the hydride phase should increase in width with H content due to the distribution of hydrogen concentrations in the hydride phase of the inhomogeneous sample.

Whatever its cause, it must be allowed for when ΔH_{H_2}(plat) values are obtained from van't Hoff plots of p_f or p_d, i.e., p_f and p_d should be evaluated at the same r value at various temperatures.

It should be mentioned that when sloping plateaux are present, the slope for p_f appears greater than that for p_d. However, if $\Delta\mu_{H_2}$, i.e. $\log p$, instead of p or $p^{1/2}$, is plotted against r, the apparent differences in slope for hydride formation and decomposition become smaller or disappear entirely.

d) Decomposition

The thermodynamics of three condensed phase equilibria has been discussed in Sect. 3.2.2. Such a three phase equilibrium will arise if the IMC breaks down in a hydriding situation to a component metal hydride and a further IMC. This complete equilibrium CE is possible only when there is sufficient atomic mobility of the metallic components. This will always occur at high temperatures but CE in IMCHs seems to be able to occur quite often at temperatures well below those which might be expected from a consideration of the properties of the pure IMC. It seems that mobility is induced by the hydriding-dehydriding pseudo-binary phase transformation and this influences the ability of the system to switch over from PE to CE at unexpectedly low temperatures as compared to those where significant diffusion of the metal atoms is expected to occur.

As long as the temperatures are low and the hydrogen gas used is extremely pure, decomposition of most IMCHs should not be a serious problem [3.34], especially since thermodynamic investigations on IMCHs are generally carried out with IMCs which have been cycled only a few times. The decomposition of IMCHs is discussed in Sect. 3.5.4 from a theoretical viewpoint and from a practical viewpoint in Chap. 5 of Volume II of this work.

e) Amorphization

It has been discovered recently that amorphization of some IMCs can occur during hydriding [3.45, 46]. This represents intermediate behavior between CE and PE and, unless its role is known, its presence can lead to erroneous interpretation of thermodynamic data for hydriding. It appears that the amorphous hydride which is formed must be quite stable because upon removal of the hydrogen the sample crystallizes, i.e. the hydrogen stabilizes the amorphous structure. *Aoki* et al. [3.47, 48] have found that hydrogen-induced amorphization occurs in most of the RNi$_2$ IMCs where R is a rare earth. The detailed mechanism of the amorphization process is not understood as yet.

This seems the appropriate place to mention that the hydrogen absorption characteristics of amorphous (glassy) and crystalline IMCs have been compared, e.g. for the case of TiCu [3.49]. No evidence has been found for a plateau pressure region in the isotherms of the amorphous materials. It is not known if there is a plateau region when hydrogen is gradually added to an IMC which undergoes the amorphization reaction. Models for the absorption of hydrogen by amorphous materials have been proposed by *Kirchheim* [3.50] and *Griessen* [3.51].

3.4 Results for Representative Systems

It is not possible to review all of the available thermodynamic data for IMCHs. Most of these data have been derived from plateau pressures and their temperature dependence. *Griessen* and *Riesterer*, in Chap. 6 of this volume, and *Buschow* et al. [3.1] have tabulated many of these results. For a few IMCHs more detailed thermodynamic data are available for the single phase regions and from reaction calorimetry in both the single and two phase regions. It is these results upon which we will concentrate here. In addition we have selected some IMCHs as representatives of a given type of behavior, e.g., $ErFe_2$–H has been chosen to illustrate multi-plateau behavior and ZrV_2–H to illustrate continuous solid solution behavior over a wide range of hydrogen contents. (In this review the plateau thermodynamic properties for hydride formation and decomposition will be given with reference to the formation reaction, i.e., ΔH(plat) will be negative for both processes – see (3.32) and (3.33).)

3.4.1 $LaNi_5$ and Some $LaNi_5$-Based Systems

Three calorimetric investigations on activated $LaNi_5$–H have given results for ΔH_{H_2}(plat) which are in substantial agreement (Table 3.1). The available calorimetric data for activated $LaNi_5$ in the dilute phase are in qualititive agreement only, i.e., there is agreement that $|\Delta H_{H_2}|_\alpha > |\Delta H_{H_2}(\text{plat})|$. The differences in the results of the different investigators [3.13, 14, 52] reflect the different annealing and evacuation conditions employed on the samples. The large values observed for $|\Delta H_{H_2}|_\alpha$ are consistent with the non-ideal (not obeying Sievert's Law) pressure-composition behavior observed in this region. This non-ideality has been attributed to trapping effects originating from the activation treatment and, possibly, from some non-stoichiometry in the starting material [3.35, 38]. A discussion of the isotope effects for $LaNi_5$–H(D) for the plateau region is given in Sect. 3.5.2.

Table 3.1. Results for ΔH_{H_2}(plat) for activated $LaNi_5$–H

ΔH_{H_2}(plat, f) [kJ mol^{-1} H_2]	ΔH_{H_2}(plat, d) [kJ mol^{-1} H_2]	Ref.
-29.4 ± 0.8	-29.8 ± 1.2	[3.13]
-32.3 ± 0.06	-31.8 ± 0.10	[3.52]
-31.6 ± 0.2	-31.4 ± 0.2	[3.14]

No calorimetric studies have been carried out in the dilute phase using well-annealed $LaNi_5$ samples. One p-r-T study in this region [3.39] gave $\Delta H^{\circ}_{H_2} = -10.8$ kJ per mole H_2, but the results on which this value was based did not intersect the origin when plotted as $p^{1/2}$ versus r, i.e., there appears to be some trapping effect present, although the authors [3.39] argued that this did not affect their thermodynamic results because the traps were saturated. In another p-r-T study [3.38] at higher temperatures and smaller H concentrations, a value of $\Delta H^{\circ}_{H_2} = -32.0$ kJ per mole H_2 was obtained. Although these results apparently obeyed Sievert's Law, it is possible that the measurements still contained a contribution due to traps since the agreement with Sieverts' law was only over very small hydrogen contents.

Ohlendorf and *Flotow* [3.22, 52] evaluated the entropies of $LaNi_5H_{0.36}$ and $LaNi_5H_{6.39}$ from measurements of the heat capacities between 5 and 300 K. ΔS_{H_2}(plat) calculated from the equation

$$\Delta S_{H_2}(\text{plat}) = \frac{2}{b-a}[S(b) - S(a)] - S^{\circ}_{H_2} \quad (3.34)$$

was found to be in good agreement with ΔS_{H_2}(plat) obtained from plateau pressure measurements.

The thermodynamic properties of $LaNi_5$ can be altered by the partial substitutional replacement of either the La or Ni. Investigations have concentrated upon the variation of the H capacity and the plateau properties as a function of the type and amount of the substituted atom. Reviews of these results have been given elsewhere [3.53–55]. The only calorimetric measurements are those of *Hubbard* et al. [3.14], who investigated $La(Ni_{4.5}Al_{0.5})$–H and $La(Ni_4Al)$–H. They found that ΔH_{H_2}(plat, f) increased in exothermicity upon substitution of Al for Ni: from -38 to -46 kJ per mole H_2 for $La(Ni_{4.5}Al_{0.5})$ and $La(Ni_4Al)$, respectively. The former value agrees well with the result of *Mendelsohn* et al. [3.55], obtained from decomposition plateau pressure measurements. Generally, it seems that substitution affects the value of ΔH_{H_2}(plat) more than that of ΔS_{H_2}(plat).

3.4.2 TiFe

Reilly and *Wiswall* [3.56] found that two hydride phases form from cubic TiFe: the β-phase ($TiFeH_2$) and the γ-phase ($TiFeH_{1.9}$). Table 3.2 summarizes the results of *Reilly* and *Wiswall* [3.56], and *Wenzl* and *Lebsanft* [3.11]. The latter authors obtained their plateau enthalpies calorimetrically for activated TiFe. The values found by *Wenzl* and *Lebsanft* [3.11] for the α/β plateau are somewhat smaller in magnitude than those found by *Reilly* and *Wiswall* [3.56] but they confirm that the $\beta-\gamma$ transformation is more exothermic than the $\alpha-\beta$ transition, which is somewhat unusual for a multi-plateau system.

The results of *Reilly* and *Wiswall* [3.56] suggest that the β-phase co-existing with the α-phase is $TiFeH_{1.4}$ during hydride formation, and $TiFe_{1.0}$ during

Table 3.2. Thermodynamic data for the TiFe–H system

	ΔH_{H_2}(plat, d) [kJ mol^{-1} H_2]	ΔH_{H_2}(plat, f) [kJ mol^{-1} H_2]	ΔS_{H_2}(plat, d) [J K^{-1} mol^{-1} H_2]	Ref.
α/β	-28.12	–	-106.2	[3.56]
β/γ	-33.30	–	-103.6	[3.56]
α/β	-24.0 ± 0.8	-23.8 ± 1.0	–	[3.11]
β/γ	-33.8 ± 5.0	-32.0 ± 3.6	–	[3.11]

decomposition. *Schefer* et al. [3.57], whose isotherms for the TiFe–D_2 system resembled those found by *Reilly* and *Wiswall* for hydrogen, introduced the terminology β_1 for TiFe$D_{1.0}$ and β_2 for TiFe$D_{1.4}$. *Schefer* et al. reported different distributions of the D atoms in the β_1 and the β_2 phases within the same orthorhombic structure and *Reidinger* et al. [3.58] have characterised the two β structures by x-ray diffraction. However, it is possible that hysteresis may be responsible for the apparent existence of the two β-phases. It is known for systems with only one hydride phase that $b_f > b_d$. For a system exhibiting two plateaux, this hysteresis effect could give rise to the appearance of different phases during hydride formation and decomposition. It is possible to estimate, from the desorption isotherms of *Reilly* and *Wiswall* in the pure β-phase region, that $(\partial \mu_{H_2}/\partial r)_T \approx 10000$ J. If this value is combined with a hysteresis free energy of 1720 J per mole H_2 [3.11], then a value of $\Delta r = 0.17$ is obtained, which is the order of magnitude of the observed difference in compositions of the β "phases". In Chap. 4 of this text more details of the structural aspects of TiFe and its hydrides are given.

Reilly et al. [3.59] found that the γ-phase can be formed directly from the α-phase if the activated TiFe is annealed at 800°C prior to the absorption isotherm determination. The plateau pressure $p_f(\alpha - \gamma)$ was found to be approximately the same as that for $p_f(\alpha - \beta)$ for unannealed samples. *Goodell* et al. [3.60] have found that isothermal cycling of TiFe 148 times leads to greatly increased plateau pressures for the $\beta - \gamma$ transformation.

Because *Wenzl* and *Lebsanft* [3.11] employed an activated sample, their results for ΔH_{H_2} in the metal-rich phase were found to be large in magnitude, reflecting trapping at defect sites introduced during activation. In contrast, *Welter* et al. [3.37] found that ΔH_{H_2} was endothermic (+21.6 kJ per mole H_2) for the annealed material.

Zhirnova and *Mogutnov* [3.61] investigated the metal-rich phase by p-r-T methods in TiFe, presumably unactivated, at elevated temperatures (673 to 1273 K). Their results, described by the equation

$$\frac{-\Delta H_{H_2}}{R} = 4452 + 2.10T - 6.84 \times 10^{-3} T^2 \tag{3.35}$$

give a value of $\Delta H_{H_2} = -37.2$ kJ per mole H_2 at 298 K but yield positive values above 975 K. Their results are therefore intermediate between the values with no trapping [3.37] and those with trapping [3.11], although the extrapolated value at 298 K seems unlikely, in that it is more exothermic than ΔH_{H_2}(plat).

3.4.3 ZrV_2 and Some $ZrMn_2$-Based Systems

ZrV_2, a Laves phase with the C15 structure, does not form a separate hydride phase at temperatures at or above 298 K at the H_2 pressures normally encountered. It does, however, dissolve large amounts of H in solid solution. The most complete thermodynamic data for the system are those of *Pebler* and *Gulbransen* [3.62], although it should be noted that these results may have been affected by the presence of a second phase in the starting material. Their results for ΔH_{H_2} and ΔS_{H_2} are shown in Fig. 3.6, where a minimum in ΔS_{H_2} at $r \simeq 0.45$ is seen to be present. The stoichiometry of $ZrMn_2$, a Laves phase with the C14 structure, can be altered from $ZrMn_{1.8}$ to $ZrMn_{3.8}$ [3.63] with concomitant changes in plateau pressures and H capacity. *Sinha* and *Wallace* [3.64] have shown that Mn replaces Zr in the hyper-stoichiometric $ZrMn_{2+x}$. Both Zr and Mn can also be partially replaced by other elements [3.65–70].

The $ZrMn_{2+x}$–H system exhibits only one plateau. The interesting feature, however, is that the "plateau" is sloping, with the slope increasing with increasing x [3.71]. Thermodynamic data, obtained by both p-r-T and calorimetric methods, exist only for this plateau region [3.63, 72, 73]. The values reported for $ZrMn_{2.8}$ at $r=1$ are given in Table 3.3. The discrepancy between these values was attributed [3.72, 73] to irreversibility effects, which give rise to uncertainties in the values obtained from van't Hoff plots in the two-phase region. However, provided attention is paid to aliquot size and the effect of hysteresis, irreversibility should not give rise to such discrepancies. A re-determination [3.74] of the plateau properties for the $ZrMn_{2+x}$–H systems for $x=0.2$ to 1.5, using both p-r-T and calorimetric methods has shown no

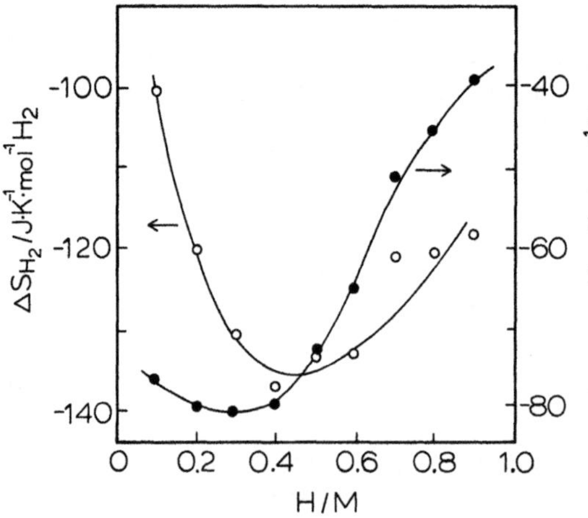

Fig. 3.6. Plots of ΔH_{H_2} and ΔS_{H_2} against hydrogen content for the ZrV_2–H system [3.62]

Table 3.3. Thermodynamic data for $ZrMn_{2.8}$ at $r=1$

Method and Ref.	ΔH_{H_2}(plat, d) [kJ mol^{-1} H_2]	ΔS_{H_2}(plat, d) [J K^{-1} mol^{-1} H_2]
p-r-T [3.63]	−17.8	−50.6
Cal. [3.72] + p-r-T	−29.9 (cal)	−91.3 (cal + p-r-T)

significant difference in the ΔH_{H_2}(plat) values obtained by the two methods. It seems, therefore, that the values obtained by p-r-T, which are given in Table 3.3 may be questioned for reasons other than lack of reversibility.

As pointed out by *Wallace* et al. [3.63], discrepancies also exist between results obtained by different investigators for the plateau pressures themselves on ostensibly the same $ZrMn_{2+x}$–H system. For example, *van Essen* and *Buschow* [3.71] report $p_d = 0.1$ atm at 323 K for $ZrMn_2$, whereas *Shaltiel* et al. [3.75] report 0.007 atm. *Majorowski* and *Flanagan* [3.74] find a value closer to that of the former investigation. There is a corresponding discrepancy between the results of *Wallace* et al. [3.63] and *van Essen* and *Buschow* [3.71] for the $ZrMn_{2.8}$–H system. These discrepancies do not appear to be caused by the effects of annealing. Whilst *van Essen* and *Buschow* did find a marked effect of annealing the IMC at 1050 °C on the subsequent isotherms, the effects on the plateau pressure are not as big as the differences between different investigations.

3.4.4 $ErFe_2$

Although $ErFe_2$, like ZrV_2, possesses a C15 Laves phase structure, its behavior towards H differs markedly. $ErFe_2$–H forms five different hydride phases, as illustrated by the results of *Kierstead* [3.76] in Fig. 3.7. No structural

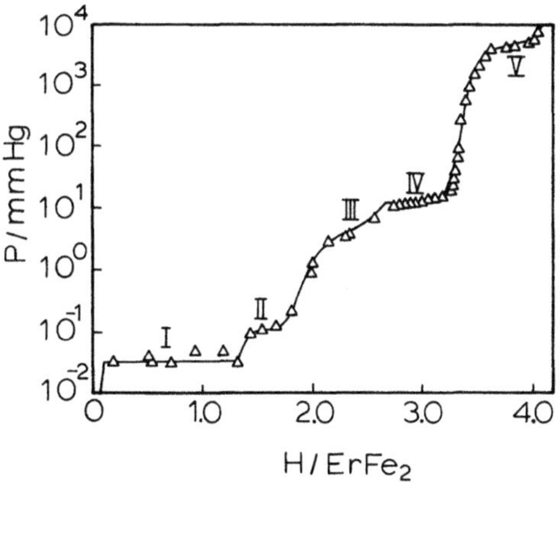

Fig. 3.7. Plot of the equilibrium hydrogen pressures against hydrogen content for the $ErFe_2$–H system at 293 K [3.76]. The five plateau regions are indicated by the roman numerals

transformations are involved for the first four hydrides which form but, when the hydride corresponding to $ErFe_2H_4$ forms, there is a rhombohedral distortion of the unit cell. *Kierstead* has determined the thermodynamic properties from p-r-T measurements on this system in the temperature range 273 to 353 K [3.76]. The values of ΔH_{H_2}(plat) for the first four plateaux are all of a similar magnitude, -47.9 ± 3.0 kJ per mole H_2, so that the different $\Delta \mu_{H_2}$(plat) values for these four plateaux are determined principally by the different ΔS_{H_2}(plat) values.

3.4.5 Mg-Based IMCs

These IMCs constitute a rather special group from the viewpoint of their potential for hydrogen storage due to their favorable ratios of weight of hydrogen to weight of metal in the IMCHs. Another unusual aspect of some of these Mg-based IMCHs is that the IMCH can be more stable than its decomposition products. Both MgH_2 itself and the Mg-based IMCHs are rather stable and whether decomposition of the IMCH to MgH_2 and another IMC or to a pure metal component occurs or not depends upon small differences in the stabilities. The similar stabilities of the hydride phases can be appreciated from the values of ΔH_{H_2}(plat) for MgH_2 and Mg_2NiH_4 which are -70.0 kJ per mole H_2 [3.77] and -64.4 kJ per mole H_2 [3.78], respectively. *Reilly* [3.79] has shown that at 562 K

$$Mg_2Cu(s) + 2H_2(g) \rightarrow Mg_2CuH_4; \quad \Delta G^{\circ}(\text{plat}) > 1.7 \text{ kJ per mole } H_2$$

$$2Mg_2Cu(s) + 3H_2(g) \rightarrow MgCu_2(s) + 3MgH_2(s);$$

$$\Delta G^{\circ}(\text{plat}) = 1.7 \text{ kJ per mole } H_2$$

whereas

$$Mg_2Ni(s) + 2H_2(g) \rightarrow Mg_2NiH_4(s); \quad \Delta G^{\circ}(\text{plat}) = 1.0 \text{ kJ per mole } H_2$$

$$2Mg_2Ni(s) + 3H_2(g) \rightarrow MgNi_2 + 3MgH_2(s);$$

$$\Delta G^{\circ}(\text{plat}) = 2.1 \text{ kJ per mole } H_2 .$$

It can be seen that the IMCH is the thermodynamically favored state for Mg_2Ni but not for Mg_2Cu. Mg_2Al_3 has also been shown to decompose upon hydriding to MgH_2 and Al [3.80].

Didisheim et al. [3.81] have recently prepared Mg_2FeH_6, which is a very hydrogen-to-metal rich IMCH, from the reaction of Mg/Fe mixtures with hydrogen at 773 K and 60 bar. Unfortunately this IMCH does not seem well suited to hydrogen storage because of its high stability which may result from the presence of $[FeH_6]^{-4}$ ions in the structure as shown by the authors from physical measurements and neutron diffraction.

3.5 Microscopic Thermodynamics

3.5.1 Configurational Models

Configurational models attempt to give an interpretation of the composition-dependent thermodynamic behavior of solutions in terms of simple interactions between the component atoms. The other contributions to the partition function are considered independently. Since exact calculations of the configurational partition function are probably impossible in a complex systems such as IMCHs, gross simplifying assumptions must be made in the calculations. As will be discussed below, these assumptions may result in physically unrealistic models of IMCHs.

a) Site and Interaction Energies

In strictly pairwise model calculations on ternary metal-hydrogen solutions it is necessary to know the metal-hydrogen and H–H interaction potentials. A modified approach, which has been used by most workers in treating alloy-H systems, is to use the local environment model [3.82, 83], in which the various possible sites for H absorption are allocated "site energies", $V_0(j)$, where j denotes a particular type of interstitial site. In its simplest form, the local environment model assumes that the number of different kinds of metal atoms which constitute the first co-ordination shell surrounding the interstitial site determine the value of $V_0(j)$. Then if j represents a tetrahedral site with two A atoms and two B atoms in the first co-ordination shell, the site energy is $V_0([A_2B_2])$. The reference state for the various $V_0(j)$ is most usefully taken as H_2(g, 1 atm).

In applying the model to finite H concentrations the H–H interaction can be considered pairwise with interaction energies V_i, where i is the ith nearest neighbor. V_i is taken as zero at infinite separation. In principle, it should be possible to predict by a priori methods the preferred sites for H occupation and the values of $V_0(j)$ and V_i required in the model calculations. A few attempts have been made along these lines and some useful results have emerged. *Switendick* [3.84], for example, has shown that $V_i \rightarrow +\infty$ when the H–H separation is < 2.1 Å, i.e., there is a hard core repulsion inside this radius. This calculation is in agreement with experimental observations.

For the most part, however, information on $V_0(j)$ and V_i has come from a knowledge of the preferred sites for occupation by H, the H capacity and the thermodynamic properties, although, with this more empirical approach, difficulties can arise in distinguishing between the relative importance of $V_0(j)$ and V_i in contributing to the observations. Multi-plateaux, for example, can be explained by assuming either equal or different $V_0(j)$ for the different sub-lattices (see below).

Westlake [3.85] has used a completely geometrical method to interpret IMH structures. He postulates that there is a minimum hole size for H occupation ($r_h > 0.33$ Å) in pure metals or IMCs, although hydride formation does not occur until $r_h > 0.40$ Å. Because of the lattice expansion accompanying H absorption, sites initially too small may become available for H occupation at higher H concentrations. Although *Westlake* does not make the claim, it might be expected, on the basis of this postulate, that the larger the hole size the lower $V_0(j)$, so that, in compounds where the A atom is larger than the B atom, $V_0([A_2B_2]) < V_0([AB_3]) < V_0([B_4])$. A further postulate of *Westlake* is that $V_i \to \infty$ at separations of less than 2.1 Å and is zero outside this hard core radius. In the case of the $[A_2B_2]$ sites in some IMCs possessing the C15 structure this hard core radius may extend out as far as the 6th nearest neighbor sites. Using these assumptions concerning $V_0(j)$ and V_i, *Westlake* has been able to interpret many observed structural features of IMCHs [3.86, 87]. *Yartys* et al. [3.88] have given a rather similar interpretation of IMH structures to that of *Westlake*. A slightly different view, however, has been presented by *Somenkov* and *Irodova* [3.89] for the C15 Laves phases who interpret the observed structures as implying that, when the lattice size is small, the $[A_2B_2]$ sites are preferred but, in large lattices, $V_0(j)$ for the $[AB_3]$ becomes lower than that for the $[A_2B_2]$ sites. The crossover occurs at a lattice parameter of $\simeq 7.9$ Å in these compounds.

Shinar et al. [3.90] have attempted to estimate the actual magnitudes of the $V_0(j)$ by using an "imaginary binary compound" concept. It is assumed that the fraction of a H atom which is bonded to an A atom is determined by the fraction of A atoms in the first shell of metal atoms surrounding the H atom. Thus, for example, the site energy per H atom for $[A_2B_2]$ sites containing n H atoms in the compound AB_2 is given by the sum of the enthalpies of formation of $AH_{n/2}$ and $B_2H_{n/2}$ divided by n. Using this procedure the $V_0(j)$ values vary with H concentration. The enthalpies of formation of the imaginary binary compounds are obtained using *Miedema's* method (see Sect. 3.5.3a), although *Westlake* [3.85] has noted that this does not appear to have been applied correctly by *Jacob* and *Shaltiel* [3.91]. *Buschow* et al. [3.1] have also pointed out that any estimation of the $V_0(j)$ should somehow take into account the stability of the IMC as well as the binary hydrides. In the case of a Nb–Pd IMC, for example, large negative values of the $V_0(j)$ would be predicted from the imaginary binary hydrides and thereby significant IMH stability, whereas the reverse stability rule would predict, correctly, high $V_0(j)$ values and hence an unstable hydride.

Whilst the approximate range of the hard core part of V_i is known, there is little information available on the longer range behavior of the H–H interaction. In IMCHs where phase separation occurs, thermodynamics requires that there must be a lowering of ΔH_{H_2} with increasing r. In the pairwise interpretation this means that there must be an effective H–H attractive interaction outside the hard core radius. However, it is possible that, as in the case of pure metals [3.92], an indirect interaction is primarily responsible, i.e., the lowering of ΔH_{H_2} comes from the lattice expansion on H absorption, which could be interpreted as implying that the $V_0(j)$ values decrease with r.

In principle, once $V_0(j)$ and V_i are known they can be used in the calculation of the configurational thermodynamic properties. From the foregoing it seems that, at present, the information on $V_0(j)$ and V_i is more qualitative than quantitative so that they are used as fitting parameters in the thermodynamic models. The models which have been used in attempting to interpret the experimental results on IMCHs and which are discussed in the next few sections are of varying degrees of sophistication.

b) Hard Core, Constant $V_0(j)$ Models

The simplest model is to assume that all the $V_0(j)$ are equal and that V_i is zero outside a cut-off radius but infinite inside that radius. The hard core assumption means that the H–H attractive interaction, required for phase separation, is assumed to be indirect. Only the configurational entropy has to be calculated. The calculation of the hard core configurational entropy associated with H occupation on a set of sites of the same energy is straightforward only if there is no blocking, i.e., only when the hard core radius does not extend beyond the nearest neighbor interstitial site. In the case of no site blocking:

$$-S_{\mathrm{H}}/R=\ln\left(\frac{\theta}{1-\theta}\right) \tag{3.36}$$

$$-S_M/R=\beta\ln(1-\theta), \tag{3.37}$$

where β is the number of interstitial sites per metal atom.

It is worth emphasising that there is an effect on $\Delta\mu_M$ as well as on $\Delta\mu_{\mathrm{H}_2}$ as a result of this configurational entropy, even though the metal lattice itself has not undergone any configurational change (cf. *Pasturel* et al. [3.93]).

When blocking of near neighbor sites occurs, the calculation of the hard core entropy is more complex. In the case of blocking without overlap the entropy can be calculated exactly by simply substituting zr/β for θ in the above equations, where z is the number of near neighbor sites blocked. However, a better calculation of the entropy allows for the overlap of the blocked sites. This calculation is not straightforward so that Monte Carlo methods must be used to calculate how z is reduced with increasing H concentration [3.94], although *Boureau* [3.95] has given an approximate method for calculating S_{H}.

In the case of IMCHs, an improved model allows for the presence of the sites of different environment and $V_0(j)$ which form different sub-lattices in the ordered IMC. The calculation of the configurational entropy is now straightforward only if the sub-lattices are assumed to be independent, if the differences in $V_0(j)$ are neglected and if it is assumed that no blocking takes place. With these assumptions:

$$\frac{-S_m}{R}=\sum_i g_i[\theta_i\ln\theta_i+(1-\theta_i)\ln(1-\theta_i)], \tag{3.38}$$

where θ_i is the fraction of type i sites occupied, g_i is the degeneracy of the type i sites and S_m is the integral configurational entropy.

Wallace et al. [3.96] used (3.38) in an attempt to give support to one particular structural model for $LaNi_5H_6$ by comparing calculated and experimental entropies. However, as *Achard* et al. [3.97] correctly pointed out, the calculation of the configurational entropy in an IMCH is more complex than this since it is necessary to allow for the blocking effect within and between the different sub-lattices. They attempted to do this by using the following equation for the thermodynamic probability, Ω:

$$\Omega=\prod_i \frac{(N_i^{\mathrm{avail}})!}{n_i!(N_i^{\mathrm{avail}}-n_i)!}, \tag{3.39}$$

where n_i is the number of H atoms on type i sites and N_i^{avail} is the number of available sites of this type.

Pasturel et al. [3.86] calculated N_i^{avail} from:

$$N_i^{\mathrm{avail}}=f_iN_i-\sum_j f_{ij}n_j, \tag{3.40}$$

where f_i and f_{ij} are blocking factors for the type i sites due to the presence of H atoms on type i and j sites, respectively, N_i is the total number of type i sites and n_j is the number of H atoms on type j sites. Although (3.39) and (3.40) are an improvement over (3.38) they are still inadequate, in that they are correct only for the case of blocking without overlap and, more importantly, are restricted to mixing on sub-lattices of the same energy.

As indicated previously, phase separation at low temperatures can be accounted for in terms of the preceding hard core entropy calculations if the attractive H–H interaction is assumed indirect. For example, in the simplest case of small hard core mixing on sites of equal $V_0(j)$ the hydrogen potential is given by:

$$\tfrac{1}{2}\Delta\mu_{H_2}=\Delta H_H^{\circ}-T\Delta S_H^{\circ}+RT\ln\left(\frac{\theta}{1-\theta}\right)+W\theta, \tag{3.41}$$

where W is the interaction coefficient. When W is constant and negative, phase separation will occur at temperatures $<-W/4R$. However, when W is constant the isotherms and phase separation envelope are symmetric about $\theta=1/2$. A linear composition dependence of W permits asymmetry about $\theta=1/2$.

In this same approximation the plateau properties are given by [3.6]:

$$\Delta H_{H_2}(\mathrm{plat})=\Delta H_{H_2}^{\circ}+W/2 \tag{3.42}$$

$$\Delta S_{H_2}(\mathrm{plat})=\Delta S_{H_2}^{\circ}. \tag{3.43}$$

The solvus compositions, a and b, can also be obtained from (3.41) (see [3.6]).

c) Multiple Plateaux Models

An interesting feature of some IMCHs is the occurrence of multiple plateaux as the H concentration changes. Even more interesting is that these multiple phase changes occur sometimes when only the one type of site is being occupied. A good example of the latter is the $ErFe_2$ system where, at least at the lower H levels, only the $[A_2B_2]$ sites are occupied [3.76].

The simplest way to explain the multi-step isotherms is to assume that the interstitial sites in the IMC can be considered as independent sub-lattices possessing different $V_0(j)$ values. The distribution of H atoms between the different sub-lattices must then be calculated. *Jacob* et al. [3.98] assumed a Boltzmann distribution for calculating the distribution of H atoms on the different sub-lattices at room temperature from a knowledge of $V_0(j)$ values calculated by the method of *Shinar* et al. [3.90]. However, the H atoms are partitioned within and between the different types of sites according to a Fermi and not a Boltzmann distribution. At the large H concentrations of interest in IMCHs large discrepancies will exist between distributions calculated from these two distribution functions.

Since the hydrogen potential must be equal on each of the sub-lattices, equations similar to (3.41) can be written for each sub-lattice. These equations incorporate the Fermi distribution of the H atoms.

In the absence of H–H interactions, the configurational ΔH_{H_2} is given by [3.99]:

$$\Delta H_{H_2} = 2\sum_j V_0(j) p_j^* , \quad \text{where} \tag{3.44}$$

$$p_j^* = \frac{f_j\{\exp[(V_0(j) - \frac{1}{2}\Delta\mu_{H_2})/kT]\}/\{1 + \exp[(V_0(j) - \frac{1}{2}\Delta\mu_{H_2})/kT]\}^2}{\sum_j \{\exp[(V_0(j) - \frac{1}{2}\Delta\mu_{H_2})/kT]\}/\{1 + \exp[(V_0(j) - \frac{1}{2}\Delta\mu_{H_2})/kT]\}^2}, \tag{3.45}$$

where f_j is the fraction of type j sites.

In the presence of indirect H–H interactions the term $2W\theta$ can be added to ΔH_{H_2} obtained from (3.44).

Kierstead [3.100] has used this model for calculating the multi-step isotherms in some IMCHs although he did not use (3.44) and (3.45) for the partial energy. Figure 3.8 illustrates, for the two-sub-lattice case, how $\Delta\mu_{H_2}$, ΔH_{H_2}, and ΔS_{H_2} vary with H concentration for selected $V_0(j)$ and W values. The experimental results of *Kierstead* for the $DyCo_3$–H system [3.101] exhibit features similar to those shown in Fig. 3.8.

Equation (3.41) provides four adjustable parameters for each sub-lattice for the fitting of experimental results, namely, ΔH°, ΔS°, W, and β, the number of interstitial sites per metal atom (which is needed to relate r to θ). If allowance is made for a linear temperature dependence of ΔH°, ΔS°, and W, the number of parameters is increased to seven, and if a linear dependence of W on θ is also introduced, the parameter count is increased to eight. Thus if there are m such

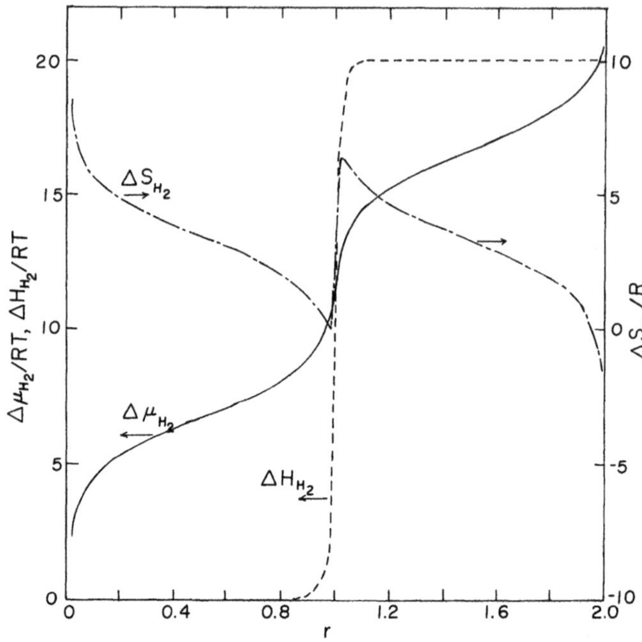

Fig. 3.8. $\Delta\mu_{H_2}$, ΔH_{H_2}, and ΔS_{H_2} plotted as a function of hydrogen content for a two sub-lattice model where $V_0(1)=0$, $V_0(2)/RT=10$ and $W=0$. The latter has been chosen for simplicity; suitable negative values of W will lead to loops in the $\Delta\mu_{H_2}$–r relationship, i.e., to horizontal plateaux

sub-lattices then between 4m and 8m parameters become available for fitting to experimental results. With such a plethora, it is not surprising that the model can be made to fit experimental results quite well. The fitting may not be unique, however, as shown by the fact that *Kierstead* first [3.101] interpreted his experimental results for $DyCo_3$ using a 4 sub-lattice/6 parameter model but subsequently [3.102] interpreted these same results using a 3 sub-lattice/8 parameter model. Both involve a total of 24 parameters. As was indicated by *Kierstead* himself, the value of this type of approach is more to give a representation of the data rather than to suggest that such models bear any close relation with physical reality.

According to the multiple site model discussed above, multi-step isotherms are explained in terms of the different $V_0(j)$ for the different types of interstitial site. However, in several IMCs the H atoms occupy only one type of site but multiple plateaux are still observed. $ErFe_2$ is a case in point. A model due to *Rees* [3.103] attempted to explain how multi-step isotherms could arise from H absorption on only one type of site. In this model the emphasis is placed on V_i rather than $V_0(j)$, i.e., the occupation of one site is envisaged as creating adjacent sites of higher energy. Subsequent occupation of these new types of site creates a further set of still higher energy and so on. It is worth noting that the calculation of the partition function for this model is difficult and was not done correctly by *Rees* (see [3.94]). *Kierstead* [3.104] has used the *Rees* model for interpreting the multi-plateau behavior of some IMCHs as an alternative to the multiple sub-lattice model. Again, however, the number of adjustable parameters is large.

From the above discussion it seems that multi-step isotherms can be interpreted as arising principally from either the effect of different $V_0(j)$ or from the effect of V_i. In actual IMCHs it is likely that both factors play a role with the relative importance varying from system to system. It is possible that the relative magnitudes of ΔS_{H_2}(plat) for the different plateaux in one IMCH may be of value in distinguishing between the relative importance of $V_0(j)$ and V_i. This can be illustrated by considering a two-plateaux system. For the two-sub-lattice model then $\Delta S_{H_2}(\text{plat}) \approx \Delta S^{\circ}_{H_2}$ for each plateau. However, where the steps in the isotherm are due to a repulsive V_i then, taking the simplest assumption of ideal mixing

$$\Delta S_{H_2}(\text{plat}) = \frac{2}{b-a}\left[\Delta S_m(b) - \Delta S_m(a)\right] \tag{3.46}$$

$$= \Delta S^{\circ}_{H_2} - \frac{2R}{b-a}\left[b \ln b + (y-b)\ln(y-b) - a \ln a - (y-a)\ln(y-a)\right]. \tag{3.47}$$

The effect of (3.47) is to make ΔS_{H_2}(plat) more negative as the average H concentration, $\bar{r} = (a+b)/2$, increases. These different predictions regarding ΔS_{H_2}(plat) from the two models are shown schematically in Fig. 3.9.

As discussed in Sect. 3.4.4, ΔH_{H_2}(plat) is approximately constant for the different plateaux in the $ErFe_2$–H system. However, as can be seen in Fig. 3.10, ΔS_{H_2}(plat) steadily becomes more negative with the average H concentration. If y is taken as 5 for the occupation of equivalent $[A_2B_2]$ sites in the C15 structure and $\Delta S^{\circ}_{H_2}$ is taken as $-104\ \mathrm{JK^{-1}}$ per mole H_2 [3.39] the calculated values of ΔS_{H_2}(plat) shown in Fig. 3.10 are obtained. Qualitatively, the calculated and experimental results agree, confirming the expectancy that, where the same interstices are being occupied, ΔS_{H_2}(plat) should become more negative in multi-plateau systems with increasing $\bar{r}$. On the other hand, if different kinds of interstices are being occupied as $\bar{r}$ increases then, to first order, ΔS_{H_2}(plat) will remain approximately constant. For the $LaCo_5$–H system ΔS_{H_2}(plat) increases slightly in passing from the first plateau to the second. This seems to be consistent with the isotope effect (see Sect. 3.5.2), which indicates a looser binding in the second hydride phase and, therefore, a less negative value of ΔS_{H_2}(plat). For the $R\mathrm{Co}_3$–H systems (R = Dy, Tb, Lu, etc.) with the $PuNi_3$ structure *Kierstead* [3.105] finds that ΔS_{H_2}(plat) does not vary from system to system or for different plateaux in the same system. This again indicates that different types of sites may predominate for each plateau region, a conclusion which is consistent with structural information [3.106].

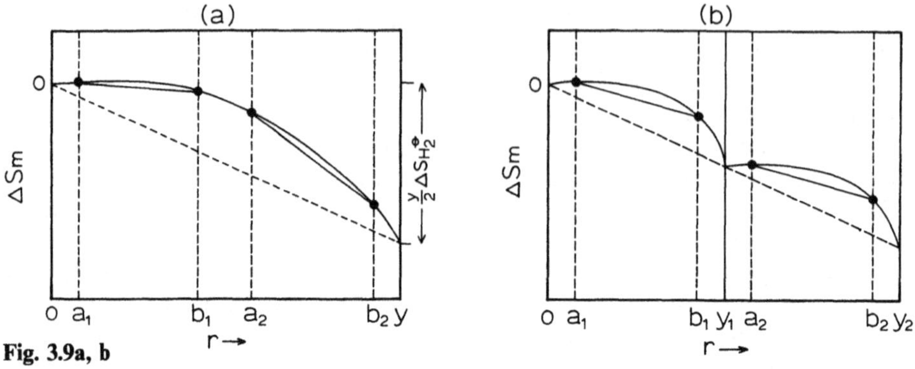

Fig. 3.9a, b

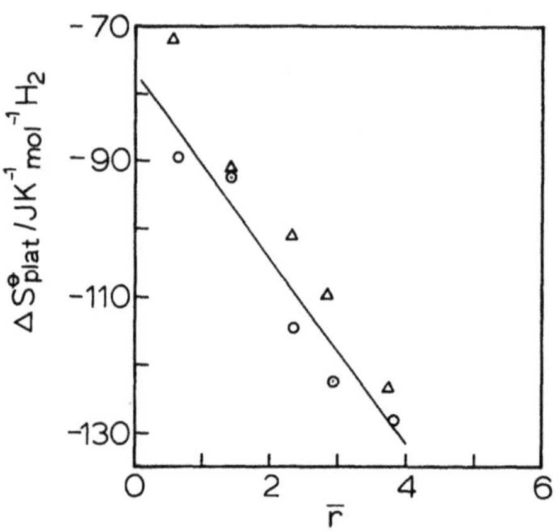

Fig. 3.9a, b. A schematic comparison of ΔS_m against hydrogen content for two multi-plateaux models: (**a**) single sub-lattice (Rees model [3.103]) and (**b**) two sub-lattice model. The values of ΔS_{H_2}(plat) can be seen to decrease markedly for model (**a**) but not for model (**b**), see (3.46)

◀ **Fig. 3.10.** ΔS_{H_2}(plat) plotted against the average hydrogen contents of the plateaux for the $ErFe_2$–H system. (○), experimental data from *Kierstead* [3.76] and (△), calculated results from (3.47)

3.5.2 Non-Configurational Contributions

Strictly speaking configurational models incorporate any non-configurational contributions to the partition function in the adjustable site energy values and assume that they are independent of composition. It is of interest, however, to have some knowledge of the electronic and, particularly, the vibrational contributions.

Little is known about the lattice dynamics of IMCHs. Some information is available from heat capacity measurements [3.22, 23], inelastic neutron scattering measurements [3.33, 107] and from measurements of the isotope solubility ratio. The neutron scattering measurements of *Shapiro* et al. on the optic modes in $FeTiH_x$ [3.107] reveal rather complicated spectra which are strongly dependent on both temperature and H concentration. Because of the difficulties in preparing single crystal hydrides, no measurements of the coherent scattering from IMCs containing deuterium have been made and, consequently, no theoretical interpretations of the lattice dynamics of IMCHs have been attempted.

In spite of the obvious limitations, it is possible to calculate the expected results for the isotope effect at constant H(D) concentration expressed as the

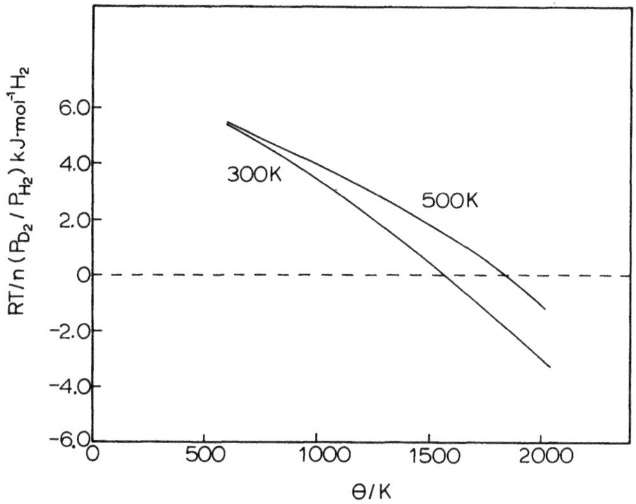

Fig. 3.11. Plot of $RT\ln[p_{D_2}/p_{H_2}]$ for conditions of infinite dilution against θ_H where $\theta_H = h\nu/k$ for 300 and 500 K from *Oates* and *Flanagan* [3.108]

isotope ratio, IR $= RT\ln(p_{D_2}/p_{H_2})$, by using the Einstein oscillator model and by assuming $\theta_H = \sqrt{2}\theta_D$. Some calculated results are shown in Fig. 3.11 [3.108]. The values of the optical frequencies obtained in this way do not agree exactly with those obtained from neutron scattering in the case of pure metal-H systems and reasons for the discrepancy have been suggested [3.108]. Thus the main value of measurements of the IR is in the observation of trends. It can be seen from Fig. 3.11 that in the region of room temperature, the IR changes from positive to negative values as the optic mode frequency increases. The IR can also provide information on the occupation of octahedral versus tetrahedral sites. Thus it is known from inelastic neutron scattering measurements on pure metal-H systems that $\theta_H \simeq 1200$–1800 K for H in tetrahedral sites, whereas $\theta_H \simeq 650$–1000 K for H in octahedral sites. As can be seen from Fig. 3.11, these differences are sufficient to have a significant effect on the IR near to room temperature.

From the viewpoint of a theoretical understanding, the IRs are best determined at infinite dilution. Unfortunately, this is not feasible for IMCHs, because most dilute phase data are affected by trapping of H at defects. Consequently, the only reliable IR information for IMCHs refers to the plateau region. Although a full analysis of the isotope effect of the plateau region is quite complex, generally it can be shown that the IR is somewhat greater for the plateau region than for the infinitely dilute case. The differences are, however, not expected to be great and thus similar conclusions regarding the type of site occupation, etc. can be drawn.

Figure 3.12 shows some results for the IR in different IMCHs. *Biris* et al. [3.109] found that the IR for $LaNi_5$ changed from positive to negative values

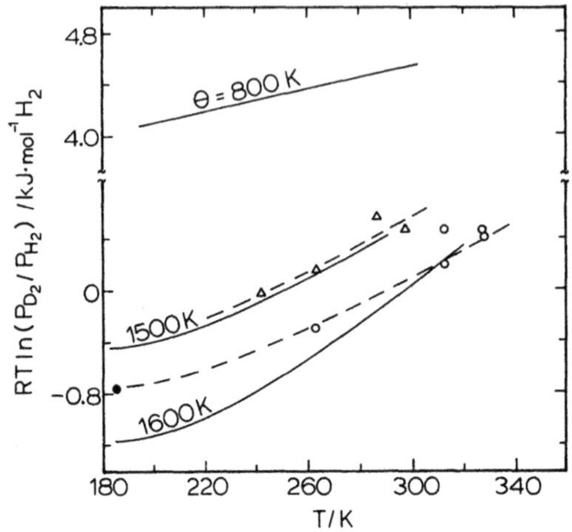

Fig. 3.12. Plot of $RT\ln(p_{D_2}/p_{H_2})$ against temperature for infinite dilution at selected values of θ_H. △, $La_{0.4}Ce_{0.6}Ni_5$ [3.111]; (○), $LaNi_5$ [3.109] and (●), $LaNi_5$ [3.111]. The experimental values all refer to the plateau pressures

near room temperature as shown in Fig. 3.12. *Andreyev* et al. [3.110] extended this work to 195 K and, as also shown in Fig. 3.12, their results are consistent with those of *Biris* et al. [3.109].

Dayan and *Dariel* [3.111] reported what they considered to be an anomalous isotope effect for the $(La_{0.4}Ce_{0.6})Ni_5$ system. Their data are also shown in Fig. 3.12, where it can be seen that these results are those to be expected for tetrahedral site occupation.

There are also a few results for the IR in multi-plateaux systems. *Kuijpers* [3.112] found, for $LaCo_5$, that IR = 410 J per mole H_2 and IR = 2000 J per mole H_2 for the first and second plateaux, respectively. These values are both within the range expected for occupation of tetrahedral interstices and suggest a looser binding of H to the sites which are being occupied in the second plateau region, as is also expected. However, it should be noted that, in the $CaNi_5$ system, the opposite behavior has been observed [3.113] in that the IR decreases as the system progresses to a higher plateau.

3.5.3 Phase Transitions

Until fairly recent times, the phase relations in pure metal–H systems appeared relatively simple but it is now realised that this is not the case. Even the phase diagram of the Pd–H system, previously thought to be the simplest of all and well established, is now known to be rather more complex and remains in some doubt [3.114]. Partial phase diagrams for IMCHs are only just becoming available [3.89] and much remains to be done in this area. Nevertheless, it is already clear that they exhibit many of the features observed in the diagrams for pure metal–H systems. Figure 3.13, for example, shows the phase diagram for the $ZrCr_2$–D system. It can be seen that, similar to many pure metal–H phase diagrams, there is a "gas/liquid" $\alpha-\alpha'$ phase transition, with order/disorder reactions occurring at lower temperatures.

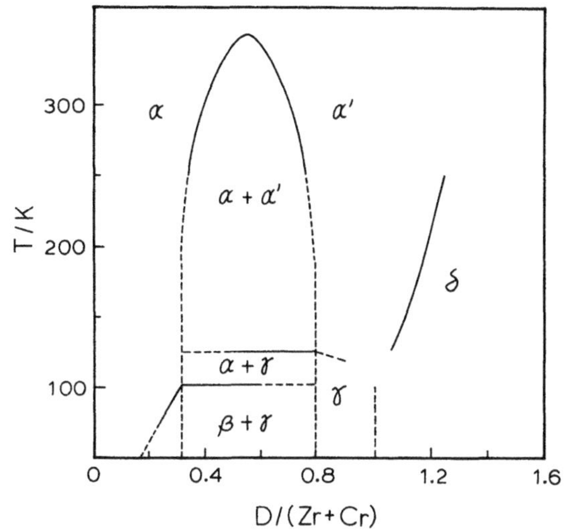

Fig. 3.13. The phase diagram for the $ZrCr_2$–D system. The solid lines represent regions established experimentally by diffraction techniques and the dashed lines represent extrapolated behavior as given by *Somenkov* and *Irodova* [3.89]

a) The $\alpha-\alpha'$ Transition

Yartys et al. [3.88] consider that this "gas/liquid" condensation transition, as opposed to a "gas/solid" structural transformation, occurs in most IMCHs at temperatures where they are likely to be used technologically. Thus the observed plateau pressures and solvus compositions will usually pertain to this type of transformation. It also seems probable that, just as in the case of pure metals [3.92], the condensation arises, principally, from the indirect H–H attractive interaction resulting from the lattice expansion associated with the introduction of H into the host lattice. The direct repulsive interactions, which operate at short range, will influence the position of the critical composition through their influence on the configurational entropy. Another effect of these short range repulsions is to give rise to short range order, which has been observed in neutron diffuse scattering in the ZrV_2–D system [3.88].

$\Delta\mu_{H_2}$(plat) and the solvus compositions for the $\alpha-\alpha'$ transition should follow from a full knowledge of the configurational and non-configurational partition functions in the single phase regions. However, in view of the fact that our understanding of these, at present, is unsatisfactory for predictive purposes, more direct methods have been suggested for estimating the technologically important plateau properties. All rely on the approximate relation (3.28) between $\Delta\mu_{H_2}$(plat) and $\Delta G_f^{\circ}(MH_y)$. The reader is referred to Chap. 6 of this volume for details of the various methods which have been suggested for the estimation of ΔH_{H_2}(plat). However, a method proposed by the present authors [3.10] for the direct estimation of $\Delta\mu_{H_2}$(plat) is discussed briefly in Sect. 3.5.4.

b) Order/Disorder Transitions

Various interstitial superstructures, formed by H atom rearrangement and characterised by a wide range of wave vectors, $\boldsymbol{k}$, have already been found in

IMCHs [3.81]. In the ZrV_2–D system, for example, ordering with $\boldsymbol{k}=(1/2\ 1/2\ 1/2)$, (1 0 0), 0, and (1 0 0) has been found at MD_2, MD_3, MD_4, and MD_6, respectively [3.89]. It appears that superstructures with $\boldsymbol{k}=0$ or (1/2 1/2 1/2) are found at higher temperatures and over wider ranges of stoichiometry than those with the other wave vectors [3.88].

The simplest type of ordering, orientation ordering, occurs when $\boldsymbol{k}=0$. However, even in this case, because of the different types and complex arrangements of the sites available in IMCs, calculations of the superstructures are quite involved. The theory of ordering on simple lattices has been developed, by amongst others, *Khachaturyan* [3.115] and, recently, *Irodova* [3.116] has used *Khachaturyan's* theory to predict the structures which may appear due to orientational ordering on the $[A_2B_2]$ sites in C 15 Laves phases. The 96 sites per unit cell form 24 interpenetrating fcc Bravais lattices. *Irodova* showed that the matrix symmetry restricts ordered structures to nine types. When the further restriction of the blocking of first and second nearest neighbors is imposed, only two of the nine remain and only one of these permits the composition AB_2H_4 to be achieved. *Gufan* and *Shirokov* [3.117] have adopted a slightly different approach and showed that ordering at the composition AB_2H_4 is possible in 13 ways which differ in symmetry but which can be reduced to 8 if it is assumed that the H–H interaction decreases fairly slowly with increasing separation. In all cases the superstructure has a tetragonal symmetry and belongs to the space group C_{4h}^6. It is clear that such comparisons of the theoretical predictions of all possible superstructures in a system with those observed experimentally will undoubtedly lead to a better understanding of H–H interactions in IMCHs.

3.5.4 Complete and Partial Equilibrium

As indicated in Sect. 3.2.2, $\Delta\mu_{H_2}$(plat) for the decomposition reactions under CE can be calculated approximately from ΔG° for the reaction (3.29). The accuracy of this approximation will be similar to that in using (3.28), i.e., it will be better at low temperatures where the mutual solubilities of the metal and hydride phases are lower. Fig. 3.14 shows the CE phase diagram for the La–Ni–H system calculated on the basis of this approximation and should be compared with Fig. 3.5, which shows the PE phase diagram for the same system. The various $\Delta\mu_{H_2}$(plat) for CE are lower than for the equivalent $\Delta\mu_{H_2}$(plat) for PE. It has been suggested [3.10, 118] that the differences between these $\Delta\mu_{H_2}$(plat) values for CE and PE will be fairly small so that those for PE can be estimated with a fair degree of confidence from the calculated ones for CE. This means of calculating the plateau pressures in pseudo-binary IMCHs provides an alternative to the methods discussed in Chap. 6.

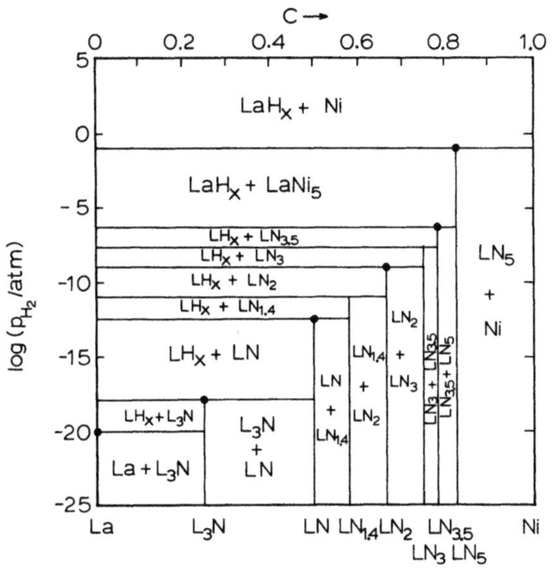

Fig. 3.14. The complete equilibrium phase diagram for the La–Ni–H system at 373 K from *Oates* and *Flanagan* [3.10]. The filled small circles represent the hydrogen pressures where the various phases are expected to decompose

3.5.5 Hysteresis

General accounts of hysteresis have been given recently by *Flanagan* and *Clewley* [3.24] and *McKinnon* [3.25]. *Sinha* and *Wallace* [3.119] have also recently developed a model of hysteresis based on elastic misfit between the hydride and the dilute phase. However, this model seems to be at odds with the fundamental point that hysteresis represents the conversion of work to heat, i.e., hysteresis must involve irreversible effects and therefore cannot be due to elastic deformation.

Hysteresis is believed to arise mainly from the irreversible plastic work which is needed for the accommodation of the dilute phase/hydride phase misfit during both hydride formation and decomposition [3.120]. In the initial activation process the expansion cannot be accommodated by dislocations and mechanical disruption occurs (brittle fracture) in most IMCs whereas in pure metal–H systems the expansion can generally be accommodated by punching out of misfit dislocations [3.121]. Following the first hydriding/dehydriding cycle in IMCs, further expansions and contractions during cycles of hydriding and dehydriding are presumably accommodated largely by dislocation creation. After the first hydriding/dehydriding cycle the dislocation density does not increase markedly and this implies that the dislocations which are created are subsequently annihilated leading to a constant, large dislocation density.

Park and *Flanagan* [3.122] recently observed an important effect of the interface velocity on the subsequent, time-independent plateau pressures of the $LaNi_5$–H system. Rapid, initial interface velocities lead to smaller values of p_f and larger values of p_d as compared, in both cases, to the effect of small interface velocities. This effect provides an explanation for the previously unexplained results of *Goodell* et al. [3.60], who found that dynamic isotherms differed from static isotherms for several IMCHs. It turns out that the former isotherms were determined under conditions which corresponded to slow interface velocities and the latter to rapid, initial interface velocities. The knowledge of the existence of this effect is important because otherwise serious errors can be introduced into thermodynamic parameters for the plateau regions. For example, if values of p_f are determined under fast interface conditions at one temperature and slow interface conditions at another temperature, erroneous data will be obtained for the plateau thermodynamic parameters.

Park and *Flanagan* [3.123] offered an explanation for this effect based on local equilibrium considerations at the interface, i.e., at the interface $\mu_A = \mu'_A$, $\mu_B = \mu'_B$, and $\mu_H = \mu'_H$ whereas in the bulk only the last of these equalities holds. The nature of the local equilibrium established depends upon the initial interface velocity and this, in turn, affects the final value of $\Delta\mu_{H_2}$(plat). If their explanation is correct, this will lead to another form of hysteresis in multi-component solid–H systems. The fact that decomposition of IMCHs can occur at relatively low temperatures as a result of repeated cycling supports the model because it indicates enhanced mobility of the metal atoms at the moving interface. *Hillert* [3.124] has pointed out that such a hysteresis will also occur in the transformation of austenite containing alloying elements but his concern was with the transformation kinetics rather than the stable, time-independent plateau pressures of interest here.

3.6 Conclusions

Until recently most "thermodynamic" investigations of IMCHs have been restricted to plateau pressure measurements in the vicinity of room temperature, i.e., the minimum information which is required for the technological applications of these materials. When models have been used for the interpretation of the results they have usually been too facile to be applicable to the quite complex systems which are involved.

More thorough investigations of both the thermodynamic properties in the single phase regions and of the phase diagrams are, however, now being conducted and it is hoped that a better understanding of the thermodynamic properties will ensue from these experimental studies. On the theoretical side the complex structures of IMCHs, with their multitude of possible environments for H atoms means that models for the configurational and vibrational contributions to the partition function will also have to become more sophisticated.

Considerable improvements in both experimental and theoretical investigations can be anticipated in the next few years. The same rapid progress which has been made in the case of pure metal–H systems in the last decade can be predicted with confidence.

Acknowledgements. TBF is grateful to the National Science Foundation and WAO to the Australian Research Grants Scheme for financial support of their research on metal hydrides.

References

3.1 K.H.J. Buschow, P.C.P. Bouten, A.R. Miedema: Rep. Prog. Phys. **45**, 937 (1982)
3.2 P.S. Rudman, G. Sandrock: Annu. Rev. Mater. Sci. **12**, 271 (1982)
3.3 D.G. Ivey, D.O. Northwood: J. Mater. Sci. **18**, 321 (1981)
3.4 H. Oesterreicher: Appl. Phys. **24**, 169 (1981)
3.5 A. Hultgren: Trans. ASM **39**, 915 (1947)
3.6 W.A. Oates, T.B. Flanagan: J. Mater. Sci. **16**, 3235 (1981)
3.7 M.L. McGlashan: Chemical Thermodynamics, Vol. 1. Special Periodical Report, Chemical Society (London) 1973, Chap. 1
3.8 D. Ohlendorf, H.E. Flotow: J. Less-Common Met. **73**, 2937 (1980)
3.9 W.E. Wallace, D. Ohlendorf, H.E. Flotow: J. Less-Common Met. **79**, 157 (1981)
3.10 W.A. Oates, T.B. Flanagan: Mater. Res. Bull. **19**, 1397 (1984)
3.11 H. Wenzl, E. Lebsanft: J. Phys. F **10**, 2147 (1980)
3.12 J.J. Murray, M.L. Post, J.B. Taylor: J. Less-Common Met. **73**, 33 (1980)
3.13 B.S. Bowerman, C.A. Wulff, T.B. Flanagan: Z. Phys. Chem. **116**, 197 (1979)
3.14 W.N. Hubbard, P.L. Rawlins, P.A. Connick, R.E. Stedwell, P.A.G. O'Hare: J. Chem. Thermo. **13**, 785 (1983)
3.15 J.J. Murray, M.L. Post, J.B. Taylor: J. Less-Common Met. **90**, 65 (1983)
3.16 M.L. Post, J.J. Murray, J.B. Taylor: Int. J. Hydrogen Energy **9**, 137 (1984)
3.17 D.M. Nace, J.G. Aston: J. Am. Chem. Soc. **79**, 3619 (1957)
3.18 G. Boureau, O. Kleppa: J. Chem. Thermo. **9**, 543 (1977)
3.19 T.L. Hill: J. Chem. Phys. **17**, 520 (1949)
3.20 A.L. Shilov, L.N. Padurets, M.E. Kost: Russ. J. Phys. Chem. **57**, 338 (1983)
3.21 M.H. Mintz, D. Hiershler, Z. Hadary: J. Less-Common Met. **48**, 241 (1976)
3.22 D. Ohlendorf, H.E. Flotow: J. Chem. Phys. **73**, 2937 (1980)
3.23 H. Wenzl, S. Pietz: Solid State Comm. **33**, 1163 (1980)
3.24 T.B. Flanagan, J.D. Clewley: J. Less-Common Met. **83**, 127 (1982)
3.25 W.R. McKinnon: J. Less-Common Met. **91**, 293 (1983)
3.26 N.A. Scholtus, W.K. Hall: J. Chem. Phys. **39**, 868 (1963)
3.27 C.E. Lundin, F.E. Lynch: In *Hydrides for Energy Storage*, ed. by A.F. Andresen, A.J. Maeland (Pergamon, Oxford 1978) p. 395
3.28 T.B. Flanagan, B.S. Bowerman, G.E. Biehl: Scripta Metall. **14**, 443 (1980)
3.29 T.B. Flanagan, W.A. Oates: J. Less-Common Met. **92**, 131 (1983)
3.30 T. Kuji, W.A. Oates, B.S. Bowerman, T.B. Flanagan: J. Phys. F **13**, 1785 (1983)
3.31 A. Percheron-Guegan, C. Lartigue, J.C. Achard, P. Germi, F. Tasset: J. Less-Common Met. **74**, 1 (1980)
3.32 T.B. Flanagan, T. Schober, H. Wenzl: to be published
3.33 R. Hempelmann, D. Richter, G. Eckold, J.J. Rush, J.M. Rowe, M. Montoya: J. Less-Common Met. **104**, 1 (1984)
3.34 P.D. Goodell: J. Less-Common Met. **99**, 1 (1984)
3.35 T.B. Flanagan, C.A. Wulff, B.S. Bowerman: J. Solid State Chem. **32**, 321 (1980)
3.36 K.H.J. Buschow: Mater. Res. Bull. **19**, 935 (1984)
3.37 J.-M. Welter, G. Arnold, H. Wenzl: J. Phys. F **13**, 1773 (1983)
3.38 J.F. Lynch, J.J. Reilly: J. Less-Common Met. **87**, 225 (1982)
3.39 T.B. Flanagan, N.B. Mason, G.E. Biehl: J. Less-Common Met. **91**, 107 (1983)
3.40 T.B. Flanagan, G.E. Biehl: J. Less-Common Met. **82**, 385 (1981)
3.41 T.B. Flanagan, S. Majorowski, J.D. Clewley, C.N. Park: J. Less-Common Met. **103**, 93 (1984)
3.42 G.D. Sandrock: In *Hydrides for Energy Storage*, ed. by A.F. Andresen, A.J. Maeland (Pergamon, Oxford 1978) p. 353
3.43 W.A. Oates, T.B. Flanagan: Scripta Metall. **17**, 983 (1983)
3.44 D.G. Ivey, D.O. Northrup: Scripta Metall. **19**, 1319 (1985)
3.45 Y.L. Yeh, K. Samwer, W.L. Johnson: Appl. Phys. Lett. **42**, 242 (1983)
3.46 K. Samwer, Y.L. Yeh, W.L. Johnson: J. Non-Cryst. Solids **61–62**, 631 (1984)

3.47 K. Aoki, A. Horata, T. Matsumoto: Proc. 4th Int. Conf. on Rapidly Quenched Metals, Sendai, August 24–28, 1981, Japan Inst. of Metals, 1982, p. 1649
3.48 K. Aoki, T. Yamamoto, T. Matsumoto: Scripta Metall. **21**, 27 (1987)
3.49 A.J. Maeland, L.E. Tanner, G.G. Libowitz: J. Less-Common Met. **74**, 279 (1980)
3.50 R. Kirchheim: Acta Metall. **30**, 1069 (1982)
3.51 R. Griessen: Phys. Rev. **B 127**, 7575 (1983)
3.52 J.J. Murray, M.L. Post, J.B. Taylor: J. Less-Common Met. **80**, 210 (1981)
3.53 C.E. Lundin, F.E. Lynch, C.B. Magee: J. Less-Common Met. **56**, 19 (1977)
3.54 G. Busch, L. Schlapbach, A. Seiler: In *Hydrides for Energy Storage*, ed. by A.F. Andresen, A.J. Maeland (Pergamon, Oxford 1978) p. 263
3.55 M.H. Mendelsohn, D.M. Gruen, A.E. Dwight: J. Less-Common Met. **63**, 193 (1979)
3.56 J.J. Reilly, R. Wiswall: Inorg. Chem. **13**, 218 (1974)
3.57 J. Schefer, P. Fischer, W. Halg, F. Stucki, L. Schlapbach, A.F. Andreson: Mater. Res. Bull. **14**, 1281 (1979)
3.58 F. Reidinger, J.F. Lynch, J.J. Reilly: J. Phys. F **12**, 149 (1982)
3.59 J.J. Reilly, J.R. Johnson, J.F. Lynch, F. Reidinger: J. Less-Common Met. **89**, 55 (1983)
3.60 P.D. Goodell, G.D. Sandrock, E.L. Huston: J. Less-Common Met. **73**, 135 (1980)
3.61 V.V. Zhirnova, B.M. Mogutnov: Russ. J. Phys. Chem. **56**, 1313 (1982)
3.62 A. Pebler, E.A. Gulbransen: Trans. Met. Soc. AIME **239**, 1593 (1967)
3.63 F. Pourarian, H. Fujii, W.E. Wallace, V.K. Sinha, H.K. Smith: J. Phys. Chem. **85**, 3105 (1981)
3.64 V.K. Sinha, W.E. Wallace: unpublished results
3.65 H. Oesterreicher, H. Bittner: Mater. Res. Bull. **13**, 83 (1978)
3.66 H. Fujii, F. Pourarian, V.K. Sinha, W.E. Wallace: J. Phys. Chem. **85**, 3112 (1981)
3.67 V.K. Sinha, F. Pourarian, W.E. Wallace: J. Phys. Chem. **86**, 4952 (1982)
3.68 F. Pourarian, W.E. Wallace: J. Less-Common Met. **91**, 223 (1983)
3.69 V.K. Sinha, W.E. Wallace: J. Less-Common Met. **91**, 229 (1983)
3.70 A. Suzuki, N. Nishimiga, S. Ono: J. Less-Common Met. **89**, 283 (1983)
3.71 R.M. van Essen, K.H.J. Buschow: Mater. Res. Bull. **15**, 1149 (1980)
3.72 A.T. Pedziwiatr, R.S. Craig, W.E. Wallace, F. Pourarian: J. Solid State Chem. **46**, 336 (1983)
3.73 F. Pourarian, V.K. Sinha, W.E. Wallace, A.T. Pedziwiztr, R.S. Craig: Proc. Int. Symp. on Electronic Structure and Properties of Hydrogen in Metals, Richmond, VA, USA (Plenum, New York 1983) p. 385
3.74 S. Majorowski, T.B. Flanagan: to be published
3.75 D. Shaltiel, I. Jacob, D. Davidov: J. Less-Common Met. **53**, 117 (1977)
3.76 H. Kierstead: J. Less-Common Met. **70**, 199 (1980)
3.77 B. Vigeholm: J. Less-Common Met. **89**, 136 (1983)
3.78 J.J. Reilly, R.H. Wiswall: Inorg. Chem. **7**, 2254 (1968)
3.79 J.J. Reilly: In *Hydrides for Energy Storage*, ed. by A.F. Andresen, A.J. Maeland (Pergamon, Oxford 1978) p. 301
3.80 M.H. Mintz, Z. Gavra, G. Kimmel: J. Less-Common Met. **74**, 263 (1980)
3.81 J.-J. Didisheim, P. Zolliker, K. Yvon, P. Fischer, J. Schefer, M. Gubelmann, A.F. Williams: Inorg. Chem. **23**, 1953 (1984)
3.82 C. Wagner: Acta Metall. **21**, 1297 (1973)
3.83 R.B. McLellan: Scripta Metall. **16**, 745 (1982)
3.84 A. Switendick: Z. Phys. Chem. N.F. **117**, 89 (1979)
3.85 D.G. Westlake: J. Less-Common Met. **91**, 1 (1983)
3.86 D.G. Westlake: J. Less-Common Met. **91**, 275 (1983)
3.87 D.G. Westlake: J. Mater. Sci. **19**, 316 (1984)
3.88 V.A. Yartys, V.V. Burnasheva, K.N. Semenenko: Russ. Chem. Revs. (Engl. trans.) **52**, 299 (1983)
3.89 V.A. Somenkov, A.V. Irodova: J. Less-Common Met. **101**, 481 (1984)
3.90 J. Shinar, I. Jacob, D. Davidov, D. Shaltiel: In *Hydrides for Energy Storage*, ed. by A.F. Andresen, A.J. Maeland (Pergamon, Oxford 1978) p. 337

3.91 I. Jacob, D. Shaltiel: J. Less-Common Met. **65**, 117 (1979)
3.92 G. Alefeld: Ber. Bunsenges. Physik. Chem. **76**, 746 (1972)
3.93 A. Pasturel, B. Brion, P. Hicter, A. Percheron-Guégan, J.C. Achard: J. Less-Common Met. **86**, 19 (1982)
3.94 W.A. Oates, P.T. Gallagher, J.A. Lambert: Trans. Met. Soc. AIME **245**, 47 (1969)
3.95 G. Boureau: J. Phys. Chem. Solids **42**, 743 (1981)
3.96 W.E. Wallace, H.E. Flotow, D. Ohlendorf: J. Less-Common Met. **79**, 157 (1981)
3.97 J.C. Achard, C. Lartigue, A. Percheron-Guegan, J.C. Mathieu, A. Pasturel, F. Tasset: J. Less-Common Met. **79**, 161 (1981)
3.98 I. Jacob, J.M. Bloch, D. Shaltiel, D. Davidov: Solid State Commun. **35**, 155 (1980)
3.99 W.A. Oates, R. Ramanathan: Proc. 2nd Internat. Congr. on Hydrogen in Metals, Paris 1977 (Pergamon, Oxford 1978) p. 2A11
3.100 H.A. Kierstead: J. Less-Common Met. **71**, 303 (1980)
3.101 H.A. Kierstead: J. Less-Common Met. **73**, 61 (1980)
3.102 H.A. Kierstead: J. Less-Common Met. **84**, 253 (1982)
3.103 A.L.G. Rees: Trans. Faraday Soc. **50**, 335 (1954)
3.104 H.A. Kierstead: J. Less-Common Met. **75**, 267 (1980)
3.105 H.A. Kierstead: J. Less-Common Met. **96**, 133 (1984)
3.106 B.D. Dunlap, P.J. Viccaro, G.K. Shenoy: J. Less-Common Met. **74**, 75 (1980)
3.107 S.M. Shapiro, F. Reidinger, J.F. Lynch: J. Phys. F **12**, 1869 (1982)
3.108 W.A. Oates, T.B. Flanagan: J. Chem. Soc., Farad. Trans., Part I, **73**, 407 (1977)
3.109 A. Biris, R.V. Bucur, P. Ghete, E. Indreea, D. Lupu: J. Less-Common Met. **49**, 477 (1976)
3.110 B. Andreyev, V. Shitikov, E. Magomedbekov, A. Shafiev: J. Less-Common Met. **90**, 161 (1983)
3.111 D. Dayan, M.P. Dariel: Mater. Res. Bull. **16**, 137 (1981)
3.112 F. Kuijpers: Philips Res. Rep. Suppl. No. 2 (1973)
3.113 G.D. Sandrock, J.J. Murray, M.L. Post, J.B. Taylor: Mater. Res. Bull. **17**, 887 (1982)
3.114 R.A. Bond, D.K. Ross: J. Physics F **12**, 597 (1982)
3.115 A.G. Khachaturayan: *Theory of the Structural Transformations in Solids* (Wiley, New York 1983)
3.116 A.V. Irodova: Soviet Phys.-Solid State **22**, 1493 (1980)
3.117 Y.M. Gufan, V.B. Shirokov: Soviet Phys.-Solid State **23**, 1992 (1981)
3.118 W.A. Oates, T.B. Flanagan: Metall. Trans. **16**A, 139 (1985)
3.119 V.K. Sinha, W.E. Wallace: J. Less-Common Met. **91**, 239 (1982)
3.120 T.B. Flanagan, B.S. Bowerman, G.E. Biehl: Scripta Metall. **14**, 443 (1980)
3.121 H.K. Birnbaum, M.L. Grossbeck, M. Amano: J. Less-Common Met. **49**, 357 (1976)
3.122 C.N. Park, T.B. Flanagan: J. Less-Common Met. **94**, L1 (1982)
3.123 C.N. Park, T.B. Flanagan: Scripta Metall. **18**, 683 (1984)
3.124 M. Hillert: *The Mechanism of Phase Transformations in Crystalline Solids* (Inst. Metals, London 1969) p. 231

4. Crystal and Magnetic Structures of Ternary Metal Hydrides: A Comprehensive Review

Klaus Yvon and Peter Fischer

With 4 Figures and 3 Tables

The crystal and magnetic structures of hydrides and deuterides of binary metal compounds as determined by neutron diffraction analysis are reviewed. Emphasis is placed on the distribution of hydrogen (deuterium) atoms in the metal atom host structure. At room temperature these distributions are characterized by thermal disorder and preferential occupation of large interstices which are surrounded by atoms belonging to hydride forming elements. The metal atom arrangements are usually similar to those of the binary metal compounds, except that their lattices are generally expanded and their symmetries reduced. At low temperatures the hydrogen (deuterium) atoms tend to order, thus inducing structural phase transitions and further symmetry lowering. Efforts to rationalize observed hydrogen (deuterium) site occupancies and maximum hydrogen (deuterium) contents in terms of empirical models are reviewed. The influence of hydrogenation (deuteration) on the magnetic structures is discussed.

4.1 Introduction

Most ternary metal hydrides[1] derive from binary metal compounds which absorb hydrogen by filling interstices of their metal atom network. The resulting hydrogen concentrations per unit volume are often higher than that of liquid hydrogen, which makes ternary metal hydrides attractive candidates for hydrogen storage (see Chap. 6, Vol. II). One of the major factors limiting hydrogen storage capacity is the reluctance of hydrogen to occupy all available interstices in the metal atom network. In order to investigate this behaviour detailed structure data are necessary. Precise atomic coordinates are also a prerequisite for energy-band calculations (see Chap. 5).

In this chapter the presently known (January 1986) crystal and magnetic structures of ternary metal hydrides are reviewed. Only those compounds are discussed for which hydrogen atom positions have been determined by neutron diffraction experiments. Compared to other methods which provide structural

1 Throughout this article the terms hydrogen and hydride(s) are also used for the terms deuterium and deuteride(s), respectively. This choice should not lead to major confusion because significant structural differences between ternary metal hydrides and deuterides have not so far been reported.

information, such as x-ray and electron diffraction, diffuse and inelastic neutron scattering, nuclear magnetic resonance, Mössbauer, Raman and infra-red spectroscopy, neutron diffraction is unique as it gives direct and precise information on the atomic coordinates of hydrogen atoms. Moreover it yields direct and detailed information on the magnetic structures and order phenomena.

In contrast to a previous critical review [4.1], emphasis in this review is placed on completeness. The only major restriction concerns the type of hydrides covered; apart from a few exceptions only metallic hydrides which derive from binary metal compounds are included, i.e. compounds in which different metal atom constituents occupy different crystallographic equipoints.

Therefore, the following classes of hydrides are not included (most recently studied representatives with known hydrogen atom distributions are given in parentheses): elemental hydrides (β_1-V_2H and β-V_2D [4.2], La(Ce)D_3 [4.3], Ce(Pr)$D_{1.95}$ [4.4], YbD_2 [4.5], β-UH(D)$_3$ [4.6], PuD_x [4.7]), hydrides which derive from pseudo-binary alloys ($Ti_{1-y}V_yD_x$ [4.8]) or amorphous alloys ($ZrNiD_{1.8}$ and $Zr_2NiD_{4.3}$ [4.9], $Cu_xTi_{1-x}(H,D)_y$ [4.10]), saline hydrides (BaLiH(D)$_3$ [4.11]), non-metallic complex hydrides (K_2ReH_9 [4.12]), ternary hydrides containing only one metal constituent (D(H)Nb_6I_{11} [4.13], ZrBrD [4.14], $TbBrD_2$ [4.15]), ternary metal hydroxides ($Na_2Pt(OD)_6$ [4.16]), hydride oxides ($DAl_{11}O_{17}$ [4.17], $Ni(OD)_2$ [4.18]), hydrogen bronzes (H(D)TaO_3 [4.19]), hydride carbides (nitrides, oxides) with close-packed metal atom arrangements (TiC_xH_y [4.20], $ZrO_{0.4}D_{0.1}$ [4.21]), and ternary hydrides which generally do not derive from stable binary metal compounds (Sr_2RuD_6 [4.22], K_2PtD_4 [4.23]).

Previous structure reviews of metal hydrides have mainly focussed on elemental hydrides [4.24–27], in particular those of group V transition elements and Pd [4.28], and actinides [4.29], transition metal hydride complexes [4.30], hydrides of compounds between transition metals and p-elements [4.31], and hydrides of selected binary metal compounds [4.1, 32–35]. Structural aspects of ternary metal hydrides have also been discussed in general reviews [4.36–42] and contributions to international symposia on the properties and applications of metal hydrides [4.43].

This review is organized as follows. In Sect. 4.2 experimental aspects such as sample preparation, neutron diffraction setup and structure refinement methods are discussed. In Sect. 4.3 a comprehensive list of ternary metal hydride structures [4.44–113] is given and their characteristic features are discussed. Emphasis is placed on those features which are of relevance for the understanding of hydrogen sorption properties, such as the distribution, the number and the type of metal atom interstices occupied by hydrogen. In Sect. 4.4 attempts to rationalize the observed hydrogen site occupancies and maximum hydrogen contents are reviewed. Most attempts have been based on two empirical models of which one emphasized the importance of geometrical restrictions [4.114] and the other nearest neighbour interactions [4.115, 116]. Work based on the former model was reviewed recently [4.114]. Section 4.5

concerns magnetic ordering of ternary metal hydrides as determined by neutron scattering. Section 4.6 presents a conclusion.

Atomic coordinates (including error estimates) and magnetic order parameters (magnetic moments and ordering temperatures as determined by neutron diffraction) for selected ternary metal hydride structures are summarized in Appendix 4.A and 4.B, respectively. Most of these data cannot be retrieved from existing crystallographic data files. Interatomic distances and structural drawings can be derived from these data by using standard crystallographic computer programs and thus are not given unless they are essential for the understanding of the text.

4.2 Experimental Method

4.2.1 Sample Preparation

Most ternary metal hydrides discussed in this review were prepared by a two-step procedure which consisted of the synthesis of the binary metal compound (mainly by arc melting or inductive heating) and subsequent hydrogenation (see Chap. 2). Exceptions are Mg_2FeD_6 [4.66] and Mg_2CoD_5 [4.65] which were directly synthesized from the elements by a sintering technique. Most hydrides were stable under atmospheric pressure, while others were kept under high deuterium pressure ($FeTiD_x$ [4.76, 79]), or were "poisoned" by air and gases such as CO, and SO_2 to prevent desorption ($FeTiD_x$ [4.77, 117], $LaNi_4AlD_{4.1}$ [4.53], $LaNi_5D_6$ [4.88]). Hydride formation sometimes required activation of the binary metal compound ($FeTiD_x$ [4.73–79, 117]). For some systems, several hydrogenation-dehydrogenation cycles were necessary to obtain homogeneous, fully hydrided samples (Mg_2NiH_4 [4.59, 61, 63, 118]). For others hydrogenation led to a segregation into various binary metal compounds and/or elemental hydrides (for example: $5La_7Ni_3+48D_2\rightarrow3LaNi_5+32LaD_3$ [4.82]), or to the formation of amorphous alloys (for example: $La_2Ni_7D_x$ [4.32 and references therein, 4.37]). The quality of the non-hydrogenated samples was often checked by x-ray powder diffraction methods, but rarely by more sensitive methods such as microprobe analysis or optical metallography ($LaNi_5$ [4.35]). Thus the presence of impurity phases was usually only detected after completion of the neutron diffraction experiment (ZrV_2D_x [4.48–50]). For some systems single-phase samples were not obtained because of the difficulty in synthesizing homogenous binary metal compounds, such as Mg_2Ni which usually contains several percent of Mg and $MgNi_2$ phases unless it is prepared by gas-phase reaction [4.119]. In view of the relatively large sample size required for neutron diffraction experiments (typically 1 cm^3), sample inhomogeneity is a major factor limiting the accuracy of the structural results. Another factor is the loss of crystallinity during hydrogenation.

The hydrogen content of the samples was usually measured prior to (or after) the diffraction experiment. Due to sample problems and the limitations of the models used for structure refinement (Sect. 4.2.3), it often differed significantly (by up to 10%, for example in $ZrV_2D_{4.5}$ [4.50] and γ-$TiFeD_{1.9}$ [4.76], see remarks in Appendix 4.A) from that determined from the diffraction experiment (Sect. 4.3.2).

4.2.2 Neutron Diffraction

Except for β-Mg_2NiD_4 [4.120–123], refinements of the metal atom substructures by x-ray diffraction have not been performed, although such experiments would have been desirable in order to detect lattice distortions and reduce the number of atomic parameters for structure refinements by neutron diffraction. Thus most neutron diffraction data were analyzed with a view to determining positional co-ordinates of both hydrogen (deuterium) and metal atoms. Structure investigations by electron diffraction were only rarely performed as for example on the system TiFe-H(D) [4.124], and on Mg_2NiH_4 [D. Noréus, 1985, private communication]. These studies did not yield atomic coordinates.

Most neutron diffraction experiments have been performed with nuclear steady-state reactors by using medium resolution two-axis diffractometers, crystal monochromators (usually graphite, or Ge) yielding wavelengths between typically 0.9 and 2.6 Å, and single detectors. Only a few studies have been performed with high-flux reactors such as that at the Institut Laue-Langevin (ILL) in Grenoble (diffractometer D1B, and high-resolution angle-dispersive diffractometer D1A [4.125]), using a multidetector system [4.126], or with pulsed neutron sources such as the Intense Pulsed Neutron Spallation Source (IPNS) at the Argonne National Laboratory, using an energy dispersive time-of-flight (TOF) method [4.127].

Due to the lack of suitable single crystals, all studies on ternary metal hydrides reported so far have been performed on powders and thus at a certain expense of accuracy. Moreover the resolution and peak to background ratios of most diffraction patterns (if stated) are relatively low and suffer from both sample dependent properties (lack of crystallinity, microtwinning, microstrains, etc., leading to anisotropic line broadening) and structural properties (disorder, short-range order, thermal diffuse scattering, etc. leading to non-uniform background, see Sect. 4.2.3). Typical values for the angular resolution are $\Delta d/d \simeq 4\times10^{-3}-10^{-2}$ (x-ray Guinier method: $\Delta d/d \simeq 10^{-3}$) whereas the cut-off value $s=\sin(\theta/\lambda)_{max}$ varied typically between $s=0.4$ Å^{-1} and $s=0.5$ Å^{-1} (see values stated in Appendix 4.A, x-rays CuKα: $s=0.6$ Å^{-1}). High-resolution studies such as those performed on β-$LaNi_5D_x$ [4.87] with the constant-wavelength spectrometer D1A ($\Delta d/d=2\times10^{-3}$, $s=0.51$ Å^{-1}) at ILL, and on $Zr_3V_3OD_x$ [4.47] and β-Mg_2NiD_4 [4.64] ($\Delta d/d=3\times10^{-3}$, $s=1.19$ Å^{-1}) with the TOF spectrometer at IPNS are rare.

The samples are usually measured in cylindrical vanadium containers of 0.8 to 1.5 cm diameter at ambient pressure and temperature, or within quartz tubes at higher pressures and temperatures. The *in situ* variation of these parameters is easier to perform for neutron diffraction (pressures up to 120 bar, temperatures between 1.3 and 600 K) than for x-ray diffraction, which explains the scarcity of the latter studies. On the other hand the precision of the lattice parameters as determined by constant-wavelength neutron diffraction is usually lower than that determined by x-rays because of the uncertainty in wavelength calibration ($\Delta\lambda/\lambda = 3 \times 10^{-4}$). Absorption is generally not an important problem. It can be determined by transmission measurements, and the corrections applied are those for cylindrical samples for which the product of linear absorption coefficient and sample radius is usually smaller than $\mu R = 0.2$ (except for compounds containing certain nuclei such as B, Sm, Gd etc.). Polarized neutron diffraction studies which would enable separation of magnetic from nuclear, and coherent from incoherent scattering, thus yielding the most detailed information on magnetic density distributions, have not yet been performed because of a lack of single crystals.

Most neutron diffraction studies have been performed on deuterides (because of the much more favorable ratio between coherent and incoherent cross-section of deuterium compared to that of hydrogen), at ambient temperature and pressure, and at fixed deuterium concentrations. Only a few studies have been performed on hydrides (Ti_2NiH_x [4.45], $ZrV_2H_{3.7}$ [4.51], $ZrTi_{1.86}H_{3.8}$ [4.53], $ErFe_2H_{3.5}$ [4.57], Mg_2FeH_6 [4.66], Nb_3SnH [4.81], α-$Mg_2NiH_{0.3}$ [4.100], $ZrNiH_{2.98}$ [4.108], $Pd_3P_{0.8}H_{0.17}$ [4.112], $Pd_6PH_{0.39}$ [4.112]) and mixed hydride-deuterides ($TiFeH_{0.78}D_{0.22}$ [4.75], $Zr_{0.35}Ti_{0.65}H_{0.76}D_{0.24}$ [4.55]), or as a function of temperature ($ZrV_2D_{3.6}$ [4.49], $Y_6Mn_{23}D_{23}$ [4.69], γ-$TiFeD_{1.8}$ [4.78], $Ti_{1.2}Mn_{1.8}D_{3.1}$ [4.98], pressure ($TiFeD_x$ [4.76], β-$LaNi_5D_x$ [4.87], Pd_6PD_x [4.113]), and deuterium content ($Zr_3V_3OD_x$ [4.47], ZrV_2D_x [4.48], $ZrCr_2D_x$ [4.51], $ThZr_2D_x$ [4.58], $Y_6Mn_{23}D_x$ [4.67], $Ho_6Fe_{23}D_x$ [4.72]), and β-$LaNi_5D_x$ [4.82, 87]. Except for one compound (β-$La^{60}Ni_5D_{7.2}$ [4.88]), all hydrides studied contained natural metal isotope mixtures.

4.2.3 Structure Refinement

Lattice parameters, atomic coordinates, amplitudes of thermal vibrations, and hydrogen site occupancy factors have usually, but not always (see below), been refined with the Rietveld powder diffraction profile fitting method [4.128] by using computer programs of which some allowed the simultaneous treatment of several phases [4.129, 130] (for reviews of the Rietveld profile fitting method see [4.131], and for its application to neutron powder diffraction see [4.132] for angle dispersive, and [4.133] for energy dispersive setup). The main problem with this method as applied to ternary metal hydrides is the complexity of the crystal structures ($Y_6Mn_{23}D_{23}$ at 4 K has 56 positional and isotropic thermal parameters, according to [4.69]; $LaNiD_{6.7}$ has 36 atomic parameters, according to [4.87, 88]; for the number of refined parameters see values of n listed in Appendix 4.A), and the low resolution of most experimental diffraction patterns (Sect. 4.2.2). Other limiting factors are the bad crystallinity of the samples ($CeRu_2D_5$ [4.56]), the presence of impurity phases (Fe and MgD_2 in Mg_2FeD_6 [4.66], MgD_2 and $MgNi_2$ in Mg_2NiD_4 [4.59–64]), anisotropic line broadening in the diffraction patterns due to microstrains (β-$LaNi_5D_{5-7}$ [4.53, 86–88], see also $Ni(OD)_2$ [4.18]), and microtwinning (β-Mg_2NiD_4 [4.64, 123]), uneven background due to short-range order of the hydrogen atoms ($ZrV_2D_{3.6}$ [4.49]), anharmonicity (leading to multisite models such as for β-$LaNi_5D_{5-7}$ [4.87, 88]), and structural disorder between the metal atoms ($Ti_{1.2}Mn_{1.8}D_{3.1}$ [4.98]). The agreement indices after profile refinement, R_{wp} and R_I (for definitions see [4.134]) vary typically between 0.08 and 0.12, and 0.03 and 0.08, respectively, whereas the errors of the atomic positional parameters lead to uncertainties in the interatomic distances of the order of ± 0.02 Å. Due to insufficient data sets and inadequacies of the refinement models some structural parameters such as the temperature factors and hydrogen site occupancy factors have been only poorly defined. The former are often constrained by the assumption of an overall temperature factor. The latter are usually not constrained although they are not always consistent with the total hydrogen content as determined by desorption measurements (see remarks in Appendix 4.A).

A controversy exists with respect to the reliability of the estimated standard deviations (e.s.d.'s) as derived by the profile fitting method (for recent work see [4.135] and references therein). Experimental evidence from cubic metal hydride structures suggests [4.48, 59] that some of the e.s.d.'s have been underestimated by a factor of up to three with respect to those derived from the same data from refinements based on integrated intensities. This discrepancy could be due to model errors [4.135] affecting the calculated integrated intensities, but no satisfactory method of correcting the e.s.d.'s for these errors has been proposed as yet. A reliable determination of e.s.d.'s is of particular importance for metal hydride structures because ambiguous, or conflicting results have often been reported as to their lattice symmetries (γ-$TiFeD_2$ [4.76–79]), space group assignments (β-$LaNi_5D_{5-7}$ [4.32, 53, 83–90]), lattice dimensions (Pd_6PD_x [4.113], $ThNi_2D_{2.6}$ [4.102]), and hydrogen atom distributions (β-Mg_2NiD_4 [4.62, 64] and β'-Mg_2NiD_4 [4.59–61], see Sect. 4.3.2). On the other hand the profile refinement method yields very accurate atomic parameters provided it is based on high-resolution data, such as those of $Zr_3V_3OD_x$ [4.47] (yielding anisotropic temperature factors) and β-Mg_2NiD_4 [4.64]. The high accuracy obtained on the latter compounds is presumably a consequence of the more favorable ratio between the number of observations and number of refined atomic parameters.

Refinements based on integrated intensities have been performed e.g. for cubic β'-Mg_2NiD_4 [4.59, 61] and ZrV_2D_x [4.48], hexagonal $LaNi_5D_{5-7}$ [4.53, 83, 86, 87, 89, 90] (showing anisotropic line broadening), and orthorhombic $ZrNiH_3$ [4.108]. The main problem with this method as applied to non-cubic structures is the treatment of overlapping peaks. Hydrogen atom positions and approximate occupancy factors (without error estimates) have been derived by inspection of nuclear density maps on α-$TiFeD_{0.057}$ [4.73] and Nb_3SnH [4.81], and from a comparison between observed and simulated powder diffraction patterns for Ti_2NiH_x [4.45], HfV_2D_4 [4.52], Th_2AlD_x [4.105], and ZrNiD [4.109].

4.3 Structural Results

During the past 25 years over 70 neutron diffraction studies on more than 50 ternary metal hydride systems have been reported. A comprehensive list of these studies [4.45–113] is given in Table 4.1. The list is complete up to, and including, 1985 to the best of the authors' knowledge. Reports which were not easily accessible to the authors may have been omitted. Most of these studies were performed on compounds having the following binary metal atom ratios (typical representatives are given in parentheses): AB($TiFeD_x$, $Zr_3V_3OD_x$, $TiCuD_x$, $ZrNiH_x$), AB_2(ZrV_2D_x, $ZrMn_2D_x$, $ZrBe_2D_x$, $ThNi_2D_x$, Mg_2NiD_4), A_2B(Ti_2NiH_x, Th_2AlD_x, Zr_2PdD_x), AB_3($HoNi_3D_x$, Pd_3PD_x), A_3B(Nb_3SnH_x), A_6B_{23}($Ho_6Fe_{23}D_x$), AB_5($LaNi_5D_x$), and AB_7($Y_2Fe_{14}BD_x$), where A denotes a transition or rare-earth element, and B a transition element or a p-element. A study of the various types of constituents reveals that all hydrides investigated so far contain at least one element which is capable of forming a stable binary hydride, such as Mg, Ti, Zr, Hf, V, Pd, Th, Y, and the rare-earths. This aspect presumably constitutes a general trend for metallic hydrides since all known binary metal compounds which do not contain one of these elements are poor hydrogen absorbers [4.136, 137]. As shown in Sect. 4.3.2 this characteristic behavior is also reflected by structural features and can be explained in terms of preferential hydrogen atom occupancies (Sect. 4.4.4).

The compounds listed in Table 4.1 are grouped according to the structure types of their (hydrogen free) metal atom matrices which are listed in order of decreasing symmetry. Each hydride structure is characterized with respect to space group, lattice parameters (including those of the hydrogen-free metal host structure), and metal atom environment of the occupied H(D) atom sites (number and type of surrounding metal atoms, point group symmetry). For convenience the H(D) sites are generally listed in order of decreasing occupancy. For characteristic compounds (those printed with bold letters) explicit atomic coordinates are given in Appendix 4.A. Lattice transformation temperatures, T_l, are indicated in both Table 4.1 and Appendix 4.A. Magnetically ordered hydride structures are listed in Appendix 4.B. The latter contains information on magnetic moments and Curie (or Néel) temperatures, T_C (or T_N), if known from neutron diffraction.

The structural features of ternary metal hydrides are summarized in the next section.

4.3.1 Metal Atom Substructure

Compared to elemental metal hydride structures which are characterized by a few, relatively simple and usually highly symmetrical, metal atom arrangements

Table 4.1. Ternary metal hydrides and deuterides studied by neutron diffraction analysis[a]

Compound[b]	Metal atom substructure[c]	Space group[d]	Ref.[e]	Lattice parameters [Å][f]	Interstices occupied by H(D)[g]
$Ti_2NiH_{0.5}$	Ti_2Ni	$Fd\bar{3}m$	[4.45]	$a=11.50$	$[Ti_6]$ *8b* $(\bar{4}3m)$ $[Ni_4]$ *8a* $(\bar{4}3m)$ $[(Ti, Ni)_6]$ *16d* $(\bar{3}m)$
Ti_2NiH	Ti_2Ni	$Fd\bar{3}m$	[4.45]	$a=11.60$	$[Ti_6]$ *8b* $(\bar{4}3m)$ $[Ni_4]$ *8a* $(\bar{4}3m)$ $[(Ti, Ni)_6]$ *16d* $(\bar{3}m)$
$Ti_4Fe_2OD_{2.25}$	Ti_2Ni (filled)	$Fd\bar{3}m$	[4.46]	Not specified	$[Ti_6]$ *8a* $(\bar{4}3m)$ orig. at $\bar{4}3m$ $[Ti_3Fe]$ *192i* (1), *96g* (m)
$Zr_3V_3OD_{1.86}$	Ti_2Ni (filled)	$Fd\bar{3}m$	[4.47]	$a=12.318$	$[Zr_3V]$ *32e* $(3m)$ orig. at $\bar{4}3m$ $[Zr_2V_2]$ *192i* (1)
$Zr_3V_3OD_{2.85}$	Ti_2Ni (filled)	$Fd\bar{3}m$	[4.47]	$a=12.381$	$[Zr_3V]$ *32e* $(3m)$ orig. at $\bar{4}3m$ $[Zr_2V_2]$ *192i* (1)
$Zr_3V_3OD_{4.93}$	Ti_2Ni (filled)	$Fd\bar{3}m$	[4.47]	$a=12.526$	$[Zr_3V]$ *32e* $(3m)$ orig. at $\bar{4}3m$ $[Zr_2V_2]$ *192i* (1) $[Zr_3V]$ $96g_1$ (m) $[ZrV_3]$ $96g_2$ (m)
$ZrV_2D_{1.5}$	$MgCu_2$	$Fd\bar{3}m$	[4.48]	$a=7.60$	$[Zr_2V_2]$ *96g* (m)
$ZrV_2D_{2.8}$	$MgCu_2$	$Fd\bar{3}m$	[4.48]	$a=7.77$	$[Zr_2V_2]$ *96g* (m) $[ZrV_3]$ *32e* $(3m)$
$ZrV_2D_{3.6}$ (340 K)	$MgCu_2$	$Fd\bar{3}m$	[4.49]	$a=7.82$; $T_1=325$ K	$[Zr_2V_2]$ *96g* (m) $[ZrV_3]$ *32e* $(3m)$
(230 K)	$MgCu_2$ (def.)	$I4_1/a$	[4.49]	$a=5.484$, $c=7.902$ (at 295 K)	$[Zr_2V_2]$ *16f* (1)
$ZrV_2D_{4.9}$	$MgCu_2$	$Fd\bar{3}m$	[4.50]	$a=7.913$	$[Zr_2V_2]$ *96g* (m) $[ZrV_3]$ *32e* $(3m)$
$ZrV_2D_{2.3}$	$MgCu_2$	$Fd\bar{3}m$	[4.51]	$a=7.910$	$[Zr_2V_2]$ *96g* (m) $[ZrV_3]$ *32e* $(3m)$
$ZrV_2H_{3.7}$	$MgCu_2$	$Fd\bar{3}m$	[4.51]	$a=7.946$	$[Zr_2V_2]$ *96g* (m) $[ZrV_3]$ *32e* $(3m)$
HfV_2D_4 (340 K)	$MgCu_2$	$Fd\bar{3}m$	[4.52]	$a=7.78$	$[Hf_2V_2]$ *96g* (m) $[HfV_3]$ *32e* $(3m)$
HfV_2D_4 (77 K)	$MgCu_2$ (def.)	$I4_1/a$	[4.52]	$a=5.42$, $c=7.85$, $T_1=278$ K	$[Hf_2V_2]$ *16f* (1)

$ZrCr_2D_{2.89}$	$MgCu_2$	$Fd\bar{3}m$	[4.51]	$a=7.625$	$[Zr_2Cr_2]$ *96g (m)*
$ZrCr_2D_{3.08}$	$MgCu_2$	$Fd\bar{3}m$	[4.51]	$a=7.628$	$[Zr_2Cr_2]$ *96g (m)*
$ZrCr_2D_{3.8}$ (118 K)	$MgCu_2$	$Fd\bar{3}m$	[4.51]	$a=7.658$	$[Zr_2Cr_2]$ *96g (m)* $[ZrCr_3]$ *32e (3m)*
$ZrCr_2D_{3.5}$	$MgCu_2$	$Fd\bar{3}m$	[4.53]	$a=7.697$	$[Zr_2Cr_2]$ *96g (m)*
$ZrCr_2D_{3.8}$ (300 K)	$MgCu_2$	$Fd\bar{3}m$	[4.54]	$a=7.711$	$[Zr_2Cr_2]$ *96g (m)* $[ZrCr_3]$ *32e (3m)*
(4 K)	$MgCu_2$ (def.)	Cc	[4.54]	$a=7.720$, $c=7.636$, $T_l=255$ K, pseudotetragonal cell	$[Zr_2Cr_2]$ 8 sites *4a (1)*
$ZrTi_{1.86}D(H)_{3.8}$	$MgCu_2$	$Fd\bar{3}m$	[4.55]	$a=8.19$	$[ZrTi_3]$ *32e (3m)*
$CeRu_2D_5$ (300 K)	$MgCu_2$	$Fd\bar{3}m$	[4.56]	$a=8.383$	$[Ce_2Ru_2]$ *96g (m)* orig. at $\bar{4}3m$
(4.2 K)	$MgCu_2$	$Fd\bar{3}m$	[4.56]	$a=8.377$	$[Ce_2Ru_2]$ *96g (m)*
$ErFe_2H_{3.5}$	$MgCu_2$	$Fd\bar{3}m$	[4.57]	$a=7.83$	Site occupancies not specified
$ThZr_2D_{3.6}$	$MgCu_2$	$Fd\bar{3}m$	[4.58]	$a=9.042$	$[Th_2Zr]$ *96g (m)* $[ThZr_3]$ *32e (3m)*
$ThZr_2D_{4.8}$	$MgCu_2$	$Fd\bar{3}m$	[4.58]	$a=9.112$	$[Th_2Zr]$ *96g (m)* $[ThZr_3]$ *32e (3m)*
$ThZr_2D_{6.0}$	$MgCu_2$	$Fd\bar{3}m$	[4.58]	$a=9.151$	$[Th_2Zr]$ *96g (m)* $[ThZr_3]$ *32e (3m)*
$ThZr_2D_{6.3}$	$MgCu_2$	$Fd\bar{3}m$	[4.58]	$a=9.154$	$[Th_2Zr]$ *96g (m)* $[ThZr_3]$ *32e (3m)*

[a] For restrictions concerning compounds included in this table see introduction of text; compounds are listed in order of decreasing symmetry of hydrogen-free metal host structure (for exceptions see text).
[b] Sample composition stated by the authors; it may differ from that resulting from structure refinement; studies performed at room temperature and 1 atm D_2-pressure, unless specified differently; for compounds marked by bold letters atomic coordinates can be found in Appendix 4.A; concerning magnetic ordering see Appendix 4.B.
[c] Structure type of hydrogen-free metal atom matrix ("def." = deformed).
[d] Standard setting according to [4.44], with origin at inversion centre (if present), unless stated differently in last column; non-standard space group symbols are marked by asterisks.
[e] Most recent work only; similar work published by same authors listed under same number in list of references.
[f] Only 4 (maximal 5) significant digits are listed; lattice parameters for hydrogen-free compounds are given in Appendix 4.A. T_l = lattice transformation temperature as determined by neutron diffraction.
[g] D(H) atom environment (in square brackets), generally listed in decreasing order of site occupancy; Wyckoff positions according to [4.44] (subscripts, if given, are those stated by authors), point symmetry (in parentheses).

Table 4.1 (continued)

Compound[b]	Metal atom substructure[c]	Space group[d]	Ref.[e]	Lattice parameters [Å][f]	Interstices occupied by H(D)[g]
β′-$Mg_2NiD_{3.9}$ (553 K) (22 bar)	CaF_2	*Fm3̄m*	[4.59]	$a=6.507$, $T_1=508$ K	[Mg_4Ni] *24e (4mm)*
β′-Mg_2NiD_4 (530 K) (3 bar)	CaF_2	*Fm3̄m*	[4.60]	$a=6.508$, $T_1=509$ K	[Mg_4Ni] *24e (4mm)*
β′-Mg_2NiD_4 (573 K) (15 bar)	CaF_2	*Fm3̄m*	[4.61]	$a=6.525$, $T_1=513$ K	[Mg_2Ni_2] *48h (mm2)* [Mg_4Ni] *24e (4mm)*
β-Mg_2NiD_4	CaF_2 (def.)	*C2/m*	[4.62]	$a=6.496$, $b=6.412$, $c=6.602$, $\beta=93.2°$, $T_1=509$ K	[Mg_4Ni] *4g (2)*, 3 sites *4i (m)*
β-Mg_2NiD_4	CaF_2 (def.)	*Ia**	[4.63]	$a=13.197$, $b=6.403$, $c=6.489$, $\beta=93.23°$, $T_1=513$ K	[Not spec.] 12 sites *4a (1)*
β-Mg_2NiD_4	CaF_2 (def.)	*C2/c*	[4.64]	$a=14.342$, $b=6.404$, $c=6.483$, $\beta=113.52°$	[Mg_4Ni] 4 sites *8f (1)*
Mg_2CoD_5 (298 K) (498 K)	CaF_2 (def.) CaF_2	*P4/nmm* *Fm3̄m*	[4.65]	$a=4.463$, $c=6.593$, $T_1=488$ K $a=6.453$	[Mg_4Co] *2c (4mm)*, *8j (m)* [Mg_4Co] *24e (4mm)*
Mg_2FeD_6	CaF_2	*Fm3̄m*	[4.66]	$a=6.430$	[Mg_4Fe] *24e (4mm)*
$Y_6Mn_{23}D_{8.3}$ (580 K)	Th_6Mn_{23}	*Fm3̄m*	[4.67]	$a=12.785$	[Y_6] *4a (m3̄m)* [Y_3Mn] *32f₃ (3m)*
$Y_6Mn_{23}D_{18}$	Th_6Mn_{23}		[4.67]	$a=12.799$	[Y_6] *4a (m3̄m)* [Y_3Mn] *32f₃ (3m)* [YMn_3] *96j₁ (m)*
$Y_6Mn_{23}D_{23}$	Th_6Mn_{23}		[4.67]	$a=12.840$	[Y_6] *4a (m3̄m)* [Y_3Mn] *32f₃ (3m)* [YMn_3] *96j₁ (m)* [Y_2Mn_2] *96k₂ (m)*
$Y_6Mn_{23}D_{23}$	Th_6Mn_{23} (def.)	*I4/mmm*	[4.68]	$a=9.031$, $c=12.787$	[Y_6] *2a (4/mmm)* [Y_3Mn] *16n₃ (m)* [YMn_3] *16m (m)*, *16l (m)* [Y_2Mn_2] *32o (1)*
$Y_6Mn_{23}D_{23}$ (295 K)	Th_6Mn_{23}	*Fm3̄m*	[4.69]	$a=12.805$, $T_1=175$ K	[Y_6] *4a (m3̄m)* [Y_3Mn] *32f₃ (3m)* [YMn_3] *96j₁ (m)*

$Y_6Mn_{23}D_{23}$ (78 K)	Th_6Mn_{23} (def.)	*P4/mmm*	[4.69]	$a=9.030$, $c=12.722$	[Y_6] *1a, 1d (4/mmm)* [Y_3Mn] *$8s_3$ (m), $8t_3$ (m)* [YMn_3] *8p, 8q, $8r_2$, $8r_4$ (m)* [Y_2Mn_2] *$8s_4$ (m), $16u_1$, $16u_2$ (1)*
$Y_6Mn_{23}D_{23}$ (4 K)	Th_6Mn_{23} (def.)	*P4/mmm*	[4.69]	$a=9.030$, $c=12.722$	[Y_6] *1a, 1d (4/mmm)* [Y_3Mn] *$8s_3$ (m), $8t_3$ (m)* [YMn_3] *8p, 8q, $8r_2$, $8r_4$ (m)* [Y_2Mn_2] *$8s_4$ (m), $16u_1$, $16u_2$ (1)*
$Th_6Mn_{23}D_{16}$ (4 K)	Th_6Mn_{23} (def.)	*$P4_2/mnm$*	[4.70]	$a=9.077$, $c=12.961$, $T_l<80$ K	[Th_2Mn_3] *8i (m)* [$ThMn_3$] *$8j_3$ (m)* [Th_2Mn_2] *$16k_3$ (1)*
$Th_6Mn_{23}D_{16}$ (295 K)	Th_6Mn_{23}	*$Fm\bar{3}m$*	[4.71]	$a=12.921$, $T_l=80$ K	[Th_3Mn] *$32f_3$ (3m)* [Th_2Mn_3] *48i (mm2)*
$Th_6Mn_{23}D_{30}$ (150 K)	Th_6Mn_{23}	*$Fm\bar{3}m$*	[4.71]	$a=13.203$	[Th_3Mn] *$32f_3$ (3m)* [Th_2Mn_3] *48i (mm2)* [Th_2Mn_2] *96k (m)*
$Th_6Mn_{23}D_{16}$ (4 K)	Th_6Mn_{23} (def.)	*P4/mmm*	[4.71]	$a=9.076$, $c=12.961$, $T_l=78$ K	[Th_3Mn] *$8s_3$, $8t_3$ (m)* [$ThMn_3$] *4n (m2m), $8r_2$, $8r_3$ (m)*
$Ho_6Fe_{23}D_{1.5}$	Th_6Mn_{23}	*$Fm\bar{3}m$*	[4.72]	$a=12.022$	[Ho_6] *4a ($m\bar{3}m$)* [Ho_3Fe] *$32f_3$ (3m)*
$Ho_6Fe_{23}D_{8.2}$	Th_6Mn_{23}	*$Fm\bar{3}m$*	[4.72]	$a=12.187$	[Ho_3Fe] *$32f_3$ (3m)* [Ho_2Fe_2] *96k (m)*
$Ho_6Fe_{23}D_{12.1}$	Th_6Mn_{23} (def.)	*I4/mmm*	[4.72]	$a=8.611$, $c=12.542$	[Ho_3Fe] *16n (m)* [Ho_2Fe_2] *32o (1)* [$HoFe_3$] *$8j_1$, $8j_2$ (m2m)*
$Ho_6Fe_{23}D_{15.7}$	Th_6Mn_{23}	*$Fm\bar{3}m$*	[4.72]	$a=12.423$	[Ho_3Fe] *$32f_3$ (3m)* [$HoFe_3$] *$96j_1$, $96j_2$ (m)* [Ho_2Fe_2] *96k (m)*
$Er_6Mn_{23}D_{23}$ (360 K)	Th_6Mn_{23} (def.)	*I4/mmm*	[4.68]	Not specified	[Er_6] *2a (4/mmm)* [Er_3Mn] *16n (m)* [$ErMn_3$] *16l, 16m (m)* [Er_2Mn_2] *32o (1)*
α-$TiFeD_{0.057}$	CsCl	*$Pm\bar{3}m$*	[4.73]	$a=2.979$	[Ti_4Fe_2] *3d (4/mmm)*
β-$TiFeD_1$	CsCl	*$Pmc2_1$*	[4.74]	$a=2.956$, $b=4.543$, $c=4.388$	[Ti_4Fe_2] *2b (m), 4c (1)*
β-$TiFeD_{1.0}$ (6.5 bar)	CsCl (def.)	*$P222_1$*	[4.75]	$a=2.966$, $b=4.522$, $c=4.370$	[Ti_4Fe_2] *2a (2), 2d (2)*

Table 4.1 (continued)

Compound[b]	Metal atom substructure[c]	Space group[d]	Ref.[e]	Lattice parameters [Å][f]	Interstices occupied by H(D)[g]
β-$TiFeH_{0.78}D_{0.22}$	CsCl (def.)	$P222_1$	[4.75]	$a=2.976$, $b=4.514$, $c=4.378$	$[Ti_4Fe_2]$ *2a (2), 2d (2)*
β_2-$TiFeD_{1.37}$ (32 bar)	CsCl (def.)	$P222_1$	[4.76]	$a=3.088$, $b=4.515$, $c=4.391$	$[Ti_4Fe_2]$ *2a (2), 2d (2)*
γ-$TiFeD_{1.9}$	CsCl (def.)	$P2/m$	[4.77]	$a=4.704$, $b=2.830$, $c=4.704$, $\beta=96.97°$	$[Ti_4Fe_2]$ *1a, 1d, 1g, (all 2/m)* $[Ti_2Fe_4]$ *1f (2/m)*
γ-$TiFeD_{1.9}$ (100 bar)	CsCl (def.)	$P2/m$	[4.76]	$a=4.706$, $b=2.835$, $c=4.697$, $\beta=96.9°$	$[Ti_4Fe_2]$ *1a, 1d, 1g, (all 2/m)* $[Ti_2Fe_4]$ *1f (2/m)*
γ-$TiFeD_{1.8}$ (300 K)	CsCl (def.)	$P2/m$	[4.78]	$a=4.708$, $b=4.697$, $c=2.835$, $\gamma=97.05°$	$[Ti_4Fe_2]$ *1a, 1d, 1g, (all 2/m)* $[Ti_2Fe_4]$ *1f (2/m)*
(100 K)	CsCl (def.)	$P2/m$	[4.78]	$a=4.705$, $b=4.695$, $c=2.828$, $\gamma=97.18°$	$[Ti_4Fe_2]$ *1a, 1d, 1g, (all 2/m)* $[Ti_2Fe_4]$ *1f (2/m)*
(4.2 K)	CsCl (def.)	$P2/m$	[4.78]	$a=4.699$, $b=4.694$, $c=2.828$, $\gamma=97.18°$	$[Ti_4Fe_2]$ *1a, 1d, 1g, (all 2/m)* $[Ti_2Fe_4]$ *1f (2/m)*
γ-$TiFeD_{1.94}$ (113 bar)	CsCl (def.)	*Cmmm*	[4.79]	$a=7.029$, $b=6.233$, $c=2.835$	$[Ti_4Fe_2]$ *4e (2/m), 2a (mmm)* $[Ti_2Fe_4]$ *2c (mmm)*
β-$TiFe_{0.9}Mn_{0.1}D_{1.25}$	CsCl (def.)	*P2cm**	[4.80]	$a=2.995$, $b=4.531$, $c=4.397$	$[Ti_4(Fe, Mn)_2]$ *2a (2), 2c (m)*
γ-$TiFe_{0.9}Mn_{0.1}D_{1.72}$	CsCl (def.)	$P2/m$	[4.80]	$a=4.714$, $b=2.837$, $c=4.714$, $\beta=97.14°$	$[Ti_4(Fe, Mn)_2]$ *1a, 1d, 1g (2/m)* $[Ti_2(Fe, Mn)_4]$ *1f (2/m)*
Nb_3SnH	Cr_3Si	$Pm\bar{3}n$	[4.81]	$a=5.34$	$[Nb_4]$ *6d* ($\bar{4}m2$)
α-$LaNi_5D_{0.15}$	$CaCu_5$	*P6/mmm*	[4.82]	$a=5.025$, $c=3.988$	$[La_2Ni_4]$ *3f (mmm)*
α-$LaNi_5D_{0.4}$	$CaCu_5$	*P6/mmm*	[4.82]	$a=5.027$, $c=3.989$	$[La_2Ni_4]$ *3f (mmm)*
β-$LaNi_5D_7$	$CaCu_5$ (def.)	*P31m*	[4.83]	$a=5.387$, $c=4.273$	$[LaNi_3]$ (or $[La_2Ni_4]$) *3c (m)* $[La_2Ni_2]$ *6d (1)*
β-$LaNi_5D_{5.6}$	$CaCu_5$ (def.)	*P31m*	[4.32]	Not specified	$[LaNi_3]$ (or $[La_2Ni_4]$) *3c (m)* $[La_2Ni_2]$ *6d (1)*
β-$LaNi_5D_6$ (5 bar)	$CaCu_5$ (def.)	*P31m*	[4.84]	$a=5.411$, $c=4.293$	$[LaNi_3]$ (or $[La_2Ni_4]$) *3c (m)* $[La_2Ni_2]$ *6d (1)*
β-$LaNi_5D_{5.5}$ (4 bar)	$CaCu_5$ (def.)	*P31m*	[4.85]	$a=5.336$, $c=4.259$	$[LaNi_3]$ (or $[La_2Ni_4]$) *3c (m)* $[La_2Ni_2]$ *6d (1)*
β-$LaNi_{5.12}D_{6.7}$ (5 atm)	$CaCu_5$	*P6/mmm*	[4.86]	$a=5.399$, $c=4.290$	$[La_2Ni_2]$ *6m (mm2)* $[La_2Ni_4]$ *3f (mmm)* $[LaNi_3]$ *12n (m), 12o (m)* $[Ni_4]$ *4h (3m)*

β-$LaNi_5D_{5.0}$ (363 K) (24 bar)	$CaCu_5$ (def.)	$P6_3mc$ ($P6mm$)	[4.87]	$a=5.395$, $c/2=4.242$ superstructure with $c'=2c$	[$LaNi_3$] *12d (1)* [La_2Ni_2] 2 sites *6c (m)* [$LaNi_3$] 2 sites *6c (m)* [Ni_4] 2 sites *2b (3m)*
β-$LaNi_5D_{5.9}$ (2 bar)	$CaCu_5$ (def.)	$P6_3mc$	[4.87]	$a=5.386$, $c=8.544$	similar as above
β-$LaNi_5D_{6.4}$ (5 bar)	$CaCu_5$ (def.)	$P6_3mc$	[4.87]	$a=5.396$, $c=8.571$	similar as below
β-$LaNi_5D_{6.7}$ (19.5 bar)	$CaCu_5$ (def.)	$P6_3mc$	[4.87]	$a=5.412$, $c=8.599$	[$LaNi_3$] *12d (1)* [La_2Ni_2] 2 sites *6c (m)* [Ni_4] *2b (3m)* [$LaNi_3$] 2 sites 6c *(m)* [Ni_4] 2 sites *2b (m)*
β-$LaNi_5D_6$ (276 K)	$CaCu_5$ (def.)	$P6_3mc$	[4.88]	$a=5.388$, $c=8.559$	[La_2Ni_4] *6c (m)* [La_2Ni_2] 2 sites *6c (m)* [Ni_4] 2 sites *2b (3m)*
β-$La^{60}Ni_5D_{7.2}$	$CaCu_5$ (def.)		[4.88]	$a=5.368$, $c=8.551$	[La_2Ni_4] *6c (m)* [La_2Ni_2] 2 sites *6c (m)* [Ni_4] 2 sites *2b (3m)*
β-$LaNi_5D_6$	$CaCu_5$ (def.)	*P321*	[4.53]	$a=5.43$, $c=4.28$	[$LaNi_3$] *6g (1)* [$LaNi_2$] *6g (1)*
β-$LaNi_5D_6$	$CaCu_5$ (def.)	*P31m*	[4.89]	$a=5.43$, $c=4.28$	[$LaNi_3$] (or [La_2Ni_4]) *3c (m)* [$LaNi_2$] *6d (1)*
β-$LaNi_5D_7$	$CaCu_5$ (def.)	*Cmmm*	[4.90]	$a=5.42$, $c=4.30$ pseudo-hexagonal cell, no atomic parameters were given; presumably isostructural to orthorhombic β^{I}-$PrCo_5D_{3.6}$	
β-$CaNi_5D_{0.77}$	$CaCu_5$ (def.)	*Pmcm**	[4.91]	$a=8.603$, $b=5.081$, $c=7.856$	[Ca_2Ni_4] 2 sites *4k (m)*
β^{I}-$PrCo_5D_{3.6}$	$CaCu_5$ (def.)	*Cmmm*	[4.92]	$a=8.868$, $b=5.341$, $c=4.083$	[Pr_2Co_2] *4h (2mm)* [Pr_2Co_4] *4e (2/m)*
β^{II}-$PrCo_5D_{2.9}$	$CaCu_5$ (def.)	*Im2m**	[4.92]	$a=8.832$, $b=5.254$, $c=8.108$	[Pr_2Co_2] *8e (1)* [Pr_2Co_4] 2 sites *4c (m)*
β^{I}-$LaCo_5D_{3.35}$	$CaCu_5$ (def.)	*Cmmm*	[4.92]	$a=8.973$, $b=5.413$, $c=4.088$ (for hydride)	[La_2Co_4] *4e (2/m)* [La_2Co_2] *4h (2mm)*
β^{II}-$NdCo_5D_{2.8}$	$CaCu_5$ (def.)	*Im2m**	[4.92]	$a=8.805$, $b=5.245$, $c=8.086$	[Nd_2Co_2] *8e (1)* [Nd_2Co_4] 2 sites *4c (m)*
β^{III}-$CeCo_5D_{2.55}$	$CaCu_5$ (def.)	*Cccm*	[4.92]	$a=8.728$, $b=5.105$, $c=8.146$	[Ce_2Co_2] *8g (2)* [Ce_2Co_4] *4e (2/m)*

Table 4.1 (continued)

Compound[b]	Metal atom substructure[c]	Space group[d]	Ref.[e]	Lattice parameters [Å][f]	Interstices occupied by H(D)[g]
$LaNi_{4.5}Mn_{0.5}D_{6.6}$	$CaCu_5$	*P6/mmm*	[4.86]	$a=5.392$, $c=4.296$	$[La_2(Ni, Mn)_2]$ *6m (mm2)* $[La(Ni, Mn)_3]$ *12n (m), 12o (m)* $[(Ni, Mn)_4]$ *4h (3m)*
$LaNi_4MnD_{5.9}$	$CaCu_5$	*P6/mmm*	[4.86]	$a=5.406$, $c=4.329$	$[La_2(Ni, Mn)_2]$ *6m (mm2)* $[La(Ni, Mn)_3]$ *12n (m), 12o (m)* $[(Ni, Mn)_4]$ *4h (3m)*
$LaNi_3Mn_2D_{5.95}$	$CaCu_5$	*P6/mmm*	[4.86]	$a=5.451$, $c=4.344$	$[La_2(Ni, Mn)_4]$ *3f (mmm)* $[La_2(Ni, Mn)_4]$ *3f (mmm)* $[La_2(Ni, Mn)_2]$ *6m (mm2)* $[La(Ni, Mn)_3]$ *12n (m)*
$LaNi_4AlD_{4.8}$	$CaCu_5$	*P6/mmm*	[4.86]	$a=5.313$, $c=4.242$	$[La_2(Ni, Al)_2]$ *6m (mm2)* $[La(Ni, Al)_3]$ *12n (m)*
$LaNi_4AlD_{4.1}$	$CaCu_5$ (def.)	*P321*	[4.53]	$a=5.30$, $c=4.24$	$[LaNi_2(Ni, Al)]$ *$6g_1$ (1)* $[La_2(Ni, Al)_2]$ *$6g_2$ (1)*
$LaNi_{4.5}Al_{0.5}D_{4.5}$ (298 K)	$CaCu_5$	*P6/mmm*	[4.93]	$a=5.341$, $c=4.236$	$[La_2(Ni, Al)_2]$ *6m (mm2)* $[La_2Ni_2(Ni, Al)]$ *6i (2mm)*
(77 K)	$CaCu_5$	*P6/mmm*	[4.93]	$a=5.331$, $c=4.236$	$[La_2(Ni, Al)_2]$ *6m (mm2)* $[La_2Ni_2(Ni, Al)]$ *6i (2mm)*
$LaNi_{4.5}Al_{0.5}D_{5.4}$	$CaCu_5$	*P6/mmm*	[4.94]	$a=5.353$, $c=4.263$	$[La_2(Ni, Al)_2]$ *6m (mm2)* $[La(Ni, Al)_3]$ *12n (m), 12o (m)* $[(Ni, Al)_4]$ *4h (3m)*
$LaNi_{4.5}Si_{0.5}D_{4.3}$	$CaCu_5$	*P6/mmm*	[4.94]	$a=5.328$, $c=4.077$	$[La_2(Ni, Si)_2]$ *6m (mm2)* $[La(Ni, Si)_3]$ *12n (m)* $[(Ni, Si)_4]$ *4h (3m)*
$LaNi_4CuD_{5.07}$	$CaCu_5$	*P6/mmm*	[4.94]	$a=5.374$, $c=4.190$	$[La_2(Ni, Cu)_2]$ *6m (mm2)* $[La(Ni, Cu)_3]$ *12n (m)* $[(Ni, Cu)_4]$ *4h (3m)*
$LaNi_4CoD_4$	$CaCu_5$	*Cmmm*	[4.95]	$a=5.319$, $b=9.213$, $c=4.070$	$[La_2(Ni, Co)_4]$ *4e (2/m)* $[La(Ni, Co)_3]$ *4h (2mm)*
$ZrMn_2D_{3.0}$	$MgZn_2$	*$P6_3/mmc$*	[4.96]	$a=5.391$, $c=8.748$	$[Zr_2Mn_2]$ *12k (m), 6h (mm2)* *24l (1), 6h (mm2)*

$ZrMoFeD_{2.6}$	$MgZn_2$	$P6_3/mmc$	[4.53]	$a=5.420$, $c=8.826$	$[Zr_2(Mo, Fe)_2]$ *6h (mm2)*, *24l (1)*
$TiMe_{1.87}D_{2.36}$ $Me=V_{0.4}Mn_{0.6}$	$MgZn_2$	$P6_3/mmc$	[4.97]	$a=5.229$, $c=8.581$	$[TiMe_3]$ *4f (3m)* $[Ti_2Me_2]$ *6h (mm2)*, *24l (1)*, *12k (m)*
$Ti_{1.2}Mn_{1.8}D_{3.1}$ (240 K)	$MgZn_2$	$P6_3/mmc$	[4.98]	$a=5.271$, $c=8.579$	$[Ti_2Mn_2]$ *12k (m)*, 2 sites *6h (mm2)* $[Ti_2(Ti, Mn)_2]$ *24l (1)*
(190 K)	$MgZn_2$	$P6_3/mmc$	[4.98]	$a=5.258$, $c=8.577$	$[Ti_2Mn_2]$ *12k (m)*, 2 sites *6h (mm2)* $[Ti_2(Ti, Mn)_2]$ *24l (1)*
(170 K)	$MgZn_2$	$P6_3/mmc$	[4.98]	$a=5.257$, $c=8.576$	$[Ti_2Mn_2]$ *12k (m)*, 2 sites *6h (mm2)* $[Ti_2(Ti, Mn)_2]$ *24l (1)*
(80 K)	$MgZn_2$	$P6_3/mmc$	[4.98]	$a=5.254$, $c=8.572$	$[Ti_2Mn_2]$ *12k (m)*, 2 sites *6h (mm2)* $[Ti_2(Ti, Mn)_2]$ *24l (1)*
α-$Mg_2NiD_{0.3}$	Mg_2Ni	$P6_222$	[4.99]	$a=5.23$, $c=13.43$	$[Mg_4Ni_2]$ sites *6h (2)*, 2 sites *12k (1)*, *6j (2)* $[Mg_4Ni_4]$ site *6f (2)*
α-$Mg_2NiH_{0.3}$	Mg_2Ni	$P6_222$	[4.100]	$a=5.232$, $c=13.404$	$[Mg_2]$ *3c (222)*
$ZrBe_2D_{1.49}$	AlB_2	$P6/mmm$	[4.101]	$a=3.714$, $c=3.470$	$[Zr_3Be_2]$ *2c ($\bar{6}m2$)*
$ThNi_2D_{2.6}$	AlB_2 (def.)	Not spec.	[4.102]	$a=4.320$, $b=7.466$, $c=3.995$ (orthorhombic), or $a=4.405$, $b=5.338$, $c=4.260$, $\beta=111.24°$ (monoclinic)	$[Th_3Ni_2]$ Not specified $[Th_2Ni_2]$ Not specified
$ThNi_2D_2$	AlB_2 (def.)	Not spec.	[4.102]	$a=3.870$, $b=6.715$, $c=3.951$	$[Th_3Ni_2]$ Not specified $[Th_2Ni_2]$ Not specified
$HoNi_3D_{1.8}$	$PuNi_3$	$R\bar{3}m$	[4.53]	$a=4.991$, $c=26.12$	$[HoNi_3]$ *6c (3m)* $[Ho_2Ni_2]$ *18h (m)*
$LaNi_3D_2$	$PuNi_3$	$R\bar{3}m$	[4.32]	$a=4.94$, $c=25.37$	Not specified "octahedral" sites $[La_2Ni_4]$, $[LaNi_5]$, $[Ni_6]$
$Pd_{15}P_2D_{0.46}$ (500 kPa)	$Pd_{15}P_2$	$R\bar{3}$	[4.103]	$a=7.117$, $c=17.128$	$[Pd_4]$ *18f (1)*
Th_2AlD_2	Al_2Cu	$I4/mcm$	[4.105]	$a=7.702$, $c=6.230$	$[Th_4]$ *16l (m)*
Th_2AlD_3	Al_2Cu	$I4/mcm$	[4.105]	$a=7.676$, $c=6.383$	$[Th_4]$ *16l (m)*
Th_2AlD_4	Al_2Cu	$I4/mcm$	[4.105]	$a=7.629$, $c=6.517$	$[Th_4]$ *16l (m)*
$Zr_2PdD_{1.70}$	$MoSi_2$	$I4/mmm$	[4.106]	$a=3.321$, $c=11.606$	$[Zr_4]$ *4d ($\bar{4}m2$)*
$Zr_2PdD_{1.96}$	$MoSi_2$		[4.106]	$a=3.372$, $c=11.438$	$[Zr_4]$ *4d ($\bar{4}m2$)*

Table 4.1 (continued)

Compound[b]	Metal atom substructure[c]	Space group[d]	Ref.[e]	Lattice parameters [Å][f]	Interstices occupied by H(D)[g]
$Y_2Fe_{14}BD_{3.5}$	$Nd_2Fe_{14}B$	Not spec. *P4₂/mnm*	[4.107] [Fruchart, 1986, private commun.]	Not spec.	$[Y_3Fe]$ *8j (m)* $[Y_2Fe_2]$ 2 sites *16k (1)*, *4e (2mm)*
$TiCuD_{0.96}$	TiCu(Ti)	*P4/nmm*	[4.104]	$a=3.020$, $c=6.837$ (X-ray values)	$[Ti_4]$ *2b* $(\bar{4}m2)$
$ZrNiH_{2.98}$	CrB	*Cmcm*	[4.108]	$a=3.53$, $b=10.48$, $c=4.30$	$[Zr_3Ni]$ *8f (m)* $[Zr_3Ni_2]$ *4c (m2m)*
ZrNiD	CrB (def.)	*P1* or *P-1*	[4.109]	$a=3.367$, $b=10.313$, $c=4.063$ $\alpha=89.18°$, $\beta=89.60°$, $\gamma=89.43°$	$[Zr_4]$ *4c (mcm)* for *Cmcm*
$ZrCoD_3$	CrB	*Cmcm*	[4.110]	$a=3.53$, $b=10.48$, $c=4.30$	$[Zr_3Co]$ *8f (m)* $[Zr_3Co_2]$ *4c (m2m)*
$LaNiD_{3.7}$	CrB	*Cmcm*	[4.53]	$a=3.98$, $b=11.94$, $c=4.87$	$[La_3Ni]$ *8f (m)* $[La_3Ni_2]$ *4c (m2m)* $[Ni_2]$ *4b (2/m)*
$Pd_3P_{0.8}D_{0.15}$ (500 kPa)	Fe_3C	*Pnma*	[4.111]	$a=5.718$, $b=7.545$, $c=5.130$	$[Pd_5]$ *8d (1)*
$Pd_3P_{0.8}H_{0.17}$ (500 kPa)	Fe_3C	*Pnma*	[4.112]	$a=5.714$, $b=7.536$, $c=5.127$	$[Pd_5]$ *8d (1)*
$Pd_6PD_{0.15}$ (200 kPa)	Pd_6P	*P2₁/c*	[4.113]	$a=5.685$, $b=9.461$, $c=8.218$, $\beta=110.41°$	$[Pd_6]$ *2d* $(\bar{1})$ $[Pd_5]$ *4e (1)*
$Pd_6PD_{0.22}$ (500 kPa)	Pd_6P	*P2₁/c*	[4.113]	$a=5.686$, $b=9.463$, $c=8.219$, $\beta=110.43°$	$[Pd_6]$ *2d* $(\bar{1})$ $[Pd_5]$ *4e (1)*
$Pd_6PD_{0.26}$ (750 kPa)	Pd_6P	*P2₁/c*	[4.113]	$a=5.687$, $b=9.465$, $c=8.220$, $\beta=110.43°$	$[Pd_6]$ *2d* $(\bar{1})$ $[Pd_5]$ *4e (1)*
$Pd_6PH_{0.39}$ (800 kPa)	Pd_6P	*P2₁/c*	[4.112]	$a=5.685$, $b=9.464$, $c=8.210$, $\beta=110.43°$	$[Pd_6]$ *2d* $(\bar{1})$ $[Pd_5]$ 2 sites *4e (1)*

(mostly face- or body-centered cubic), ternary metal hydride structures show a great variety of complex metal atom arrangements of various (usually low) symmetries. Apart from the usual expansion and distortion (see below) most arrangements are similar to those of the corresponding hydrogen free binary metal compounds. Thus ternary metal hydride structures can be conveniently classified according to the structure types of their corresponding (undistorted) binary metal compounds as shown in the following list (structure types in italics; representative hydrides in parentheses; same sequence as in Table 4.1):

Cubic structure types: *Ti_2Ni* and *filled Ti_2Ni* (also called η-carbide) *type* (Ti_2NiH_x, $Ti_4Fe_2OD_x$, $Zr_3V_3OD_x$); *$MgCu_2$ type* (often called cubic Laves phase) (ZrV_2D_x, HfV_2D_4, $ZrCr_2D_x$, $ZrTi_{1.86}D(H)_{3.8}$, $CeRu_2D_5$, $ErFe_2H_{3.5}$, $ThZr_2D_x$); *CaF_2 type* (β, β'-Mg_2NiD_4, Mg_2CoD_5, Mg_2FeD_6); *Th_6Mn_{23} type* ($Y_6Mn_{23}D_x$, $Th_6Mn_{23}D_x$, $Ho_6Fe_{23}D_x$); *CsCl type* (α-$TiFeD_{0.057}$, $\beta(\beta')$-$TiFeD_{1(1.4)}$, γ-$TiFeD_{\simeq 2}$, $TiFe_{0.9}Mn_{0.1}D_x$) and *Cr_3Si-type* (Nb_3SnH).

Hexagonal structure types: *$CaCu_5$ type* ($LaNi_5D_{5-7}$, and substitutional derivatives $La(Ni_{1-x}B_x)_5D_x$, with $B=Mn$, Al, Si, Cu, Co; $RCo_5D_{\simeq 3}$, with $R=Pr$, La, Nd, Ce; β-$CaNi_5D_{0.77}$), *$MgZn_2$ type* (also called hexagonal Laves phase) ($ZrMn_2D_3$, $ZrMoFeD_{2.6}$, $Ti(V_{0.4}Mn_{0.6})_{1.87}D_{2.36}$, $Ti_{1.2}Mn_{1.8}D_{3.1}$); *Mg_2Ni type* (α-$Mg_2NiD_{0.3}$); *AlB_2 type* ($ZrBe_2D_{1.49}$, $ThNi_2D_x$); *$PuNi_3$ type* ($HoNi_3D_{1.8}$, $LaNi_3D_2$); and *$Pd_{15}P_2$ type* ($Pd_{15}P_2D_{0.46}$).

Tetragonal structure types: *TiCu(Ti)(=B11) type* ($TiCuD_{0.96}$); *$CuAl_2$ type* (Th_2AlD_x); *$MoSi_2$ type* (Zr_2PdD_x), and *$Nd_2Fe_{14}B$ type* ($Y_2Fe_{14}BD_{3.5}$).

Orthorhombic structure types: *CrB type* ($ZrNiD(H)_x$, $ZrCoD_3$, $LaNiD_{3.7}$), and *Fe_3C type* ($Pd_3P_{0.8}D_x$).

Monoclinic structure types: *Pd_6P type* (Pd_6PD_x).

Upon hydrogenation most binary metal compounds undergo a lattice expansion (between 2 and 3A^3 per H atom [4.138]); for example $LaNi_5$ whose lattice expands by up to 25 vol% at the composition $LaNi_5D_7$. The exceptionally strong expansion of $CeRu_2$ (37% for $CeRu_2D_5$) is due to a hydrogen induced electronic transition as shown by XPS measurements [4.139]. A lattice contraction upon hydrogenation has so far only been observed for $ThNi_2$ (−2.2% for $ThNi_2D_2$). For some hydride structures the space group symmetries at room temperature are identical to those of the binary metal compounds, such as for cubic $Zr_3V_3OD_x$, $x=1.86$, 2.85, 4.93; ZrV_2D_x, $x=1.5$, 2.8, 4.9; and $Ho_6Fe_{23}D_x$, $x=1.5$, 8.2, 15.7; for hexagonal $ZrMn_2D_{3.0}$; tetragonal Th_2AlD_4; orthorhombic $LaNiD_{3.7}$, and monoclinic Pd_6PD_x. For others these symmetries differ – see hydrides marked "def." (deformed) in second column of Table 4.1 – e.g. β-$TiFeD_1$, γ-$TiFeD_2$, β^{I}-$PrCo_5D_{3.6}$, β^{II}-$NdCo_5D_{2.8}$, β^{III}-$CeCo_5D_{2.55}$, β-$LaNi_5D_7$, β-$CaNi_5D_{0.77}$, $ThNi_2D_x$, and ZrNiD, where hydrogenation leads to a lowering of lattice symmetry but does not induce a major reconstruction of the metal atom substructure. Only a few hydrides with reconstructed metal atom arrangements, or hydrides which do not derive from stable binary metal compounds have been reported to date (see below). For some hydride structures the lattice symmetries depend upon hydrogen content

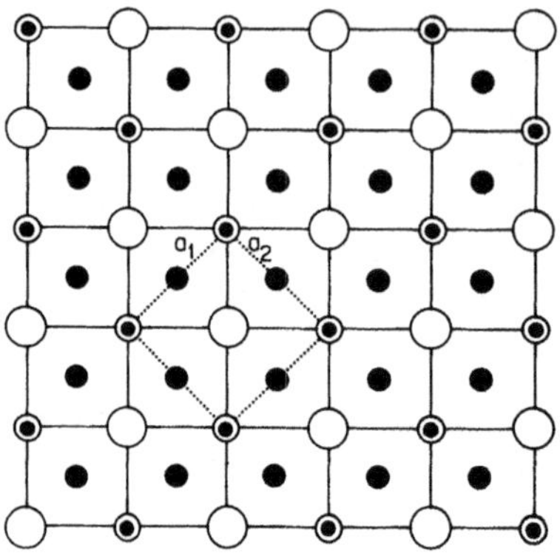

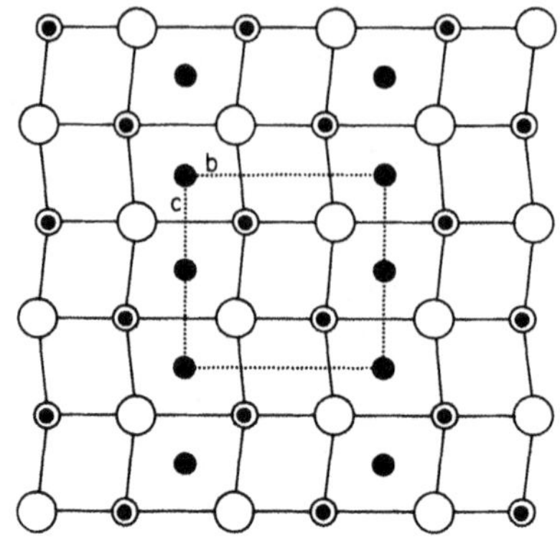

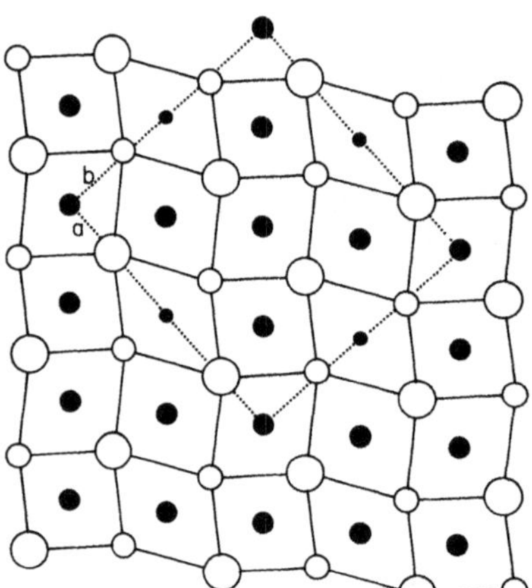

Fig. 4.1. Projections of the crystal structure of the α, β, and γ-hydrides of TiFe. Large open circles Ti at hight 1/2, small open circles Fe at height 0, small filled circles D at 1/2, large filled circles D at 0. Unit cell marked by dotted lines

($Ho_6Fe_{23}D_x$, ZrV_2D_x, $TiFeD_x$, $LaNi_5D_x$) and upon temperature (HfV_2D_4, $ZrV_2D_{3.6}$, $Y_6Mn_{23}D_{23}$, $Th_6Mn_{23}D_{16}$). In general the symmetry decreases as a function of hydrogen content ($TiFeD_x$, see below) and increases as a function of temperature ($Y_6Mn_{23}D_{23}$, $ZrV_2D_{3.6}$, Mg_2NiH_4).

Until now only a few systems that contain two (or more) well-defined hydride phases have been structurally characterized. An example is the TiFe–D system which contains a cubic (solid solution) phase of approximate composition $TiFeD_{0.06}$ (called α), one (or possibly two) orthorhombic phases of approximate composition $TiFeD_{\sim 1}$ (one called β [4.74, 75] or β_1 [4.75], and the other β_2 [4.76]), and a hydrogen-richer monoclinic [4.76–79] or probably orthorhombic phase [4.79]) of approximate composition $TiFeD_{\sim 2}$ (called γ). As shown in Fig. 4.1 the distortion of the overall metal atom substructures increases as a function of the overall hydrogen content. Another example is the $LaNi_5$–D system which contains a hexagonal (solid solution)

α-phase of approximate composition $LaNi_5D_{0.4}$ [4.82], a trigonal [4.32, 33, 83–85, 89] or hexagonal [4.86–88] β-phase of approximate composition $LaNi_5D_{5.5}$ and a hydrogen-rich hexagonal [4.87, 88] or orthorhombic [4.90] phase of approximate composition $LaNi_5D_7$.

For some systems (β-$LaNi_5D_{\simeq 6}$, β- and γ-$FeTiD_{1-2}$, β-Mg_2NiD_4) the space group symmetries are not yet fully established. For β-$LaNi_5D_6$ five different space groups have been proposed (*P6/mmm* [4.86], *P6₃mc* [4.87, 88], *P31m* [4.83—85, 89], *P321* [4.53], and *Cmmm* [4.1, 90]). Of these only one (*Cmmm*) was capable of explaining the observed anisotropic line broadening, but was judged unlikely from refinements on integrated intensities [4.87]. The other space groups required the assumption of microstrains [4.86, 87] or particle size effects [4.88]. A weak diffraction line previously not included in structure refinements was interpreted [4.87, 88] by doubling the hexagonal cell along *c*, and by assuming the (non-centrosymmetric) space group $P6_3mc$. For β-$FeTiD_x$ various orthorhombic space groups have been proposed [4.74–76]. For γ-$FeTiD_2$ the published monoclinic structure [4.76–78] has been described in terms of an orthorhombic space group with fewer parameters [4.79]. The structure of the hydrogen storage material β-Mg_2NiD_4 at room temperature was originally described as tetragonal [4.118]. Later it was described [4.122, 140, 141] as orthorhombic (space groups $P2_12_12$ [4.122] or $P222_1$ [4.140]) and monoclinic (space groups *C2/m* [4.62], *Ia* (or *Cc*) [4.63, 121, 122], or *Cm* [4.120]. Recently an x-ray structure refinement which took into account anisotropic line broadening due to microtwinning has suggested [4.123] that its structure is monoclinic with space group *C2/c*. Based on this model, hydrogen atom positions were later successfully derived from high-resolution neutron diffraction data [4.64]. The presence of microtwinning in this compound has been confirmed by electron diffraction [D. Noréus, private communication].

An example of a well-characterized hydride structure in which hydrogenation induces significant shifts of the metal atoms while the lattice expands isotropically is cubic $Zr_3V_3OD_x$ [4.47]. Reconstructed metal atom substructures were found for the following hydrides: cubic $ZrTi_{1.86}D_{3.8}$ which derives from a hexagonal close-packed alloy; orthorhombic $ZrCoD_3$ which derives from a cubic compound with CsCl structure, and monoclinic Mg_2NiH_4 which derives from a hexagonal compound with complex structure. On the other hand tetragonal Mg_2CoH_5 and cubic Mg_2FeH_6 do not derive from stable intermetallic compounds. They were included in this review because of their close structural and bonding relationships to Mg_2NiH_4, and because they have provided insight into the factors which limit maximum hydrogen content in this type of compound (Sect. 4.4.3). Diffraction evidence for reconstructed metal atom arrangements also exists for other metal hydrides (for a recent list see [4.42]). An interesting example is $TiCr_{1.8}H_{5.3}$ for which a disordered CaF_2-type metal host structure was proposed [4.142] while the hydrogen-free alloy has a $MgCu_2$-type structure. However this compound has not been fully characterized.

4.3.2 Distribution of Hydrogen Atoms

In contrast to elemental hydride structures which are often ordered, ternary metal hydride structures are usually disordered at room temperature. Consequently their local hydrogen atom configurations around the metal atoms are unknown. Examples of compounds with ordered hydrogen atom distributions at room temperature and thus known hydrogen configurations are Mg_2FeD_6, Mg_2CoD_5, and Mg_2NiD_4, as shown in Fig. 4.2. In the first two of these the transition metal atoms are surrounded respectively by octahedral (as in Sr_2RuD_6 [4.22]), and square-pyramidal (as in Sr_2IrD_5 [4.143]), hydrogen atom configurations. In Mg_2NiD_4 they are surrounded by a tetrahedral configuration, as shown from analysis of high-resolution neutron diffraction data [4.64], in contrast to the square-planar [4.62] and disordered [4.63] configurations reported previously from analysis of medium- and low-resolution data. Examples of compounds with partially or nearly ordered hydrogen atom distributions at room temperature are $TiCuD_{0.96}$, $LaNiD_{3.7}$, ZrNiD, $ZrNiH_{2.98}$, Th_2AlD_4, γ-$TiFeD_2$, and $ZrV_2D_{3.6}$.

One of the most characteristic and intriguing structural features of ternary metal hydrides are their preferential hydrogen site occupancies. As can be seen from Table 4.1, the hydrogen atoms exclusively occupy those interstices of the metal atom network which allow them to be in contact with at least one hydride forming element such as Mg, Ti, Zr, Hf, V, Ni, Pd, Th, Y, or rare-earths. Moreover, in structures which contain several types of interstices those surrounded by a larger fraction of hydride-forming elements, or those surrounded by an element forming a more stable binary hydride, appear to be preferentially occupied. For example in the ZrV_2D_x system the $[Zr_2V_2]$ holes are populated at lower overall deuterium concentrations than the $[ZrV_3]$ holes, and in the hydrides of TiFe the $[Ti_4Fe_2]$ holes are filled at lower overall hydrogen concentrations than the $[Ti_2Fe_4]$ holes. In the $Zr_3V_3OD_x$ system most of the hydrogen atoms first fill the $[Zr_3V]$ and $[Zr_2V_2]$ holes, while the

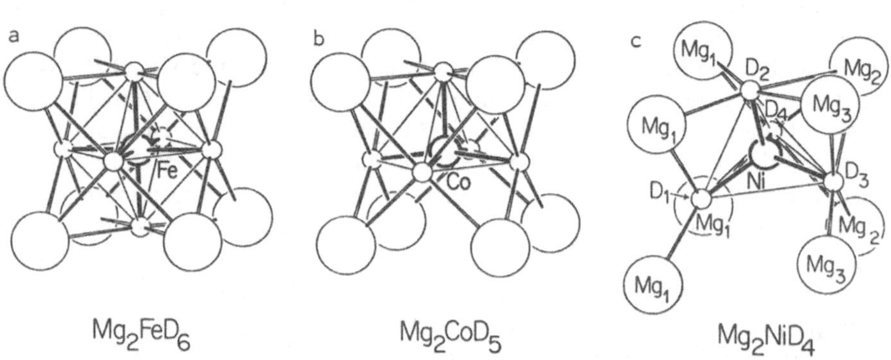

Fig. 4.2. Transition metal environments in Mg_2FeD_6, Mg_2CoD_5, and Mg_2NiD_4 (from [4.64]). Large circles Mg, small circles D

$[ZrV_3]$ and $[V_4]$ interstices mostly remain empty even at high hydrogen contents. In $LaNi_5D_x$ most of the hydrogen atoms occupy the $[La_2Ni_2]$ (or $[La_2Ni_4]$) and $[LaNi_3]$ holes, while the $[Ni_4]$ holes are only filled at high hydrogen concentrations. Preferential hydrogen site occupancy was also observed with amorphous alloys such as $ZrNiD_{1.8}$ [4.9] for which neutron diffraction analysis has shown occupation of tetrahedral $[Zr_4]$ holes. The above trends correlate with macroscopic sorption properties and constitute the main basis for a model [4.115] which was used to rationalize hydrogen site occupancies and relative stabilities in terms of attractive nearest-neighbour interactions (Sect. 4.4.4). According to another model [4.114] the hydrogen site occupancies can also be interpreted in geometrical terms by assuming that hydrogen prefers to occupy large interstices and avoids small interstices, and by correlation effects between hydrogen atoms (Sect. 4.4.1).

The metal atoms surrounding the hydrogen atom sites form polyhedra of various shapes and symmetries (for number and type of metal atoms, and point group symmetries see last column in Table 4.1). The polyhedra which occur most frequently are deformed tetrahedra ("tetrahedral sites") and octahedra ("octahedral sites"). The hydrogen sites within these polyhedra usually have low symmetry (mostly 2, *m* or 1).

Tetrahedral sites are occupied in hydrides belonging to the following structure types (a representative compound and the nature of metal atom environment are given in parentheses): Ti_2Ni (Ti_2NiH, $[Ni_4]$) and filled Ti_2Ni($Zr_3V_3OD_x$, $[Zr_3V]$, $[Zr_2V_2]$, $[ZrV_3]$); $MgCu_2$(ZrV_2D_x, $[Zr_2V_2]$, $[ZrV_3]$); Th_6Mn_{23}($Y_6Mn_{23}D_x$, $[Y_3Mn]$, $[Y_2Mn_2]$, $[YMn_3]$); Cr_3Si(Nb_3SnH_x, $[Nb_4]$); $CaCu_5$(β-$LaNi_5D_x$, $[La_2Ni_2]$, $[LaNi_3]$, $[Ni_4]$); $MgZn_2$($ZrMn_2D_3$, $[Zr_2Mn_2]$); $PuNi_3$($HoNi_3D_{1.8}$, $[HoNi_3]$, $[Ho_2Ni_2]$); $CuAl_2$(Th_2AlD_x $[Th_4]$); $MoSi_2$(Zr_2PdD_x, $[Zr_4]$); TiCu(Ti) ($TiCuH_x$ $[Ti_4]$); $Nd_2Fe_{14}B$ ($Y_2Fe_{14}BD_x$, $[Y_3Fe]$, $[Y_2Fe_2]$) and CrB($ZrNiH_3$, $[Zr_3Ni]$).

Octahedral sites are occupied in: Ti_2Ni(Ti_2NiH, $[Ti_6]$); Th_6Mn_{23}($Y_6Mn_{23}D_x$, $[Y_6]$); CsCl($\beta(\beta')$-$FeTiD_x$, $[Ti_4Fe_2]$; γ-$FeTiD_2$, $[Ti_4Fe_2]$, $[Ti_2Fe_4]$); $CaCu_5$($LaNi_5D_x$, $[La_2Ni_4]$); and Pd_6P(Pd_6PD_x, $[Pd_6]$).

Square-pyramidal sites are occupied in: CaF_2(Mg_2NiD_4, $[Mg_4Ni]$; Mg_2CoD_5, $[Mg_4Co]$; Mg_2FeD_6, $[Mg_4Fe]$); Fe_3C($Pd_3P_{0.8}D_x$, $[Pd_5]$); and Pd_6P(Pd_6PD_x $[Pd_5]$).

Triangular, or triangular bi-pyramidal sites are occupied in the hydrides $Th_6Mn_{23}D_{16}$($[Th_2Mn_3]$), $LaNi_5D_6$($[LaNi_2]$, according to [4.89]), $ZrBe_2D_{1.49}$ ($[Zr_3Be_2]$), and $ZrNiH_{2.98}$($[Zr_3Ni_2]$), whereas

linear hydrogen coordination was found in $LaNiD_{3.7}$($[Ni_2]$) and α-$Mg_2NiH_{0.3}$($[Mg_2]$).

In the discussed ternary metal hydrides no example of double occupancy of a hole such as that found [4.14] in the halide ZrBrD exists as yet. Due to a lack of resolution, the exact locations of the hydrogen atoms within the coordination polyhedra are not always known. A typical example is $LaNi_5D_6$ in which the deuterium atoms occupying the octahedral $[La_2Ni_4]$ interstices appear to be displaced from the centre in various directions, depending on the

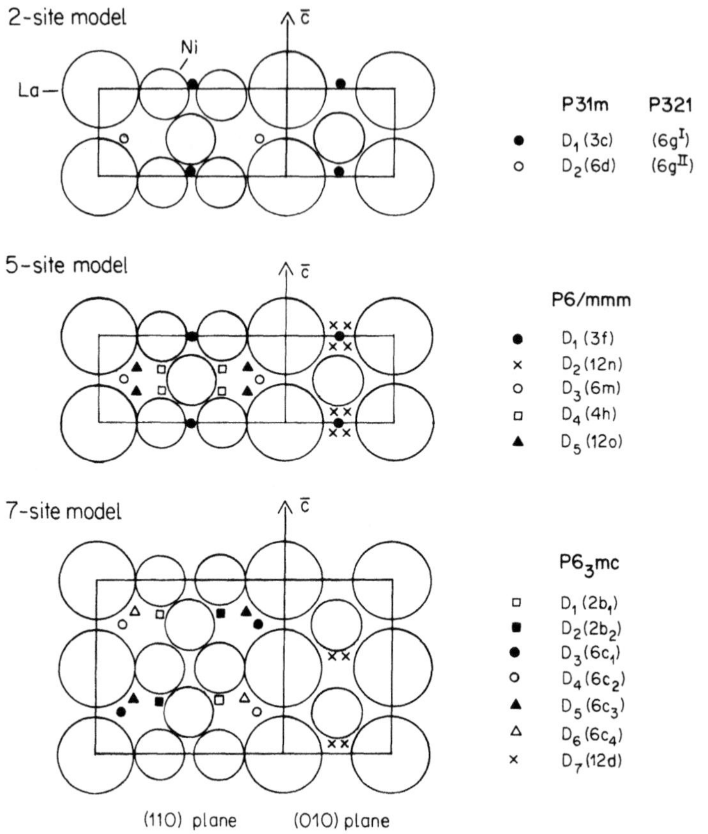

Fig. 4.3. Deuterium atom sites and space group symmetries of various multi-site models proposed for the structure of β-$LaNi_5D_x$ (adapted from [4.87]; Wyckoff positions [4.44] in parentheses)

model used for structure refinement (Fig. 4.3). Conflicting results with respect to the deuterium atom displacements and site occupancies have also been reported for other systems such as $Y_6Mn_{23}D_{23}$ [4.67–69], β-$TiFeD_1$ [4.74–76], and $ZrV_2D(H)_{3.6}$ [4.49–51]. Some of these ambiguities could be due to anharmonic thermal motion of the H atoms ($Y_6Mn_{23}D_{23}$, β-$TiFeD_1$) or a reduction of cell symmetry ($ZrV_2H_{3.6}$) due to hydrogen atom ordering. An interesting example of an ambiguous hydrogen atom distribution is the disordered β'-Mg_2NiD_4 modification for which two models with respectively square-pyramidal [Mg_4Ni] [4.59, 60] and square-planar [Mg_2Ni_2] [4.61] hydrogen site coordination have been reported which both allow an equally good fit to the diffraction data. The latter model was considered less likely [4.144] in view of its unusually short magnesium-hydrogen distances (Sect. 4.3.3), and electronic properties [4.145].

In disordered metal hydrides the hydrogen atoms undergo large amplitude thermal vibrations (rms. values of up to 0.1 Å at room temperature, see *B* values in Appendix 4.A). Residual nuclear density in Fourier difference maps ($LaNi_5D_x$ [4.35]) and temperature dependent site occupancy factors ($ZrV_2D_{3.6}$ [4.49]) suggest that these vibrations are strongly anharmonic. Anharmonicity is presumably one of the factors responsible for the systematic difference between the hydrogen content found by diffraction methods and that measured by desorption (see remarks in Appendix 4.A).

In some compounds structural phase transitions occur below room temperature (see T_1 values given in Table 4.1). These transitions are generally due to ordering of the hydrogen atoms which leads to a reduction of lattice symmetry. Examples are $ZrV_2D_{3.6}$, HfV_2D_4, $Y_6Mn_{23}D_{23}$, and $Th_6Mn_{23}D_{16}$. Compounds with ordered hydrogen atom distributions at room-temperature and disordered distributions at high temperatures are β-Mg_2NiD_4 and Mg_2CoD_5. Their high temperature structures resembles those of Mg_2FeD_6(K_2PtCl_6 type), and K_2PtD_4 [4.23].

4.3.3 Interatomic Distances

In view of relatively low precision (typically ± 0.02 Å, see Sect. 4.2.3) the interatomic distances of the ternary metal hydride structures characterized to date reveal only qualitative trends.

Metal-Metal Distances. At the hydrogen-rich phase limit these distances are longer by about 0.1 Å, on the average, than those in the corresponding binary metal compound. This suggests that the metal-metal bonds are weakened during hydride formation. Examples of the reverse behavior are found in $Zr_3V_3OD_{1.38}$, $Zr_2PdD_{1.98}$, and $Y_6Mn_{23}D_x$, in which some metal-metal distances (V–V, Zr–Zr, and Mn–Mn, respectively) are shortened by about 0.2 Å with respect to those in the hydrogen-free compounds.

Metal-Hydrogen Distances. These correspond closely to those in the corresponding elemental hydrides, except that they may be anomalously shortened due to partial occupancy of the hydrogen atom sites. For example some Zr–D distances in $Zr_3V_3OD_{1.86}$([Zr–D] = 1.94 Å) are shorter than those in binary ZrD_2([Zr–D] = 2.09 Å), while some metal-hydrogen distances in β'-Mg_2NiD_4([Ni–D] = 1.49(3) Å) according to [4.59] and α-$FeTiD_{0.057}$([Fe–D] = 1.49 Å) are close to the sum of covalent radii ([Ni–H] = 1.47 Å, [Fe–H] = 1.49 Å). Such short distances may not be characteristic for all hydrogen atom sites belonging to a particular crystallographic location because of possible local distortions in their metal atom environments, as was suggested [4.146] for α-$TiFeD_{0.057}$, and found [4.147] in binary transition metal hydrides (for example in NbH_x). On the other hand, the unusually short metal-hydrogen distances reported in an early structure model of $LaNi_5D_7$([Ni–D] = 1.17 Å [4.83]), and those recently reported for disordered β'-Mg_2NiH_4([Mg–D] = 1.66 Å [4.61]) are presumably a consequence of limited data sets ($LaNi_5D_7$) and deficiencies of the structure models (β'-Mg_2NiD_4) used for structure refinement. The Ni–D distance in the former compound is smaller than the atomic radius of the metal atom (r[Ni] = 1.24 Å) whereas the Mg–D distance reported for the latter compound is significantly shorter than that in binary MgD_2([Mg–D] = 1.95 Å), and that reported [4.59] previously for β'-Mg_2NiH_4([Mg–D] = 2.31 Å) by using a different structure model. The conflicting results obtained for β'-Mg_2NiD_4 from diffraction data of similar resolution were analyzed in [4.144] and found to be mainly due to the particular ratio between the neutron scattering lengths of the metal nuclei ($b_{Ni} \simeq 2b_{Mg}$). Short metal-hydrogen distances in ordered (or nearly ordered) hydrides occur in $ZrNiH_{2.98}$([Zr–H] = 1.95 Å), Mg_2CoD_5([Co–D] = 1.52 Å) and Mg_2FeD_6([Fe–D] = 1.56 Å). The distances in the two latter hydrides, and also those in $FeTiD_{0.057}$, are of particular interest because they involve metal atoms which do not belong to elements forming stable hydrides (Sect. 4.4.4). Most metal-hydrogen distances in ternary metal hydride structures can be rationalized by 12-fold coordinated metal atom radii [4.148] and a fixed H atom radius of 0.40 Å [4.114]. However, descriptions in terms of normal-valence cation radii (or crystal radii [4.149]) and fixed H atom radii of 1.1 Å [4.150] (or 1.27 Å [4.24]) such as those used [4.24, 150] to rationalize metal-hydrogen distances in binary metal hydrides presumably fit these distances equally well.

Hydrogen-Hydrogen Distances. A characteristic feature of ternary metal hydride structures is their minimum hydrogen atom separation. In ordered hydrides this separation was never found to be smaller than 2.1 Å, a limit already noticed [4.151] in binary metal hydrides. Apparent exceptions are Th_2AlD_4([D–D] = 1.79 Å according to [4.105], see also [4.31]), and β-Mg_2NiD_4([D–D] = 1.75 Å, according to [4.62]) which could be due to the low resolution of the reported diffraction data (Th_2AlD_4) and the inadequacy of the model used for structure refinement (β-Mg_2NiD_4). Well-documented examples of hydrogen-hydrogen distances shorter than 2.1 Å have so far only been reported for (non-metallic) complex hydrides such as K_2ReH_9([D–D] = 1.87(2) Å [4.12]).

In disordered hydrides partially occupied hydrogen atom sites are almost always separated by less than 2.1 Å. Typical examples are cubic ZrV_2D_5 and hexagonal $ZrMn_2D_3$ in which neighboring D atom sites are separated by about 1.3 Å. As shown in Fig. 4.4 these sites are connected to form quasi-infinite three-dimensional networks, thus providing possible diffusion paths through the structures. Since the occupancies of the hydrogen atom sites are usually lower than 50% (see Appendix 4.A), a minimum hydrogen atom separation of about 2.1 Å was assumed [4.52, 114] to be effective also in these and other disordered hydrides. A compound showing hydrogen-hydrogen contacts which are possibly shorter than 2.1 Å is $Y_6Mn_{23}D_{23}$ for which full occupancy (without e.s.d) of a site ($32f_3$) which is only 1.3 Å away from another partially occupied site ($96k_1$) was reported in one study [4.67], and 62% occupancy in another

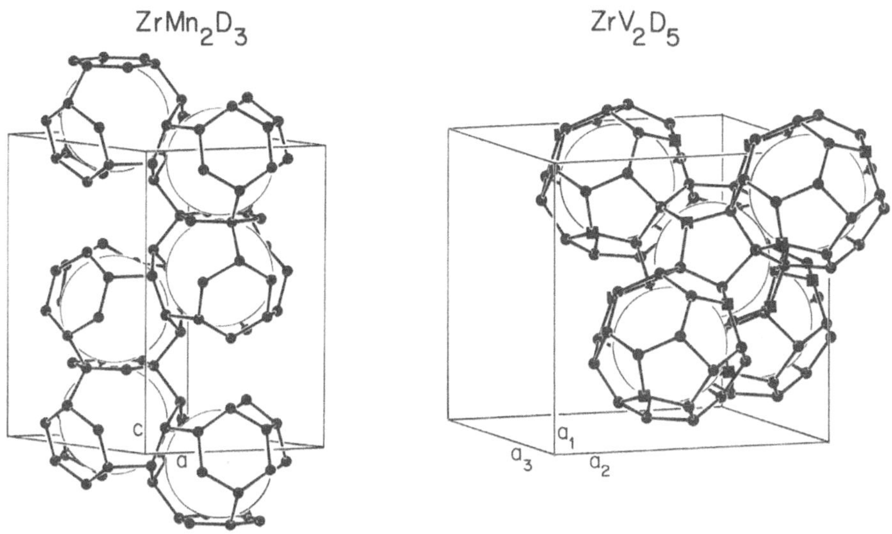

Fig. 4.4. Distribution of hydrogen atom sites in hexagonal $ZrMn_2D_3$ and cubic ZrV_2D_5 (from [4.1]). Large circles Zr, small filled circles and squares D atom sites. The V and Mn atoms are omitted. Solid lines indicate possible diffusion paths in the structure

study [4.69]. Experimental support for the assumption of a minimum separation between hydrogen atoms in disordered hydrides came mainly from the observation of diffuse intensity peaks close to $\lambda/(2\sin\theta)=2$ Å in the neutron diffraction patterns of hydrides with $MgCu_2$ type metal atom substructure such as HfV_2D_4 [4.52], ZrV_2D_x [4.48, 50], $ZrCr_2D_{3.8}$ [4.51, 54], $CeRu_2D_5$ [4.56], the disordered modifications of Mg_2NiH_4 [4.59, 61] and Mg_2CoD_5 [4.65], and of $Ti_{1.2}Mn_{1.8}D_{3.1}$ [4.98]. No such peaks were found in the patterns of the ordered low-temperature phases of HfV_2D_4 [4.52], $ZrV_2D_{3.6}$ [4.49], and $ZrV_2H_{3.7}$ [4.51]. In these last hydrides, the shortest separations between occupied hydrogen atom sites are about 2.1 Å, compared to 1.3 Å found in the corresponding disordered modifications. Diffuse neutron diffraction intensity peaks were also reported for $ThNi_2D_x$ [4.102].

Hydrogen–Non-Metal Distances. Except for $Ti_4Fe_2OD_{2.25}$([O–D] = 1.68 Å [4.46]) and $ZrO_{0.4}D_{0.1}$([O–D] not specified in [4.21] but presumably smaller than 2.0 Å) these distances are larger than 2.0 Å (examples: $Zr_3V_3OD_x$: [O–D] = 2.01 Å, Pd_6PD_x: [P–D] = 2.94 Å, $Pd_3P_{0.8}D_{0.15}$: [P–D] = 3.51 Å). For the last compound, an unusually small hydrogen–non-metal distance is found ([P–D] = 0.58 Å) which was attributed [4.111] to the presence of defects on both deuterium and phosphorus sites.

4.4 Stability Aspects

Many attempts have been made to rationalize the stabilities and maximum hydrogen contents of ternary metal hydrides (see Chap. 6). Electronic band-structure calculations have so far only been performed [4.145, 152–155] for hydrides having simple and highly symmetric crystal structures such as cubic $TiFeD_x$, β'-Mg_2NiD_4, Mg_2FeD_6, ZrV_2D_x, and tetragonal Mg_2CoD_5 [see Chap. 5]. While these calculations were most valuable for the understanding of electronic properties, their level of sophistication did not allow heats of formation, relative site stabilities, or maximum hydrogen contents to be derived. The results for $TiFeD_x$ suggest [4.152] that its stability is mainly controlled by a lowering of empty metal states by the hydrogen potential, similar to the behavior suggested [4.151] for elemental metal hydrides. The results on ZrV_2D_x were claimed [4.155] to be consistent with the experimentally observed hydrogen site occupation (mainly $[Zr_2V_2]$ interstices). The calculations on β'-Mg_2NiD_4 were performed [4.145] on two disordered structure models of which one [4.59, 60] assumed octahedral and the other [4.61] linear surroundings of nickel atoms by hydrogen. The results and comparison with XPS spectra [4.145] favor the former model. However in view of recent structure work [4.64] on the ordered modification of β-Mg_2NiD_4, new calculations based on the actual hydrogen atom distribution (tetrahedral configuration around nickel) and a reinterpretation of the XPS spectra are desirable. The calculations performed [4.154] on Mg_2FeD_6 predict semiconducting behavior and confirm that the previously postulated band-filling effect [4.59, 66] is an important factor limiting hydrogen capacity.

Various semi-empirical attempts to rationalize enthalpies of hydride formation and maximum hydrogen contents have been reported. Some are based on the inverse relationship found [4.136] between the thermal stability of metal hydrides and the heats of formation of the corresponding binary metal compounds ("rule of reversed stability"), while others are based on correlations between the heats of formation of ternary metal hydrides and the stabilities of the corresponding elemental hydrides [4.156], or some suitable electronic parameters of the hydrogen-free metal host structure [4.157]. The influence of entropy effects have been discussed [4.158] for $LaNi_5D_x$ and isostructural hydrides [4.159], while the relative importance of geometric and electronic contributions to the thermodynamic properties have been evaluated [4.160] for body-centered-cubic metal hydrides. Correlations were pointed out between atomic volume contraction during binary metal compound formation and the stability [4.161] and maximum hydrogen content [4.162] of the corresponding ternary metal hydride. A model for calculating enthalpies of hydride formation has been proposed [4.163] which takes into account both electronic and geometrical factors. For some systems, the maximum hydrogen content can be rationalized by decomposing the ternary metal hydride into elemental hydrides, such as [4.37]:

$$LaNi_5H_{5.5} = LaH_3 + 5NiH_{0.5},$$

or by decomposing the ternary metal hydride into hydrides of structurally related binary metal compounds. For example [4.164]:

$$AB_3H_{x+y}(\text{PuNi}_3\text{type}) = \tfrac{1}{3}[AB_5]H_x(\text{CaCu}_5\text{type}) + \tfrac{2}{3}[AB_2]H_y(\text{MgZn}_2\text{type}).$$

A multiplateau theory with consideration of hydrogen interactions has been proposed [4.165] which allows experimental isotherms to be fitted.

Apart from a few exceptions the above methods are all structure independent. Their success is presumably due to the relatively small contribution of structure-dependent terms to the total energy of hydride crystals, as confirmed by theoretical [4.166] and semi-empirical [4.163, 167] considerations, and by the crystal chemistry of binary metal compounds [4.168]. Consequently these methods fail to rationalize (or predict) structure-dependent features such as relative stabilities of potentially occupied hydrogen atom sites and maximum hydrogen content. Attempts to rationalize these two features have mainly been based on empirical considerations, such as the availability of interstitial holes of sufficient size, electrostatic repulsion acting between hydrogen atoms or between hydrogen and non-metal atoms, electronic factors such as band filling, and attractive nearest-neighbor interactions between hydrogen and metal atoms. These attempts are summarized in the next section.

4.4.1 Interstitial Hole Size

Correlations between hydrogen equilibrium pressures and intersticial hole size in AB, AB_2, and AB_5 compounds [4.169] and cell volume in AB_5 compounds [4.170] suggest that hydrogen prefers to occupy large interstices. Hole sizes were defined [4.114, 171] by the largest sphere which can be inserted into the holes. The radii of the metal atom spheres were either derived [4.171] self-consistently by assuming that these spheres touch each other, or taken [4.114] from a tabulated set of 12-fold coordinated metal atom radii [4.148 and 4.168, p. 151]. Only holes for which the radius of the inscribed sphere exceeded a threshold value of about 0.40 Å were found to be occupied [4.114]. This value of the radius constituted one of the two major parameters of the so-called "geometrical model" [4.114] which was used to rationalize hydrogen site occupancies and maximum hydrogen contents (Sect. 4.4.2). Its application showed that the requirement of minimum hole size was fulfilled for nearly all metal hydride structures studied so far by neutron diffraction analysis (for example: $Zr_3V_3OD_x$ [4.172], $TiFeD_x$ [4.146], $ZrBe_2D_x$ [4.173], $Y_6Mn_{23}D_x$ [4.174], Sr_2RuD_6 [4.175], $LaNi_5D_x$ [4.176], $Pd_3P_{1-x}H_y$ [4.177], ZrV_2D_x and $ZrMn_2D_x$ [4.138], Ti_2NiD_x [4.178]). A compound for which this requirement was not fulfilled was β'-Mg_2NiH_4, suggesting [4.175] a possible deficiency in its structure model. For hydrides of Th_6Mn_{23}-type alloys, the occupancy of a particular site (site $96j_1$, see Appendix 4.A) has been investigated [4.179] along the lines of this model and found to be critical for the tetragonal lattice distortion. However, there also exist hydride structures containing holes of sufficient size which are nonetheless not occupied by hydrogen, such as some of the [Mg_4Co] holes in Mg_2CoD_5, some of the [Mg_4Ni] holes in β-Mg_2NiD_4, the [Zr_3Pd] and [Zr_2Pd_2] holes in $Zr_2PdD_{1.70}$, the [M_6] holes (M=Th, Ho) in $M_6Mn_{23}D_{16}$ and $Ho_6Fe_{23}D_x$ (x=12.1, 15.7), the triangular-prismatic [Pd_6] holes in Pd_3P_{1-x} and Pd_6P, and some of the tetrahedral [Zr_2M_2] holes (M=V, Mn, Ti) in disordered ZrM_2D_3, ordered $ZrV_2D_{3.6}$ [P. Zolliker, private communication], and $ZrTi_{1.86}D_{3.8}$. For some compounds, repulsive interactions between the D atoms are presumably responsible for the non-occupancy of these holes (Sect. 4.4.2). For others (such as $Pd_3P_{1-x}H_y$), repulsive interactions between hydrogen and non-metal atoms have been invoked [4.177]. There also exist compounds which absorb little hydrogen under normal conditions, such as $LaPt_5$ and $ThNi_5$ (both $CaCu_5$ type [4.180]) whose metal atom substructures presumably contain holes which are sufficiently large to be filled by hydrogen atoms. In these compounds other factors apart from interstitial volume appear to govern hydrogen site occupancy. Thus the requirement of minimum hole size may be a neccessary but not a sufficient criterion for hydrogen occupancy, and the maximum hydrogen content calculated by purely geometrical considerations may be higher than that observed (see below).

4.4.2 Electrostatic Repulsion

The observation of a minimum hydrogen atom separation of about 2.1 Å has been interpreted in terms of electronic effects [4.151], electrostatic repulsion acting between the hydrogen atoms [4.50, 114], blocking effects [4.181, 182], or antibonding effects [4.150]. This separation corresponds to about twice the pseudo-fluorine-like hydrogen radius of 1.1 Å used [4.150] to rationalize metal-hydrogen distances in elemental hydrides, and is consistent with anion-like behavior of hydrogen. Anion-like behavior of hydrogen in ternary metal hydrides was also suggested from a study of their magnetic properties [4.183]. The existence of a minimum hydrogen atom separation in metal hydride structures correlates with empirical considerations which suggest [4.137] that the maximum hydrogen content corresponds to similar volume concentrations of hydrogen. The 2.1 Å separation constituted the second major parameter of the geometrical model which was used [4.114] to rationalize hydrogen site occupancies and hydrogen capacities. According to this model, the maximum hydrogen content of a particular metal structure corresponds to the maximum number of spheres which can be packed into interstitial space without violating the conditions of minimum sphere size (0.4 Å) and sphere separation (2.1 Å), regardless of the nature of the metal atoms. Satisfactory agreement with observed hydrogen site occupancies and maximum hydrogen contents was observed for most systems investigated by this model, e.g. ZrV_2D_6 and $ZrMn_2D_{2.75}$ [4.138], $LaNi_5D_6$ [4.31, 176], $Zr_3V_3OD_x$ [4.172], $Y_6Mn_{23}D_{23}$ [4.174], $Pd_3P_{1-x}H_y$ [4.177], $ZrBe_2D_{1.5}$ [4.173], Ti_2NiD_x, and $Ti_4Fe_2OD_x$ [4.178]. In contrast, the agreement is less satisfactory for compounds for which such an analysis has not yet appeared in the literature, such as Mg_2CoD_5 and β-Mg_2NiD_4. For both structures the model predicts a maximum hydrogen content of six atoms per formula unit [P. Zolliker, 1985; private communication]. Furthermore, predictions arising from this model depend on the availability and accuracy of experimentally determined lattice parameters and metal atom positions [4.178]. Thus *a priori* estimates of maximum hydrogen contents based on this model are not always possible, as is the case for $LaCo_5D_x$($x=9$ at 1500 atm. [4.184]), for which structural information at high deuterium pressure is not yet available. The model also cannot predict the onset of structural phase transitions due to ordering of the hydrogen atoms (ZrV_2D_x), and the occurrence of lattice distortions, as was discussed [4.179] for $Y_6Mn_{23}D_x$ and related deuterides. Successful attempts to rationalize maximum hydrogen content by the related exclusion principle [4.181] were reported [4.67] for hydrides containing mainly tetrahedral interstices such as those having the $MgCu_2$, $MgZn_2$, and Th_6Mn_{23} type metal host structures. The previously proposed "blocking model" [4.182] has been mainly applied to binary hydride structures for which it allows the calculation of configurational entropies of interstitial atoms. Evidence for repulsive interactions between hydrogen and other non-metallic *p*-elements such as oxygen and phosphorus have been reported [4.31, 172, 177] for $Zr_3V_3OD_x$, $Pd_3P_{0.8}D_{0.15}$, Pd_6PD_x, and other compounds not covered by this review such as hydride-oxides (Ti_2OH_x [4.186]) or hydride-carbides ($ZrC_{1-x}D_y$ [4.187]), which were discussed in a previous review [4.31]. Short oxygen-deuterium distances have to date only been reported for $Ti_4Fe_2OD_{2.25}$([O–D]$=1.68$ Å [4.46]) and $ZrO_{0.4}D_{0.1}$ with *hcp* host structure ([O–D] not stated, D atoms filling tetrahedral holes next to octahedral holes occupied by O atoms, [4.21]). Repulsive interactions between hydrogen and metallic *p*-elements such as Al and Sn have been found to influence the hydrogen content in the substitutional series $La(Ni,Al)_5D_x$ [4.31, 170, 185], and in Nb_3SnH_x [4.31, 81]. On the other hand the adverse influence of Al on the maximum hydrogen content in the former series was interpreted in geometrical terms by invoking [4.176] the larger atomic size of Al compared to that of Ni, and in terms of electronic factors by invoking a non-availability of *d*-band bonding [4.180] (see below).

4.4.3 Electronic Factors

The particular bonding situation in non-metallic hydrides such as Mg_2NiH_4, Mg_2CoH_5, and Mg_2FeH_6 allows their maximum hydrogen contents to be rationalized in terms of purely electronic factors. Their crystal structures

(Fig. 4.2) are considered [4.59, 65, 66] to be formally built by divalent Mg^{2+} cations and tetravalent complex anions $[NiH_4]^{4-}$, $[CoH_5]^{4-}$, and $[FeH_6]^{4-}$, respectively. Since the latter anions all satisfy the 18-electron rule [4.188] the hydrogen content of the above compounds was assumed to be determined by band filling effects [4.59]. Subsequent bandstructure calculations [145, 154] confirm this view, but also stress the partially covalent nature of the magnesium-hydrogen interaction [4.145]. The usefulness of this approach in predicting new ternary metal hydride systems was initially demonstrated by the discovery of Mg_2FeH_6 and $MgCoH_5$ which were synthesized [4.65, 66] several years after their possible existence was suggested [4.59]. Mg_2FeH_6 is of particular interest because of its unusually high volume efficiency for hydrogen storage (more than twice that of liquid hydrogen) and its low material cost. For Mg_2NiH_4 predictions from the 18-electron rule prompted a re-investigation of its low-temperature structure for which an unusual square-planar configuration of Ni by hydrogen atoms had been proposed [4.62]. A high-resolution neutron diffraction study recently confirmed [4.64] that the Ni atoms had the expected [4.59, 188] tetrahedral configuration (Fig. 4.2).

Band-filling arguments were used to rationalize [4.180, 189] maximum hydrogen contents in Ni-based $CaCu_5$ type hydrides and electronic factors have been invoked [4.190] to explain the variation of hydride stability in the substitutional series $Zr(V_{1-x}Cr_x)_2H_y$.

4.4.4 Attractive Nearest-Neighbor Interactions

Preferential hydrogen occupancy of metal atom interstices that are surrounded by hydride forming elements have been interpreted in terms of a model [4.115] that assumes the existence of attractive interactions between hydrogen and its nearest-metal neighbors ("imaginary binary hydride model" [4.191, 192]). Relative stabilities of potentially occupied hydrogen atom sites are estimated by calculating local formation enthalpies of suitably partitioned imaginary binary hydrides, and by assuming a Fermi or Boltzmann type energy distribution [4.116, 194]. The calculations are based on available structure data and empirical constants given by *Miedema* [4.167] which have previously been used [4.137] to rationalize the stability of binary metal hydrides. Justification for this approach comes mainly from empirical considerations such as the crystal chemistry of intermetallic compounds, and less from theoretical calculations [4.155]. Good qualitative agreement between observed and calculated hydrogen site occupancies has been observed for $MgCu_2$-, and $MgZn_2$-type hydrides [4.48, 191, 192] (for a review of the results on these compounds see [4.193]), and also for Th_6Mn_{23}-type hydrides [4.192] (except for $Y_6Mn_{23}D_{23}$ [4.69]), for $LaNi_5D_x$ [4.159] and for $ZrNiD_x$ [4.192, 194]. Quantitative agreement was observed [4.48] for ZrV_2, which could be fortuitous [4.114, 195]. Generally speaking, the imaginary binary hydride model succeeds in predicting relative site occupancies for all those compounds and interstices which are also successfully rationalized by the geometrical model. On the other hand, it failed for those interstices which could not be rationalized by the geometrical model such as the empty [M_6] holes (M = Th, Ho] in $Th_6Mn_{23}D_{16}$ and $Ho_6Fe_{23}D_x$ (x=12.1, 15.7) and some empty tetrahedral [Zr_2M_2] holes (M = V, Mn) in disordered $ZrMn_2D_3$ and ordered $ZrV_2D_{3.6}$. It is also unable to explain why some compounds such as $ZrPd_2$, $LaPt_5$, and $ThNi_5$ are bad hydrogen absorbers [I. Jacob, 1986; private communication]. This parallel behaviour was suggested [4.1] to be a consequence of a correlation which exists between atomic size and the formation enthalpies of binary metal hydrides, i.e. elements with large atomic size tend to have larger enthalpies of hydride formation than those with small atomic size. In contrast to the geometrical model, the imaginary binary hydride model did not allow rationalization of the maximum hydrogen content. For ZrV_2D_x this failure has been attributed [4.48] to the neglect of geometrical restrictions for hydrogen site occupancies, i.e. minimum hydrogen separation due to repulsive interactions, entropy effects [4.196] similar to those investigated in β-$LaNi_5D_{6.5}$ [4.158], and the weakening of metal-metal bonds due to lattice expansions. The last of these effects appears to be the structural basis for "the rule of reversed stability" [4.136] which suggests that the weakening of the metal-metal bonds occurs at the expense of metal-hydrogen bonds.

4.5 Magnetic Ordering

The influence of hydrogen absorption on the magnetic properties of intermetallic compounds is reviewed in Chap. 7 and previously in [4.42]. Typical hydrogen storage systems such as TiFe and $LaNi_5$ as well as their hydrides do not show long-range magnetic order at low temperatures (e.g. $TiFeH_x$ [4.76]). Those ternary metal hydrides for which long-range magnetic order was found from neutron diffraction experiments are summarized in Appendix 4.B. The list gives information on the magnetic structures including ordered magnetic sublattice moments and the corresponding ferro- (or ferri-) and antiferromagnetic ordering temperatures, T_C and T_N, respectively. Usually these temperatures agree well with those determined from bulk magnetic measurements. The most extensive neutron diffraction work concerns cubic Laves phase compounds such as $RFe_2H(D)_x$ (R = Ho, Er, Tm) [4.57], and $A_6B_{23}H(D)_x$ systems (A = Y, Th, Ho, Er, B = Mn, Fe; for a review see [4.197]). In $RCo_5H(D)_x$ systems the magnetic moment of cobalt decreases as the hydrogen content is increased [4.42, 92]. In ferrimagnetic $ErFe_2$(H, D) with Curie temperature T_C = 574 K the magnetic moments of the Er atoms at low temperatures show nearly their full free-ion value of $9\mu_B$ and they are coupled antiparallel to the moments of the Fe atoms of magnitude $1.6\mu_B$. Upon deuteration to $ErFe_2D_{3.5}$ the Curie temperature decreases to 440 K. While the Fe moment remains unchanged upon deuteration the Er moment at saturation is reduced to $4.3\mu_B$. It vanishes at 295 K as the temperature is increased, presumably due to the weaker exchange interactions between Er ions compared to those between the Fe ions [4.57]. This system also shows a small isotope effect ($ErFe_2H_{3.5}$: $T_C = 450$ K). More dramatic changes of magnetic properties due to hydrogenation were found for A_6M_{23} systems. For example the Pauli paramagnet Th_6Mn_{23} changes on deuteration to a ferrimagnet with a Curie temperature of $T_C = 329$ K at the composition $Th_6Mn_{23}D_{30}$ [4.71]. The ferrimagnetic compound Y_6Mn_{23} ($T_C = 486$ K) changes on deuteration to an antiferromagnetic compound with $T_N = 175$ K at the composition $Y_6Mn_{23}D_{23}$ [4.69]. Of particular interest is also the superconducting compound $CeRu_2$ exhibiting short-range ferromagnetic fluctuations. It shows an electronic transition (often considered as a valence change between Ce^{4+} and Ce^{3+}) upon hydriding. Cubic $CeRu_2D_4$ is non-superconducting and orders antiferromagnetically below $T_N \approx 1.4$ K [4.56].

4.6 Conclusions

The structures of ternary metal hydrides differ greatly from those of elemental metal hydrides in terms of their diversity and complexity. Their lattices are considerably expanded and often distorted with respect to those of the corresponding binary metal compounds but their overall metal atom arrangements do not change much as a function of hydrogenation. The lattice expansion usually (but not always) leads to a weakening of the metal-metal bonds at the expense of metal-hydrogen bonds. So far, no evidence exists for the formation of hydrogen-hydrogen bonds. The hydrogen atoms occupy predominantly tetrahedral and octahedral interstices in the metal atom network, are mobile at room temperature, and are usually disordered over several different sites which are connected to form three-dimensional diffusion paths through the structures. The number of available interstices in the structures generally greatly exceeds the number of hydrogen atoms absorbed. At low temperatures the hydrogen atoms tend to order thus inducing structural phase transitions. The interstices which are preferentially occupied are large, and are surrounded by at least one atom of a binary hydride forming element. Their occupancies can be rationalized in terms of geometrical factors (atomic size), attractive nearest-neighbor interactions (metal-hydrogen bond formation), repulsive second-nearest-neighbor interactions (mainly between the hydrogen atoms), and electronic factors (band-filling effects), thus providing a qualitative understanding at the atomic level of those factors which limit hydrogen sorption capacities. Clearly, more neutron and x-ray investigations, possibly on single crystals, more sophisticated structure models (taking into account effects due to anharmonicity and microtwinning etc.), and diffraction data of higher resolution are desirable in order to gain a quantitative understanding. This is also true for the study of magnetically ordered hydrides, in particular with respect to investigating correlations between chemical and magnetic ordering. Finally the detection of significant isotope effects (hydrogen versus deuterium) on both static and dynamic properties of ternary hydrides would be of interest.

4.A Appendix

Structural data for ternary metal hydrides and deuterides studied by neutron diffraction analysis[a]

Compound[b]	Ref.[c]	Space group[d]	Lattice parameters[e]	Atom parameters[f]					
$Zr_3V_3OD_{4.93}$	[4.47]	$Fd\bar{3}m$ (227)*	$a=12.5260(1)$	Zr	*48f*	x	0	0	$x=0.1942(1)$, $B=0.32(2)$
			$\{a=12.1703(1)\}$	V1	*16d*	5/8	5/8	5/8	$B=0.3$
				V2	*32e*	x	x	x	$x=0.830(1)$, $B=0.3$
				O	*16c*	1/8	1/8	1/8	$B=0.30(4)$
				D1	*32e*	x	x	x	$x=0.9120(1)$, PP$=0.998(13)$, B_{ij}
				D2	*192i*	x	y	z	$x=0.7490(3)$, $y=0.8616(2)$, $z=0.9429(2)$, PP$=0.221(3)$, B_{ij}
				D3	*96g*	x	x	z	$x=0.906(3)$, $z=0.280(4)$, PP$=0.027(7)$, $B=1.4(10)$
				D4	*96g*	x	x	z	$x=0.968(2)$, $z=0.657(3)$, PP$=0.025(5)$, $B=0.7(7)$
				$R_p=0.019$, $R_{F^2}=0.092$, $n=35$, $s=1.19$ (711 *hkl*)					

Remarks: High-resolution TOF data, refinement of anisotropic temperature factors with $s=0.56$; refinements also performed for $Zr_3V_3OD_{2.85}$ and $Zr_3V_3OD_{1.86}$; sequence of filling of D atom sites as a function of overall D content: D1, D2, D3+D4

[a] Only those compounds which are marked by bold letters in Table 4.1 are listed here; each structure contains results of one refinement only unless conflicting results were reported by different workers; the selection choice was guided mainly by the quality of the diffraction data and structure refinement, the priority of publication, etc.; the order in which the compounds are listed is the same as that of Table 4.1.

[b] Composition stated by the authors; the D-content usually corresponds to that determined by desorption measurements.

[c] Most recent work only; similar work published by same authors listed under same number in list of references.

[d] Standard setting according to [4.44], with origin at center of inversion, if present; space group symbols marked by asterisks indicate non-standard settings, or settings with origin away from center of inversion; sequence number of space group in parentheses.

[e] Figures in curly brackets indicate lattice parameters of the hydrogen-free compound ("n.s." = not specified).

[f] Atom symbol, Wyckoff symbol (subscripts, if present, are those stated by authors) and first equipoint listed in [4.44], positional coordinates (x, y, z), population parameter PP (only stated if different from unity), and isotropic thermal parameter B ($Å^2$) (or mean-square displacements $\langle u^2 \rangle = B/8\pi^2$ $Å^2$), followed by agreement indices of profile refinement, R_p, R_{wp}, R_I, R_F, R_{F^2}, or R (for definitions see [4.128, 134], number of refined parameters n ("n.s." = not specified), s = maximum value of $(\sin\theta)/\lambda$ [$Å^{-1}$], and number of contributing reflections; figures in parentheses indicate errors of the least significant digits.

4.A Appendix (continued)

Compound[b]	Ref.[c]	Space group[d]	Lattice parameters[e]	Atom parameters[f]					
$ZrV_2D_{4.9}$	[4.50]	$Fd\bar{3}m$ (227)	a=7.913(2)	Zr	8a	1/8	1/8	1/8	B=0.89(9)
			{a=7.44}	V	16d	1/2	1/2	1/2	B=0.89(9)
				D1	96g	x	x	z	x=0.609(3), z=0.8743(5), PP=0.249(4), B=0.89(9)
				D2	32e	x	x	x	x=0.2739(2), PP=0.387(4), B=0.89(9)
				R_{wp}=0.079, R_I=0.024, n=6 (atomic), s=0.40 (13 *hkl*)					
		Remarks: Calculated D content $ZrV_2D_{4.5}$; for refinements by other workers see [4.48, 4.49, 4.51]; Below 325 K $ZrV_2D_{3.6}$ is tetragonal due to D atom ordering [4.49]; for temperature dependence of lattice parameters, see [4.198].							
HfV_2D_4 (77 K)	[4.52]	$I4_1/a$ (88)	a=5.42(–)	Hf	4a	0	1/4	1/8	B=0.45(–) at 340 K
			c=7.85(–)	V	8d	0	0	1/2	B=0.20(–)
			{a=7.37}	D1	16f	x	y	z	x=0.182(–), y=−0.068(–), z=0.060(–), B=1.63(–)
				R_I=0.066, n=n.s. (6 atomic), s=0.44 (36 *hkl*)					
		Remarks: Structure not refined, no e.s.d.'s given; D atoms positions obtained by analysis of integrated intensites; becomes cubic at T_I=278 K (a=7.78 A at 300 K; D atoms disordered) with marked hysteresis; refinement for isostructural $ZrV_2D_{3.6}$ as a function of temperature reported in [4.49]; NMR evidence for two types of V sites in tetragonal $HfV_2H_{3.9}$ (T_I=296 K) reported in [4.199]							
β'-$Mg_2NiD_{3.9}$ (553 K) (22 bar)	[4.59]	$Fm\bar{3}m$ (225)	a=6.507(2)	Ni	4a	0	0	0	B=1.3(2)
			{a, c=n.s.}	Mg	8c	1/4	1/4	1/4	B=1.3(2)
				D	24e	x	0	0	x=0.229(1), PP=0.69(2), B=7.6(3)
				R_I=0.056, n=4 (atomic), s=0.40 (11 *hkl*)					
		Remarks: Refinements on both profile and integrated intensities (see also [4.60]); significantly different D atom distribution proposed in [4.61]; unusually large B value for D confirmed in [4.60]; for anisotropic B_{ij} see [4.60, 61]; D atoms order below T_I=508(5) K							
β'-Mg_2NiD_4 (573 K) (15 bar)	[4.61]	$Fm\bar{3}m$ (225)	a=6.5247(3)	Ni	4a	0	0	0	B=2.5(1)
			{a, c=n.s.}	Mg	8c	1/4	1/4	1/4	B=3.10(2)
				D1	48h	x	x	0	x=0.215(4), PP=0.312(4), B_{ij}
				D2	24e	x	0	0	x=0.195(4), PP=0.043(3), B=5.0(–)
				R_I=0.083, n=0 (atomic), s=0.54 (21 *hkl*)					

Remarks: Type of refinement not specified, presumably on integrated intensities; disordered D atom distribution differing significantly from those reported in [4.59, 4.60]; Mg–D1 distance of 1.66 Å unusually short; consistency index $R=0.083$ not significantly different from that for occupancy of mainly *24l* sites ($R=0.086$); for critical evaluation of structure refinements see [4.144]

β-Mg_2NiD_4	[4.62]	C2/m (12)	$a=6.496(1)$	Ni1	*2a*	0	0	0	
			$b=6.412(2)$	Ni2	*2d*	0	1/2	1/2	
			$c=6.602(2)$	Mg	*8j*	*x*	*y*	*z*	$x=0.295(2)$, $y=0.275(2)$, $z=0.245(2)$
			$\beta=93.23(2)°$	D1	*4g*	0	*y*	0	$y=0.223(4)$
			{*a*, *c*=n.s.}	D2	*4i*	*x*	0	*z*	$x=0.253(4)$, $z=0.119(4)$
				D3	*4i*	*x*	0	*z*	$x=0.420(5)$, $z=0.716(4)$
				D4	*4i*	*x*	0	*z*	$x=0.272(4)$, $z=0.487(4)$

$R_F=0.12$, n=n.s. (10 positional), $s=0.54$

Remarks: Ordered, square-planar D atom configuration around Ni; a disordered configuration at room temperature was reported in [4.61, 63]; a tetrahedral D atom configuration was reported in [4.64]; the D3–D4 distances of 1.75(4) Å are unusually short; the atom positions correspond to those of a substructure only, superstructure cell doubled along *c*

β-Mg_2NiD_4	[4.64]	*C2/c* (15)	$a=14.342(5)$	Ni	*8f*	*x*	*y*	*z*	$x=0.1194(6)$, $y=0.231(1)$, $z=0.083(1)$, $\langle u^2\rangle=-0.004(2)$
			$b=6.404(1)$						
			$c=6.483(1)$	Mg1	*8f*	*x*	*y*	*z*	$x=0.265(1)$, $y=0.483(2)$, $z=0.075(2)$, $\langle u^2\rangle=-0.002(2)$
			$\beta=113.52(4)$						
				Mg2	*4e*	0	*y*	1/4	$y=0.014(3)$, $\langle u^2\rangle=-0.002(-)$
				Mg3	*4e*	0	*y*	1/4	$y=0.513(3)$, $\langle u^2\rangle=-0.002(-)$
				D1	*8f*	*x*	*y*	*z*	$x=0.211(1)$, $y=0.300(3)$, $z=0.304(3)$, $\langle u^2\rangle=0.032(4)$
				D2	*8f*	*x*	*y*	*z*	$x=0.136(1)$, $y=0.316(2)$, $z=0.881(2)$, $\langle u^2\rangle=0.013(2)$
				D3	*8f*	*x*	*y*	*z*	$x=0.011(1)$, $y=0.287(2)$, $z=0.054(2)$, $\langle u^2\rangle=0.013(-)$
				D4	*8f*	*x*	*y*	*z*	$x=0.131(1)$, $y=0.995(2)$, $z=0.082(2)$, $\langle u^2\rangle=0.013(-)$

$R_F=0.044$, $R_{wp}=0.138$, $n=24$ (atomic), $s=0.56$

Remarks: Metal atom parameters determined from X-ray profile refinement [4.123]; tetrahedral D atom configuration, in contrast to planar configuration reported previously [4.62]; line broadening due to microtwinning investigated by X-ray diffraction [4.123], see also [4.121]

Mg_2CoD_5	[4.65]	*P4/nmm* (129)	$a=4.463(4)$	Co	*2c*	1/4	1/4	*z*	$z=0.256(2)$, $B=0.4(2)$
			$c=6.593(6)$	Mg1	*2a*	3/4	1/4	0	$B=0.04(4)$
				Mg2	*2b*	3/4	1/4	1/2	$B=0.04(-)$
				D1	*2c*	1/4	1/4	*z*	$z=0.497(2)$, $B=1.44(9)$
				D2	*8j*	*x*	*x*	*z*	$x=0.4879(4)$, $z=0.2257(3)$, $B=1.39(5)$

$R_{wp}=0.12$, $R_F=0.08$, $n=8$ (atomic), $s=0.77$ (178 *hkl*)

continued on p. 122

4.A Appendix (continued)

Compound[b]	Ref.[c]	Space group[d]	Lattice parameters[e]	Atom parameters[f]					

Remarks: Ordered; square-pyramidal D atom distribution; main impurity MgD_2; structure transforms above $T_1=488(5)$ K into a disordered, cubic modification; structure similar to that reported for the low-temperature modification of Sr_2IrD_5 (space group *I4/mmm* according to [4.143], $T_1 \simeq 205$ K)

Compound[b]	Ref.[c]	Space group[d]	Lattice parameters[e]	Atom parameters[f]					
Mg_2FeD_6	[4.66]	$Fm\bar{3}m$ (225)	$a=6.430(1)$	Fe	*4a*	0	0	0	$B=0.15(7)$
				Mg	*8c*	1/4	1/4	1/4	$B=0.5(1)$
				D	*24e*	x	0	0	$x=0.2420(7)$, $B=1.6(1)$
				$R_p=0.27$, $R_F=0.067$, $n=4$ (atomic), $s=0.73$ (40 *hkl*)					

Remarks: Ordered, octahedral D atom configuration around Fe, isostructural to Sr_2RuD_6 (K_2PtCl_6 type) [4.22]; preliminary low-resolution data also on hydride Mg_2FeH_6

Compound[b]	Ref.[c]	Space group[d]	Lattice parameters[e]	Atom parameters[f]					
$Y_6Mn_{23}D_{23}$ (295 K)	[4.69]	$Fm\bar{3}m$ (225)	$a=12.805(*)$ {a=n.s.}	Y	*24e*	x	0	0	$x=0.205(*)$, $B=0.82(*)$
				Mn1	*4b*	1/2	1/2	1/2	$B=0.47(*)$
				Mn2	*24d*	0	1/4	1/4	$B=0.47(*)$
				Mn3	$32f_1$	x	x	x	$x=0.179(*)$, $B=0.47(*)$
				Mn4	$32f_2$	x	x	x	$x=0.372(*)$, $B=0.47(*)$
				D1	*4a*	0	0	0	PP=1.000(–), $B=6.35(*)$
				D2	$32f_3$	x	x	x	$x=0.100(*)$, PP=0.625(–), $B=1.37(*)$
				D3	$96j_1$	0	y	z	$y=0.169(*)$, $z=0.373(*)$, PP=0.417(–), $B=4.95(*)$
				D4	$96k_1$	x	x	z	$x=0.161(*)$, $z=0.049(*)$, PP=0.292(–), $B=2.28(*)$
				$R_{wp}=0.106$, $n=18$ (atomic), $s=0.55$					

Remarks: (*) standard deviations of positional parameters were stated to be "generally less than ±2 in the last digit"; those of the thermal parameters were approximately 30%; no standard deviation of occupancy factors were given; temperature factors of D1 and D3 are anomalously large; occupancies of $32f_3$ (D2) and $96k_1$ (D4) sites are significantly different from those reported previously for $Y_6Mn_{23}D_{23}$ (PP(D2)=1.0, PP(D4)=0.18, according to [4.67]); transforms at low temperature into a tetragonal structure due to partial D atom ordering on sites D3 and D4 (see below); structure is tetragonal at room temperature according to [4.68]

Compound[b]	Ref.[c]	Space group[d]	Lattice parameters[e]	Atom parameters[f]					
$Y_6Mn_{23}D_{23}$ (4 K)	[4.69]	*P4/mmm* (123)	$a=9.030(*)$ $c=12.722(*)$ {a=n.s.}	Y1	*2g*	0	0	z	$z=0.213(*)$, $B=0.72(*)$
				Y2	*2h*	1/2	1/2	z	$z=0.285(*)$, $B=0.72(*)$
				Y3	*4j*	x	x	0	$x=0.204(*)$, $B=0.72(*)$
				Y4	*4k*	x	x	1/2	$x=0.297(*)$, $B=0.72(*)$
				Mn1	*1b*	0	0	1/2	$B=0.01(*)$

				Mn2	*1c*	1/2	1/2	0	$B=0.01(*)$
				Mn3	*2e*	0	1/2	1/2	$B=0.01(*)$
				Mn4	*2f*	0	1/2	0	$B=0.01(*)$
				Mn5	$8r_1$	x	x	z	$x=0.255(*)$, $z=0.747(*)$, $B=0.01(*)$
				Mn6	$8s_1$	x	0	z	$x=0.360(*)$, $z=0.182(*)$, $B=0.01(*)$
				Mn7	$8t_1$	x	1/2	z	$x=0.145(*)$, $z=0.317(*)$, $B=0.01(*)$
				Mn8	$8s_2$	x	0	z	$x=0.275(*)$, $z=0.387(*)$, $B=0.01(*)$
				Mn9	$8t_2$	x	1/2	z	$x=0.249(*)$, $z=0.137(*)$, $B=0.01(*)$
				D1	*1a*	0	0	0	PP=1.00(–), $B=7.80$
				D2	*1d*	1/2	1/2	1/2	PP=1.00(–), $B=7.90$
				D3	$8s_3$	x	0	z	$x=0.219(*)$, $z=0.093(*)$, PP=0.59(–), $B=0.25$
				D4	$8t_3$	x	1/2	z	$x=0.305(*)$, $z=0.398(*)$, PP=0.66(–), $B=0.02$
				D5	*8p*	x	y	0	$x=0.230(*)$, $y=0.438(*)$, PP=0.50(–), $B=2.30$
				D6	*8q*	x	y	1/2	$x=0.304(*)$, $y=0.045(*)$, PP=0.38(–), $B=1.80$
				D7	$8r_2$	x	x	z	$x=0.141(*)$, $z=0.331(*)$, PP=0.75(–), $B=2.83$
				D8	$8r_4$	x	x	z	$x=0.360(*)$, $z=0.132(*)$, PP=0.88(–), $B=2.40$
				D9	$8s_4$	x	0	z	$x=0.324(*)$, $z=0.041(*)$, PP=0.31(–), $B=1.08$
				D10	$16u_1$	x	y	z	$x=0.066(*)$, $y=0.177(*)$, $z=0.141(*)$, PP=0.22(–), $B=2.67(*)$
				D11	$16u_2$	x	y	z	$x=0.296(*)$, $y=0.392(*)$, $z=0.327(*)$, PP=0.50(–), $B=0.20(*)$
				$R_{wp}=0.086$, $n=56$ (atomic)+4(magnetic), $s=0.55$					
		Remarks: For (*) and (–) see remarks above; B's restricted; refinement also performed on data collected at 78 K; temperature factors of D1 and D2 anomalously large; two *8r* and one *8t* sites which are occupied in the cubic high-temperature structure are empty; $T_1=175$ K							
α-$TiFeD_{0.057}$	[4.73]	$Pm\bar{3}m$ (221)	$a=2.979(1)$ {a=n.s.}	Ti	*1b*	1/2	1/2	1/2	$B=0.57(4)$
				Fe	*1a*	0	0	0	$B=0.57(4)$
				D	*3d*	1/2	0	0	PP=0.019(–), $B=0.7(1.0)$
				$R_I=0.018$, n=n.s. (3 atomic), $s=0.63$ (11 *hkl*)					
β-$TiFeD_{1.0}$ (6.5 bar)	[4.75]	$P222_1$ (17)	$a=2.966(6)$, $b=4.522(10)$ $c=4.370(9)$ {$a=2.975$}	Ti	*2d*	1/2	y	1/4	$y=0.757(3)$, $B=0.56(8)$
				Fe	*2c*	0	y	1/4	$y=0.2941(8)$, $B=0.56(8)$
				D1	*2a*	x	0	0	$x=0.00(–)$, PP=0.88(1); $B=0.56(8)$
				D2	*2d*	1/2	y	1/4	$y=0.30(1)$, PP=0.12(1), $B=0.56(8)$
				$R_{wp}=0.175$, $R_I=0.062$, n=n.s. (6 atomic), $s=0.40$ (27 *hkl*)					

continued on p. 124

4.A Appendix (continued)

Compound[b]	Ref.[c]	Space group[d]	Lattice parameters[e]	Atom parameters[f]					
		Remarks: Also called β_1 modification refinements in same space group performed by other workers [4.74, 4.80]; refinements in space groups $Pmc2_1$ (26), *P2cm* (28), *Pmcm* (51) reported in [4.74, 4.75, 4.80]; according to [4.76] another more D-rich β_2-$TiFeD_{1.4}$ phase with same structure exists; refinement also performed on mixed hydride-deuteride $TiFeH_{0.78}D_{0.22}$							
γ-$TiFeD_{1.9}$	[4.77]	*P2/m* (10)	$a=4.704(1)$	Ti	*2n*	*x*	1/2	*z*	$x=0.294(17)$, $z=0.229(19)$, $B=0.3(1)$
			$b=2.8301(5)$	Fe	*2m*	*x*	0	*z*	$x=0.203(3)$, $z=0.721(3)$, $B=0.3(1)$
			$c=4.704(1)$	D1	*1a*	0	0	0	$B=1.6(4)$
			$\beta=96.97(5)$	D2	*1d*	1/2	0	0	$B=1.6(4)$
			$\{a=2.980\}$	D3	*1g*	1/2	0	1/2	$B=1.6(4)$
				D4	*1f*	0	1/2	1/2	PP=0.85(5), $B=1.6(4)$
				$R_I=0.10$, $n=15$ (7 atomic), $s=0.50$					
		Remarks: Sample poisoned by CO; refinements performed by other workers [4.76, 4.78–80]; for temperature dependence of structural parameters see [4.78]; space group *Cmmm* proposed in [4.79] (see below)							
γ-$TiFeD_{1.94}$ (113 bar)	[4.79]	*Cmmm* (65)	$a=7.029(4)$	Ti	*4h*	*x*	0	1/2	$x=0.223(2)$, $B=0.22(6)$
			$b=6.233(4)$	Fe	*4i*	0	*y*	0	$y=0.2887(7)$, $B=0.22(-)$
			$c=2.835(1)$	D1	*4e*	1/4	1/4	0	PP=1.00(2), $B=1.4(1)$
			$\{a=2.975\}$	D2	*2c*	1/2	0	1/2	PP=0.91(2), $B=1.4(-)$
				D3	*2a*	0	0	0	PP=0.99(2), $B=1,4(-)$
				$R_I=0.085$, $R_{wp}=0.16$, $n=7$ (atomic), $s=0.59$					
		Remark: Calculated D content $TiFeD_{1.94(3)}$							
Nb_3SnH	[4.81]	$Pm\bar{3}n$ (223)	$a=5.34(-)$	Nb	*6c*	1/4	0	1/2	$B=0.72(-)$
			$\{a=5.29(-)\}$	Sn	*2a*	0	0	0	$B=0.72(-)$
				D	*6d*	1/4	1/2	0	PP=0.33(–), $B=0.72(-)$
				R=n.s., n=n.s., s=n.s. (10 *hkl*)					
		Remarks: H atom located on Fourier difference map							
β-$LaNi_5D_{5.5}$ (4 bar)	[4.85]	*P31m* (157)	$a=5.336(1)$	La	*1a*	0	0	*z*	$z=0(-)$, $B=4.5(2)$
			$c=4.259(1)$	Ni1	*2b*	1/3	2/3	*z*	$z=0.980(5)$, $B=2.1(1)$
			$\{a, c$=n.s.$\}$	Ni2	*3c*	*x*	0	*z*	$x=0.499(1)$, $z=0.496(5)$, $B=2.0(1)$
				D1	*3c*	*x*	0	*z*	$x=0.470(1)$, $z=0.092(5)$, $B=1.8(1)$

				D2	8d	x	y	z	$x=0.189(2)$, $y=0.804(3)$, $z=0.495(4)$, PP$=0.42(4)$, $B=2.3(2)$

$R_I=0.083$, $n=$n.s. (14 atomic), $s=0.75$

Remarks: Two-D-site model; temperature factor of La anomalously large; atomic parameters significantly different from those reported by other workers [4.32, 4.84]; structure refinements based on same space group reported in [4.32, 4.83, 4.84, 4.89]; structure refinements based on other space groups reported in [4.53, 4.86–4.88]

β-$LaNi_{5.12}D_{6.7}$ (5 atm)	[4.86]	*P6/mmm* (191)	$a=5.399(2)$	La	*1a*	0	0	0	PP$=0.89(-)$, $B=4.0(6)$
			$c=4.290(2)$	Ni1	*2c*	1/3	2/3	0	PP$=0.95(-)$, $B=1.8(3)$
			$\{a=5.016\}$	Ni2	*3g*	1/2	0	1/2	$B=1.0(2)$
			$\{c=3.983\}$	D1	3f	1/2	0	0	PP$=0.21(3)$, $B=6(2)$
				D2	*4h*	1/3	2/3	z	$z=0.369(1)$, PP$=0.13(2)$, $B=1(1)$
				D3	*6m*	x	$2x$	1/2	$x=0.136(3)$, PP$=0.32(2)$, $B=2.3(9)$
				D4	*12n*	x	0	z	$x=0.455(3)$, $z=0.117(5)$, PP$=0.18(1)$, $B=1.3(1)$
				D5	*12o*	x	$2x$	z	$x=0.204(3)$, $z=0.354(5)$, PP$=0.11(1)$, $B=2(1)$

$R=0.073$, $n=$n.s. (24 atomic), $s=0.72$

Remarks: Five-D-site model; refinement on integrated intensities; a small fraction of Ni (0.03 per formula unit) was distributed on sites *6l* (x, $2x$, 0; etc. $x=0.287$) and *2e* (0, 0, z; etc. $z=0.313$); temperature factor of La anomalously large; anisotropic line broadening in diffraction patterns; orthorhombic distortion for $LaNi_5D_7$ suggested in [4.90]; nominal composition: $LaNi_5D_{6.6}$; refinements based on other space groups reported in [4.32, 4.53, 4.83–4.85, 4.87, 4.90]

β-$LaNi_5D_{6.7}$ (19.5 bar)	[4.87]	*P6₃mc* (186)	$a=5.412(2)$	La	2a	0	0	z	$z=0(-)$, $B=1.0(5)$
			$c=8.599(3)$	Ni1	*2b*	1/3	2/3	z	$z=0(-)$, $B=2.9(4)$
			$\{a=5.015(5)\}$	Ni2	*2b*	1/3	2/3	z	$z=1/2(-)$, $B=2.9(4)$
			$\{c=3.97(1)\}$	Ni3	*6c*	x	$-x$	z	$x=1/2(-)$, $z=1/4(-)$, $B=1.2(4)$
				D1	*2b*	1/3	2/3	z	$z_1=0.826(6)$, PP$=0.75(7)$, $B=1.3(5)$
				D2	*2b*	1/3	2/3	z	$z_2=z_1+1/2$, PP$=0.08(1)$, $B=1.3(-)$
				D3	*6c*	x	$-x$	z	$x_3=0.110(8)$, $z_3=1/4(-)$, PP$=0.80(15)$, $B=1.3(-)$
				D4	*6c*	x	$-x$	z	$x_4=x_3$, $z_4=z_3+1/2$, PP$=0.08(2)$, $B=1.3(-)$
				D5	*6c*	x	$-x$	z	$x_5=0.178(4)$, $z_5=0.289(4)$, PP$=0.30(2)$, $B=1.3(-)$
				D6	*6c*	x	$-x$	z	$x_6=x_5$, $z_6=z_5+1/2$, PP$=0.18(2)$, $B=1.3(-)$
				D7	*12d*	x	y	z	$x=0.476(10)$, $y=0(-)$, $z=0.057(2)$, PP$=0.25(-)$, $B=1.3(-)$

$R_w=0.017$, $n=16$ (atomic), $s=0.51$ (58 *hkl*)

Remarks: Seven-D-site model having doubled *c*-axis with respect to previously models; overall temperature factor B(D)$=1.3(5)$; model contains 36 free atomic parameters of which 19 were constrained during refinement; refinement performed on integrated intensities; refinement in same space group also reported in [4.88]; for refinements in other space groups see list above and below; D7–D7 and D3–D5 distances are anomalously short: 0.3 Å and 0.7 Å, respectively

4.A Appendix (continued)

Compound[b]	Ref.[c]	Space group[d]	Lattice parameters[e]	Atom parameters[f]					
β-$LaNi_5D_6$	[4.53]	*P321*(150)	$a=5.43(-)$	La	*1a*	0	0	0	
			$c=4.28(-)$	Ni1	*2d*	1/3	2/3	*z*	$z=0.04(-)$
			{*a*, *c*=n.s.}	Ni2	*3f*	*x*	0	1/2	$x=0.471(-)$
				D1	*6g*	*x*	*y*	*z*	$x=0.553(-)$, $y=0.499(-)$, $z=0.102(-)$, PP=0.5(–)
				D2	*6g*	*x*	*y*	*z*	$x=0.181(-)$, $y=0.364(-)$, $z=0.595(-)$, PP=0.5(–)
				$R=0.083$, n=n.s. (10 atomic), s=n.s.					
		Remarks: B values not stated, for structure refinements based on other space groups see above							
β-$CaNi_5D_{0.77}$	[4.91]	*Pmcm* (51)*	$a=8.603(1)$	Ca1	*2f*	1/2	*y*	1/4	$y=0.499(4)$, $B=0.5(-)$
			$b=5.081(1)$	Ca2	*2e*	0	*y*	1/4	$y=0.052(4)$, $B=0.5(-)$
			$c=7.856(2)$	Ni1	*4k*	*x*	*y*	3/4	$x=0.340(2)$, $y=0.039(3)$, $B=0.5(-)$
			{*a*, *c*=n.s.}	Ni2	*4k*	*x*	*y*	1/4	$x=0.835(2)$, $y=0.547(3)$, $B=0.5(-)$
				Ni3	*2c*	0	1/2	1/2	$B=0.5(-)$
				Ni4	*2b*	1/2	0	0	$B=0.5(-)$
				Ni5	*8l*	*x*	*y*	*z*	$x=0.250(3)$, $y=0.253(3)$, $z=0.0133(5)$, $B=0.5(-)$
				D1	*4k*	*x*	*y*	3/4	$x=0.247(4)$, $y=0.721(5)$, PP=0.64(1), $B=1.0(-)$
				D2	*4k*	*x*	*y*	1/4	$x=0.25(3)$, $y=0.77(4)$, PP=0.12(1), $B=1.0(-)$
				$R=0.069$, $n=15$ (atomic), $s=0.41$					
		Remarks: Fixed temperature factors for metal atoms ($B=0.5$) and D atoms ($B=1.0$); possibility of coherent [010] twinning indicated							
β[I]-$PrCo_5D_{3.6}$	[4.92]	*Cmmm* (65)	$a=8.868(-)$	Pr	*2a*	0	0	0	$B=1.2(2)$
			$b=5.341(-)$	Co1	*4g*	*x*	0	0	$x=0.366(2)$, $B=1.2(-)$
			$c=4.083(-)$	Co2	*2c*	1/2	0	1/2	$B=1.2(-)$
			{$a=5.022(-)$}	Co3	*4f*	1/4	1/4	1/2	$B=1.2(-)$
			{$c=3.989(-)$}	D1	*4e*	1/4	1/4	0	PP=0.98(10), $B=4.7(2)$
				D2	*4h*	*x*	0	1/2	$x=0.1375(7)$, PP=0.98(6), $B=0.9(2)$
				$R=0.053$, $n=7$ (atomic), $s=0.35$					
		Remark: Temperature factor of D1 anomalously large							
$ZrMn_2D_{3.0}$	[4.96]	$P6_3/mmc$ (194)	$a=5.391(3)$	Zr	*4f*	1/3	2/3	*z*	$z=0.0661(8)$, $B=1.4(1)$
			$c=8.748(8)$	Mn1	*2a*	0	0	0	$B=1.4(-)$

			{$c=8.276$}	D1	*24l*	x	y	z	$x=0.042(2)$, $y=0.325(2)$, $z=0.562(1)$, PP$=0.179(4)$, $B=1.4(-)$
				D2	*12k*	x	$2x$	z	$x=0.456(1)$, $z=0.632(1)$, PP$=0.376(4)$, $B=1.4(-)$
				D3	*6h*	x	$2x$	1/4	$x=0.463(2)$, PP$=0.312(1)$, $B=1.4(-)$
				D4	*6h*	x	$2x$	1/4	$x=0.202(9)$, PP$=0.052(2)$, $B=1.4(-)$
				$R_p=0.107$, $R_I=0.040$, $n=$n.s. (14 atomic), $s=0.40$					
		Remarks: Calculated D content $ZrMn_2D_{2.75}$; order-disorder transformation at $T_1=230(10)$ K according to [4.200]							
$ZrBe_2D_{1.49}$	[4.101]	*P6/mmm* (191)	$a=3.7139(6)$	Zr	*1a*	0	0	0	
			$c=3.4704(8)$	Be	*2d*	1/3	2/3	1/2	
			{a, $c=$n.s.}	D	*2c*	1/3	2/3	0	PP$=0.75(-)$
				$R=0.040$, $n=$n.s. (4 atomic), $s=$n.s.					
		Remark: No B values given							
$HoNi_3D_{1.8}$	[4.53]	$R\bar{3}m$ (166)	$a=4.991(-)$	Ho	*3a*	0	0	0	
			$c=26.12(-)$	Ho	*6c*	0	0	z	$z=0.138(-)$
			{a, $c=$n.s.}	Ni1	*3b*	0	0	1/2	
				Ni2	*6c*	0	0	z	$z=0.336(-)$
				Ni3	*18h*	x	$-x$	z	$x=0.494(-)$, $z=0.081(-)$
				D1	*18h*	x	$-x$	z	$x=0.142(-)$, $z=0.071(-)$, PP$=0.64(-)$
				D2	*6c*	0	0	z	$z=0.775(-)$, PP$=0.78(-)$
				$R=0,067$, $n=$n.s. (9 atomic), $s=$n.s.					
		Remarks: No B values given; D atom distribution differs from that proposed for $LaNi_3D_2$ [4.32]							
Th_2AlD_4	[4.105]	*I4/mcm* (140)	$a=7.629(-)$	Th	*8h*	x	$x+1/2$	0	$x=0.162(-)$
			$c=6.517(-)$	Al	*4a*	0	0	1/4	
			{$a=7.62$}	D	*16l*	x	$1/2+x$	z	$x=0.368(-)$, $z=0.137(-)$
			{$c=5.86$}	$R=$n.s., $n=$n.s., $s=$n.s.					
		Remarks: D atoms were placed by trial and error; no structure refinement was performed; D–D distance of 1.7 Å anomalously short; only ordered D atom distributions were tested							
$Zr_2PdD_{1.70}$	[4.106]	*I4/mmm* (139)	$a=3.3207(2)$	Pd	*2a*	0	0	0	$B=0.6(-)$
			$c=11.6058(9)$	Zr	*4e*	0	0	z	$z=0.3566(1)$, $B=0.4(-)$
			{a, $c=$n.s.}	D	*4d*	0	1/2	1/4	$B=1.0(-)$
				$R_p=0.070$, $n=$n.s. (4 atomic), $s=$n.s.					
		Remark: Refinement also for $Zr_2PdD_{1.96}$							

4.A Appendix (continued)

Compound[b]	Ref.[c]	Space group[d]	Lattice parameters[e]	Atom parameters[f]					
$TiCuD_{0.96}$	[4.104]	*P4/nmm* (129)	$a=3.020(-)$	Ti	*2c*	1/4	1/4	*z*	$z=0.6751(9)$, $B=0.65(2)$
			$c=6.837(-)$	Cu	*2c*	1/4	1/4	*z*	$z=0.0933(5)$, $B=0.65(-)$
			{*a*, *c*=n.s.}	D	*2b*	3/4	1/4	1/2	PP=0.90(1), $B=0.65(-)$
				$R_p=0.128$, $R_I=0.088$, $n=4$ (atomic), $s=0.76$					
$ZrNiH_{2.98}$	[4.108]	*Cmcm* (63)	$a=3.53(-)$	Zr	*4c*	0	*y*	1/4	$y=0.140(1)$, $B=4.3(4)$
			$b=10.48(-)$	Ni	*4c*	0	*y*	1/4	$y=0,430(1)$, $B=1.1(2)$
			$c=4.30(-)$	H1	*4c*	0	*y*	1/4	$y=0.956(2)$, $B=3.1(5)$
			{*a*, *b*, *c*=n.s.}	H2	*8f*	0	*y*	*z*	$y=0.298(1)$, $z=0.507(4)$, $B=4.6(5)$
				R_p=n.s., *n*=n.s. (9 atomic), $s=0.46$					
		Remarks: Refinement in integrated intensities; same data later also refined in [4.110]							
ZrNiD	[4.109]	*Cmcm* (63)	$a=3.367(2)$	Zr	*4c*	0	*y*	1/4	$y=-0.140(-)$
			$b=10.313(4)$	Ni	*4c*	0	*y*	1/4	$y=-0.429(-)$
			$c=4.063(2)$	D	*4c*	0	*y*	1/4	$y=0.2152(-)$
		{*a*, *b*, *c*=n.s.}	$\alpha=89.18(3)$						
			$\beta=89.60(5)$						
			$\gamma=89.43(3)$						
				R=n.s., *n*=n.s., *s*=n.s.					
		Remarks: Atomic positions described in space group *Cmcm*, lattice parameters are those of a triclinic lattice; no structure refinement was performed							
$LaNiD_{3.7}$	[4.53]	*Cmcm* (63)	$a=3.98(-)$	La	*4c*	0	*y*	1/4	$y=0.131(-)$
			$b=11.94(-)$	Ni	*4c*	0	*y*	1/4	$y=0.418(-)$
			$c=4.87(-)$	D1	*4c*	0	*y*	1/4	$y=0.912(-)$, PP=1.0(−)
			{$a=3.81$}	D2	*8f*	0	*y*	*z*	$y=0.309(-)$, $z=0.483(-)$, PP=0.95(−)
			{$b=10.53$}	D3	*4b*	0	1/2	0	PP=0.80(−)
			{$c=4.37$}						
				R_p=n.s., *n*=n.s. (7 atomic), *s*=n.s.					
		Remark: B values and type of refinement not specified							
$Pd_3P_{0.8}D_{0.15}$ (500 kPa)	[4.111]	*Pnma* (62)	$a=5.7182(4)$	Pd1	*8d*	*x*	*y*	*z*	$x=0.1791(3)$, $y=0.0627(2)$, $z=0.3316(3)$, $B=0.80(3)$
			$b=7.5448(6)$	Pd2	*4c*	*x*	1/4	*z*	$x=0.0249(4)$, $z=0.8671(5)$, $B=1.25(6)$

			$c=5.1304(4)$	P	$4c$	x	1/4	z	$x=0.8845(10)$, $z=0.4592(11)$, $B=1.50(14)$			
			$\{a=5.7019\}$	D	$8d$	x	y	z	$x=0.379(6)$, $y=0.178(4)$, $z=0.003(6)$, $B=2.8(7)$			
			$\{b=7.5366\}$									
			$\{c=5.1191\}$									

$R_{wp}=0.071$, $R_{F^2}=0.044$, $n=22$, $s=0.46$

Remarks: Structure refinement also performed on hydride $Pd_3P_{0.8}H_{0.17}$ [4.112]; results not significantly different from those of the deuteride

									x	y	z	B
$Pd_6PD_{0.26}$ (750 kPa)	[4.113]	$P2_1/c$ (14)	$a=5.6870(4)$									
			$b=9.4654(7)$	Pd1	$4e$	x	y	z	0.9612(20)	0.8602(6)	0.9391(10)	0.68(2)
			$c=8.2196(8)$	Pd2	$4e$	x	y	z	0.5307(22)	0.1357(7)	0.0716(11)	0.68(–)
			$\beta=110.431(9)°$	Pd3	$4e$	x	y	z	0.7750(19)	0.8690(8)	0.5726(11)	0.68(–)
			$\{a=5.6740\}$	Pd4	$4e$	x	y	z	0.7154(18)	0.1318(6)	0.4395(11)	0.68(–)
			$\{b=9.4409\}$	Pd5	$4e$	x	y	z	0.8599(20)	0.5794(6)	0.7481(11)	0.68(–)
			$\{c=8.2100\}$	Pd6	$4e$	x	y	z	0.6250(24)	0.4173(7)	0.2560(11)	0.68(2)
			$\{\beta=110.414\}$	P	$4e$	x	y	z	0.8761(22)	0.2614(5)	0.7481(13)	0.53(9)
				D1	$2d$	1/2	0	1/2	PP=0.169(6)			3.8(5)
				D2	$4e$	x	y	z	0.311(10)	0.083(8)	0.226(9)	3.8(–)
									PP=0.091(6)			

$R_{wp}=0.053$, $R_I=0.026$, $n=36$, $s=0.45$

Remarks: Structure model with $\beta=110.43°$ cannot be discriminated from one with $\beta=110.12°$; refinements also performed on $Pd_6PD_{0.15}$ (200 kPa) and $Pd_6PD_{0.22}$ (500 kPa); occupancy of D1 and D2 sites change differently as a function of pressure; structure of the hydride $Pd_6PH_{0.39}$ [4.112] is not significantly different from that of the deuteride; a third hydrogen position (D3 at $4e$ with $x=0.25(5)$, $y=0.01(3)$, $z=0.97(2)$) is occupied at 800 kPa H_2 pressure.

4.B Appendix

Magnetic structures of ternary metal hydrides and deuterides studied by neutron diffraction

Compound[a]	Ref.	Magnetic ordering[b]
$CeRu_2D_5$	[4.56]	Antiferromagnetic; ferromagnetic (111) planes stacked in a $++--$ sequence along [111] $\mu_{Ce}=0.6(2)\ \mu_B$ at 120 mK
$HoFe_2D_{3.5}$	[4.57]	Ferrimagnetic ("fanned"), $T_C=T_C^{Ho}=295$ K $\mu_{sat}^{Ho}=5.3(4)\ \mu_B$, $\mu_{sat}^{Fe}=-1.9(3)\ \mu_B$ *Remarks:* Temperature dependence of magnetic moments determined
$ErFe_2D_2$	[4.57]	Ferrimagnetic ("fanned"); $T_C^{Er}=369$ K $\mu_{sat}^{Er}=7.5\ \mu_B$, $\mu_{sat}^{Fe}=-1.8\ \mu_B$ *Remark:* Temperature dependence of magnetic moment of Er determined
$ErFe_2D(H)_{3.5}$	[4.57]	Ferrimagnetic ("fanned"), $T_C=440$ K (deuteride), 450 K (hydride) $T_C^{Er}=295$ K (deuteride), 300 K (hydride) $\mu_{sat}^{Er}=4.3(4)\ \mu_B$ (deuteride), $4.9(4)\ \mu_B$ (hydride) $\mu_{sat}^{Fe}=-1.6(3)\ \mu_B$ (deuteride), $-1.5(3)\ \mu_B$ (hydride) *Remark:* Temperature dependence of magnetic moments determined
$TmFe_2H_{3.5}$	[4.57]	Ferrimagnetic ("fanned"), $T_C^{Tm}=250$ K $\mu_{sat}^{Tm}=3.8(4)\ \mu_B$ *Remark:* Temperature dependence of magnetic moment of Tm determined
$Y_6Mn_{23}D_{23}$	[4.69]	Antiferromagnetic, $T_N=172(7)$ K 4 K 78 K $\mu_{Mn1}=-\mu_{Mn2}=3.68\ \mu_B$, $3.60\ \mu_B$ $\mu_{Mn9}=-\mu_{Mn8}=1.60\ \mu_B$, $2.08\ \mu_B$ *Remark:* Zero ordered magnetic moments at other Mn sites
$Th_6Mn_{23}D_{30}$	[4.71]	Ferrimagnetic at 4 K $\mu_{Mn1}=-3.5(-)\ \mu_B$, $\mu_{Mn2}=0.8(1)\ \mu_B$, $\mu_{Mn3}=1.8(1)\ \mu_B$, $\mu_{Mn4}=1.5(1)\ \mu_B$
$Ho_6Fe_{23}D_{1.5}$	[4.72]	Ferrimagnetic 4 K 295 K $\mu_{Ho}=-10.2(2)\ \mu_B$ $-5.7(2)\ \mu_B$ $\mu_{Fe1}=2.2\ \mu_B$ $2.2\ \mu_B$ $\mu_{Fe2}=0.9(2)\ \mu_B$ $1.2(2)\ \mu_B$

[a] Listed in same order as in Table 4.1; see also Appendix 4.A

[b] μ = ordered magnetic moment in units of Bohr magneton, μ_B; μ_{sat} = ordered magnetic moment at saturation, i.e. for temperature T close to zero; negative signs indicate opposite directions of magnetic moments; $T_{N(C)}$ = Néel (Curie) temperature determined by neutron diffraction; T_C^R = ordering temperature of rare-earth sublattice; values within parentheses indicate estimated standard deviations

4.B Appendix (continued)

Compound[a]	Ref.	Magnetic ordering[b]
		$\mu_{Fe3} =$ 1.3(2) μ_B 1.4(2) μ_B $\mu_{Fe4} =$ 2.0(2) μ_B 1.9(2) μ_B
$Ho_6Fe_{23}D_{8.2}$	[4.72]	Ferrimagnetic, 4 K 78 K 295 K $\mu_{Ho} = -9.1(2)\,\mu_B$ $-8.3(2)\,\mu_B$ $-4.2(2)\,\mu_B$ $\mu_{Fe1} =$ 2.2 μ_B 2.2 μ_B 2.2 μ_B $\mu_{Fe2} =$ 1.8(2) μ_B 1.8(2) μ_B 1.6(2) μ_B $\mu_{Fe3} =$ 0.9(2) μ_B 0.8(2) μ_B 0.8(2) μ_B $\mu_{Fe4} =$ 2.1(2) μ_B 2.1(2) μ_B 1.9(2) μ_B
$Ho_6Fe_{23}D_{12.1}$	[4.72]	Ferrimagnetic, 4 K 295 K $\mu_{Hoa(8e)} = -10.2(2)\,\mu_B$ $-4.1(2)\,\mu_B$ $\mu_{Hob(16h)} = -10.2(2)\,\mu_B$ $-4.1(2)\,\mu_B$ $\mu_{Fe1(4b)} =$ 2.0(2) μ_B 2.2(–) μ_B $\mu_{Fe2a(8c)} =$ 2.5(2) μ_B 2.0(2) μ_B $\mu_{Fe2b(16f)} =$ 2.5(2) μ_B 2.3(2) μ_B $\mu_{Fe3(16n_1)} =$ 1.1(2) μ_B 1.1(2) μ_B $\mu_{Fe4(16n_2)} =$ 2.5(2) μ_B 2.0(2) μ_B *Remark:* Ho moment reorientation from [100] to [101] between 295 and 4 K, Fe1 and Fe3 moments reorient similarly whereas Fe2 and Fe4 moments remain within the basal plane.
$Ho_6Fe_{23}D_{15.7}$	[4.72]	Ferrimagnetic, 4 K 295 K $\mu_{Ho} = -10.2(2)\,\mu_B$ $-4.3(2)\,\mu_B$ $\mu_{Fe1} =$ 2.2(–) μ_B 2.2(–) μ_B $\mu_{Fe2} =$ 2.4(2) 2.3(2) μ_B $\mu_{Fe3} =$ 2.0(2) μ_B 1.8(2) μ_B $\mu_{Fe4} =$ 2.2(2) μ_B 1.9(2) μ_B
$Er_6Mn_{23}D_{23}$	[4.68]	Ferrimagnetic, $\mu_j \parallel c$; at 150 K: $\mu_{Er(4e)} = -2.5(3)\,\mu_B$, $\mu_{Mn(2b)} = -2.7(3)\,\mu_B$ $\mu_{Er(8h)} = 1.2(3)\,\mu_B$, $\mu_{Mn(4c)} = 3.4(2)\,\mu_B$ $\mu_{Mn(8f)} = 0$ $\mu_{Mn(16n_1)} = 0.8(2)\,\mu_B$ $\mu_{Mn(16n_2)} = 0.9(2)\,\mu_B$
β^I-$PrCo_5D_{3.6}$	[4.92]	Ferromagnetic, $\mu_j \parallel c$; at 300 K: $\mu_{Co} = 0.83(4)\,\mu_B$ 4.2 K 77 K $\mu_{Pr} = 1.6(1)\,\mu_B$ 0.50(4) μ_B $\mu_{Co1} = 0.12(8)\,\mu_B$ 0.98(6) μ_B $\mu_{Co2,3} = 0.79(6)\,\mu_B$ 1.28(5) μ_B
β^{II}-$PrCo_5D_{2.9}$	[4.92]	Ferromagnetic, $\mu \parallel c$; at 300 K: $\mu_{Co} = 1.05(5)\,\mu_B$
β^I-$LaCo_5D_{3.35}$	[4.92]	Ferromagnetic, $\mu \parallel c$; at 300 K: $\mu_{Co} = 1.14(5)\,\mu_B$
β^{III}-$CeCo_5D_{2.55}$	[4.92]	Ferromagnetic, $\mu \parallel c$; at 300 K: $\mu_{Co} = 0.98(3)\,\mu_B$
$ZrMn_2D_3$	[4.200]	Antiferro- or ferrimagnetic, $T_{N(C)} = 175(5)$ K
$Ti_{1.2}Mn_{1.8}D_{3.1}$	[4.98]	Ferromagnetic, $\mu_{Mn(6h)}$ perpendicular to c, at 80 K: $\mu_{Mn} = 0.31\,\mu_B$

Acknowledgements. The authors thank Professors S. Rundqvist and L. Schlapbach for valuable suggestions, and Drs. Y. Andersson, D. Fruchart, I. Jacob, and P. Zolliker for a careful reading of the manuscript. Errors which may remain in the text and tables are the sole responsibility of the authors.

Support by the Swiss National Energy Research Foundation (NEFF) and the Swiss National Science Foundation is gratefully acknowledged.

References

4.1 K. Yvon: J. Less-Common Met. **103**, 53 (1984)
4.2 Y. Noda, T. Kajitani, M. Hirabayashi, S. Sato: Acta Crystallogr. C **41**, 1566 (1985)
4.3 R.A. Alikhanov, V.I. Buzin, N.I. Kulikov, M.E. Kost, W. Sikora, L.S. Smirnov: J. Less-Common Met. **101**, 291 (1984)
4.4 R.R. Arons, J.W. Cable: Sol. State Commun. **55**, 835 (1985)
4.5 B. Lebech, N. Hessel-Andersen, S. Steenstrup, A. Schroeder Pedersen: Acta Cryst. C **39**, 1475 (1983
P. Fischer, J. Schefer, K. Tichy, R. Bischof, E. Kaldis: J. Less-Common Met. **94**, 151 (1983)
4.6 W. Bartscher, A. Boeuf, R. Caciuffo, J.M. Fournier, W.F. Kuhs, J. Rebizant, F. Rustichelli: Solid State Commun. **53**, 423 (1985)
4.7 W. Bartscher, A. Boeuf, R. Caciuffo, J.M. Fournier, J.M. Haschke, L. Manes, J. Rebizant, F. Rustichelli, J.W. Ward: Physica **130B**, 530 (1985)
4.8 H. Nagel, R.S. Perkins: Z. Metallkd. **66**, 362 (1975)
4.9. H. Kaneko, T. Kajitani, M. Hirabayashi, M. Ueno, K. Suzuki: J. Less-Common Met. **89**, 237 (1983)
4.10 B. Rodmacq, Ph. Mangin, A. Chamberod: J. Phys. F: Met. Phys. **15**, 2259 (1985)
4.11 A.J. Maeland, A.F. Andresen: J. Chem. Phys. **48**, 4660 (1968)
4.12 S.C. Abrahams, A.P. Ginsberg, K. Knox: Inorg. Chem. **3**, 558 (1964)
S.C. Abrahams, K. Knox: J. Phys. **25**, 461 (1964)
4.13 A.N. Fitch, S.A. Barrett, B.E.F. Fender, A. Simon: J. Chem. Soc. Dalton Trans. **4**, 501 (1984)
4.14 S.D. Wijeyesekera, J.D. Corbett: Solid State Commun. **54**, 657 (1985)
4.15 H.J. Mattausch, A. Simon, K. Ziebeck: J. Less-Common Met. **113**, 149 (1985)
4.16 G. Bandel, M. Müllner, M. Trömel: Z. Anorg. Allg. Chem. **453**, 5 (1979)
4.17 J.M. Newsam, A.K. Cheetham, B.C. Tofield: Solid State Chem. **60**, 214 (1985)
4.18 C. Greaves: J. Appl. Cryst. **18**, 48 (1985)
C. Greaves, M.A. Thomas: Acta Cryst. B **42**, 51 (1986)
4.19 M.T. Weller, P.G. Dickens: J. Solid State Chem. **58**, 164 (1985)
4.20 I.S. Latergaus, V.T. Em, I. Karimov, D.Ya. Khvatinskaya, S.K. Dolukhanyan: Inorg. Mat. **20**, 1420 (1984)
4.21 S. Mukawa, T. Kajitani, M. Hirabayashi: J. Less-Common Met. **103**, 19 (1984)
4.22 R.O. Moyer, Jr., C. Stanitski, J. Tanaka, M.I. Kay, R. Kleinberg: J. Solid State Chem. **3**, 541 (1971)
4.23 W. Bronger, G. Auffermann, P. Müller: J. Less-Common Met. **116**, 9 (1986)
4.24 G.G. Libowitz: *The Solid-State Chemistry of Binary Metal Hydrides* (W.A. Benjamin, Inc., New York, Amsterdam 1965)
4.25 W.M. Mueller, J.P. Blackledge, G.G. Libowitz: *Metal Hydrides* (Academic, New York, London 1968)
4.26 P. Eckerlin, H. Kandler: in *Landolt-Börnstein, Vol. 6* Structure Data of Elements and Intermetallic Phases, ed. by K.-H. Hellwege, A.M. Hellwege (Springer, Berlin, Heidelberg 1971) pp. 138–174
4.27 R.R. Arons et al.: In *Landolt-Börnstein III*, Vol. 12c (Springer, Berlin, Heidelberg 1982) pp. 372–405

4.28 H. Asano, M. Hirabayashi: Z. Phys. Chem. NF **114**, 1 (1979)
H. Asano, M. Hirabayashi: In *Metal Hydrides*, ed. by G. Bambakidis (Plenum, New York 1981) pp. 81–103
M. Hirabayashi, H. Asano: ibid., pp. 53–80
4.29 J.W. Ward: Physica **130B**, 510 (1985)
4.30 R.G. Teller, R. Bau: In *Structure and Bonding*, Vol. 44, pp. 1–82 (Springer, Berlin, Heidelberg, New York 1981); see also
R. Bau, M.Y. Chiang, D.M. Ho, S.G. Gibbins, T.J. Emge, T.F. Koetzle: Inorg. Chem. **23**, 2823 (1984)
4.31 S. Rundqvist, R. Tellgren, Y. Andersson: J. Less-Common Met. **101**, 145 (1984)
4.32 A.F. Andresen: In *Hydrides for Energy Storage*, ed. by A.F. Andresen, A.J. Maeland (Pergamon, Oxford 1978) pp. 61–72
4.33 A.F. Andresen: J. Less-Common Met. **88**, 1 (1982)
4.34 K.N. Semenenko, V.V. Burnasheva: J. Less-Common Met. **105**, 1 (1985)
4.35 A. Percheron-Guégan, C. Lartigue, J.C. Achard: J. Less-Common Met. **109**, 287 (1985)
4.36 G.D. Sandrock: In *Hydrides for Energy Storage*, ed. by A.F. Andresen, A.J. Maeland (Pergamon, Oxford 1978) pp. 353–393
4.37 H. Oesterreicher: Appl. Phys. **24**, 169 (1981)
4.38 G.K. Shenoy, B.D. Dunlap, P.J. Viccaro, D. Niarchos: Adv. Chem. Series No. 194, ed. by J.G. Stevens, G.K. Shenoy, Chap. 23, pp. 501–521 (1981)
4.39 H. Buchner: *Energiespeicherung in Metallhydriden* (Springer, Wien, New York 1982)
4.40 K.H.J. Buschow, P.C.P. Bouten, A.R. Miedema: Rep. Prog. Phys. **45**, 937 (1982)
4.41 D.G. Ivey, D.O. Northwood: J. Mater. Sci. **18**, 321 (1983)
4.42 K.H.J. Buschow: In *Handbook on the Physics and Chemistry of Rare Earths*, Vol. 6, Chap. 47, ed. by K.A. Gschneider, Jr., L. Eyring (Elsevier, Amsterdam 1984)
4.43 *Proceedings of the International Symposium on the Properties and Applications of Metal Hydrides:* J. Less-Common Met. Vols. **73** and **74** (1980), **88** (1982); **89** (1983); **101**, **103**, and **104** (1984); **129**, **130**, and **131** (1987)[1]
4.44 *International Tables for Crystallography*, Vol. A (D. Reidel, Dordrecht 1983)
4.45 H. Buchner, M.A. Gutjahr, K.-D. Beccu, H. Säufferer: Z. Metallkde. **63**, 497 (1972)
4.46 C. Stioui, D. Fruchart, A. Rouault, R. Fruchart, E. Roudaut, J. Rebiére: Mat. Res. Bull. **16**, 869 (1981)
4.47 F.J. Rotella, H.E. Flotow, D.M. Gruen, J.D. Jorgensen: J. Chem. Phys. **79**, 4522 (1983)
4.48 J.-J. Didisheim, K. Yvon, P. Fischer, D. Shaltiel: J. Less-Common Met. **73**, 355 (1980)
4.49 J.-J. Didisheim, K. Yvon, P. Fischer, P. Tissot: Solid State Commun. **38**, 637 (1981)
4.50 J.-J. Didisheim, K. Yvon, D. Shaltiel, P. Fischer, P. Bujard, E. Walker: Solid State Commun. **32**, 1087 (1979)
4.51 D. Fruchart, A. Rouault, C.B. Shoemaker, D.P. Shoemaker: J. Less-Common Met. **73**, 363 (1980); Phys. Status Solidi a **57**, K119 (1980)
4.52 A.V. Irodova, V.P. Glazkov, V.A. Somenkov, S.Sh. Shilstein: J. Less-Common Met. **77**, 89 (1981); Sov. Phys. Solid State **22**, 45 (1980)
4.53 V.A. Yartys', V.V. Burnasheva, K.N. Semenenko, N.V. Fadeeva, S.P. Solov'ev: Int. J. Hydrogen Energy **7**, 957 (1982)
V.A. Yartys', V.V. Burnasheva, S.E. Tsirkunova, E.N. Kozlov, K.N. Semenenko: Sov. Phys. Crystallogr. **27**, 148 (1982)
V.V. Burnasheva, N.V. Fadeeva, K.N. Semenenko, S.P. Solov'ev, V.A. Yartys': Acta Crystallogr. A **37**, C-182 (1981)
4.54 A.V. Irodova, O.A. Lavrova, G.V. Laskova, L.N. Padurets: Sov. Phys.-Sol. State **24**, 22 (1982)
4.55 N.F. Miron, V.I. Shcherbak, V.N. Bykov, V.A. Levdik: Sov. Phys.-Crystallogr. **16**, 266 (1971)
4.56 D. Fruchart, F. Vaillant, A. Rouault, A. Benoit, J. Flouquet: J. Less-Common Met. **101**, 285 (1984)

[1] These references are only partly considered in this review.

D. Fruchart, F. Vaillant, E. Roudaut, A. Nemoz, X.G. Tessema: Phys. Status Solidi a **65**, K19 (1981)
4.57 J.J. Rhyne, S.G. Sankar, W.E. Wallace: In *The Rare-Earths in Modern Science and Technology*, Vol. 1, ed. by G.J. McCarthy, J.J. Rhyne (Plenum, New York 1978) pp. 63–68;
J.J. Rhyne, G.E. Fish, S.G. Sankar, W.E. Wallace: J. de Phys. **40**, C5–209 (1979)
G.E. Fish, J.J. Ryne, S.G. Sankar, W.E. Wallace: J. Appl. Phys. **50**, 2003 (1979)
G.E. Fish, J.J. Rhyne, T. Brun, P.J. Viccaro, D. Niarchos, B.D. Dunlap, G.K. Shenoy, S.G. Sankar, W.E. Wallace: In *The Rare-Earths in Modern Science and Technology*, Vol. 2, ed. by G.J. McCarthy, J.J. Rhyne, H.B. Silber (Plenum, New York 1980) pp. 569–570
4.58 W. Bartscher, J. Rebizant, A. Boeuf, R. Caciuffo, F. Rustichelli, J.M. Fournier, W.F. Kuhs: Proc. Actinides 85, Aix-en-Provence, Sept. 1985. J. Less-Common Met. **121**, 455 (1986)
4.59 K. Yvon, J. Schefer, F. Stucki: Inorg. Chem. **20**, 2776 (1981)
J. Schefer, P. Fischer, W. Hälg, F. Stucki, L. Schlapbach, J.J. Dididsheim, K. Yvon, A.F. Andresen: J. Less-Common Met. **74**, 65 (1980)
4.60 D. Noréus, L.G. Olsson: J. Chem. Phys. **78**, 2419 (1983)
4.61 J.L. Soubeyroux, D. Fruchart, A. Mikou, M. Pezat, B. Darriet, P. Hagenmuller: Mat. Res. Bull. **19**, 969 (1984)
B. Darriet, J.L. Soubeyroux, M. Pezat, D. Fruchart: J. Less-Common Met. **103**, 153 (1984); see also [4.63, 99]
4.62 D. Noréus, P.-E. Werner: J. Less-Common Met. **97**, 215 (1984)
4.63 J.L. Soubeyroux, D. Fruchart, A. Mikou, M. Pezat, B. Darriet: Mat. Res. Bull. **19**, 1119 (1984)
4.64 P. Zolliker, K. Yvon, J.D. Jorgensen, F. Rotella: Inorg. Chem. **25**, 3590 (1986)
4.65 P. Zolliker, K. Yvon, P. Fischer, J. Schefer: Inorg. Chem. **24**, 4177 (1985)
4.66 J.-J. Didisheim, P. Zolliker, K. Yvon, P. Fischer, J. Schefer, M. Gubelmann, A.F. Williams: Inorg. Chem. **23**, 1953 (1984)
J.-J. Didisheim, K. Yvon, P. Fischer, J. Schefer, M. Gubelmann, A.F. Williams: Z. Kristallogr. **162**, 61 (1983)
4.67 M. Commandre, D. Fruchart, A. Rouault, D. Sauvage, C.B. Shoemaker, D.P. Shoemaker: J. de Phys. **40**, L639 (1979)
4.68 C. Crowder, B. Kebe, W.J. James, W. Yelon: In *The Rare-Earths in Modern Science and Technology*, Vol. 3, ed. by G.J. McCarthy, H.B. Silber, J.J. Rhyne (Plenum, New York 1982) pp. 473–476; for work on $Er_6Mn_{23}D_{23}$ see: Amer. Inst. Phys. Conf. Proc. **89**, 318 (1982)
4.69 K. Hardman-Rhyne, J.J. Rhyne, E. Prince, C. Crowder, W.J. James: Phys. Rev. B **29**, 416 (1984)
4.70 K. Hardman, J.J. Rhyne, E. Prince, H.K. Smith, S.K. Malik, W.E. Wallace: In *The Rare-Earths in Modern Science and Technology*, Vol. 3, ed. by G.J. McCarthy, H.B. Silber, J.J. Rhyne (Plenum, New York 1982) pp. 477–478
K. Hardman, J.J. Rhyne, H.K. Smith, W.E. Wallace, S.K. Malik: J. Appl. Phys. **52**, 2070 (1981)
4.71 K. Hardman-Rhyne, H.K. Smith, W.E. Wallace: J. Less-Common Met. **96**, 201 (1984)
K. Hardman, J.J. Rhyne, K. Smith, W.E. Wallace: J. Less-Common Met. **74**, 97 (1980)
4.72 J.J. Rhyne, K. Hardman-Rhyne, H.K. Smith, W.E. Wallace: J. Less-Common Met. **94**, 95 (1983)
4.73 P. Thompson, F. Reidinger, J.J. Reilly, L.M. Corliss, J.M. Hastings: J. Phys. F: Met. Phys. **10**, L57 (1980)
4.74 P. Thompson, M.A. Pick, F. Reidinger, L.M. Corliss, J.M. Hastings, J.J. Reilly: J. Phys. F: Met. Phys. **8**, L75 (1978)
4.75 P. Fischer, W. Hälg, L. Schlapbach, F. Stucki, A.F. Andresen: Mat. Res. Bull. **13**, 931 (1978)

G. Busch, L. Schlapbach, F. Stucki, P. Fischer, A.F. Andresen: In *Hydrogen Energy System*, ed. by T.N. Veziroglu, W. Seifritz (Pergamon, Oxford 1979), Vol. 5, pp. 2615–2636; Int. J. Hydrogen Energy **4**, 29 (1979)
4.76 J. Schefer, P. Fischer, W. Hälg, F. Stucki, L. Schlapbach, A.F. Andresen: Mat. Res. Bull. **14**, 1281 (1979)
L. Schlapbach, A. Seiler, F. Stucki, P. Zürcher, P. Fischer, J. Schefer: Z. Phys. Chem. NF **117**, 205 (1979)
4.77 P. Thompson, J.J. Reilly, F. Reidinger, J.M. Hastings, L.M. Corliss: J. Phys. F: Met. Phys. **9**, L61 (1979)
4.78 W. Schäfer, G. Will, T. Schober: Mat. Res. Bull. **15**, 627 (1980)
W. Schäfer, E. Lebsanft, A. Bläsius: Z. Phys. Chem. NF **115**, 201 (1979)
see also [4.124]
4.79 P. Fischer, J. Schefer, K. Yvon, L. Schlapbach, T. Riesterer: J. Less-Common Met **129**, 39 (1986)
4.80 D. Fruchart, M. Commandré, D. Sauvage, A. Rouault, R. Tellgren: J. Less-Common Met. **74**, 55 (1980)
4.81 L.J. Vieland, A.W. Wicklund, J.G. White: Phys. Rev. B **11**, 3311 (1975)
4.82 A. Furrer, P. Fischer, W. Hälg, L. Schlapbach: In *Hydrides for Energy Storage*, ed. by A.F. Andresen, A.J. Maeland (Pergamon, Oxford 1978) pp. 73–82; see also [4.84]
4.83 A.L. Bowman, J.L. Anderson, N.G. Nereson: Proceedings of the 10th Rare-Earth Research Conference, Carefree (1973) pp. 485–489
4.84 P. Fischer, A. Furrer, G. Busch, L. Schlapbach: Helv. Phys. Acta **50**, 421 (1977)
G. Busch, L. Schlapbach, W. Thoeni, Th. v. Waldkirch, P. Fischer, A. Furrer, W. Hälg: Proc. 2nd Int. Congress on Hydrogen in Metals (Paris 1977) 1D7, pp. 1–8
P. Fischer, W. Hälg, L. Schlapbach, Th. von Waldkirch: Helv. Phys. Acta **51**, 4 (1978); see also [4.82]
4.85 D. Noréus, L.G. Olsson, P.-E. Werner: J. Phys. F: Met. Phys. **13**, 715 (1983)
4.86 A. Percheron-Guégan, C. Lartigue, J.C. Achard, P. Germi, F. Tasset: J. Less-Common Met. **74**, 1 (1980)
C. Lartigue, A. Percheron-Guégan, J.C. Achard, F. Tasset: In *The Rare-Earths in Modern Science and Technology*, Vol. 2, ed. by G.J. McCarthy, J.J. Rhyne, H.B. Silber (Plenum, New York 1980)
C. Lartigue, A. Percheron-Guégan, J.C. Achard, F. Tasset: J. Less-Common Met. **75**, 23 (1980)
J.C. Achard, F. Givord, A. Percheron-Guégan, J.L. Soubeyroux, F. Tasset: J. de Phys. **40**, C5–218 (1979)
4.87 C. Lartigue, A. Percheron-Guégan, J.C. Achard, J.L. Soubeyroux: J. Less-Common Met. **113**, 127 (1985)
4.88 P. Thompson, J.J. Reilly, L.M. Corliss, J.M. Hastings, R. Hempelmann: J. Phys. F: Met Phys. **16**, 679 (1986)
4.89 V.V. Burnasheva, V.A. Yartys', N.V. Fadeeva, S.P. Solov'ev, K.N. Semenenko: Sov. Phys. Dokl. **23**, 97 (1978)
4.90 A.V. Irodova, M.E. Kost, L.N. Padurets, V.A. Somenkov, E.I. Sokolova, S.Sh. Shil'shtein: Russ. J. Inorg. Chem. **26**, 166 (1981)
4.91 L.D. Calvert, B.M. Powell, J.J. Murray, Y. le Page: J. Sol. State Chem. **60**, 62 (1985)
4.92 F.A. Kuijpers, B.O. Loopstra: J. Phys. Chem. Solids **35**, 301 (1974); J. de Phys. **32**, Cl 657 (1971)
F.A. Kuijpers: Philips Res. Repts. Suppl. No. 2 (1973)
4.93 C. Crowder, W.J. James, W. Yelon: J. Appl. Phys. **53**, 2637 (1982)
4.94 J.C. Achard, A.J. Dianoux, C. Lartigue, A. Percheron-Guégan, F. Tasset: In *The Rare-Earths in Modern Science and Technology*, Vol. 3, ed. by G.J. McCarthy, H.B. Silber, J.J. Rhyne (Plenum, New York 1982) pp. 481–486
4.95 E. Gurewitz, H. Pinto, M.P. Dariel, H. Shaked: J. Phys. F **13**, 545 (1983)
4.96 J.-J. Didisheim, K. Yvon, D. Shaltiel, P. Fischer: Solid State Commun. **31**, 47 (1979)

4.97 H.W. Mayer, K.M. Alasafi, O. Bernauer: J. Less-Common Met. **88**, L7 (1982)
4.98 D. Fruchart, J.L. Soubeyroux, R. Hempelmann: J. Less-Common Met. **99**, 307 (1984)
4.99 J.L. Soubeyroux, D. Fruchart, A. Mikou, M. Pezat, B. Darriet: Mat. Res. Bull. **19**, 895 (1984); see also [4.61]
4.100 D. Noréus, P.-E. Werner: Acta Chem. Scand. A **36**, 847 (1982)
4.101 A.F. Andresen, K. Otnes, A.J. Maeland: J. Less-Common Met. **89**, 201 (1983)
4.102 A.F. Andresen, H. Fjellvåg, A.J. Maeland: J. Less-Common Met. **103**, 27 (1984)
4.103 Y. Andersson, S. Rundqvist, R. Tellgren, T.B. Flanagan: Z. Phys. Chem. NF **145**, 43 (1985)
4.104 A. Santoro, A. Maeland, J.J. Rush: Acta Cryst. B **34**, 3059 (1978)
4.105 J. Bergsma, J.A. Goedkoop, J.H.N. van Vucht: Acta Crystallogr. **14**, 223 (1961)
J.H.N. van Vucht: Philips Res. Repts. **18**, 35 (1963)
4.106 A.J. Maeland, E. Lukacevic, J.J. Rush, A. Santoro: J. Less-Common Met. **129**, 77 (1987)
4.107 D. Fruchart, P. Wolfers, P. Vulliet, A. Yaouanc, R. Fruchart, P.L'Heritier: In *Proc. of the E. C. Workshop Meeting on* Nd–Fe *Permanent Magnets*, Oct. 1984, Brussels, ed. by I.V. Mitchell (1985)
4.108 S.W. Peterson, V.N. Sadana, W.L. Korst: J. de Phys. **25**, 451 (1964)
4.109 D.G. Westlake, H. Shaked, P.R. Mason, B.R. McCart, M.H. Mueller, T. Matsumoto, M. Amano: J. Less-Common Met. **88**, 17 (1982)
4.110 A.V. Irodova, V.A. Somenkov, S.Sh. Shil'shtein, L.N. Padurets, A.A. Chertkov: Sov. Phys. Crystallogr. **23**, 591 (1978)
4.111 Y. Andersson, S. Rundqvist, R. Tellgren, J.O. Thomas, T.B. Flanagan: J. Sol. State Chem. **32**, 321 (1980)
4.112 Y. Andersson, S. Rundqvist, R. Tellgren: J. Sol. State Chem. **52**, 327 (1984)
4.113 Y. Andersson, S. Rundqvist, R. Tellgren, J.O. Thomas, T.B. Flanagan: Acta Crystallogr. B **37**, 1965 (1981)
4.114 D.G. Westlake: J. Less-Common Met. **91**, 1 (1983)
4.115 I. Jacob, D. Shaltiel, D. Davidov, I. Miloslavski: Solid State Commun. **23**, 669 (1977)
I. Jacob, D. Shaltiel: J. Less-Common Met. **65**, 117 (1979)
4.116 I. Jacob, J.M. Bloch, D. Shaltiel, D. Davidov: Solid State Commun. **35**, 155 (1980)
4.117 J.J. Reilly, R.H. Wiswall, Jr.: Inorg. Chem. **13**, 218 (1974)
4.118 J.J. Reilly, R.H. Wiswall, Jr.: Inorg. Chem. **7**, 2254 (1968)
4.119 M.L. Post, J.J. Murray, G.J. Despault, J.B. Taylor: Mat. Res. Bull. **20**, 337 (1985)
4.120 D. Noréus, P.-E. Werner: Mat. Res. Bull. **16**, 199 (1981)
4.121 S. Ono, H. Hayakawa, A. Suzuki, K. Nomura, N. Nishimiya, T. Tabata: J. Less-Common Met. **88**, 63 (1982)
4.122 H. Hayakawa, Y. Ishido, K. Nomura, H. Uruno, S. Ono: J. Less-Common Met. **103**, 277 (1984)
4.123 P. Zolliker, K. Yvon, C. Bärlocher: J. Less-Common Met. **115**, 65 (1986)
4.124 T. Schober, W. Schäfer: J. Less-Common Met. **74**, 23 (1980), see also [4.78]
4.125 A.W. Hewat, I. Bailey: Nucl. Instrum. Methods **137**, 463 (1976)
4.126 R. Allemand, J. Bourdel, E. Roudaut, P. Convert, K. Ibel, J. Jacobe, J.P. Cotton, B. Farnoux: Nucl. Instrum. Methods **126**, 29 (1975)
W. Hälg, H. Heer, J. Schefer, P. Fischer, B. Bron, A. Isacson, M. Koch: Helv. Phys. Acta **57**, 741 (1984);
P. Fischer: Neutron Diffraction Newsletter, Int. Union of Cryst. (1985) p. 15
4.127 J.D. Jorgensen, F.J. Rotella: J. Appl. Cryst. **15**, 27 (1982)
4.128 H.M. Rietveld: J. Appl. Crystallogr. **2**, 65 (1969)
4.129 P.-E. Werner, S. Salomé, G. Malmros, J.O. Thomas: J. Appl. Crystallogr. **12**, 107 (1979)
4.130 T.G. Worlton, J.D. Jorgensen, R.A. Beyerlein, D.L. Decker: Nucl. Instrum. Methods **137**, 331 (1976)
4.131 A. Albinati, B.T.M. Willis: The Rietveld method, publication MPD-NBS-136, Materials Physics Division, AERE Harwell (1980); J. Appl. Crystallogr. **15**, 361 (1982)

4.132 A.W. Hewat: Proc. of Symposium on Accuracy in Powder Diffraction, Gaithersburg, MD (1979), National Bureau of Standards Special Publication **567**, 111 (1980); see also: C.R.A. Carlow (ed.): *High Resolution Powder Diffraction*, Mat. Science Forum (Trans Tech. Publ., Switzerland) Vol. **9**, pp. 69–80 (1986)
4.133 R.B. Von Dreele, J.D. Jorgensen, C.G. Windsor: J. Appl. Crystallogr. **15**, 581 (1982)
4.134 R.A. Young, E. Prince, R.A. Sparks: J. Appl. Crystallogr. **15**, 357 (1982)
4.135 H.G. Scott: J. Appl. Crystallogr. **16**, 159 (1983)
4.136 H.H. van Mal, K.H.J. Buschow, A.R. Miedema: J. Less-Common Met. **35**, 65 (1974) A.R. Miedema, K.H.J. Buschow, H.H. van Mal: J. Less-Common Met. **49**, 463 (1976)
4.137 P.C.P. Bouten, A.R. Miedema: J. Less-Common Met. **71**, 147 (1980)
4.138 D.G. Westlake: J. Less-Common Met. **90**, 251 (1983)
4.139 J. Osterwalder, T. Riesterer, L. Schlapbach, F. Vaillant, D. Fruchart: Phys. Rev. B **31**, 8311 (1985)
4.140 Z. Gavra, M.H. Mintz, G. Kimmel, Z. Hadari: Inorg. Chem. **18**, 3595 (1979)
4.141 J.P. Darnaudery, M. Pezat, B. Darriet, P. Hagenmuller: Mat. Res. Bull. **16**, 1237 (1981)
4.142 J.R. Johnson, J.J. Reilly, F. Reidinger, L.M. Corliss, J.M. Hastings: J. Less-Common Met. **88**, 107 (1982)
4.143 J. Zhuang, J.M. Hastings, L.M. Corliss, R. Bau, Chiau-Yu Wei, R.O. Moyer, Jr.: J. Sol. State Chem. **40**, 352 (1981)
4.144 P. Zolliker, K. Yvon: Mat. Res. Bull. **21**, 415 (1986)
4.145 M. Gupta, E. Belin, L. Schlapbach: J. Less-Common Met. **103**, 389 (1984)
4.146 D.G. Westlake: J. Mater. Sci. **19**, 316 (1984)
4.147 H. Behr, H. Metzger, J. Peisl: J. Less-Common Met. **88**, 159 (1982)
4.148 E.T. Teatum, K.A. Gschneidner, J.T. Waber: Los Alamos Natl. Lab., Los Alamos, New Mexico, Rept. No. LA-4003 (1968); see also [Ref. 4.168, p. 151]
4.149 R.D. Shannon: Acta Crystallogr. A **32**, 751 (1976)
4.150 J.D. Corbett, H.S. Marek: Inorg. Chem. **22**, 3194 (1983)
4.151 A.C. Switendick: Z. Phys. Chem. NF **117**, 89 (1979); B.K. Rao, P. Jena: Phys. Rev. B **31**, 6726 (1985)
4.152 M. Gupta: J. Less-Common Met. **88**, 221 (1982)
4.153 D.A. Papaconstantopoulos, A.C. Switendick: J. Less-Common Met. **88**, 273 (1982)
4.154 M. Gupta: J. Less-Common Met. **103**, 325 (1984)
4.155 B.M. Klein, W.E. Pickett: J. Less-Common Met. **88**, 231 (1982)
4.156 W.A. Oates, T.B. Flanagan: Mat. Res. Bull. **19**, 1397 (1984)
4.157 R. Griessen, A. Driessen, D.G. de Groot: J. Less-Common Met. **103**, 235 (1984)
4.158 J.C. Achard, C. Lartigue, A. Percheron-Guégan, J.C. Mathieu, A. Pasturel, F. Tasset: J. Less-Common Met. **79**, 161 (1981)
4.159 J. Shinar, D. Shaltiel, D. Davidov, A. Grayevsky: J. Less-Common Met. **60**, 209 (1978)
4.160 R. Burch, N.B. Mason: J. Less-Common Met. **63**, 57 (1979)
4.161 F.L. Carter: J. Less-Common Met. **74**, 245 (1980)
4.162 A. Yoshikawa, T. Matsumoto, K. Yagisawa: J. Less-Common Met. **88**, 73 (1982)
4.163 A.L. Shilov, M.E. Kost, N.T. Kuznetsov: J. Less-Common Met. **105**, 221 (1985)
4.164 B.D. Dunlap, P.J. Viccaro, G.K. Shenoy: J. Less-Common Met. **74**, 75 (1980)
4.165 H.A. Kierstead: J. Less-Common Met. **96**, 141 (1984)
4.166 A.R. Williams, J. Kübler, C.D. Gelatt, Jr.: Phys. Rev. B **19**, 6094 (1979)
4.167 A. Miedema, P.F. de Châtel, F.R. DeBoer: Physica **100** B, 1 (1980)
4.168 W.B. Pearson: *The Crystal Chemistry and Physics of Metals and Alloys* (Wiley, New York 1980)
4.169 C.E. Lundin, F.E. Lynch, C.B. Magee: J. Less-Common Met. **56**, 1,9 (1977)
4.170 M.H. Mendelsohn, D.M. Gruen, A.E. Dwight: J. Less-Common Met. **63**, 193 (1979); Nature (London) **269**, 45 (1977)
4.171 C.B. Magee, J. Liu, C.E. Lundin: J. Less-Common Met. **78**, 119 (1981)
4.172 D.G. Westlake: J. Chem. Phys. **79**, 4532 (1983)
4.173 D.G. Westlake: Mat. Res. Bull. **18**, 1409 (1983)

4.174 D.G. Westlake: J. Mat. Science **18**, 605 (1983); and Scripta Metall. **16**, 1049 (1982)
4.175 D.G. Westlake: Sol. State Chem. **53**, 130 (1984)
4.176 D.G. Westlake: J. Less-Common. Met. **91**, 275 (1983)
4.177 D.G. Westlake: J. Less-Common Met. **103**, 203 (1984); J. Less-Common Met. **107**, 189 (1985)
4.178 D.G. Westlake: J. Less-Common Met. **105**, 69 (1985)
4.179 G.A. Stewart, D C. Creagh: J. Phys. F **15**, 1639 (1985)
4.180 Y. Chung, T. Takeshita, O.D. McMasters, K.A. Gschneidner, Jr.: J. Less-Common Met. **74**, 217 (1980)
4.181 D.P. Shoemaker, C.B. Shoemaker: J. Less-Common Met. **68**, 43 (1979)
4.182 W.A. Oates, J.A. Lambert, P.T. Gallagher: Trans. Met. Soc. AIME **245**, 47 (1969)
4.183 W.E. Wallace: J. Less-Common Met. **88**, 141 (1982)
4.184 J.F. Lakner, F.S. Uribe, S.A. Steward: J. Less-Common Met. **72**, 87 (1980)
4.185 W.E. Wallace, E.B. Boltich: J. Sol. State Chem. **33**, 435 (1980)
4.186 H. Goretzki: Naturwiss. **54**, 163 (1967)
4.187 K. Yvon, H. Nowotny, R. Kieffer: Monatsh. Chem. **98**, 2164 (1967)
4.188 D.M.P. Mingos, J.C. Hawes: In *Structure and Bonding*, Vol. 63 (Springer, Berlin, Heidelberg 1986) pp. 1–63
4.189 K.A. Gschneidner, Jr., T. Takeshita, Y. Chung, O.D. McMasters: J. Phys. F. Met. Phys. **12**, L1 (1982)
4.190 M.H. Mendelsohn, D.M. Gruen: J. Less-Common Met. **78**, 275 (1981)
4.191 I. Jacob, A. Stern, A. Moran, D. Shaltiel, D. Davidov: J. Less-Common Met. **73**, 369 (1980)
4.192 I. Jacob: Sol. State Commun. **40**, 1015 (1981)
I. Jacob, J.M. Bloch, E. Gurevich: J. Less-Common Met. **88**, 25 (1982)
4.193 D. Shaltiel: J. Less-Common Met. **73**, 329 (1980)
4.194 I. Jacob, J.M. Bloch: Sol. State Commun. **42**, 541 (1982)
4.195 T. Riesterer: J. Less-Common Met. **103**, 219 (1984)
4.196 D.M. Gruen, M. Mendelsohn: J. Less-Common Met. **55**, 149 (1977)
4.197 C.E. Crowder, W.J. James: J. Less-Common Met. **95**, 1 (1983)
4.198 C. Geibel, W. Goldacker, H. Keiber, V. Oestreich, H. Rietschel, H. Wühl: Phys. Rev. B **30**, 6363 (1984)
4.199 M.Yu. Belyaev, A.V. Skripov, V.N. Kozhanov, A.B. Stepanov, E.V. Galoshina: Solid State Commun. **48**, 1049 (1983)
4.200 J.-J. Didisheim, P. Fischer: J. Less-Common Met. **103**, 267 (1984)

5. Electronic Properties

Michèle Gupta and Louis Schlapbach

With 25 Figures and 2 Tables

5.1 Introduction

The dissolution of hydrogen in a metal lattice and the formation of a metal hydride perturbs considerably the electrons and phonons of the host metal. Accordingly the study of the electronic properties of a hydrogen-metal system is not only of fundamental interest to understand the hydrogen-metal interaction in the bulk and at the surface, but also sheds light on technologically important parameters such as the heat of formation and heat conductivity and to some extent also the phenomena of hydrogen embrittlement (decohesion) and catalysis.

Until the late sixties, research on electronic properties of metal hydrides involved mainly experimental work [5.1]. The first theoretical investigations started in the early fifties with the pioneering work of Friedel in which he studied the screening of a hydrogen impurity and the heats of solution of *H* in noble metals [5.2]. However, up to the early seventies, oversimplified models were often used to describe the bonding mechanisms of *H* in transition metals [5.3]. The first ab initio band structure calculations on binary hydrides performed by *Switendick* [5.4] gave a new impetus and stimulated a lot of experimental and theoretical work. From the theoretical results on concentrated binary hydrides several important features have emerged and the hydrides are no longer considered, as they were in the early days, as interstitial alloys but rather as real compounds. The role of the chemical bonding has been emphasized and the crude rigid-band model in its protonic or anionic form is no longer being safely applied to the electronic structure of the host metal to interpret the experimental data on hydrides. For about ten years both theoretical and experimental work was limited to hydrides of elemental metals, e.g. Pd, Ti, rare earth metals and Th [5.4, 5]. Only a few years ago the first results on the electronic structure of hydrides of intermetallic compounds elaborated by band structure calculations (FeTi) [5.6] and experimentally by photoelectron spectroscopy ($ZrMn_2$) [5.7] and x-ray emission spectroscopy (crystalline and amorphous alloys of Zr and V) [5.8–10] became available. Recent reviews on theoretical [5.6] and experimental work on crystalline [5.11] and amorphous [5.12] materials have been given.

After a brief description of the theoretical and experimental methods given in Sect. 5.2, we shall essentially review in this chapter the progress made since the appearance of the two volumes on Hydrogen in Metals [5.13] of this series.

Since then, the work on binary hydrides mainly of transition and rare earth metals has been pursued. For stoichiometric compounds new results concerning Fermi surfaces, superconductivity, and prediction of various observables using realistic wavefunctions have been obtained. An effort towards a self-consistent treatment of the crystal potentials has been made and some conclusions have been obtained concerning the problem of sign of the charge transfers. Also, theoretical studies of stabilities and of preferential occupation of sites by hydrogen atoms by means of total energy calculations have started to appear. The recent results on binary hydrides are described in Sect. 5.3 while Sect. 5.4 is devoted to the very recent theoretical and experimental investigations of the electronic properties of intermetallic hydrides. The essential experimental and theoretical results are gathered in Tables 5.1 and 5.2. The consequences of the electronic properties of hydrides for heats of formation, superconductivity, and maximum hydrogen content will be discussed while some features relevant to surface electronic structure, cohesion, and embrittlement will be briefly mentioned.

Four different effects are in general relevant for understanding the changes in the electronic structure of the host metal due to the H absorption: (i) The generally observed expansion (or in some special cases, the contraction) of the lattice often accompanied by a change in the crystal structure results in a modification of the symmetry of the states and in the band widths. (ii) The attractive potential of the proton affects those metal wavefunctions which have a finite density at the H site and leads to the so-called metal-hydrogen bonding band below the metal *d*-band. Furthermore it can pull below the Fermi energy E_F some metal states which were located above E_F in the pure metal. (iii) The additional H–H interactions in the hydrides which have more than one H atom per unit cell lead to new features in the lower portion of the density of states. (iv) The presence in the unit cell of the additional electrons brought by the H atoms and the inbalance between the additional number of electrons and the number of new electron states shifted to below the Fermi energy of the host metal results in a shift of the Fermi level.

We shall underline the similarities and differences observed between the electronic structure of binary and ternary hydrides. One of the goals of the analysis of electronic properties of intermetallic compounds and their hydrides is to understand why some compounds of non hydride forming components form a stable ternary hydride e.g. Fe–Ni alloys [5.14] or vice versa why a compound of hydride forming components does not form a ternary hydride e.g. $ZrPd_2$ [5.15].

5.2 Theoretical and Experimental Methods: General Features

5.2.1 Theoretical Methods

In this section we shall briefly remind the reader of the most salient features of some of the methods used in the study of the electronic structure of hydrogen in metals and indicate references for further details concerning the formalisms and the numerical approaches.

The Local Density Functional Approximation. Since a wide variety of theoretical studies ranging from the problem of hydrogen as an impurity in jellium to the treatment of stoichiometric transition metal hydrides, rest on the local density functional (LDF) formalism developed by *Hohenberg*, *Kohn* and *Sham* [5.16], we shall first summarize the main ideas associated with this method. The basis of the LDF is that the ground-state energy E of an interacting N particle system in an external potential is a unique functional of the electron density $\varrho(\boldsymbol{r})$ and that the ground-state electron density is the one which minimizes the energy functional $E[\varrho(\boldsymbol{r})]$. The application of the variational principle leads to the self-consistent solution of a set of N one-particle Schrödinger equations

$$-\nabla^2 + V_{\text{eff}}[\varrho(\boldsymbol{r}), \boldsymbol{r}]\,\Psi_i(\boldsymbol{r}) = \varepsilon_i \Psi_i(\boldsymbol{r}) \qquad (5.1)$$

from which the wave function Ψ_i and energy eigenvalues ε_i are generated. The effective one-electron potential V_{eff} can be expressed in terms of the external potential, the average electrostatic potential and the exchange and correlation potential $V_{\text{xc}}[\varrho(\boldsymbol{r})]$ which is the functional derivative of the energy functional $E_{\text{xc}}[\varrho(\boldsymbol{r})]$. In the LDF, the exchange correlation energy E_{xc} for the interacting inhomogeneous electron gas is approximated by:

$$E_{\text{xc}}[\varrho(\boldsymbol{r})] \simeq \int \varrho(\boldsymbol{r})\varepsilon_{\text{xc}}[\varrho(\boldsymbol{r})]d\boldsymbol{r}\,, \qquad (5.2)$$

where $\varepsilon_{\text{xc}}[\varrho(\boldsymbol{r})]$ is the exchange and correlation energy per particle for an interacting but homogeneous electron gas having the local density $\varrho(\boldsymbol{r})$. Various forms of the exchange and correlation potential can be used.

Within the LDF, a self-consistent calculation can be set up (i) by constructing an initial guess of the charge density, for example for a solid by an overlap of atomic charge densities corresponding to a given configuration or the renormalized atom model; (ii) calculating the effective crystal potential $V_{\text{eff}}[\varrho(\boldsymbol{r}), \boldsymbol{r}]$; (iii) solving the Schrödinger equation (5.1); (iv) calculating the new charge density by summing the one-electron densities over the lowest occupied states and returning to (i) with an admixture of the old and the new charge densities until a satisfactory convergence criterion is reached. The total energy of the system is then given by:

$$E = \sum_{n\,\text{occ}} \varepsilon_n - \tfrac{1}{2}\int d^3r \int d^3r' \frac{\varrho(\boldsymbol{r})\varrho(\boldsymbol{r}')}{|\boldsymbol{r}-\boldsymbol{r}'|} + \int d^3r\varrho(\boldsymbol{r})\,\{\varepsilon_{\text{xc}}[\varrho(\boldsymbol{r})] - \mu_{\text{xc}}[\varrho(\boldsymbol{r})]\}\,, \qquad (5.3)$$

where the second term corrects for the double counting of the electron Coulomb self-interaction and the third term accounts for the exchange correlation energy within the LDA approximation.

a) Stoichiometric Hydrides

Energy Bands. The methods which have been mostly used for studying the electronic structure of stoichiometric hydrides are the augmented plane wave (APW) and Korringa-Kohn-Rostoker (KKR) methods [5.17] in which the wavefunctions are expanded in a set of energy-dependent partial waves. The crystal potential is often assumed to have a muffin-tin (MT) shape: it has a spherical symmetry inside touching spheres surrounding the atomic centres and is constant in the remaining interstitial region. This approximation is known to work rather well for closed – packed structure such as the fcc structure often encountered in metal hydrides. For more open structures one can also introduce the spherical approximation around vacant interstitial sites – as done sometimes for the fluorite structure dihydrides – or use the so called warping corrections to account for the departure of the potential in the interstitial region from a constant value. The warping corrections can become rather important for intermetallic hydrides such as e.g. $FeTiH_x$ and Mg_2NiH_4 for which the non-overlapping MT spheres occupy less than 50% of the unit cell volume.

Other band structure techniques such as the LCAO method [5.17] which is based on an expansion of the wavefunctions on a set of energy-independent fixed basis functions and on the use of general crystal potentials have not been very much applied to the study of metal hydrides. The LCAO method has the advantage – unlike the APW and KKR schemes – of leading to energy-independent Hamiltonian matrix elements. However, it suffers from the problem of necessarily limited basis sets and from the difficulty in the evaluation of multicentre integrals.

Linear methods of band theory such as the linear augmented plane wave (LAPW) or the linear muffin-tin orbital method (LMTO) [5.18] on the other hand, combine the advantage of the use of energy-independent fixed basis functions with those of the partial waves. They have been shown to be very successful in the treatment of compounds having complex structure; their use for the study of metal hydrides is just starting [5.19].

Density of States (DOS). Improvements on the accuracy of the DOS and Fermi energy calculations have been made since earlier work on metal hydrides [5.20]; the histogram method with coarse mesh is no longer used. The energy eigenvalues calculated ab initio in a discrete mesh of $\boldsymbol{k}$ points in the Brillouin Zone (BZ) [typically at 89 points in the $(1/48)^{\text{th}}$ wedge of the irreducible fcc BZ] are then interpolated using symmetrized plane-wave expansions or symmetrized Fourier series expansions. The BZ integration can then be performed by analytical methods. It is widely admitted that one of the most satisfactory schemes is the so-called linear-energy tetrahedron method [5.21]. Accuracies typically of the order of a few tenths of an meV are obtained using these techniques. Other analytical procedures such as the so-called Quad scheme have also been used [5.22]; this usually differs by a few percent from the tetrahedron method.

The energy band interpolation can also be performed by the LCAO method. The Hamiltonian matrix elements are expressed in terms of the tight-binding energy integrals of Slater and Koster (SK) [5.23]. The numerical accuracy achieved is often inferior to that of the above mentioned techniques; it depends in particular upon the number of basis orbitals included in the calculation, on the range of the interactions included (nearest neighbours or next nearest neighbours) and on the number of ab initio $\boldsymbol{k}$ points used in the fitting procedure. However, the interest of the SK scheme resides in its physical meaning; the energy integrals obtained as parameters are extremely useful for building tight-binding Hamiltonians which allow the treatment of a variety of electronic structure problems [5.24].

Decomposed DOS and Charge Transfer Analysis. Decompositions of the DOS into site and angular momentum components are now available for most of the metal hydrides; they shed light on the bonding characteristics and are extremely useful in interpretating many experimental data such as the x-ray emission spectra for example. The KKR and APW wavefunctions are from their definition perfectly suited for an angular momentum analysis inside the MT spheres. This decomposition depends, however, on the a priori partition of the real space into non-overlapping atomic spheres and further assumptions have to be made for the interstitial charge. The weights of the angular momentum decomposed DOS can for example be obtained by continuing the radial integrations beyond the MT to a Wigner-Seitz sphere. DOS analysis in terms of LCAOs on the other hand suffer from the ill-defined spatial extension of the atomic orbitals; if these functions are quite extended a sizeable portion of the charge is sometimes located in the neighbouring MT sphere. The peculiarities of the different schemes should be kept in mind when numerical values of sensitive qualities such as charge transfer are discussed as done in the following sections. Charge transfer definitions are still a matter of debate; it is however our opinion that in the discussion of these effects the most meaningful quantity is the charge associated with spatially localized states such as the transition metal *d*-electron charge.

Total Energy Calculations. Besides the large amount of theoretical work on the cohesive energies of transition metals using simplified schemes [5.25–28], total energy calculations obtained from ab initio band structure studies by means of (5.3) or equivalent expressions, are now available for a variety of transition metals (TM) [5.29] and transition metal compounds [5.30]. However metal hydrides have not been systematically investigated in this respect. *Gelatt* et al. [5.31] first proposed a theoretical explanation of the trends in the heats of formation of binary TM hydrides. More recently total energy calculations from first principles have been performed for binary hydrides of the end of the series: Co to Cu and Ru to Ag by *Williams* et al. [5.32], *Methfessel* and *Kübler* [5.33] using the augmented spherical wave (ASW) method, and for monohydrides of Nb by *Ho* et al. [5.34] within the pseudopotential method. Detailed work on binary hydrides has been recently reported by *Switendick* [5.35] who calculated total energies as a function of lattice constant for PdH, NiH, TiH_2, and PdH_2 using self-consistent APW band structure calculations. No such calculations are however yet available for hydrides of intermetallic compounds.

The calculated heat of formation ΔH of say – a binary metal hydride is defined from the reaction of the metal M with gaseous hydrogen

$$M+\tfrac{1}{2}xH_2 \rightarrow MH_x + \Delta H_{MH_x}\,. \tag{5.4}$$

Thus ΔH_{MH_x} is obtained by substrating from the total energy of the metal hydride E_{MH_x} the total energy of the corresponding pure metal E_M and the total energy per hydrogen atom of an H_2 molecule E_{H_2}

$$\Delta H_{MH_x} = E_{MH_x} - E_M - \tfrac{1}{2}xE_{H_2}\,. \tag{5.5}$$

We wish to point out the difficulty of obtaining accurate theoretical predictions since the measured values of ΔH_{MH_x} for binary TM hydrides are in absolute value smaller than 1 eV/H atom while the first two terms in (5.5) are several orders of magnitude larger.

In order to obtain accurate absolute values of ΔH_{MH_x} even the theoretical method involved in the calculation of E_{H_2} becomes important. For reasons of consistency, the same type of theoretical treatment should be used to calculate each of the three terms of (5.5). The problems encountered here are similar to those found in the computation of cohesive energies of solids where one subtracts the total energy of the atoms from that of the corresponding solid. Large absolute errors are known to be made for example in the atomic energies with the use of the LDF. However, due to cancellation effects, the accuracy of the computed cohesive energies usually turns out to largely exceed the absolute accuracies for the atomic or solid calculations. On the basis of the success obtained in the calculation of cohesive energies for TM, one can expect that the use of (5.5) in conjunction with band structure studies will correctly reproduce trends and provide a microscopic understanding of the heats of formation of hydrides.

Gelatt et al. [5.31] proposed that the main contributions to the heat of formation of a metal monohydride are essentially: (i) The metal-hydrogen bonding contribution which results in the lowering of metal states having an *s* symmetry at the H sites. This term is approximated by the energy lowering of the lowest band of the metal upon formation of the hydride; this band accommodates two electrons. (ii) The increase in the binding of the metal *d*-bands due to the attractive potential of the protons. This term is taken to be the change in the average energy of the *d*-states located above the lowest metal band; these states are assumed to accommodate $n-2$ electrons where n is the number of valence electrons of the pure metal. (iii) The shift in the Fermi energy of the metal due to the presence of an addition electron in the hydride. It is assumed here that the protonic rigid-band model is correct since this term is approximated by the Fermi energy of the metal. (iv) Coulomb repulsions due to the increased charge density at the H sites lead to a decrease in the heat of formation. These different contributions have been calculated by *Gelatt* et al. [5.31] using the renormalized atom theory. The trends in the variation of the experimental curves are reproduced and ascribed essentially to the variation of the first and third contribution. The analysis of the total energy calculations of *Williams* et al. [5.32]have essentially confirmed that the general upward trend in the enthalpy curve as Z increases in a TM series is due to a decrease of the M–H bonding contribution while it is the variation of the metal chemical potential which explains the dip in the ΔH curve observed for Ni and Pd.

A very different simplified model has been proposed by *Nørskov* et al. [5.36] to evaluate the heat of solution of H in bulk metals as well as chemisorption of H on metal surfaces, trapping of H at defects etc. by means of the effective medium theory. One calculates the binding energy of an H atom immersed in an homogeneous electron gas which has a density equal to the average value of the host metal density around the H interstitial site. Interactions with the metal core electrons, hybridization between H *s*-electrons and transition metal *d*-electrons and a correction term for the inhomogeneity of a real metal are also introduced. Using this model, the general trends in the heat of solution of H in 3*d*, 4*d* and 5*d* TM can be explained mostly by the variation of the first term. Besides the nonnegligible role of the hybridization contribution, the large host metal electron densities in the middle of the TM series – also associated with the maximum in cohesive energies – appear to be responsible for the observed maximum in the heat of solution of H. Another interesting feature of the model of *Nørskov* et al. [5.36] is the proposed explanation of the lattice expansion observed in transition metals around the H sites and the lattice contraction in alkali metals, in terms of the existence of an optimum host electron density which leads to a minimum in the binding energy curve of a H atom in an electron gas. The idea of the existence of an optimum host electron density provides also a ground for the empirical criterion of *Westlake* [5.37] of a hydrogen minimum hole size of 0.38 Å.

Daw et al. [5.41] showed that the problem of decohesion and hydrogen embrittlement can be treated by calculating the embedding energy of a H atom in a transition metal in an effective medium theory.

The need for a rational understanding of the observed trends and the possibility of making predictions about metal hydride stabilities is particularly acute for hydrides of intermetallic compounds of technological interest. In view of the scarcity of theoretical work for these materials of complex crystal structures, numerous empirical or semi-empirical models have been proposed. *Griessen* and *Driessen* [5.38] have recently further simplified and proposed an empirical expression for the heat of formation of binary hydrides ΔH which depends linearly upon the difference $\Delta E = E_F - E_S$ between the Fermi energy E_F and the centre of the lowest conduction band of the host metal E_S (see Chap. 6). This empirical relation has then been extended to ternary metal hydrides by using results on TM alloys obtained from the tight-binding coherent potential approximation (CPA) model of *Cyrot* and *Cyrot-Lackmann* [5.39]. Another powerful scheme for predicting the properties of ternary hydrides has been proposed by *Miedema* et al. [5.40] following his work on transition metal alloys. However it has been shown [5.30] that the microscopic picture in which Miedema's work is based overemphasizes the properties of the constituents rather than the bonding properties of the compounds.

b) Nonstoichiometric Hydrides

Several approaches have been used for studying the electronic structure of nonstoichiometric hydrides. The best one to date is based on the coherent potential approximation (CPA) within the KKR method [5.42]. We want to mention however that in older work using the virtual crystal approximation (VCA) the crystal potential of the disordered binary system is simply approximated by the crystal potential of an effective ordered system by taking a concentration weighted average of the atomic potentials for a two component system. This approach is obviously too crude since the VCA potential is real and periodic and thus leads to real eigenvalues and infinite lifetimes of the associated eigenstates. In the average t-matrix approximation (ATA) of Korringa, single-site potentials are complex leading to complex eigenvalues and finite lifetimes. In the ATA it is the scattering t-matrix at a given site which is taken as the concentration weighted average of the single scattering t-matrices [5.43]. The CPA was first introduced by *Soven* [5.44] and has been successfully applied to the study of various alloy systems. In this approximation one assumes an effective ordered system by locating on every site some effective potential which is determined by requiring that the replacement at a given site of the coherent potential by the true site potential produces no further scattering on the average. From the averaged Green's function of the system $G(\boldsymbol{r}, \boldsymbol{r}', E)$, quantities such as e.g. the densities of states and the charge densities can be obtained. CPA calculations using the tight-binding method have been performed for binary TM hydrides and alloys [5.45, 46]. Recently, the CPA in its KKR form has been used to study alloys such as Pd–Ag–H for which the disorder in both the metal and the hydrogen sublattices have been taken into account [5.42]. In that case the CPA medium consists of two sublattices of effective sites.

c) H Impurity in Metals

A substantial number of theoretical studies of a hydrogen impurity in simple metals have appeared since the mid 70 s; most of them are based on the jellium model. These self-consistent calculations of the screening of the proton in an electron gas using the LDF approximation have shown the importance of nonlinear effects as emphasized in the pioneering work of *Friedel* [5.2]. The effect of the lattice ions is introduced using approximate treatments such as local pseudopotential theory [5.17] or the spherical solid model [5.47]. Other theoretical treatments of H in simple metals using supercell calculations, KKR scattering theory have also been used. Studies of heats of solution of H, determinations of hydrogen site occupancies, activation energies for overcoming potential barriers as involved in classical diffusion mechanisms, and calculations of local mode vibration frequencies are available; attempts to study lattice relaxation effects around the H impurity in simple metals have also been made. The corresponding theoretical work on simple metals has been recently reviewed [5.48]. The treatment of H as an impurity in TM is much more complex than in simple metals due to the presence of the metal *d*-electrons. The amount of theoretical work in this field has started to increase: Ab initio calculations such as the KKR Green's function method [5.49], the large supercell [5.20, 50] or cluster models [5.51, 52] have been used as well as parametrized tight-binding Hamiltonians [5.53].

d) Calculation of Other Observables

From these theoretical studies on metal hydrides and H in metals, accurate information can be obtained concerning properties related to the Fermi energy: Fermi surface geometries, densities of states at E_F which can be related, in the independent particle model, to the Pauli susceptibility χ_P and the electronic specific heat coefficient γ. Details concerning the nature of the bonding, the charge transfer and to a lesser extent total energies and predictions of vibrational frequencies are becoming increasingly available.

Using quantities obtained from ab initio band structure work such as the single-site scattering phase shifts at the Fermi energy δ_l, the partial DOS of angular character l, n_l, evaluated at site κ, and the total DOS at E_F for one spin direction $N_\uparrow(E_F)$, one can obtain an estimate of the electron-phonon coupling constant λ and thus the superconducting transition temperature T_c. Following the work of *McMillan* [5.54] for transition metals, an approximate expression of λ has been proposed for compounds with a large mass ratio between the constituent atoms

$$\lambda \sim \lambda_{\mathrm{Metal}} + \lambda_{\mathrm{H}} = \frac{\eta_{\mathrm{Metal}}}{M_{\mathrm{Metal}}\langle\omega^2\rangle_{\mathrm{ac}}} + \frac{\eta_{\mathrm{H}}}{M_{\mathrm{H}}\langle\omega^2\rangle_{\mathrm{optic}}}, \tag{5.6}$$

where η is the "electronic contribution", M the atomic mass and $\langle\omega^2\rangle$ the second moment of the renormalized phonon frequencies defined by McMillan. Using the rigid ion approximation which ignores renormalization effects due to lattice vibrations, *Gaspari* and *Gyorffy* [5.55] have shown that:

$$\eta_\kappa \sim \frac{E_F}{N_\uparrow(E_F)\pi^2} \sum_l 2(l+1)\sin^2(\delta^\kappa_{l+1} - \delta^\kappa_l)\frac{n^\kappa_l(E_F)n^\kappa_{l+1}(E_F)}{n^{\kappa(1)}_l(E_F)n^{\kappa(1)}_{l+1}(E_F)}. \tag{5.7}$$

This expression will be used in the following sections to comment on the superconducting properties of metal hydrides.

5.2.2 Experimental Methods

In order to understand the electronic structure of metal hydrides, properties such as the total and partial (*s*, *p*, *d*, *f*) densities of states at E_F, the position, shape and width of the occupied and empty energy bands and their $\boldsymbol{k}$-dependence, the charge transfer and the transitions from metallic to nonmetallic behaviour and related energy gaps should be measured and compared to theory. Essentially all the experimental methods which are used to look at the electronic properties of metals and alloys can also be applied to metal hydrides. However, severe restrictions may be introduced by the shape of the samples. Hydrides of intermetallic compounds are easily available in the form of powder, in some cases as thin films or amorphous ribbons, but so far, never in the form of polycrystalline or single crystal bulk samples. Furthermore the rather low thermodynamic stability of metal hydrides clearly limits investigations at high temperature and at low pressure (ultrahigh vacuum, UHV) according to the pressure–composition isotherms.

The Pauli contribution to the magnetic susceptibility, and the electronic specific heat coefficient γ are proportional to the density of states at E_F for a free electron gas. In real metals electron-electron and electron-phonon interactions cause large deviations and render quantitative evaluations uncertain. Magnetic precipitates in the bulk or at the surface of intermetallic compounds can cause large errors in the evaluation of susceptibility data [5.56]. Whereas electrical resistivity measurements still yield very valuable results for binary hydrides, e.g. rare earth hydrides, only very few measurements have been performed on hydrides of intermetallic compounds because of the difficulties in making contacts to the samples. Contactless measurements of Q values or capacitances in resonance circuits allow the study of the variation of the resistivity with temperature and thus to the identification of metallic or semiconducting behaviour.

Nuclear magnetic resonance (NMR), a well-known method to study H diffusion, also provides information on the electronic structure [5.57]. The hyperfine fields at the site of a nucleus are affected, among others things, by conduction electrons which cause a Knight shift; this can be measured by NMR, Mössbauer spectroscopy [5.58] and muon spin rotation spectroscopy (μSR) [5.59, 60]. With NMR, the Knight shift at the nucleus of the host metal and also that of the proton can be measured. The time dependence of the hyperfine field is related to a relaxation mechanism and contributes a conduction electron component ($1/\tau_1$) to the spin-lattice relaxation rate. ($\tau_1 \times T$), where T is the absolute temperature, is determined by the *s–p–d* conduction band density of states at E_F. However, relations between electronic structure, relaxation rate, and Knight shift are not simple and often electronic specific heat and magnetic susceptibility are also needed for a full interpretation of the NMR data.

The "classical" methods mentioned so far have the advantage of requiring only a reasonably simple experimental set up which nonetheless usually allows measurements over a wide range of temperature and hydrogen pressure. However, these methods probe the electronic structure around E_F only and do not give direct information on hydrogen induced bonding and antibonding bands etc.

The de Haas–van Alphen effect, i.e. the quantum oscillatory part of the magnetization of a metal in an external magnetic field, probes the effective electron mass and the curvature of the Fermi surface and is very well suited to measure small changes in the electronic structure of dilute H-metal systems [5.61, 62] Compton scattering of high energy x-rays, i.e. the inelastic scattering of a photon with electrons, probes the electron momentum distribution of the host metal and its variation due to hydride formation. H_2 partial pressures up to the mbar level are possible [5.63]. Positron annihilation has also been used to study the Fermi surface of TM hydrides [5.64] as well as their momentum space properties [5.65].

Among the most powerful methods to test the electronic structure of elemental metals, intermetallic compounds, glassy and crystalline alloys and their hydrides are the spectroscopic methods of electron and x-ray photo-emission [5.66, 67] inverse photo-emission [5.68] and also x-ray emission and absorption [5.10, 69]. In photoelectron spectroscopy (PES) monoenergetic radiation from an UV lamp (UPS, $h\nu = 10$–100 eV), from a tunable synchrotron radiation source, or from an x-ray source (XPS), e.g. Mg K_α ($h\nu = 1253$ eV) or AlK_α ($h\nu = 1486$ eV), is used to excite and emit electrons from a sample by the photoelectric effect. An electron energy analyzer yields count-rate versus binding energy curves which are called energy distribution curves or photo-electron spectra. With UPS, the valence band and possibly high lying core levels are probed with an energy resolution $\Delta E \approx 0.1$ eV whereas with XPS, valence band and core levels down to 1.5 keV are accessible ($\Delta E \approx 0.5$–1.5 eV). The mean free path of the photoelectrons in the solid determines the probing depth. It varies between ≈ 0.5 nm for kinetic energies $E_K \approx 50$ eV to ≈ 3 nm for $E_K \approx 1.5$ keV. The small probing depth requires ultrahigh vacuum instruments to prevent contamination of the surface. Though hydrogen partial pressure up to 10^{-5} mbar is possible, pressure is the most serious limitation of PES in studying metal hydrides.

A valence band photoelectron spectrum resembles a one-electron density of states curve, but they are not in fact identical, for many reasons: The spectrum represents an excited state (one electron is missing) and many-body effects provide screening of the photohole, the emission of electrons of different quantum numbers and of different elements is enhanced or diminished by cross-section effects and the limited instrumental resolution broadens the spectra. Screening and correlation lead to particularly strong and interesting effects in the PES spectra of the rare earths and actinides.

Within some approximations, photoelectron spectra yield directly the position and width of the occupied bands, the variation of the emission at E_F accounts for changes in the DOS at E_F and the core-level shifts indicate the charge transfer and/or changes in hybridization. The use of photons of various energies allows, to some extent, the evaluation of partial DOS owing to the energy dependence of the cross-section. Soft x-ray emission spectroscopy (SXES) is even better at distinguishing the partial DOS by selecting various valence band-core level transitions. To look at the bulk electronic properties of hydrides XPS has significant advantages over UPS: the larger probing depth and the analysis of the core levels of all components make possible controls of deviations of hydrogen concentration and of alloy composition at the surface.

The major methods for studying unoccupied states above E_F are inverse photoemission (also called bremsstrahlung-isochromat spectroscopy, BIS) [5.68, 70] and x-ray absorption (XAS) particularly in the form of x-ray absorption near-edge structure (XANES) [5.71]. In XANES monoenergetic photons of variable energy are absorbed and excite bound electrons below E_F to empty states above E_F. The structure near the absorption edge reflects the empty, local, partial DOS, but unfortunately only with rather low energy resolution.

If single crystal samples are available, the dispersion $E(\boldsymbol{k})$ of the occupied and empty energy bands can be measured by angle resolved photoemission and inverse photoemission, respectively, (ARUPS, ARBIS).

In optical spectroscopies the complex dielectric function is probed by measurements of absorption or reflectivity. The dielectric function is directly related to the electronic band structure. Features in the dielectric function can be related to electric dipole interband transitions [5.72].

5.3 Results for Binary Hydrides

5.3.1 Hydrides of Pd and Ni

a) Energy Bands and Densities of States

Stoichiometric palladium hydride has been the most thoroughly investigated of all the binary hydrides. All the calculations are in qualitative agreement with the results first obtained by *Switendick* [5.20] so we shall summarize only briefly the main results. An improvement of the densities of states curves has been achieved over the earlier calculations due to a better sampling of the BZ and due to the use of more accurate integration techniques. However, a dispersion of the theoretical results still exists concerning the position and width of the metal-hydrogen bonding bands, the metal *d*-bands, the value of the DOS at E_F, and the values of charge transfer. Some examples are given in Table 5.1. An effort towards a self-consistent treatment has been made but here too differences are observed between different self-consistent calculations due to the different treatment of the exchange term of the crystal potential and due to numerical accuracy problems.

As shown in Fig. 5.1, the main differences between the calculated DOS of pure Pd metal and that of PdH can be summarized as follows [5.20, 35, 73–75].

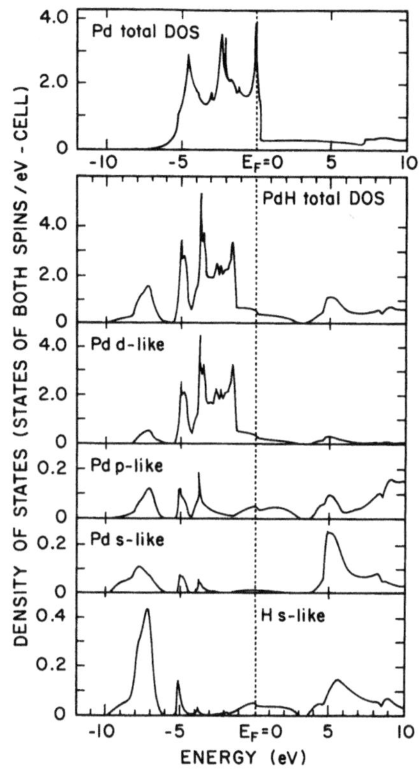

Fig. 5.1. Slater-Koster density of states for Pd (top, from [5.94]) and total and partial wave analysis of the DOS of PdH [5.35]

(i) The appearance below the metal *d*-bands of metal-hydrogen bonding states which, as discussed later, have been observed by several spectroscopic techniques. These states are mostly already filled in pure Pd; in the hydride they are lowered in energy by the H potential and hybridize strongly with H *s*-states. However, metal states of *p* symmetry which are empty in pure Pd, also participate in this metal-hydrogen bonding band. They bring about 0.24 electrons below E_F. (ii) The filling of the metal *d*-bands. The Fermi energy of PdH is found to lie in the metal *sp*-band in a region of low DOS. This explains the drastic reduction of the electronic specific heat and Pauli susceptibility upon hydrogenation. (iii) The lowering of empty metal states below E_F results in a Fermi level position in the hydride which is necessarily closer to the *d*-bands than predicted by the protonic rigid-band model. Assuming that the metal *d*-bands are not affected around E_F, only $1-0.24=0.76$ electrons should be accommodated above the Fermi level of the metal. It is interesting to notice that experimentally the hydrogen desorbs easily for concentrations larger than $x=0.7$; this seems to correspond to the complete filling of the 0.36 holes in the *d*-bands, if one takes into account the 0.24 additional electrons at low energy. For $x \geqq 0.7$ the Fermi level of the hydride lies in the metal *sp*-band which has a low DOS and it is energetically not favourable to shift the Fermi level towards

Table 5.1. Valence band and core level daty of binary hydrides

Sample composition	Method[a]	Conduction band width [eV][b]	H-bonding band [eV][c]	Core level binding energy and (core level shift) [eV][d]	Remarks	Ref.
Pd	T SK-APW	5.7				[5.94]
Pd	E XPS	4.8		$3d_{5/2}$ 335.3		[5.77–79]
Pd	E UPS	3.4				[5.78]
PdH	T SCAPW	5.5	7.4			[5.35]
PdH	T APW	5.0	7.4			[5.74]
$PdH_{0.6-0.8}$	E XPS	4.2	7.5–8.0	$3d_{5/2}$ 335.5 (0.2±0.1)	Less asymmetric CL CL satellite disappears	[5.77–79]
$PdH_{0.6-0.8}$	E UPS	2.8	7.5			[5.78]
$PdH_{0.6}$	E SXES		7.4			[5.80]
NiH	T APW	4.2	6.6–11.9			[5.83]
NiH_x	E ARUPS	2.2	8–14		Ion implanted	[5.85]
Ti	E XPS			3p 32.7 $2p_{3/2}$ 454.0		[5.180, 190]
TiH_2	T SCAPW	2.9	4,5, 5.2			[5.99]
$TiH_{1.5}$	E UPS	1.7	5.1 (7.3)	3p 33.3 (0.7)		[5.105]
$TiH_{1.8}$	E SXES		7			[5.8]
$TiH_{1.9}$	E XPS			$2p_{3/2}$ 454.5 (0.5)		[5.190]
$TiH_{0.9-1.8}$	E XPS		7	(0.2–0.6)		[5.107]
Zr	E XPS			$4p_{3/2}$ 27.1 $3d_{5/2}$ 178.8		[5.180, 190]
ZrH	T LAPW	3.4	5.5, 6.5, 7.3			[5.115]
ZrH_2	T APW	2.7	5.7, 6.7			[5.96]
	T SCAPW	3.6	5.0, 6.0			[5.99]
$ZrH_{1.6}$	E UPS	2.0	5.3 (7.2)	$4p_{3/2}$ 28.1 (1.0)		[5.105]
$ZrH_{1.7}$	E XPS	2.6	7	$4p_{3/2}$ 28.0 (1.0) $3d_{5/2}$ 179.1 (0.7)		[5.106]
$ZrH_{1.9}$	E UPS	2.1	5.3 (7.8)			[5.105]
	E XPS			$3d_{5/2}$ 179.6 (0.7)		[5.190]

$ZrH_{0.5-1.9}$	E SXES		7				[5.108]
ZrH_x	E XPS		3.4	$4p_{3/2}$	28.9 (1.4)	Ion implanted	[5.109]
				$3d_{5/2}$	179.6 (0.6)		
Hf	E XPS			$4f_{7/2}$	14.2		[5.180]
				$5p_{3/2}$	29.9		
HfH_2	T SCAPW	4.2	5.7, 6.7				[5.99]
$HfH_{1.6}$	E UPS	2.2	5.5 (7.8)	$4f_{7/2}$	14.7 (0.5)		[5.105]
				$5p_{3/2}$	30.3 (0.4)		
Nb	E XPS			$4p_{3/2}$	30.8		[5.180]
				$3d_{5/2}$	202.3		
NbH	T PsP	4.3	6.4, 8			β-phase	[5.34]
		4.4	7			γ-phase	
$NbH_{1.0}$	E UPS	2.7	5.5, 7.5	$4p_{3/2}$	30.7 (0)		[5.118]
NbH_2	T	4	7.0, 8.1				[5.96]
NbH_x	E XPS		4.6	$4p_{3/2}$	31.7 (0.7)	Ion implanted	[5.109]
				$3d_{5/2}$	203.2 (0.8)		
V	E XPS			3p	37.1		[5.180]
VH	T APW	5.4	8.2, 9.5				[5.4]
$VD_{0.7}$	E SXES		6, 8				[5.119]
$VH_{1.0}$	E UPS	1.8	5.5, 7.5	3p	37.1 (0)		[5.118]
VH_2	T SCAPW	4.1	5.8				[5.46]
Ta	E XPS			$5p_{3/2}$	32.7		[5.180]
$TaH_{0.8}$	E UPS	2.1	5.3, 7.7	$5p_{3/3}$	32.7 (0)		[5.118]
Sc	E UPS	1.6					[5.145]
$ScH_{0.2}$	E UPS	1.6	5.0	3p	28.9		[5.145]
ScH_2	T KKR	2.1	4.3				[5.151]
$ScH_{1.85}$	E UPS	1.5	5.7	3p	29.5 (0.6)		[5.137]
Y	E XPS	2.5		$3d_{5/2}$	155.7		[5.144]
				$3p_{3/2}$	299.3		
$YH_{0.15}$	E UPS	1.6	5.0				[5.145]
YH_2	T KKR	2.1	3.9, 4.5				[5.151]
$YH_{1.73}$	E UPS	1.9	5.5	$4p_{3/2}$	24.2		[5.137]

Table 5.1 (continued)

Sample composition	Method[a]	Conduction band width [eV][b]	H-bonding band [eV][c]	Core level binding energy and (core level shift) [eV][d]	Remarks	Ref.
$YH_{2.1}$	E XPS		5.6	$3d_{5/2}$ 156.2 (0.5) [0.5]		[5.144]
				$3p_{3/2}$ 300.1 (0.8)		
YH_3	E XPS		6.0	$3d_{5/2}$ 157.5 (1.8) [0.5]		[5.144]
				$3p_{3/2}$ 301.0 (1.7)		
La	E UPS	1.3		$5p_{3/2}$ 16.9		[5.191]
	E XPS	1.8		$5p_{3/2}$ 17.2		[5.191]
				$3d_{5/2}$ 835.8		
LaH_2	T APW	1.5	4.5, 5.7			[5.148]
LaH_2	T SCAPW	1.7	3.5, 4.6			[5.149]
$LaH_{1.98}$	E UPS	1.1	4.3 (5.7)	$5p_{3/2}$ 17.6 (0.7)		[5.138]
$LaH_{2.2}$	E XPS	2	4.5	$5p_{3/2}$ 17.6 (0.4)		[5.140, 70]
				$3d_{5/2}$ 837.1 (1.3) [≈1]		
$LaH_{2.48}$	E UPS		4.7 (6.7)	$5p_{3/2}$ 18.5 (1.6)		[5.138]
LaH_3	T APW	–	3.8, 4.7, 5.4			[5.148]
LaH_3	T SCAPW	–	3.9, 4.9, 5.5			[5.149]
$LaH_{2.89}$	E UPS	–	4.9 (7.0)	$5p_{3/2}$ 18.5 (1.6)		[5.138]
$LaH_{2.9}$	E XPS	–	5.8	$5p_{3/2}$ 18.8 (1.6)		[5.139, 140, 70]
				$3d_{5/2}$ 838.7 (2.9) [≈2]		
Ce	E UPS	2.5[e]				[5.192]
	E XPS	2.6[e]		$5p_{3/2}$ 17.5		[5.140]
CeH_2	T APW		6.4, 7.0			[5.152]
$CeH_{2.1}$	E UPS	1.1	4.8			[5.142, 168]
	E XPS	1	5.1	$5p_{3/2}$ 18.5 (1.0)		[5.140, 70]
				$3d_{5/2}$ 885.7 (1.9) [≈1]		
$CeH_{2.4}$	E UPS	1	5.0–5.4	$5p_{3/2}$ 18.3		[5.141, 142]
CeH_3	T APW	–	5.8, 6.8, 7.4			[5.152]
$CeH_{2.9}$	E UPS	–	5.4	$5p_{3/2}$ 18.8		[5.168]
$CeH_{2.9}$	E XPS	–	5.4	$5p_{3/2}$ 18.7 (1.2)		[5.140, 70]
				$3d_{5/2}$ 885.9 (2.1) [≈1.3]		

Pr	E XPS	2		$5p_{3/2}$ 18.4 $3d_{5/2}$ 932.0		[5.143]
$PrH_{2.1}$	E XPS	2	5–7[f]	$5p_{3/2}$ 19.0 (0.6) $3d_{5/2}$ 933.2 (1.2)		[5.143]
PrH_3	E XPS	–	5–7[f]	$5p_{3/2}$ 19.8 (1.4) $3d_{5/2}$ 934.2 (2.2)		[5.143]
ErH_2	T APW	1.8	5.2, 6.1			[5.147]
$YbH_{2.6}$	E XPS	f		$3d_{5/2}$ 1521.9 ($4f^{14}$) and 1530 ($4f^{13}$)	mixed value	[1.49, 5.84]
Lu	E XPS			$5p_{3/2}$ 26.6 $3d_{5/2}$ 1589.7		[5.84, 193]
$LuH_{1.77}$	E UPS	1.7	5.5	$5p_{3/2}$ 27.5		[5.137]
$LuH_{2.1}$	E XPS			$3d_{5/2}$ 1590.3 (0.7)		[5.84]
Ca	E UPS	1.2				[5.181]
	E XPS			3p 25.4		[5.180]
CaH_2	T APW	–	3.7			[5.181]
	E UPS	–	5.0	3p 26.2		[5.181]
Th	E XPS			$6p_{3/2}$ 16.6		[5.180]
ThH_2	E UPS	1.9	5.0 (6.7)	$6p_{3/2}$ 17.7		[5.188]
Th_4H_{15}	E UPS	≈0.6	4.8 (3, 7)	$6p_{3/2}$ 18.6		[5.188]
U	E XPS			$4f_{7/2}$ 377.4		[5.180]
UH_3	T APW		width 7.3			[5.185]
UD_3	E XPS		6	$4f_{7/2}$ 379.1 (2.1)		[5.186]

[a] Method: T theoretical; E experimental; APW augmented plane waves; SC self-consistent; SK Slater-Koster; L linearized; PsP Pseudopotential; KKR Korringa-Kohn-Rostoker; UPS UV-photoelectron spectroscopy; XPS x-ray photoelectron spectroscopy; AR angle resolved; SXES soft x-ray emission spectroscopy.

[b] Conduction band width: E: FWHM, not corrected for instrumental resolution; T, E: in the case of an overlap between conduction band and H-bonding band, the band width is estimated from the dip in the DOS

[c] H-bonding band: centre of band or pronounced peaks, rel. to E_F

[d] (core-level shift, uncorrected), [core level shift, corrected for screening effects], most probable values relative to host metal values

[e] Including 4f-emission

[f] Overlap with 4f-emission

higher energies. (iv) The metal *d*-bands are narrowed by the 5% lattice expansion effect; however, most of the *d*-states are not affected by the metal hydrogen interaction, with the notable exception of the states located at the bottom of the band which participate in the *M*–H bonding.

In recent calculations, [5.35, 73–75] analysis of the total DOS either into its angular momentum components *s, p, d* at the metal and at the H sites, as shown in Fig. 5.1, or analysis in terms of LCAO's are now available. Such details are very important in interpreting the x-ray emission and absorption data and shed some light on the hybridization of the states in the full energy range. The nature of the low-lying bonding states for example is made very clear by means of such analysis. It has been shown that when H occupies octahedral interstices of the fcc metal lattice as in PdH, the H *s*-states interact with the metal *d*-states of e_g symmetry.

The results of calculations of the DOS of nonstoichiometric Pd hydrides [5.46] are given in Fig. 5.2, E_F is found closer to the top of the Pd *d*-bands.

Eastman et al. [5.76] observed in early room-temperature photoemission studies of Pd films exposed to hydrogen a strong additional emission at 5.5 eV, which was interpreted as the hydrogen induced band, but their spectra show no decrease of the emission at E_F. Other groups failed to see even that hydrogen induced band due to the rapid desorption of hydrogen (see introduction of [5.77]). These contradictions were eliminated by the first low-temperature photoemission studies [5.77] of (a) properly hydrogenated Pd which showed the hydrogen induced band at 8 eV, a strong decrease of the emission at E_F, a decrease of the bandwidth of $\approx 10\%$ and a very weak core-level shift, in very good agreement with theory and of (b) oxygen dosed Pd which showed the 5.5 eV emission observed in the earlier work.

Within the last couple of years the occupied and empty parts of the valence band and the core levels of properly prepared PdH_x ($x = 0.6$ to 0.8) were studied using UPS [5.77, 78], XPS [5.77–79], EELS [5.79], SXES [5.80] and low energy BIS [5.78]. As compared to pure Pd the following features were observed (Fig. 5.3a, b). The UPS, XPS and SXES spectra reveal a weak and rather broad hydrogen induced band centred at 7.5 eV to 8 eV. E_F moves out of the Pd 4*d*-band into the flat *sp*-band which is clearly seen from the disappearance of the peak at E_F in the BIS spectra (empty 4*d*-states) and from the decrease of the emission at E_F by a factor of about 3 in the UPS and XPS spectra. In agreement with the filling of the 4*d*-band and the reduction of the DOS at E_F, the shake up satellites of the Pd 3*d* and 2*p* core levels disappear and the core levels become less asymmetric. The width of the 4*d*-band decreases by almost 1 eV. The shift of the core levels is very small (0.15 ± 0.10 eV) indicating a very small, transfer of charge, if any, in agreement with theory.

A slightly larger core-level shift (0.3 eV) was found in a recent photoemission study of Pd hydride which was prepared by ion implantation technique and thus contained higher hydrogen concentration [5.81]. Compton profiles of $PdH_{0.72}$ confirm the existence of a hydrogen induced bonding band below E_F [5.63].

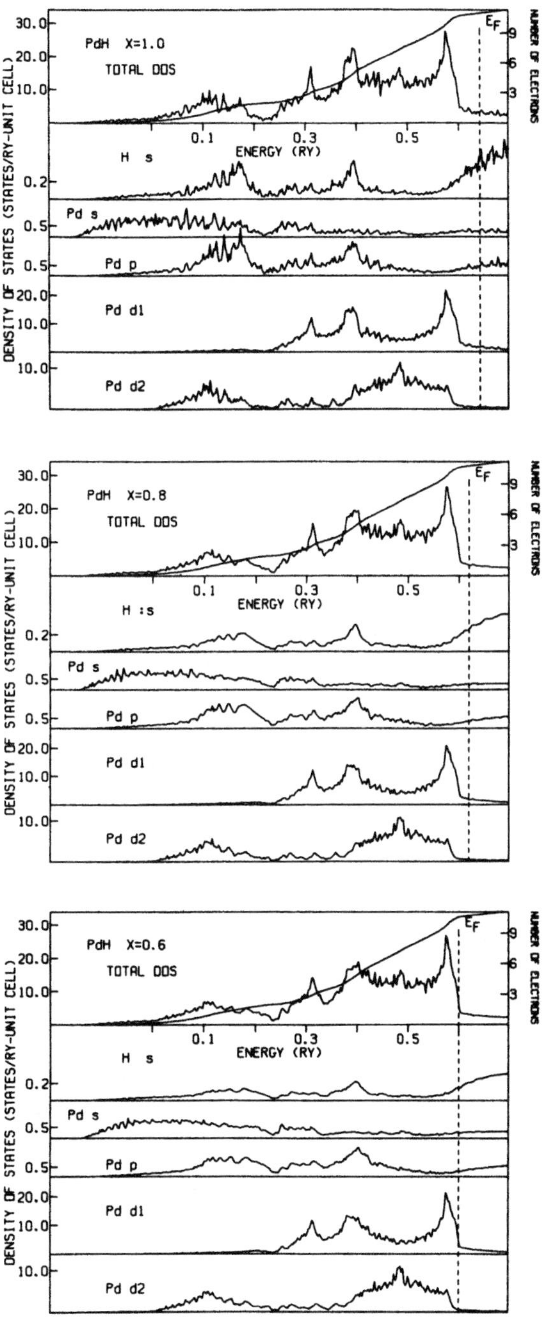

Fig. 5.2. Total and partial DOS of nonstoichiometric PdH_x [5.73] and number of electrons

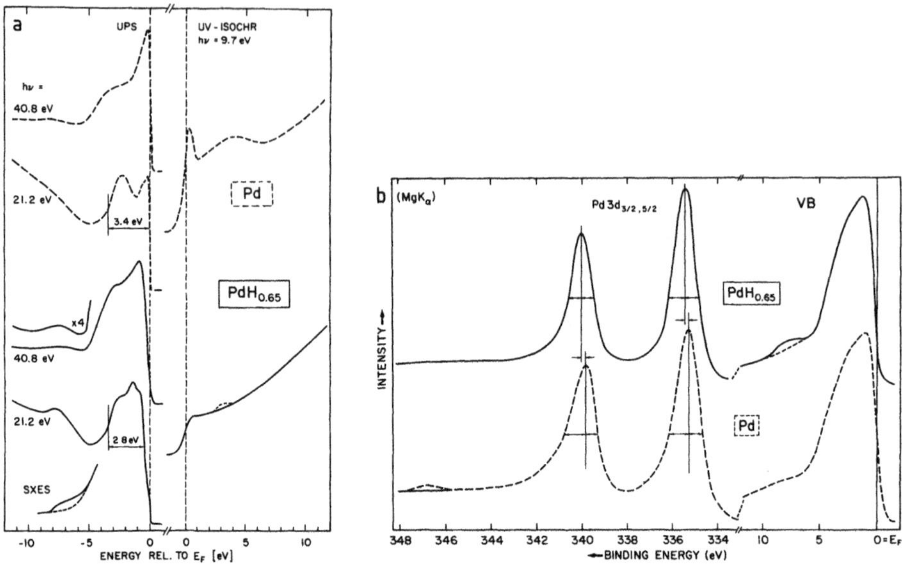

Fig. 5.3. (a) UPS and UV-isochromat (BIS) spectra of the occupied and empty part of the valence band of Pd and $PdH_{0.65}$ [5.78]. Upon hydride formation the Pd *d*-band becomes narrower and shifts away from E_F (UPS emission decreases, UV-isochromat peak at E_F disappears). Hydrogen-induced bonding band appears at 8 eV in UPS and also SXES [5.80] spectra. **(b)** XPS spectra of the valence band and of the Pd 3*d* core-levels of Pd and $PdH_{0.65}$. Upon hydride formation the width of the Pd *d*-band and the emission at E_F decrease, the hydrogen induced bonding band appears at 8 eV, the Pd 3*d* core-levels shift to 0.2 eV larger binding energy and become less asymmetric [5.78]

The only remaining controversy between theory and these spectroscopic results concerns the empty antibonding states which yield a prominent peak ≈ 5 eV above E_F in the calculated DOS, but which are not seen in the BIS spectra (see Fig. 5.3a).

H dissolves in Ni endothermally (α-phase) and nickel hydride (β-phase) is rather unstable at room temperature. High pressure hydrogenation, cathodic charging and ion implantation have been used successfully to prepare NiH_x in the composition range $0 \leq x \leq 2.0$. Low temperature ion implantation leads to a supersaturated α-phase at least up to $x = 0.85$. The magnetization decreases linearly with H concentration and reaches zero at $x = 0.8$, in agreement with the fact that the electrolytically charged β-phase is nonmagnetic ([5.82] and refs. therein.)

The main differences between the two isoelectronic compounds PdH and NiH originate mostly from the differences already existing in the metal matrices. The Fermi level of NiH [5.83] is found to be closer to the top of the *d*-bands than in PdH; thus the DOS at E_F is larger and the "*d*" character of the states is more pronounced due to the fact that the number of holes in the *d*-bands is larger in pure Ni than in Pd. This result has consequences for the

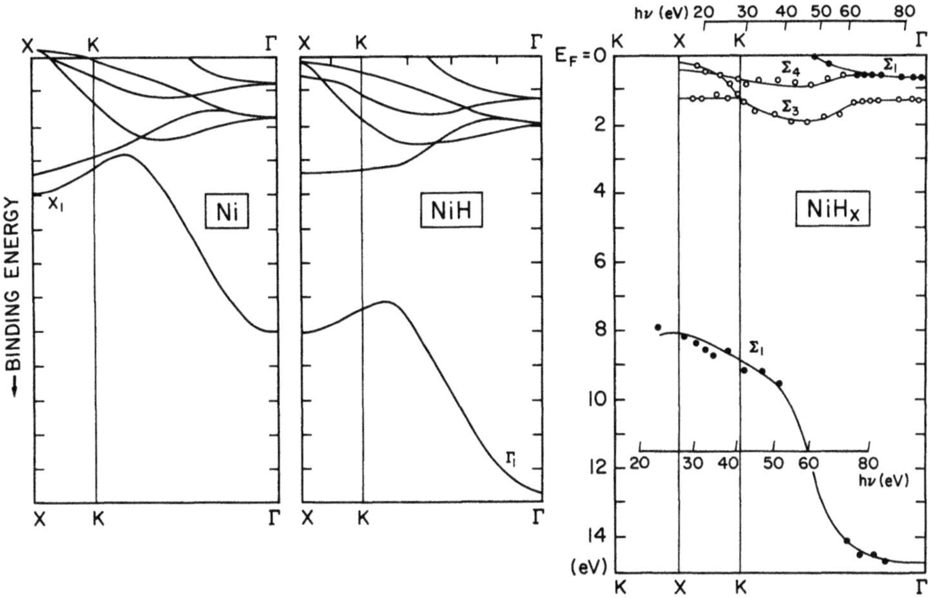

Fig. 5.4. Comparison of calculated energy bands of Ni and NiH, plotted along several high symmetry directions of the BZ [5.83], and energy bands of ion implanted NiH_x measured by angle resolved photoemission [5.85]. The metal–hydrogen interaction leads to a strong lowering of the Σ_1 branch

Fermi Surface (FS) anisotropies and plays a role in determining the strength of the electron-phonon interaction. In the hydride NiH, the low-lying metal-H bonding states have a larger tail of metal *s*-character than in PdH; this is again reminiscent of the differences in the host band structure. Photoelectron spectroscopic investigations of electrolytically charged Ni hydride have failed so far due to the fast desorption of H across a clean Ni surface even at ≈ 100 K and a hydrogen partial pressure of 10^{-5} [5.84]. However, low temperature ion implantation in situ allowed the study of the valence band of a hydrogenated Ni single crystal by angle resolved photoelectron spectroscopy along the ΓKX direction of the Brillouin zone [5.85] (Fig. 5.4). In rather good agreement with band structure calculations the band Σ_1 is strongly lowered in energy, e.g. from 3 eV in Ni to 8 eV in Ni hydride at the X_1 point. Further information was obtained from the 6 eV and 13 eV satellites of the valence band of Ni. Their intensity depends strongly on the number of 3*d*-holes. In Ni hydride the 13 eV satellite has disappeared completely and the intensity of the 6 eV satellite has decreased by a factor of two, in agreement with a filling of 3*d*-states. Quantitative conclusions cannot be drawn as the hydrogen concentration of the sample was not known.

The empty part of the valence band of $NiH_{0.85}$ was studied using XANES. The structure found in the absorption edge reflects quite nicely the structure for the *p*-projected calculated DOS [5.71] after an expansion of the theoretical energy scale by 5%, the origin of which is not yet clear. Similar energy rescaling has been tentatively justified in terms of the influence of the core hole (see [5.71]).

b) Fermi Surfaces (FS)

Since the Fermi level of stoichiometric PdH and NiH lie in the metal *sp*-band, the Fermi surfaces of the hydrides are expected to be qualitatively similar to those of the noble metals; they are multiply connected surfaces having the shape of a warped electron sphere centred at the fcc BZ centre with necks along the $L(111)$ direction [5.74, 83]. Since the Fermi level of the hydrides is closer to the top of the *d*-bands than it is for noble metals, the FS anisotropies and the *d* character of the states at E_F are expected to be larger. FS cross-sections have been calculated for PdH and NiH. Fermi surface nesting features have been invoked to explain the observed concentration-dependent short-range order of PdD_x for $0.7<x<0.78$ in the vicinity of the (1, 1/2, 0) point [5.86]. Although the exact dimensions of the nesting vector are questionable since they were obtained by applying the rigid-band model to the bands of stoichiometric PdH, this suggestion deserves further examination since it is based on more general grounds and has been applied successfully to other materials.

Experimental FS studies by de Haas–van Alphen measurements are only available for very dilute hydrides CuH_x [5.87], PdH_x and PdD_x and PdT_x [5.88]. The hydrogen concentration dependence of the Dingle temperature dT_D/dx of some high-symmetry orbits of dilute CuH_x shows that the larger values of dT_D/dx are observed for the neck orbits in the $L(111)$ directions. Indeed, this result agrees with theoretical calculations since the corresponding 4*p*-branch of metal states at the L_2 $\boldsymbol{k}$ point has an *s*-symmetry at the interstitial octahedral site and is thus expected to be strongly affected by the H potential.

The modifications of some extremal FS cross-sections of Pd in the presence of dilute amounts of H, D, and T [5.88] reveal large isotope effects for the small hole ellipsoids around the X and L points of the fcc BZ while the large FS sheet around Γ is isotope independent. This experimental result raises the question of the possible role of the zero-point motion of H (and D) on the band structure [5.62]. However, from the available FS measurements, *Bakker* et al. [5.88] estimate that in the dilute limit, the DOS at E_F should be almost isotope independent.

c) Study of Other Observables

Total Energies. Calculations of equilibrium lattice constants and heats of hydride formation as defined in (5.5) are available for PdH and NiH in the work of *Williams* et al. [5.32] and *Switendick* [5.35]. The calculated lattice constants are in good agreement to within 1%–1.5% of the experimental values. The calculated heats of formation of PdH are -50.16 kJ/(mole H) [5.32] and -40.13 kJ/(mole H) [5.35] compared to the experimental result -20.06 kJ/(mole H). For NiH the calculated value is -33.44 kJ/(mole H) compared to the positive experimental value +5.02 kJ/(mole H). Accuracy in the absolute values of ΔE_{MH} is difficult to achieve. As seen in (5.5) a result of a few tenths of an eV is expected as the difference between two large values E_{MH} and E_M each being of the order of 100 000 eV [5.35]. This is thus at the limit of the accuracy of the calculations. In addition, the absolute value of ΔE_{MH} depends sensitively on the theoretical value used for the energy of the hydrogen molecule. Nevertheless, the calculations reproduce the observed trend of the greater stability of PdH versus NiH. In this connection it has been noted by *Switendick* [5.35] that the endothermic value of NiH could be accounted for by the difference between the value for magnetic and nonmagnetic configuration (used in the calculation) of pure Ni metal. From their calculation *Williams* et al. [5.32] trace back the increased stability of PdH compared to NiH to differences in the chemical potential of the host metal.

Hyperfine Interactions. Calculations of wavefunctions and their use in the prediction of various observables have been rather scarce in the field of metal hydrides. We shall however mention here a theoretical study of the contact term of the hyperfine interaction at the proton site in PdH which gives interesting information on the *s*-character of the Fermi surface states at the proton site. The theoretical study was motivated by the interpretation of proton spin-lattice relaxation times τ_1, measured in the temperature range between 3 and 77 K where the proton does not diffuse [5.89].

The nuclear spin-lattice relaxation can be viewed as a spin-flip scattering of the conduction electrons due to the magnetic hyperfine interaction with the nuclear spins. The contact term of the relaxation rate is often written in the convenient factorized form:

$$\tau_1^{-1} = 4\pi\gamma_N^2 k_B T (N_s H_{hfs})^2 , \qquad (5.8)$$

where N_s is the partial *s*-type DOS at E_F for one spin direction and H_{hfs} the hyperfine field at the H site.

The theoretical value obtained by generating 15 000 wavefunctions at the FS leads to $(N_s H_{hfs})_{theory} = 4.4 \times 10^{15}\,\mathrm{G\,erg^{-1}}$ which is to be compared to $(N_s H_{hfs})_{exp} = (4.0 \pm 0.1) \times 10^{15}\,\mathrm{G\,erg^{-1}}$. The theoretical value is quite satisfactory since it lies within 7% of the experimental error bar; it provides valuable information about the H site *s*-character of the states at E_F. It is to be noted, as discussed below, that this feature plays a role in the value of the matrix elements of the electron–optical phonon coupling and thus on the superconducting properties of PdH. The NMR experiments [5.89] show that for non-stoichiometric compounds corresponding to $x = \mathrm{H/Pd} = 0.7$, $\tau_1 T = 70 \pm 3\,\mathrm{s\,K}$; this increase over the value $\tau_1 T = 48 \pm 2\,\mathrm{s\,K}$ of stoichiometric PdH could correspond, if other factors are equal, to a 21% decrease in the value of the partial H *s*-DOS at E_F. This trend is corroborated by theoretical calculations [5.46] on non-stoichiometric compounds. It is also reproduced by applying in the vicinity of E_F the rigid-band model to stoichiometric PdH.

Electron-Phonon Interaction and Superconductivity. The study of the electron-phonon coupling constant λ of PdH is particularly interesting in view of the superconducting properties of PdH_x (for $x > 0.8$) and of the existence of an inverse isotope effect on the superconducting transition temperature T_c [T_c(PdH) ~ 9 K while T_c (PdD) ~ 11 K] [5.90]. It seems now well-established from superconducting tunneling experiments as well as from theoretical work [5.73, 91] on λ_H that the electron-optical phonon coupling is essentially responsible for the large value of T_c in PdH. The inverse isotope effect in T_c has been ascribed to anharmonicity effects [5.73] in agreement with inelastic neutron scattering data. However, other experimental results suggest that the value of the anharmonicity is too small to account for the observed differences in T_c [5.90]. Other possible mechanisms such as the role of zero-type amplitude effects on the band structure have not yet been quantitatively investigated.

Another interesting question is why NiH, which is isoelectronic to PdH and diamagnetic, is not superconducting down to 1 K [5.92]. Indeed, calculations of the electron–optical phonon electronic term η_H defined in Sect. 2 show a 57% decrease of this term from PdH to NiH [5.91]. This is due to the decrease of the *s*-electron density at the H site for states at E_F, the Fermi level of NiH being closer to the top of the *d*-bands than in PdH. Moreover, the temperature dependence of the electrical resistivity [5.92] indicates that the optical phonon modes in NiH are much harder than in PdH, the Einstein temperature being $\Theta_E(\text{NiH}) \approx 1.8\, \Theta_E(\text{PdH})$, a factor which further contributes to the decrease of T_c. Inelastic neutron scattering data [5.93] of $NiH_{0.75}$ confirmed the higher frequencies of the optical phonons in NiH than in PdH. The spectrum presents a sharp peak around 88 meV and a broad shoulder extending up to 110 meV.

Recent calculations of λ and T_c for Pd–Ag–H alloys using the CPA [5.94] have confirmed the role played by the variation of the H *s*-type DOS at E_F on the value of η_H.

We can conclude that due to the large amount of experimental and theoretical work performed on PdH_x and to a lesser extent on NiH_x and alloys with noble metals, these systems appear to be quite well understood and provide an example of the electronic properties of transition metal monohydrides.

5.3.2 Hydrides of Ti, Zr, and Hf

The group IVB metals Ti, Zr, and Hf form stable metallic dihydrides of the CaF_2-type structure (fcc) over a wide range of concentrations [5.1]. In this structure, the H atoms occupy tetrahedral interstices of the fcc metal lattice. For hydrogen concentrations approaching $x = 2$ and below a critical temperature, a tetragonal distortion of the lattice occurs. Most experimental and theoretical studies of the electronic structure of hydrides of Ti, Zr, and Hf were performed in relation to this fcc–fct lattice distortion.

Early calculations [5.4] have shown, as proposed by *Ducastelle* et al. [5.95] from their study of the temperature and H concentration dependence of the electronic specific heat, thermoelectric power and magnetic susceptibility, that the cubic to tetragonal distortion can be ascribed to a Jahn-Teller effect. Since then, several non-self-consistent calculations for TiH_2 and ZrH_2 by *Gupta* [5.96], for TiH_x by *Kulikov* et al. [5.97] *Fujimori* and *Tsuda* [5.98] as well as self-consistent results for TiH_2, ZrH_2, HfH_2 by *Papaconstantopoulos* and *Switendick* [5.99] and studies of substoichiometric hydrides of group IVB metals [5.46] have become available. Improvements in the numerical accuracy of the DOS led to the conclusion that the Fermi level of group IVB metal cubic dihydrides falls in a peak of the DOS which araises from a flat branch of degenerate metal *d*-states in the ΓL direction of the fcc BZ. The quadratic distortion lifts this degeneracy and leads to a lowering of the Fermi level and of the DOS at E_F in agreement with a variety of experimental data mentioned previously such as electronic specific heat [5.95, 100] and positron annihilation studies [5.65]. NMR investigations of TiH_x and ZrH_x for $1.5 \leqq x \leqq 1.9$ have been performed by *Korn* [5.101], and *Bowman* et al. [5.102]. The spin-lattice relaxation rate and the Knight shift also indicate that the DOS at E_F is lowered by the fcc→fct distortion. The strong modification of the states at E_F associated with the Jahn-Teller effect is also evidenced by the sign change of the Hall coefficient from negative in the cubic phase to positive in the tetragonal phase [5.103]; the sensitivity of the FS geometry to the distortion appears from the analysis of the FS cross-sections of cubic TiH_2 [5.96]. Work function measurements [5.104] reveal a discontinuity of ~0.2 eV accompanying the fcc →fct transition of TiH_x and ZrH_x.

Wavefunction calculations and partial-wave analysis of the DOS are now available [5.46, 96, 99]. They have revealed clearly the details of the *M*–H and H–H bonding and the coupling of the H *s*-states with the metal *d*-states of the t_{2g} symmetry when H occupies the tetrahedral interstices of the fcc metal lattice. These calculations show that besides the low-lying metal-hydrogen bonding band observed in monohydrides, a second band due to H–H and metal-H interactions appears in dihydrides and leads to a double-peak structure in the DOS below the metal *d*-bands. In group IVB metal dihydrides, this second low-lying band overlaps the metal *d*-states.

The photoelectron spectra of $MH_{\approx 1.6}$ (M = Ti, Zr, Hf) obtained by *Weaver* et al. [5.105] are plotted in Fig. 5.5. The *d*-band derived emission extends within 3 eV from E_F while the hydrogen induced bands centred at ~5.5 eV below E_F show, in agreement with theory, a double-peak structure, the second peak being observed below 7 eV. Direct comparison with calculations is however hazardous since the samples correspond to $x \sim 1.6$. The spectra taken for ZrH_x, $x = 1.63$ and $x = 1.94$ reveal the change associated with the fcc→fct distortion especially around E_F and also below 7 eV. Self-consistent calculations on dihydrides [5.99] locate the H induced states at 4.9, 5.4, and 5.7 eV respectively for TiH_2, ZrH_2, and HfH_2 while non-self-consistent results [5.96] locate these states about 1 eV lower. The effect of iterating to self-consistency is a lowering of

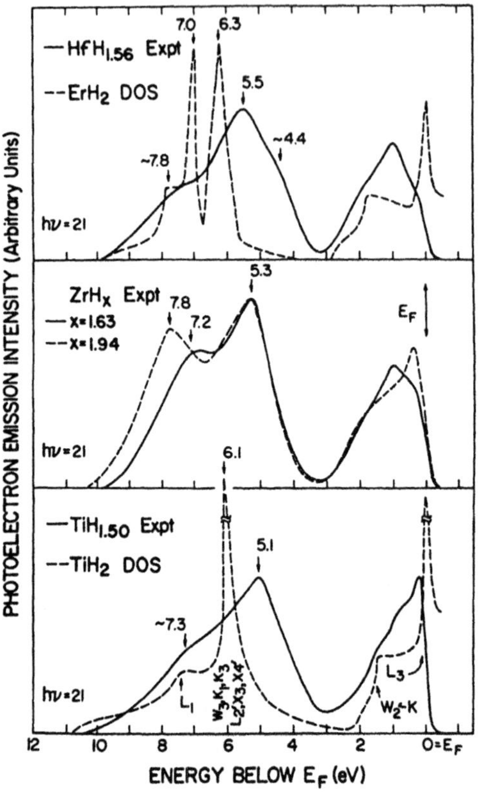

Fig. 5.5. Photoelectron spectra of $HfH_{1.56}$, $ZrH_{1.63}$, $ZrH_{1.94}$, and $TiH_{1.50}$ after background subtraction compared with the calculated DOS of ErH_2 and TiH_2. The spectra for ZrH_x reveal the change in the bandstructure associated with the fcc–fct distortion [5.105]

the d-bands and of E_F with respect to the metal-H bonding states. This lowering is about 0.9 eV for ZrH_2, but it amounts to 1.4 eV in TiH_2. A peculiar ordering of the H-H antibonding states at the BZ centre has been obtained in self-consistent calculations on TiH_2 [5.99]. This is, however, not observed in self-consistent results on other dihydrides [5.35].

An XPS analysis of the valence band of an oxygen contaminated sample of $ZrH_{1.7}$ reveals the hydrogen induced band at 7 eV [5.106]. All core levels, as measured by different groups using UPS and XPS, shift 0.5–1.0 eV to larger binding energy, consistent with charge transfer to the hydrogen site (Table 5.1). This result agrees with the trends observed in theoretical calculations. Apart from this shift, the core level spectra do not show any hydride effect. The Ti $2p$ core level was shown to shift approximately linearly with hydrogen concentration [5.107].

In soft x-ray L_3 emission spectra of ZrH_x ($x = 0.5$, 1.0, and 1.9) and of $TiH_{1.8}$, hydrogen induced bands at 7 eV developed at the expense of the emission just below E_F [5.8, 108]. The implantation of 8 keV hydrogen into Zr [5.109] results in the appearance of a distinct peak in XPS valence band spectra at 3.4 eV, which was attributed to a (Zr-4d–H-1s)-band. As the position of this peak does not agree either with band structure calculations of the dihydrides or

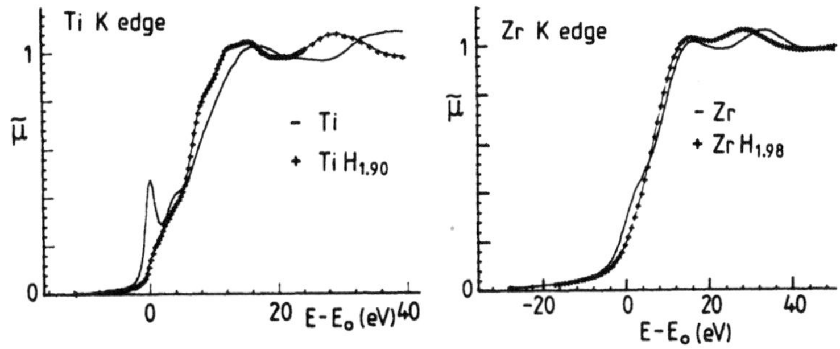

Fig. 5.6. XANES at the K edges of Ti, $TiH_{1.9}$, Zr, and $ZrH_{1.98}$ showing features of the empty part of the DOS above E_F [5.71]

with photoelectron spectra of ordinary bulk ZrH_x, the authors suggest that it could be due to different sites of the implanted hydrogen. However, upon annealing of the sample to 600 °C the peak does not disappear nor shift to the position of the hydrogen induced band at about 7 eV, but rather becomes more intense. The exposure of clean Zr to 3000 Langmuir hydrogen ($1L = 10^{-6}$ s Torr) at room temperature results in rather weak effects in photoelectron spectra which are qualitatively similar to those observed for hydrides. It does not lead to a significant hydride formation [5.110]. XANES K-edges of Ti and Zr hydrides (see Fig. 5.6) have less structure than do those of the corresponding pure metals [5.71], a fact which has not yet been explained by DOS calculations in this energy range.

Theoretical studies of the total energy of TiH_2 versus lattice constant [5.35] reproduce the existence of a minimum in the total energy curve. The lattice constant is obtained with an accuracy better than 1.5% and the heat of formation has been improved considerably over previous results. However, a detailed analysis of the theoretical calculation reveals the sensitivity of total energies to the approximation used for the treatment of core electrons. The theoretical results [5.35] range from -35.53 to -52.25 kJ/(mole H) depending on the core approximation used while experimental data range from -44.73 to -61.86 kJ/(mole H).

The electron-phonon coupling constant has been caclculated for dihydrides of group IVB metals [5.91, 111]. Since in these dihydrides E_F falls at the bottom of the TM d-bands, the value of the H s-DOS at E_F is very small, unlike in PdH and to a lesser extent NiH. Using (5.7) one obtains negligible values of the electronic term of the electron–optical phonon coupling η_H. Moreover available neutron scattering data [5.112] and a study of the temperature dependence of the phonon resistivity [5.113] indicate values of the optical phonon frequencies much larger than those in PdH. This leads to very small values of the electron–optical phonon coupling λ_H. Moreover, the electron–acoustic phonon coupling is found to be smaller than in the corresponding pure metal. Thus, the dihydrides are not found to be superconducting, in agreement with experimental data [5.90]. An increase of T_c of pure Zr ($T_c = 0.7$ K) has been reported [5.114] after implantation of small doses of H and D with values of $T_c = 3.14$ K for H and 4.65 K for D. Nevertheless this observation does not contradict previous data nor theoretical results since it concerns disordered dilute hcp phase for which the electronic structure is expected to be drastically different from that of a stoichiometric fluorite structure dihydride.

CPA calculations on substoichiometric hydrides ($x < 2.0$) indicate that the H-induced states become less sharp and decrease in height as the H concentration decreases; however, they do not shift in energy. The position of E_F on the other hand appears to shift to lower energies by about 0.27 eV from $x = 2.0$ to $x = 1.6$ for TiH_x resulting in a large reduction of $N(E_F)$ [5.46].

Theoretical studies of the electronic properties of group IV hydrides of symmetry lower than cubic remain up to date very scarce. However, recently a band structure calculation of the tetragonal γ-phase of ZrH has been performed [5.115]; substantial differences from the DOS of ZrH_2 have been obtained. A detailed analysis of the evolution of the states with the removal of the H atoms rules out fully the validity of a rigid-band behaviour applied to the bands of cubic ZrH_2.

The calculated DOS at E_F is found to be 0.88 electron states eV^{-1} (Zr atom)$^{-1}$ for the γ-phase compared to the value of 1.76 electron states eV^{-1} (Zr atom)$^{-1}$ for cubic ZrH_2 [5.35]. A comparison with magnetic susceptibility measurements [5.116] seems to indicate that the orbital contribution to the magnetic susceptibility is large in this compound.

Hydrides of Ti, Zr, and Hf are a nice example to demonstrate the effect of structural distortion on electronic properties. These effects are clearly visible in calculated DOS and in a number of observables however, the limited experimental resolution prevents detection of detailed effects in PES spectra.

5.3.3 Hydrides of Nb, V, and Ta

The group VB metals Nb, V, and Ta form monohydrides in a rather complicated diagram of ordered and disordered phases, and dihydrides [5.117]. NbH_x, e.g., exists as monohydride at 300 K in the orthorhombic β-phase and below 200 K in the pseudocubic γ-phase and as dihydride in an fcc structure (presumably CaF_2). In the monohydrides, the H atoms occupy tetrahedral interstices of the bcc Nb lattice. Band structure calculations of the monohydride were first performed for γ-NbH [5.118]. A single and rather narrow peak was obtained in the DOS of the hydrogen-metal induced states, as usually observed for other monohydrides. Photoelectron spectra of $NbH_{1.0}$, $VH_{1.0}$, and $TaH_{0.8}$ however present a double-peak structure in the hydrogen induced states at ~ 5.5 eV and ~ 7.5 eV [5.118]. In fact very early SXES investigations of $VD_{0.7}$ [5.119] already revealed a double peak at ~ 6.5 eV and ~ 8.5 eV below E_F in the $V-L_3$ emission spectra contrary to more recent SXES spectra of $NbH_{0.8}$ and $VH_{0.4}$, which show only one hydrogen induced band [5.120]. Hydrogen induced states were also observed in electron energy loss spectra of NbH_x and

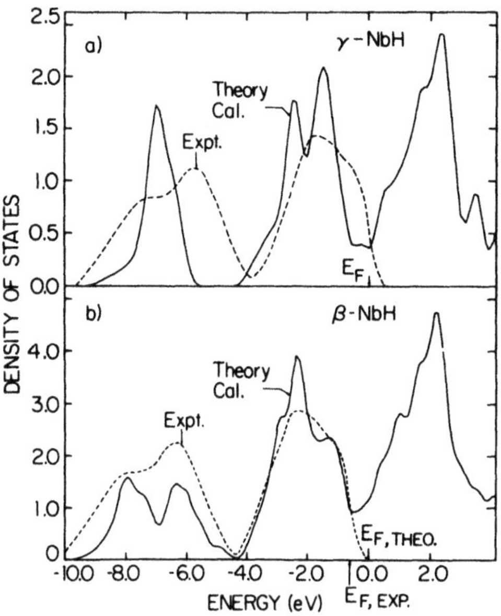

Fig. 5.7. Calculated DOS in units of [states/eV unit cell] for (*a*) γ-NbH and (*b*) β-NbH compared with photoelectron spectra. The experimental curve has been shifted in (*b*) to line up with the DOS. The DOS curve calculated for β-NbH shows the double peak structure of the hydrogen-induced bonding band, in agreement with experiment [5.34]

VH_x of unknown concentrations at 4 and 7 eV respectively [5.121] but with less structure than in the photoelectron spectra. Several explanations were proposed to account for the unusual double peak, among them the formation of a surface dihydride. However, recent first principles pseudopotential calculations [5.34] of both β- and γ-phases of NbH yield a more probable reason. As shown in Fig. 5.7 the calculated DOS of the β-phase is in good agreement with the photoemission data. Although the orthorhombic distortion was ignored in this calculation, the width and double-peak structure in the hydrogen induced states is reproduced; it has been ascribed to a splitting of the bands due to shorter distances between the H atoms in the β-phase. These results demonstrate the sensitivity of the hydrogen derived bands to the crystal structure and lattice parameters.

The empty DOS was studied by x-ray absorption for monohydrides and deuterides of V [5.122] and by XANES for $VD_{0.72}$ and $NbH_{0.93}$ [5.71]. Upward shifts of the *K*-edge and a suppression of fine structures just above the edge were observed and explained by the filling of unoccupied *d*-states by excess electrons of hydrogen and the lowering of empty *p*-states to a position below E_F, respectively.

Surprisingly, the PES spectra of $TaH_{0.8}$, Nb_4D_3, and $VH_{1.0}$ [5.118] do not show a shift of the *p* core levels, in contradiction to the theoretical calculations which indicate some charge transfer from the transition metal to the H sites. The *p*- and *d*-spectra of NbH_x of unknown composition produced by ion implantation technique show a shift of 0.8 eV [5.109].

Total energy calculations for both the β- and γ-phases of NbH have been performed by *Ho* et al. [5.34]. A good agreement with available experimental results is reported for lattice constants, bulk modulus (within 4.5%) and heats of formation. However, as indicated by the authors, the calculated heats of formation are small, a few tenths, of an eV per H atom, thus numerical accuracies and effects such as neglecting the orthorhombic distortion and contributions from zero-point vibration energy of hydrogen in the solid and in the hydrogen molecule can become critical. From total energy calculations as a function of the H atom position in the unit cell, the optical phonon frequencies have also been obtained for the first time for a transition metal hydride [5.34]. The results are in rather good agreement with neutron scattering data [5.123]; the importance of anharmonic effects was also pointed out.

The fluorite structure dihydrides NbH_2 and VH_2 have been studied theoretically [5.96, 99, 124]; an interesting behaviour is the disappearance of superconductivity in the high T_c group V metals upon formation of hydride phases [5.90] which was confined by recent heat capacity measurements [5.125]. The essential features of the bands and DOS of NbH_2 and VH_2 are very similar to those of the isostructural group IV metal dihydrides discussed previously; the Fermi levels of group V metal dihydrides fall however at higher energies since one more electron needs to be accommodated in the metal d-bands. The very small value of the H s-type DOS at E_F in conjunction with the large values of the optical mode frequencies $\omega_{opt} > 120$ meV observed by inelastic neutron scattering [5.123] lead to very small values of the electron–optical phonon coupling constant. A drastic weakening of the electron–acoustic phonon term η_{metal} between pure Nb and NbH_2 has also been obtained [5.96]. From a detailed analysis of the structural changes in the metal lattice, and from values of the DOS at E_F [5.91] this reduction in η_{metal} has been ascribed mostly to a weakening of the metal d–d interactions due to the lattice expansion rather than to a density of states effect. The observed hardening of the acoustic phonons upon hydrogen absorption [5.112] leads to a further decrease of the electron–acoustic phonon coupling and explains why the group V metal dihydrides are not superconducting down to ~ 1 K. The studies of Knight shifts and nuclear relaxation rates in NbH_2 [5.126] indicate clearly in agreement with theory the small value of $N_s(E_F)$ compared to $N_d(E_F)$. Further information concerning the variation of the DOS at E_F as a function of hydrogen content in VH_x and NbH_x with x up to 1.93 is available from the detailed low temperature heat capacity measurements [5.125]; in the absence of theoretical results on hydrides in the full concentration range covered by the experiment these results have been successfully compared with results on V metal and with data on V/Cr-alloys.

5.3.4 Hydrides of Rare Earth Metals Including Sc, Y, and La

The trivalent RE metals as well as Sc, Y, and La are known to from metallic dihydrides REH_2 and semiconducting (or poorly metallic) trihydrides REH_3 [5.127, 128], with the exception of Sc, which probably does not form a trihydride. The dihydrides are of almost ideal cubic CaF_2-type structure where the tetrahedral sites are roughly filled and the lattice parameter is increased by typically 8% as compared to the elemental metals. The subsequent occupation of octahedral sites leads to cubic BiF_3-type and hexagonal HoD_3-type structure for the trihydrides of the light (La to Nd) and heavy rare earth and Y, respectively. Ce does not seem to be an exception. The divalent RE metals Er and Yb form semiconducting dihydrides.

The phase diagram for the intermediate range REH_2 to REH_3 is not yet known in detail. A generalized diagram [5.127] shows a solid solubility range of the REH_2 phase from $x \approx 1.9$ to $x \approx 2.3$, then a range of coexistence of the REH_2 and REH_3 phases, and finally the REH_3 phase. The light RE metals show a rather extended solubility range of the REH_2 phase which ends in a tetragonally distorted phase at $REH_{2.33}$ with slightly contracted lattice parameters [5.129]. Very recently the tetragonal distortion was observed in CeH_x at room temperature from $CeH_{2.28}$ to $CeH_{2.64}$ and evidence for a miscibility gap from $CeH_{2.56}$ to $CeH_{2.64}$ was given [5.130]. In fact tetragonal distortions have been observed in many RE hydrides over the last ten years; however, at low temperature only, i.e. in the range from 200 K to 250 K [5.131–136]. They are thought to be caused by hydrogen ordering effects. Down to 4 K no further structural transitions have so far been observed. Possibly the range of concentration for which the tetragonal distortion occurs is more extended at low temperatures. Tentative phase diagrams for the range CeH_2 to CeH_3 including low temperatures can be found in [5.130, 134].

The transition of the electronic structure from rare earth metals with a $(5d6s)^3$ conduction band to metallic dihydrides and further on to nonmetallic or at most poorly metallic trihydrides, obtained in the calculations (Fig. 5.8), as

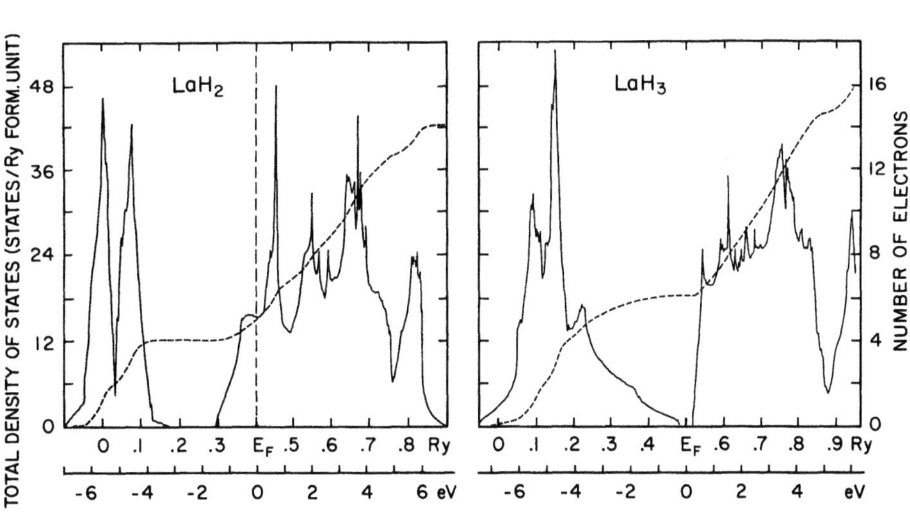

Fig. 5.8. The total DOS of LaH_2 and LaH_3 (—, *left hand scale*) and the total number of electrons (– – –, *right hand scale*). DOS units are [states of both spins/Rydberg formula unit] [5.148]

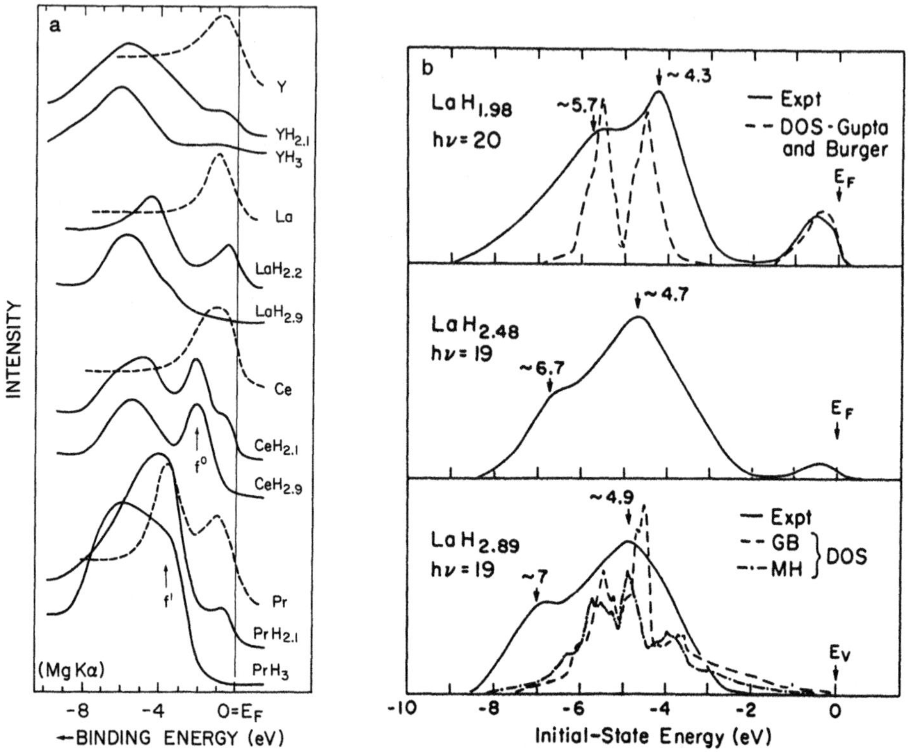

Fig. 5.9. (a) XPS spectra of the valence bands of Y, La, Ce, Pr and their dihydrides and trihydrides. The dihydrides are metallic but show reduced emission at E_F, whereas the trihydrides are non-metallic. The hydrogen-induced bands appear at 5–6 eV (4*f* emission of cerium hydrides at 2 eV; that of praseodymium overlap with the hydrogen-induced band) [5.11]. **(b)** Photoelectron spectra for $LaH_{1.98}$, $LaH_{2.48}$, and $LaH_{2.89}$ after background subtraction compared with calculated DOS for LaH_2 and LaH_3 (*Gupta* and *Burger* [5.148]) and LaH_3 (*Misemer* and *Harmon* [5.149]). The calculated bonding band centers agree with experiments but their widths are significantly narrower [5.138]

well as the appearance of the hydrogen-induced bands centred around 5–6 eV are easily seen in photoelectron spectra (XPS) of hydrides of Y, La, Ce, and Pr (Fig. 5.9a). The double-peak structure of the hydrogen-induced band, typical for dihydrides, is nicely resolved in the low photon energy spectra of $LaH_{1.98}$ (Fig. 5.9b) and $CeH_{2.1}$. The results of all valence band studies of rare earth hydrides by PES are summarized in Table 5.1 [5.11, 137–145]. From optical studies of interband transitions in ScH_{2-x}, YH_{2-x}, and LuH_{2-x} it was concluded that the octahedral site occupation is not negligible in the dihydrides [5.72].

The essential ideas put forward by Switendick from his early band structure sutides of stoichiometric YH_2, YH_3 and rare earth di- and trihydrides [5.4] have been confirmed by more recent calculations for Y_4H_9 and Y_4H_{11} [5.146],

ErH_2 [5.147], LaH_2 and LaH_3 [5.148–150] ScH_2 and YH_2 [5.151], CeH_2, CeH_3 [5.150, 152]. For all these band structure calculations the $4f$ electrons are omitted.

Due to differences in the nuclear charge of the metal atom, the low energy H induced states overlap the metal d-bands in the case of Sc and Y while a gap is observed in the case of LaH_2 and RE dihydrides. This explains why the divalent RE dihydrides are semiconductors. For trivalent RE dihydrides only one electron occupies the bottom of the metal d-bands; the DOS at E_F is much reduced from its value in the pure metal in agreement with trends observed in the electronic specific heat [5.127, 132, 136] and magnetic susceptibility data [5.127]. In the trihydrides, a third band appears at low energy leading, as observed in the photoemission spectra (Figs. 5.9a, b), to a further depletion of the RE metal d-band. Wavefunction analysis shows that the additional third band of RE trihydrides has essentially a metal d–e_g and octahedral H s-character while the first two low-lying bands gain also, besides the metal d–t_{2g} tetrahedral H s-character dominant in the dihydride, an additional metal d–e_g octahedral H s-contribution.

Most theoretical calculations on RE trihydrides indicate the presence of a gap between the three low-lying bands and the higher metal d-states, with the exception of *Kulikov* et al. [5.150, 153, 154] who obtained an overlap; their calculation, however is based on the use of a model Hamiltonian and the authors themselves warn that the results are only qualitative. The value of the energy gap in LaH_3 appears to be reduced by the use of a less strong exchange and correlation potential and by self-consistency effects. The theoretical value of the gap – if any – remains an open question since it could be further affected by spin-orbit coupling effects not fully included up to date. The use of a strong exchange potential ($\alpha=1$ in the X_α scheme) leads to a narrowing of the bands and results in a better agreement with the optical conductivity data $\sigma(\omega)$ [5.138]. These data show that the onset of interband transitions occurs at 1.1 eV, in much better agreement with the value of 1.3 eV obtained from non-self-consistent strong exchange potential calculations. The value of 1.75 eV obtained from self-consistent Hedin and Lundqvist exchange correlation potential calculation is too large. On the other hand, the strong exchange potential which narrows the valence band width by 13% from the theoretical value of 4.2 eV [5.149] leads to a poorer agreement with the photoemission experiments [5.138] which indicate a valence band width of about 6 eV. It has been argued in this connection that the premature filling of octahedral sites in LaH_2 will lead to an increase in the theoretical band width [5.149]. The narrowing of the bands lead to a DOS at E_F for LaH_2 of 1.11 states/eV-cell larger than the self-consistent value of 0.85 states/eV-cell and in better agreement with the experimental value.

The results of different band structure calculations and of measurements of specific heat [5.132, 136] susceptibility [5.127], NMR Knight shift [5.155] and photoemission [5.11, 137–145] show "in agreement" that the trihydrides are nonmetallic or at most poorly metallic.

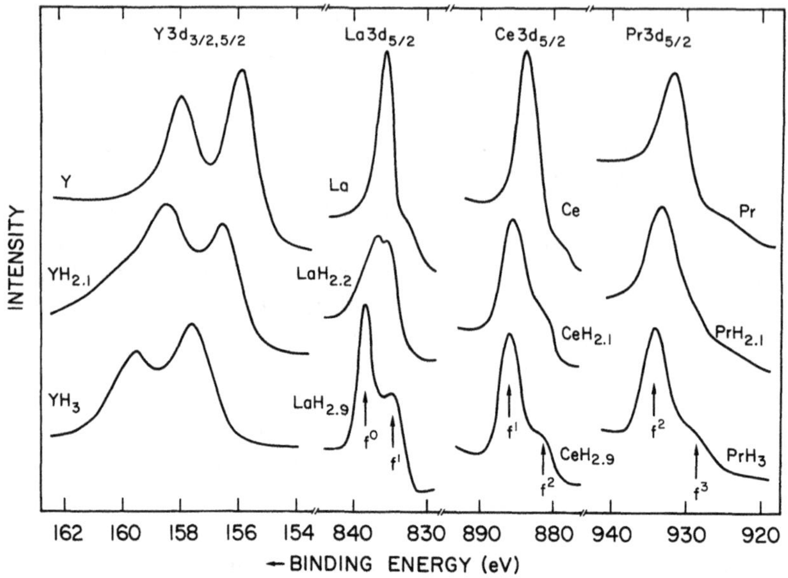

Fig. 5.10. XPS 3*d* core-levels of Y, La, Ce, and Pr and of their dihydrides and trihydrides. The evaluation of the chemical shifts from the measured core-level shifts is not straightforward as satellites indicate screening effects [5.11]

Inverse photoemission spectra of light rare earth hydrides show weak structures at 1–1.5 eV and 3–3.5 eV above E_F which can be attributed to empty (*spd*) states [5.70].

All band structure calculations indicate that a charge transfer exists from the metal to the hydrogen tetrahedral sites in the dihydrides of TM and RE metals while the octahedral site in the trihydrides can be considered as neutral. This general trend is corroborated by the experimental XPS data on the metal core-level shifts obtained upon hydrogenation [5.11]; the core-level photoemission spectra (Fig. 5.10 and Table 5.1) are all shifted to larger binding energy as compared to the elemental metals. The apparent shift from the metal to the dihydride can be larger (e.g. Ce) or smaller (e.g. La) than the shift from the dihydride to the trihydride. A quantitative determination of the chemical shift from the observed spectra is not straightforward owing to final state effects in the photoemission caused by screening of the core hole. A careful analysis agrees with a charge transfer to the tetrahedral site and no further transfer to the octahedral site [5.139, 140, 143, 144, 70, 156].

The theoretical values of charge transfers are still a matter of controversy. In the KKR and APW methods, the charge transfer is best defined as the charge difference inside a given MT sphere between the final self-consistent results and the overlapping neutral atoms case. Such a definition which of course suffers from the arbitrariness in the partition of the real space into several regions, leads to charge transfers of the order of 0.2 electrons towards the hydrogen MT spheres at the tetrahedral sites. In tight binding and LCAO analysis, the number of electrons per atomic orbital is used to define the ionicity; this definition however does not take proper care of the spatial localization of the charge; as example, such analysis leads to about 1.8 electrons in the 1*s* tetrahedral hydrogen in CeH_2 which is certainly an overestimate of the ionicity since if one looks at the interatomic separations in this compound, it appears that the 1*s* wavefunction is quite extended spatially and part of the corresponding charge would be located inside the Ce MT sphere.

The Fermi surfaces of the dihydrides are determined only by one band and are therefore much simpler than those of the corresponding metals. Ignoring relativistic effects, these surfaces are found to be multiply connected hole surfaces formed of a warped cube centred around Γ with necks in the ΓL directions [5.147, 148]. Strong nesting properties of this surface have been noticed; the nesting vectors are along the 100 directions their exact dimensions are expected to be somewhat modified by relativistic effects; nevertheless their values remain close to the magnetic ordering vectors found experimentally by polarized neutron scattering studies [5.157]. This result is already encouraging and suggests that the magnetic ordering could be explained by the existence of Fermi-surface induced divergences in the generalized susceptibility; further theoretical investigations concerning the role of relativistic effects are needed to confirm this point.

The drastic change in the characteristics of the FS states from the pure metal to the dihydride together with the strong modification of the phonons has several important consequences on the electron-phonon coupling and thus on the superconducting properties as well as on the electrical resistivity. While fcc La is a superconductor with $T_c \sim 6$ K, LaH_x with $1.8 < x < 2.36$ does not show any superconducting transition above 1 K [5.90]. The strong decrease of the electron-acoustic phonon coupling from the pure metal to the dihydride [5.158] is due (i) to a reduction of the electronic term η_{metal} which originates from the decrease in the La *d*–*d* coupling due to the lattice expansion and to a lesser extent to a reduction of the DOS at E_F (ii) to a strong increase in the acoustic phonon frequencies [5.159, 160]. The Debye temperature essentially doubles from La to $LaH_{2.03}$ [5.161]. This drastic change explains also the large decrease observed in the room temperature resistivity of the dihydrides [5.162, 163]. Further modifications of the electronic structure with hydrogen concentration induce changes in the electron-phonon interaction; shifts in the phonon frequencies have been reported from the study of infra-red and Raman spectra of CeH_x with $2 < x < 3$ [5.164].

Whereas the electronic structure of stoichiometric dihydrides and trihydrides is rather well understood, there is considerable confusion in the intermediate concentration range, particularly at concentrations approaching trihydride composition and at low temperatures. A concentration dependent metal to semiconductor transition was observed in Ce hydrides at $x \approx 2.75$. Furthermore the occurrence of a semiconductor to metal transition upon increasing temperature was suggested. Results published up to 1979 are

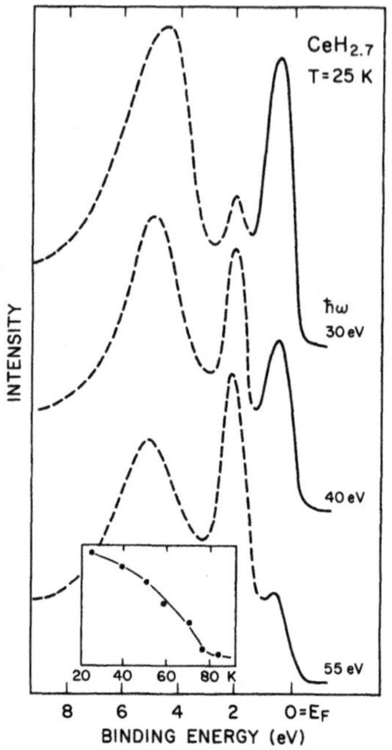

Fig. 5.11. Low temperature photoelectron spectra of $CeH_{2.7}$ for various photon energies $\hbar\omega$, showing the hydrogen-induced band at 5 eV and the 4*f*-emission at 2 eV. The intense peak at E_F (—) grows below $T = 70$ K, indicating a semiconductor to metal transition. Inset: Intensity of the peak at E_F versus temperature [5.168]

summarized in the review of *Libowitz* and *Maeland* [5.127]. Never results on specific heat and NMR Knight shift reveal poorly metallic or nonmetallic behaviour at low temperature [5.132, 155, 165]. But resistivity measurements seem to indicate metallic behaviour at low temperature and less metallic or semiconducting behaviour at $T > 200$ K [5.149]. The origin of this puzzling behaviour is not known, and the fact that the metal-to-semiconductor transition is observed already at $x \approx 2.75$ instead of $x = 3$ is not fully understood and cannot be explained by one-electron band structure calculations. Several models were proposed: *Fujimori* and *Tsuda* [5.152] describe the temperature dependent transition by a delocalized-localized transition of electrons associated with the order-disorder transition of octahedral hydrogen, which is related to the tetragonal distortion. *Kulikov* [5.153, 154] proposed that the small band overlap and the resulting nesting of hole and electron pockets of the Fermi surface lead to an excitonic insulator phase at low temperature (see also discussion in [5.149]). Recent NMR results of Ce hydrides do not confirm a temperature dependent transition. The results on the concentration dependent transition are consistent with a Mott transition [5.166, 167]. Concerning these theoretical models one should not forget that the metal-to-semiconductor transition was detected in resistivity measurements of rare earth hydrides in 1959 and 1972 [5.127], i.e. at a time when the purity of rare earth metals and the hydride preparation methods where rather limited as compared to the present time. Accordingly it can be expected that a repetition of similar measurements will eliminate most of the controversies.

Recent photoemission studies on hydrides of Ce and La revealed a surface semiconductor-to-metal transition at temperature below $T \approx 70\,K$ (Fig. 5.11) [5.168]. A depletion of octahedral-like surface hydrogen sites is thought to occur. This pushes *d*-states from the hydrogen-induced band back to E_F and results in a metallic surface dihydride. The results indicate that at room temperature the hydrogen concentration at the surface exceeds that of the bulk, whereas at low temperature the opposite seems to be the case.

Resistivity anomalies were found in the solid solution α phase of the elemental metals [5.169] and of the dihydride α phase [5.170] at temperatures around 150 K. They seem to be caused by a peculiar linear ordering tendency [5.169].

So far we have completely neglected the 4*f*-electrons, which are otherwise so important for the behaviour of many RE compounds. The atomic-like 4*f*-electron states are thought not to contribute significantly to the metal-hydrogen bonding. Valence band photoemission spectra of Ce hydrides [5.142, 168] show that the hydrogen bonding band does not contain any noticeable 4*f* contribution. The 4*f* related emission does not seem to be affected by the hydride formation. However, as a very recent analysis of 4*f* related features in core-level photoemission and inverse photoemission showed, there is significant 4*f*-conduction band hybridisation in the light rare earth hydrides [5.70] which could possibly improve the results of band structure calculations, if taken into account.

The theoretical treatment of 4*f*-electrons as band states is very delicate and controversial [5.172]; in the case of lanthanum hydrides, the 4*f*-levels are found to be 2.0 to 3.0 eV above E_F [5.149], whereas inverse photoemission shows 4*f* related features 5.2 eV above E_F [5.70].

A nice and illustrative consequence of a weak 4*f*-conduction band hybridisation is the large coefficient γ of the electronic specific heat of Ce hydrides $\gamma(CeH_{2.6}) = 110\,mJ\,mole^{-1}\,K^{-2}$ (as compared to $\gamma(LaH_{2.6}) \leqq 0.04\,mJ\,mole^{-1}\,K^{-2}$), which suggest that Ce hydrides are so-called heavy-electron compounds [5.172, 168].

Whereas in Ce hydrides the Ce 4*f*-electrons and the valency of Ce are hardly affected, hydrogen causes a valence change in Yb, which forms according to optical re-emission measurements a semiconducting $YbH_{1.94}$ (divalent, in agreement with NMR data of *Zogal* [5.173]) and metallic $YbH_{2.65}$ (more than divalent) [5.174, 175]. Very recent XPS and susceptibility studies proved that $YbH_{2.6}$ is a mixed valent metallic magnetic compound which does not order down to 1.7 K [1.49, 5.84].

For magnetic rare earths, hydrogen does not directly affect the magnetic moments. However, the decrease in the number of conduction electrons and also the large lattice expansion lead to a decrease of the strength of the RKKY indirect exchange interaction of the local moments mediated by the conduction electrons. Consequently, the magnetic transition temperatures of the dihydrides are much lower than those of the pure metals. The type of magnetic order changes with increasing hydrogen concentration. For example CeH_x orders antiferromagnetically for $x \leqq 2.0$, ferromagnetically for $2.1 < x < 2.7$ and again antiferromagnetically for $x > 2.8$ [5.134, 176, 177]. A tentative magnetic phase diagram for Ce hydrides was proposed recently by *Arons* et al. [5.178], who also reviewed magnetic ordering properties of other RE hydrides EuH_2 is a ferromagnetic semiconductor [5.179].

5.3.5 Hydrides of Ca and Mg

The alkaline earth metals Ca and Mg form transparent insulating saline dihydrides [5.1]. Though these light weight hydrides are of potential interest for hydrogen storage, very little is known about their electronic properties.

APW band structure calculations performed for CaH_2 in a simplified CaF_2 structure together with photoemission results reveal Ca–H bonding states centred 5 eV below E_F and a large band gap [5.181]. The bonding band has largely Ca *d*-, Ca *p*- and H *s*-character. The binding energy of the Ca 3*p* core level is 1.3 eV larger than in Ca metal. These results indicate that H-metal bonding in saline or covalent hydrides is not radically different from that of metallic hydrides.

From Compton scattering experiments [5.182] and from different theoretical approaches to the electronic properties [5.183] it is concluded that the valence electrons of MgH_2 are not localized as in ordinary ionic crystals. The estimated effective charges of the ions are $Mg^{1.90+}$ and $H^{0.65-}$.

5.3.6 Hydrides of Actinide Metals

APW band structure calculations for UH_3 in the crystallographically slightly different phases α and β (both are based on the A 15 structure) [5.184] give evidence of a fairly large bonding interaction between U 5*f*- and H *s*-electrons [5.185]. As usual there is a hydrogen bonding band some eV below E_F, which overlaps the U 5*f*-levels, in good agreement with photoelectron spectra of β-UD_3 [5.186]. This results in two types of 5*f*-electrons, itinerant and localized (magnetic). No photon energy dependent spectra have been measured so far to elucidate the 5*f* contribution in detail. The measured 4*f*, 6*p*, and 5*f* spectra of UD_3 are all shifted by 2 eV to larger binding energy in comparison to pure uranium. The electronic specific heat coefficient takes the values $\gamma = 33.9\,\mathrm{mJ\,mole^{-1}\,K^{-2}}$ for β-UD_3 [5.186] and $\gamma = 28.7\,\mathrm{mJ\,mole^{-1}\,K^{-2}}$ for β-UH_3 [5.187] as compared to $\gamma = 9.88\,\mathrm{mJ\,mole^{-1}\,K^{-2}}$ for U metal. This increase of γ by a factor of three upon the formation of the hydride probably does not reflect a large increase of the DOS at E_F but rather points to a weak *f*–*d* correlation effect as in heavy electron systems.

Photoelectron spectra of ThH_2 and Th_4H_{15}, performed at energies 20–35 eV with a small cross-section for *f*-electron emission, clearly show the growth of the Th 6*d*-hydrogen bonding band at the expense of the conduction band [5.188]. The conduction band emission for Th_4H_{15} is weak, but non-vanishing.

Specific heat measurements [5.189] indicate a 50% increase of γ in going from ThH_2 to Th_4H_{15}. According to the valence band spectra this increase is not caused by an enhanced DOS at E_F. The *f*-electron contribution to the H-metal bonding band has not been studied in detail. The $6p_{3/2}$ core levels shift from 16.6 eV in Th [5.180] to 17.7 eV in ThH_2 and 18.6 eV in Th_4H_{15} [5.188].

5.4 Results on Ternary Hydrides

During the last decade much progress has been made in the understanding of the electronic properties of binary transition and rare earth metal alloys. A number of theoretical band structure investigations of ordered crystalline alloys [5.41, 194, 195] as well as CPA studies or model calculations of disordered binary TM alloys [5.196–200] and numerous photoemission investigations of crystalline [5.66, 201, 202] and amorphous [5.194, 203–205] alloys have fully ruled out the validity of crude models such as the rigid-band approach and provided insight into the cohesion of these alloys. The importance of hybridization between the metal *d*-states of the two components in binary systems has been stressed. It leads to bandfilling effects which are more important than those previously ascribed to charge transfers. Indeed, XPS core-level spectroscopy [5.201] as well as theoretical results indicate that charge transfer in these alloys is much smaller than previously thought. The concept of charge transfer remains however controversial due to the ill-defined spatial extension of the charge in most theoretical calculations and to the difficulty of analyzing experimental core-level shifts.

The investigation of the electronic structure of ternary hydrides began rather recently with photoemission [5.7] and x-ray emission [5.8–10] work, often hindered by hydrogen desorption, and with theoretical calculations [5.6], made difficult by the complexity of the crystal structures and in some cases by the controversies concerning the hydrogen site occupancy in the lattice [5.206].

In what follows we shall review the results obtained for hydrides of the intermetallic compounds of the type AB_5 (e.g. $LaNi_5$) AB (FeTi), and AB_2 Laves phase compounds (ZrV_2), on Mg_2Ni and related compounds and on some amorphous alloys. Valence band and core-level data are summarized in Table 5.2 together with results of some other ternary hydrides.

5.4.1 Hydrides of AB_5 Compounds (Haucke Compounds)

Intermetallic compounds of the general formula AB_5, where A is a rare earth, Y or La and B a transition element have been investigated for the last fifteen years mainly due to their technological importance as permanent magnets ($SmCo_5$) and hydrogen storage material ($LaNi_5$).

$LaNi_5$. The intermetallic RB_5 compounds including $LaNi_5$ crystallize within the hexagonal $CaCu_5$ structure with space group *P*6/mmm.

Table 5.2. Valence band and core level data of intermetallic compounds and alloys and their hydrides

Sample composition	Method[a]	Conduction band width [eV][b]	H-bonding band [eV][c]	Core level binding energy and (core level shift) [eV][d]		Remarks	Ref.
$LaNi_5$	T SCAPW	3.4					[5.207]
	T TBRM	3.2					[5.208]
	E XPS	2.8		La $3d_{5/2}$ 836.0	Ni $2p_{3/2}$ 852.8		[5.66, 191, 201]
	E UPS	1.8					[5.213]
$LaNi_5H_7$	T TBRM	5.0	8.2				[5.208]
FeTi	T SCAPW	6.5					[5.229]
	E UPS	2.2		Ti 3p 33.2	Fe $3p_{3/2}$ 53.1		[5.234]
	E XPS			Ti $2p_{3/2}$ 454.1	Fe $2p_{3/2}$ 707.0	Fe 2p more symmetric	[5.84, 233]
FeTi+1 L H_2	E UPS		5.5				[5.234]
FeTiH	T APW	5.5	9.9				[5.6]
$FeTiH_2$	T APW	4.1	7.7, 9.3				[5.6]
Mg_2Ni	E XPS	1.7 Peak at 1.2		Mg 2p 49.2 Mg 1s 1303.2	Ni $2p_{3/2}$ 852.7		[5.84, 201]
Mg_2NiH_4	T APW	Width of *d*-states 2.7	5.1–8.2 and 0–5.1			Nonmetallic	[5.261]
	E XPS	Peak at 2.5	7, 9–10	Mg 2p 51 (2)	Ni $2p_{3/2}$ 855 (2)	Severe oxygen contamination	[5.84]
$ZrMn_2$	E XPS	4		Zr $3d_{5/2}$ 178.7	Mn $2p_{3/2}$ 638.9		[5.7, 11]
$ZrMn_2H_3$	E XPS	4	6–7	179.2 (0.5)	638.9 (0.2)		
$ZrCr_2$	E UPS	1.3				CB peak at 0.2 eV	[5.11, 269]
	E XPS	3.5		Zr $3d_{5/2}$ 178.4	Cr $2p_{3/2}$ 573.9		[5.11, 269, 270]
$ZrCr_2H_{3.8}$	E XPS	3.5	6–7	179.5 (1.1)	574.3 (0.4)		[5.11, 269]
ZrV_2	T SCAPW	3.6					[5.267]
	E UPS	1.3				CB peak at 0.8 eV	[5.11, 269]
	E XPS	3.0		Zr $3d_{5/2}$ 178.4	V $2p_{3/2}$ 512.1		[5.11, 269]
$ZrV_2H_{5.1}$	E XPS	2.5	6.5	179.8 (1.4)	512.9 (0.8)		[5.11, 269]

$ZrV_2D_{4.8}$	E SXES		6.5					[5.271]
ZrVCr	E XPS	3.6		Zr 3d 178.2	V 2p 512.1	Cr 2p 573.7		[5.269]
$ZrVCrH_{4.4}$	E XPS	3.0	6.4	179.4 (1.2)	512.8 (0.7)	574.3 (0.6)		
$ZrVCrH_{2.8}$	E SXES		≈7.0	Zr 4d–2p (0.3)				[5.272]
$ZrVCuH_{2.8}$	E SXES		≈7.5	(0.5)	V 3d–2p (0.6)			
$ZrVCoH_{3.9}$	E SXES		≈6.2	(0.6)	(0.6)			
$ZrVFeH_{3.7}$	E SXES		≈5.2	(0.5)	(0.6)			
$CeRu_2$	E XPS			Ce $3d_{5/2}$ 884.8	Ru $3d_{5/2}$ 279.7			[5.278]
$CeRu_2H_4$	E XPS		5–8	885.5 (0.7)	280.1 (0.4)			
YFe_2	E XPS	4		Y $3d_{5/2}$ 155.1	Fe $2p_{3/2}$ 706.8			[5.84]
YFe_2H	E XPS	4	6–8	156.2 (1.1)	707.2 (0.4)			
Hf_2Fe	E UPS	1.5						[5.84]
	E XPS	2		Hf $4d_{5/2}$ 210.9	Fe $2p_{3/2}$ 707.0		Fe 2p asymmetry weak	[5.279]
Hf_2FeH_4	E XPS	2	5–6	211.5 (0.6)	707.0 (0)		Increased	
ZrCo	E XPS			Zr $3d_{5/2}$ 179.1	Co $2p_{3/2}$ 778.6			[5.84]
$ZrCoH_3$	E XPS		6–8	179.7 (0.6)	778.6 (0)			
Zr_2Co	E XPS			179.0	778.7			
Zr_2CoH_2	E XPS		6–8	179.7 (0.7)	778.8 (0.1)			
Zr_2Ni	E XPS			179.0	Ni $2p_{3/2}$ 853.6			
Zr_2NiH_2	E XPS		6–8	179.6 (0.6)	853.6 (0)			
$Zr_{76}Fe_{24}$	E XPS	3.0		Zr $3d_{5/2}$ 178.6	Fe $2p_{3/2}$ 706.9		Amorphous	[5.284]
$Zr_{76}Fe_{24}H_{246}$	E XPS	3.5	6	179.4 (0.8)	707.2 (0.3)		Fe 2p asymmetry increased	
$Zr_{74}Ni_{26}$	E XPS	3.0		178.9	Ni $2p_{3/2}$ 853.6		Amorphous	[5.285]
$Zr_{74}Ni_{26}H_x$	E XPS	3.6	6–9	179.6 (0.7)	853.7 (0.1)		Amorphous	
$Zr_{67}Ni_{33}D_{100}$	E SXES		7				Amorphous	[5.10]
$Zr_{65}Pd_{35}H_{100}$	E SXES		7				Amorphous	[5.10]

[a] Method: T theoretical; E experimental; TBRM tight binding recursion method; APW augmented plane waves; SC self-consistent; UPS UV photoelectron spectroscopy; XPS x-ray photoelectron spectroscopy; SXES soft x-ray emission spectroscopy

[b] Conduction band width: E FWHM, not corrected for instrumental resolution; T from edge to edge

[c] H-bonding band: center of band or pronounced peaks, rel. to E_F

[d] Core level binding energy relative to E_F, (core level shift relative to value of the intermetallic compound)

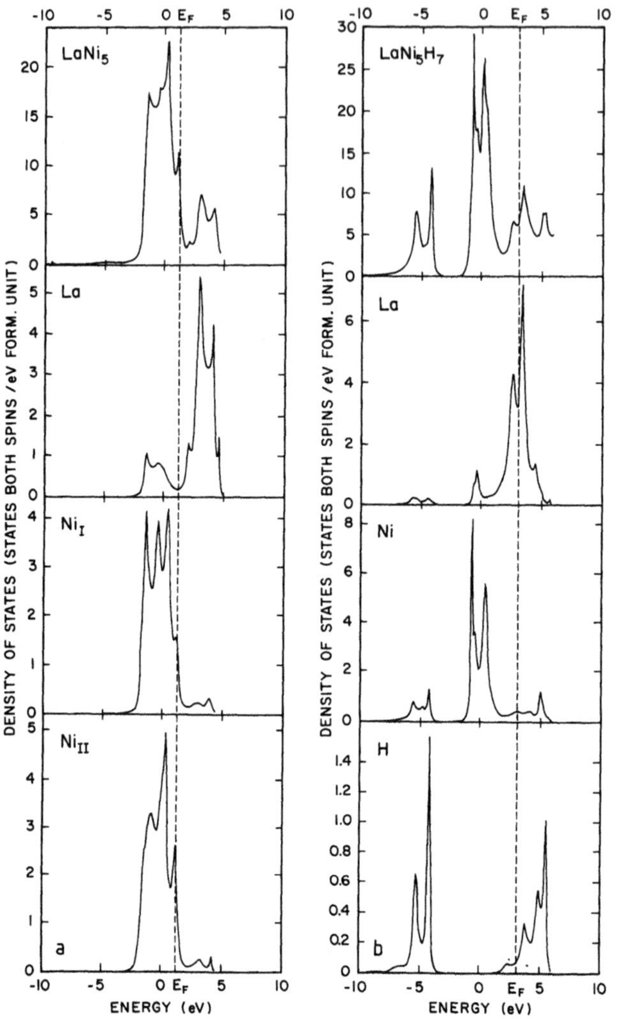

Fig. 5.12 **(a)** The total DOS of $LaNi_5$ and its character at the La, Ni_I (*basal plane*), and Ni_{II} sites (*middle plane*) [5.208]. **(b)** The total DOS of $LaNi_5H_7$, its *d*-character at a La site and at a Ni site in the basal plane and its *s*-character at a tetrahedral H site with Ni_2La_2 coordination [5.208]

The band structure of $LaNi_5$ has been studied by the spin polarized self-consistent APW method [5.207] and by the tight-binding recursion scheme [5.208]. Partial results using the LMTO method have also been reported [5.209]. The local DOS of *d*-type at the two different Ni sites and at the La site and the total DOS obtained by *Gupta* [5.208] are plotted in Fig. 5.12a. The Ni 3*d*-bands are located at lower energies than the La 5*d*-states, as expected from the relative position of the atomic *d*-levels. They give rise to a narrow and high peak in the DOS of about 3.2 eV width, while band structure calculations for paramagnetic Ni within the local exchange approximation lead to a *d*-band width of the order of 3.8 eV. We thus observe a narrowing of the Ni *d*-bands from pure fcc Ni to $LaNi_5$ which we ascribe essentially to the decrease of the Ni coordination number from 12 in pure Ni to 7.2 in $LaNi_5$, and to a lesser extent to a small increase in the Ni–Ni distances ranging from 2.46 Å to 2.51 Å in $LaNi_5$ compared to 2.45 Å in fcc Ni. The detailed features of the Ni DOS at the two different sites show of course some variation due to different atomic environments. The La 5*d*-bands centred at higher energies are also narrower than in fcc La due to the large increase in the La–La distances and to the decrease in the coordination number from fcc La to $LaNi_5$. In $LaNi_5$ each La atom is surrounded by 6 La atoms in the basal plane at distances ~5.01 Å and by 2 La atoms in the *c* direction located at ~3.98 Å. This is to be compared to $d_{La-La} = 3.75$ Å in fcc La. The La–Ni interaction is not very strong; it is responsible for the presence of smaller peaks observed in the local *d*-DOS plotted in Fig. 5.12a. The Fermi level E_F of $LaNi_5$ falls in a rapidly decreasing portion of the Ni *d*-bands which are not filled. The contribution of La *d*-states at E_F is very small.

The calculated values of the DOS are rather large, they correspond in the independent particle model and without enhancement to $\gamma = 28$ mJ (mole $LaNi_5)^{-1}$ K^{-2} [5.208] and $\gamma = 32.4$ mJ (mole $LaNi_5)^{-1}$ K^{-2} [5.207]. From heat capacity measurements of $LaNi_5$ between 1.6 and 4.2 K *Nasu* et al. [5.210] found $\gamma = 34.3$ mJ $mole^{-1}$ K^{-2} while *Takeshita* et al. [5.211] from their measurements between 1 K to 5 K obtained $\gamma = 36.5$ mJ $mole^{-1}$ K^{-2}. Data taken at higher temperatures between 5 and 300 K by *Ohlendorf* and *Flotow* [5.212] lead to $\gamma = 42.6$ mJ $mole^{-1}$ K^{-2}. The experimental results, in agreement with the calculations, point to a large value of the DOS at E_F in $LaNi_5$. This is also consistent with the large value of the magnetic susceptibility $\chi = 4.6 \times 10^{-6}$ emu/g [5.56]. It is to be noted that $LaNi_5$ is reported to be Pauli paramagnetic with strong Stoner enhancement [5.56]. Spin-polarized calculations [5.207], however, predict that $LaNi_5$ should be a weak ferromagnetic with a small spin moment of 0.69 μ_B per formula unit almost entirely due to Ni. The majority-spin 3*d* sub-band has been found to be filled unlike the minority-spin 3*d* sub-band which is well described by a rigid shift of about 0.35 eV due to exchange splitting. We should point out here the difficulty of the APW calculation due to the complexity of the crystal structure and to the self-consistency requirement imposed on the value of the magnetic moment. Valence band photoemission spectra of $LaNi_5$ [5.66, 191, 213] show the similarity of occupied bands with those of Ni, i.e. strong emission at E_F and the presence of the 6 eV satellite which both indicate that the Ni *d*-band is not completely filled (Fig. 5.13), in agreement with theory. A shoulder at 2 eV seems to be caused by the La–Ni interaction. The valence band of $LaNi_5$ is slightly narrower than that of Ni. The experimental width of ≈3.2 eV (2.4 eV FWHM), [5.66] agrees with theoretical values of 3.2 eV [5.208] and 3.4 eV [5.207].

Core-level photoemission spectra [5.191, 201] show that apart from a more intense low binding energy satellite of the La 3*d*-peak, which is ascribed to a $4f^1$ final state, the position and shape of the La 3*d* and Ni 2*p* core-level peaks do not change in $LaNi_5$ as compared to La and Ni, indicating that no significant charge transfer takes place upon formation of the alloy. In agreement, soft x-ray appearance-potential spectra do not reveal charge transfer [5.214]. Indeed, as

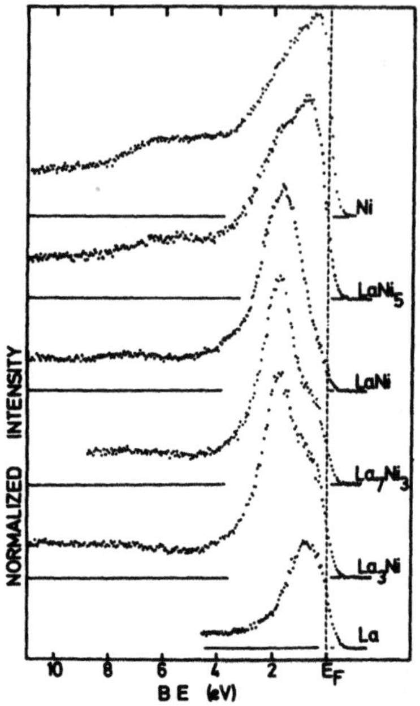

Fig. 5.13. XPS valence band spectra of La–Ni compounds [5.66]. The intensity of the 6 eV satellite and the high emission at E_F in the spectrum of $LaNi_5$ are comparable with those of Ni and indicate that the Ni *d*-band is not full

mentioned previously [5.66], the bonding in intermetallic compounds arises more from hybridization than from charge transfer. We have no explanation for the large charge transfer of 1.5 electrons from La to 5 Ni atoms calculated by *Malik* [5.207] for $LaNi_5$, while in contrast the same authors did not obtain a charge transfer in the isoelectronic compounds $GdNi_5$.

$LaNi_5$ Hydride. The hydrogen absorption is accompanied by a large lattice expansion ($\Delta a/a \approx \Delta c/c \approx 8\%$ for β-$LaNi_5H_x$ with $5<x<7$). Some controversy concerning the location of the H atoms in the lattice has developed amongst the numerous crystallographic investigations which are described in detail in the Chap. 4. However, recent neutron diffraction experiments on hydrogen-rich phases of β-$LaNi_5H_x$ with ($5 \leqq x \leqq 7$) [5.206, 215] have been interpreted in terms of the $P6_3mc$ non-centrosymmetric space group, in which the 6-fold *c*-axis of the $CaCu_5$ structure becomes a 6_3 screw axis due to the ordering of the H atoms. This ordering leads to a reduction of symmetry and a doubling of the unit cell size along the *c*-axis. The H atoms occupy octahedral sites coordinated with La_2Ni_4 neighbours and tetrahedral sites with Ni_4 and La_2Ni_2 neighbours. The Ni–H distances range from ~1.61 to ~1.68 Å and are shorter than in NiH ($d_{Ni-H} = 1.86$ Å). These crystallographic data made possible the first theoretical investigation of the electronic structure of $LaNi_5H_7$ which is the only hydride of $LaNi_5$ which has been investigated theoretically to date [5.208].

The total DOS of $LaNi_5H_7$ plotted in Fig. 5.12b shows the formation of metal-hydrogen derived states centred at ~5 eV below the centre of the Ni *d*-bands. A DOS analysis at the different atomic sites indicates that the H atoms interact mostly with Ni atoms despite the larger affinity of La for hydrogen. A similar situation occurs in FeTiH in which the Fe–H bonding is dominant [5.6]. The Ni d-bands are found to narrow from 3.2 eV in $LaNi_5$ to less than 2.5 eV in $LaNi_5H_7$. This is due in part to the large lattice expansion and to the lowering of the lowest portion of the Ni d-bands due to the Ni–H interaction. Some of the Ni d-states are also depleted from the main portion of the narrow d-bands and participate in the antibonding Ni d-, H s-states which extend over the La $5d$-bands and interact with them. This effect tends to oppose to the narrowing of the La d-bands caused by the lattice expansion.

The Fermi level of $LaNi_5H_7$ is found to fall at higher energy compared to that of $LaNi_5$. This is due in part to the presence of 7 supplementary electrons brought by H atoms in the unit cell and also to a deformation of the Ni d-bands, part of the Ni d-states being now found at higher energies in the Ni–H antibonding region. Although there are not too many of these states per Ni atom, since there are five Ni atoms in the unit cell this corresponds to a non-negligible number of electrons to be accommodated at higher energies. This effect contributes to the increase in the value of E_F. At the Ni sites the DOS at E_F in the hydrides is much smaller than that in the intermetallic compound.

The Fermi level of $LaNi_5H_7$ is found to fall in a rising portion of the DOS of the La atom. Thus the total DOS at E_F in the hydride remains large as in $LaNi_5$. The theoretical result agrees with the trend observed in the measured value of the electronic specific heat coefficient: *Ohlendorf* and *Flotow* [5.212] have observed a small decrease from $\gamma = 42.6\,\mathrm{mJ\,mole^{-1}\,K^{-2}}$ in $LaNi_5$ to $\gamma = 40.4\,\mathrm{mJ\,mole^{-1}\,K^{-2}}$ in $LaNi_5H_{6.3}$ and *Chung* et al. [5.216] found a decrease from 36.5 to $\sim 30\,\mathrm{mJ\,mole^{-1}\,K^{-2}}$. It is to be noted that the theoretical calculation [5.208] points to a large difference in the nature of the states at E_F, which are largely Ni d-like in $LaNi_5$ with no La contribution while in $LaNi_5H_7$, the La d-contribution becomes important. In $LaNi_5H_7$ the DOS at E_F also has some H s-character due to the antibonding Ni–H states. Although the Ni d-contribution per Ni atom at E_F remains smaller than the La contribution, the total Ni d-character at E_F is substantial since there are five Ni atoms per unit cell.

Magnetic susceptibility measurements reveal a strong decrease from $\chi = 4.6 \times 10^{-6}$ emu/g for $LaNi_5$ to $\approx 1.3 \times 10^{-6}$ emu/g for $LaNi_5H_6$ [5.56, 217] which is, however, caused by a sharp decrease of the Stoner enhancement factor and only a rather small decrease of the DOS at E_F.

No spectroscopic data are available for $LaNi_5$ hydride due to the fast desorption of hydrogen in vacuum. Thus an important part of the metal-hydrogen bonding features described above have not been confirmed experimentally. The theory suggests the importance of the Ni-d, H-s interaction and a much weaker La-d, H-s interaction. This is not in contradiction with Mössbauer data [5.218] which samples only the metal s-states and indicates a charge transfer from La to H and no transfer from Ni to H.

μSR Knight shift measurements [5.219] reveal $K\mu \approx 0$ for $LaNi_5$, which is a typical value for transition metals, and $K\mu = +80$ ppm for $LaNi_5H_6$ which is closer to the values for normal metals.

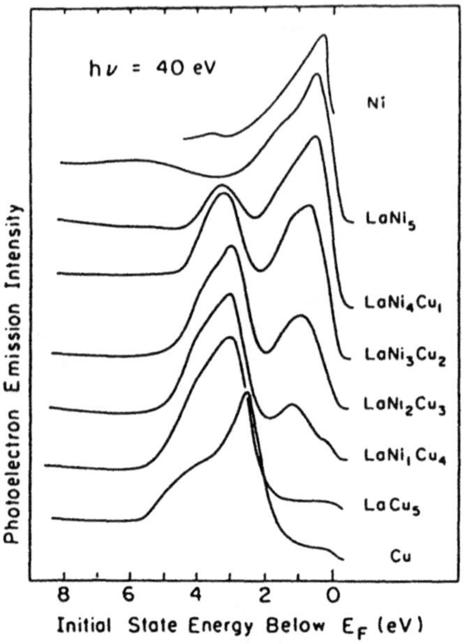

Fig. 5.14. Valence band photoelectron spectra of Ni, $LaNi_5$, $LaNi_{5-x}Cu_x$, and Cu [5.224]

Resistivity values are rather controversial; the room temperature electrical resistivity of evaporated $LaNi_5$ films (300 nm thick) was shown to be 7.6 $\times 10^{-4}\,\Omega$ cm – more than an order of magnitude larger than values measured on bulk samples of $CeNi_5$, $NdNi_5$, and $GdNi_5$ [5.220] and to decrease to 7.2 $\times 10^{-4}\,\Omega$ cm upon charging of the sample with hydrogen [5.221]. The opposite trend, i.e. an increase of the resistivity upon hydride formation was observed by *Larsen* et al. [5.222] also on thin films and by Walsh et al. [5.223] by EPR experiments on powder samples.

Effects of Substitution. The partial or total substitution of La or Ni in $LaNi_5$ is known to strongly affect the heat of hydride formation and the maximum hydrogen content. In metallic hydrides such variations should be caused by differences of the electronic structure of the host intermetallic compound. Two simple models have been proposed to correlate the hydride formation properties with the requirement of a partially empty Ni 3*d*-band either by appropriate position of E_F relative to this band (Fig. 5.14, [5.224]) or by the number of unpaired *d*-electrons [5.225]. To date neither total energy calculations nor electron spectroscopic methods are sensitive enough to detect these differences unambiguously: *Weaver* et al. [5.213] showed that the valence band photoelectron spectra of $CaNi_5$, UNi_5, $LaNi_5$, and $ThNi_5$ are nearly identical in agreement with electronic specific heat measurements [5.211], and reveal no differences which could be related to the wide range of the stability of the hydrides [5.226]. They suggest that crucial differences in the band structure, which are too subtle to appear as substantial changes in angle integrated spectra, may exist and propose angle resolved photoemission studies.

Photoemission spectra of various La–Ni intermetallics (Fig. 5.13 [5.66]) do show certain trends; however, no correlation with the trends of the heats of hydride formation has been established. Spectra of hydride forming La_7Ni_3 are rather similar to those of non hydride forming AlNi.

The heat of hydride formation depends among others on the availability of empty host metal states which can be pulled down or filled by a shift of E_F. The empty part of the density of states is clearly important and should be studied e.g. by inverse photoemission. The 6 eV satellite of the valence band spectra of Ni compounds which is a measure of the empty part of the band, gives us a hint: it is significantly more intense [5.66, 213] in the hydride forming $LaNi_5$ than in $ThNi_5$ and $CeNi_5$, which do not form stable hydrides.

5.4.2 Hydrides of FeTi

FeTi was considered for many years to be the most promising hydrogen storage material and considerable effort was made to understand its structural, electronic and activation behaviour. FeTi dissolves some hydrogen in the α-phase and forms the three distinct hydride phases $\beta_1(FeTiH_1)$, $\beta_2(FeTiH_{1.4})$ and $\gamma(FeTiH_{1.9})$. Amongst the ternary hydrides, the hydrides of FeTi were the first to be investigated theoretically. Band structure calculations are available for FeTiH, $FeTiH_2$ [5.6] and for the substoichiometric $FeTiH_{x<1}$ [5.227].

FeTi (α-Phase). This is a Pauli paramagnetic compound of cubic CsCl-type ordered structure which is isoelectronic to Cr. Systems with CsCl structure were among the first intermetallic compounds studied using band theory. Band structure calculations for ordered stoichiometric FeTi using SC–KKR formalism [5.198a, 228], SC–APW [5.229] and SC–ASW with a local density description of exchange and correlation [5.195] yield in rather good agreement a total DOS curve with E_F in the valley between occupied *d*-like bonding states derived primarily from Fe and empty *d*-like antibonding states derived primarily from Ti. The total DOS at E_F is $N(E)_F = 0.46$ states/eV and the width of the occupied band is 6.5 eV [5.229]. A transfer of 0.18 electrons from Ti to Fe is indicated [5.229]. Results of recent KKR–GF and KKR–CPA calculations [5.230] for the pseudobinary compound $TiFe_xCo_{1-x}$ agree for $x=1$ with the earlier calculations and emphasize the effect of antistructural atoms (e.g. Fe on Ti sites) on the magnetic properties. For disordered FeTi the sharp DOS features disappear, the valley around E_F becomes filled and $N(E_F)$ increases considerably [5.198a, 228].

Most of the experimentally determined parameters agree qualitatively with the results of band structure calculations: De Haas–van Alphen measurements of the Fermi surface can be identified with a hole surface around the *M* point of the BZ and an electron surface around the *X* point [5.231]. The measured hole pocket volumes are 2.7 times smaller than the calculated values. The electronic specific heat coefficient of annealed samples is small ($\gamma = 0.18$ mJ mole^{-1} K^{-2})

[5.232]. The magnetic susceptibility has a value of 3.4×10^{-6} emu/g [5.233] almost independent of temperature. Photoelectron spectra of the valence band show the main features of the band calculations [5.234]. Core-level spectra show, as compared to those of elemental Fe and Ti, no shift of the Fe 3*p*- and Fe 2*p*-levels, but shifts of ≈ 0.5 eV to larger binding energy for Ti 2*p*- and Ti 3*p*-levels were observed [5.233, 234]. The asymmetry of the Fe 2*p*-line of FeTi is by far less pronounced than in elemental Fe, in agreement with a low partial density of Fe states at E_F [5.84, 233]. From XANES studies, it is concluded that the empty *p*-like DOS for Fe and Ti sites are very similar in shape to that of elemental Ti [5.235].

Surface precipitates of Fe or Fe–Ti–O compounds formed after selective surface oxidation greatly affect the magnetic properties [5.233]. Furthermore antistructure Fe atoms on Ti sites may also influence the magnetic and electronic properties, if indeed they exist in FeTi [5.240]. However, the order parameter as measured by neutron diffraction [5.236] is $S = 1.00$; the local order around both Fe and Ti is good up to the sixth coordination shell (EXAFS) [5.235] and upon annealing, when antistructure atoms should disappear, the electronic specific heat coefficient γ increases [5.232] in contradiction to band structure calculations, which reveal a much larger $N(E_F)$ for disordered FeTi [5.228]. The electrical resistivity of FeTi samples with a residual resistivity ratio $\varrho(300\,K)/\varrho(4.2\,K) = 8$ amounts to $16.7 \pm 0.5\,\mu\Omega$ cm at room temperature. FeTi shows the largest increase of the resistivity ever observed due to the dissolution of hydrogen in the α-phase. The specific resistivity increases by $\Delta\varrho = 8.2\,\mu\Omega$ cm per 1% hydrogen in FeTi. A linear increase up to the composition $FeTiH_{0.07}$ has been observed [5.237].

$FeTi_{1.0}$ (β-Phase). Although FeTi crystallizes in the CsCl structure, β-FeTiH has an orthorhombic structure [5.238, 239] which is a distorted version of the tetragonal lattice formed from the doubling of the CsCl cell of FeTi (see Chap. 4). The H atoms occupy distorted octahedral intersitices with two iron atoms as nearest neighbours and four titanium atoms as next-nearest neighbours. The hydrogen atoms are located in chain-like positions in the planes containing the Fe atoms. It is interesting to notice that the hydrogens occupy octahedral sites in FeTiH instead of the more usual tetrahedral sites characteristic of the bcc metals, and that they form chains. Further, the H atoms are much closer to the two Fe atoms (Fe–H distance = 1.579 Å) than to the four Ti atoms (Ti–H distance = 2.163 Å), despite the greater affinity of hydrogen for Ti than for Fe.

Keeping in mind the remarks concerning the crystal structure, the band structure results [5.6] of FeTiH plotted in Fig. 5.15a can be understood (i) by a folding of the bands of FeTi due to the doubling of the size of the CsCl unit cell, (ii) the expansion of the cell in the (110) direction of the CsCl structure and lifting of the degeneracies due to the orthorhombic distortion and (iii) by the role of the hydrogen atoms in the lattice. In order to illustrate the first point we have plotted in Fig. 5.15b the bands of pure FeTi (dashed lines indicate the

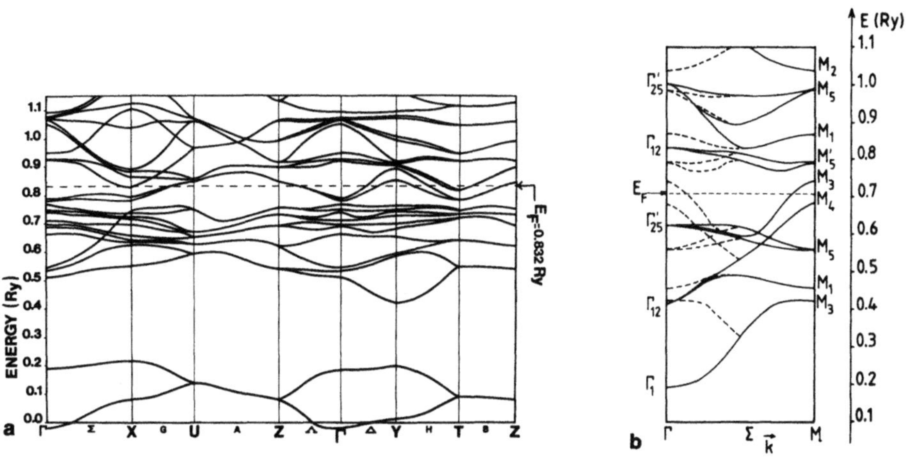

Fig. 5.15. (a) The energy bands $E(\mathbf{k})$ of β-FeTiH along several high symmetry directions of the orthorhombic BZ [5.6]. **(b)** The energy bands of FeTi along the ΓM direction of the bcc BZ. (——). The dashed lines indicate the folding of the bands due to the BZ reduction in that direction

folding of the bands) in the (110) direction of the CsCl Brillouin Zone (BZ) due to the doubling of the unit cell size. Thus, at the BZ centre of β-FeTiH we obtain the states corresponding to the points Γ and M of the BZ of pure FeTi; in the hydride, all degeneracies are lifted by the orthorhombic distortion. Since the electronic structure of FeTi in the CsCl simple cubic BZ is characterized by 10 metal d-bands overlapped and hybridized with a wider metal sp-band, these will lead to 22 bands in β-FeTiH. The comparison of Figs. 5.15a and b shows that the presence of H in the lattice leads to a drastic lowering of the lowest lying states; for example, at the BZ centre of FeTiH, the two low lying states which correspond to Γ and M_3 in pure FeTi are lowered by about 4 eV. The lowering of metal states having an s-symmetry at the interstitial site by the H potential which strongly scatters s-waves, is a feature common to all the metal hydrides, as mentioned previously for binary hydrides. The second band in Fig. 5.15a should be understood as resulting from the folding of the low lying metal-hydrogen band, due to the doubling of the size of the unit cell. A detailed analysis of the states modified by the metal-hydrogen interaction shows that additional states which are empty in FeTi appear in the metal d-bands below the Fermi energy in the hydride. This is the case for a branch of metal $4p$-states corresponding to the M_5 point of the BZ of FeTi. These states lie at about 1.5 eV above E_F in FeTi; they are lowered by about 3.4 eV by the H potential and lie below E_F in the hydride. The orthorhombic distortion leads to important splittings in the d-bands; for example, at the BZ centre, the Γ_{12} and Γ'_{25} states of the cubic BZ are split by as much as 0.25 meV in the hydride. The low lying metal-hydrogen bands are formed out of states already filled in the pure intermetallic, which have been lowered by the H potential. The width of the

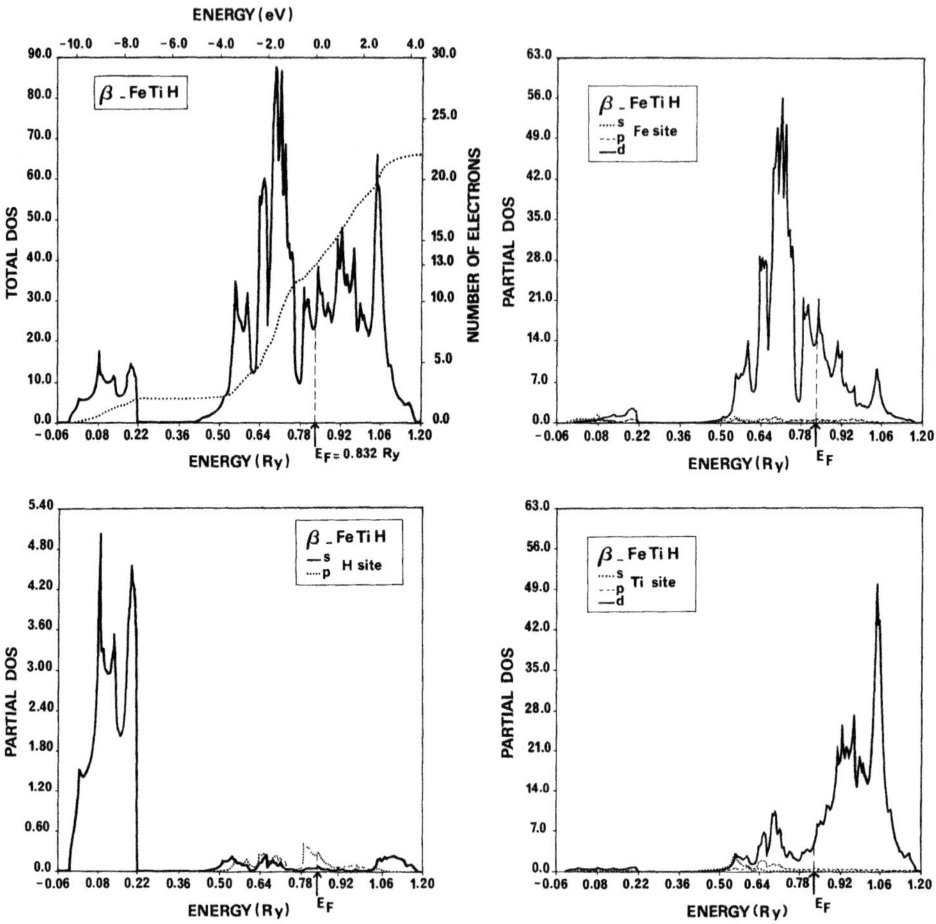

Fig. 5.16. The total DOS of β-FeTiH and its partial wave analysis inside the Fe, Ti, and H muffin-tin spheres. Units are [states of both spins/Rydberg formula unit] [5.6]

lowest metal *d*-states measured up to the energy of the valley is slightly smaller in FeTiH than in FeTi due to the lowering of the lowest portion of the Fe *d*-states by the H potential; the increase in the lattice parameter also plays a role in the slight narrowing of the *d*-bands. Since additional states which are empty in FeTi appear below E_F in the hydride, less than one electron brought by the H atom has to be accommodated at the top of the *d*-bands and thus, although E_F is shifted towards higher energies in the hydride, the protonic rigid-band model is not quantitatively correct. Furthermore, the metal *d*-bands are deformed in going from the intermetallic to its hydride.

In Fig. 5.16, we plot the DOS of FeTiH obtained from an APW calculation [5.6]; it can be characterized (i) by a structure at low energy, resulting from the metal-hydrogen interaction which is centred at 9 eV below the

Fermi level, and (ii) the metal d-bands at higher energies in which we identify, as in pure FeTi the two peaked structure characteristic of the bonding and antibonding metal states in the bcc materials. The increase by 2.3% of the Fe–Ti bond length between FeTi and FeTiH results in a weakening of the bond and in a larger overlap between the bonding and antibonding metal d-states in FeTiH, the valley being narrower in the hydride. The number of states at E_F increases substantially in going from FeTi to FeTiH; for the latter we obtained $N(E_F) = 1.76$ states of both spin/eV–FeTiH which corresponds to an unenhanced electronic specific heat coefficient $\gamma = 2.02\,\mathrm{mJ\,mole^{-1}\,K^{-2}}$.

A partial DOS analysis into its angular momentum components inside the H and metal MT spheres is plotted in Fig. 5.16. It reveals that the low lying energy states have essentially a H s- and also a Fe d-character, corresponding to states having mostly an e_g symmetry; the Ti d-contribution and the metal s- and p-components are substantially smaller. The lowest portion of the metal d-bands is essentially dominated by Fe d-states while Ti d-states have their most important contribution for energies larger than that of the valley in the DOS as in pure FeTi. However, inspite of an increase in E_F, the Fe d-character of the DOS at E_F analyzed inside the MT spheres is almost three times larger than the Ti d-contribution.

The lowering of metal states by the H potential has been recognized to be an important factor for the stability of the hydride. As indicated above, these metal states can be classified into two categories: (i) metal states already filled in pure FeTi, like those forming the two low lying metal-hydrogen bands and (ii) metal states previously empty in FeTi, like the $4p$ metal branch around the M_5 point of the FeTi, BZ, which is lowered by about 3.4 eV in the hydride. It is interesting to notice that in bcc VH *Switendick* [5.4] found that metal p-states at the N point of the bcc BZ, having an s-symmetry at the H sites, are lowered by the H potential by as much as 5 eV. The H concentration appears to be an important factor in the lowering of these metal p-states and could explain, in part, the difference in the enthalpies of formation of α- and β-phases. For binary hydrides such as PdH it has also been noticed that the H–H next nearest neighbour interactions are crucial in the lowering below E_F of the empty Pd $5p$-states.

In FeTiH, the lowering of empty metal states brings, however, only about 0.15 electrons below E_F and thus the Fermi level of the hydride is shifted towards higher energies, a factor which adversely affects the stability of the compound.

With the aim of assessing the role of the H–H interactions along the chains parallel to the (100) direction in FeTiH, we chose to break these chains by locating the two H atoms at (0, 1/4, 1/4) and (0, 3/4, 3/4) instead of (0, 1/4, 1/4) and (0, 1/4, 3/4) as in β-FeTiH. The breaking of the chains results in a shift of the low lying metal-hydrogen bands towards higher energies by 0.2 eV on the average. With this new location of the H atoms, in addition to the plane URTZ which is doubly degenerate in β-FeTiH, we obtain in the BZ another plane TYRS of doubly degenerate eigenvalues. The metal d-states and the position of the Fermi level are essentially unchanged except for the states at the point Y

which are shifted up in the new structure and become doubly degenerate like the point Z. Thus, from the study of the one-electron energy eigenvalues, we can conclude that the H–H interactions along the chains appear to play a role in the stability of the compound. Evidence for this role is given by the difference in the heats of solution between the dilute α-phase ($\Delta H = 0.11$ eV/H atom) and the β-phase ($\Delta H = 0.16$ eV/H).

We also chose to locate the H atoms in the octahedral interstices close to the Ti atoms. We find, in this case, that the low lying bands are shifted up by 0.34 eV on average. Since, as shown above, the low lying bands of β-FeTiH are mostly formed of H and Fe states, the raising of these states in the new structure can be interpreted via the decrease in the strength of the H–Fe interaction due to the new location of the H atoms close to Ti. In addition we observe sensible modifications of the metal d-bands but the final conclusion concerning the stability of the hydrides needs to await total energy calculations. Besides the role of the electronic interactions, demonstrated here by the change in the position of the low lying metal-hydrogen states, and also by the modification of the d-bands, the size of the interstice available for H may also be an important factor for the preferred site occupancy of H close to the Fe atoms.

The few available experimental results support the theoretical conclusions; rapid desorption of hydrogen and difficulties related to the need of activation has so far rendered photoelectron spectroscopic measurements of FeTi hydride impossible. The exposure of FeTi to small doses of hydrogen induces enhanced emission in the range 4–8 eV [5.234].

Both magnetic susceptibility and electronic specific heat [5.232] increase upon hydride formation. Due to magnetic precipitates in the near surface region formed in the activation process [5.233], and due to the effect of antistructure Fe atoms possibly formed upon hydride formation [5.240, 232] the estimation of $N(E_F)$ from susceptibility and specific heat data is even more difficult than usually. Values of $\chi = (7.5 \pm 1) \times 10^{-6}$ emu/g for $FeTiH_{1.0}$ [5.233] and $\gamma = 1.01$ mJ mole^{-1} K^{-2} for FeTi hydride of unknown composition [5.232] were obtained. Both values indicate a significantly higher $N(E_F)$ in the hydride, a trend which is in agreement with theory.

Information on the momentum distribution of the electrons has been obtained by *Lässer* et al. [5.241] from their Compton profile analysis of FeTi and the β-phase $FeTiH_{1.17}$. The main differences between the two profiles are observed for small values of the reduced scattering vector ($q < 2$ a.u.) which correspond to large values of the position vector in real space and can thus be ascribed to bonding effects. Realistic wavefunction calculations are needed to give a reliable interpretation of the data since crude estimates obtained from a simple scaling of the profiles using the protonic model are in very poor agreement with the experiment. Indications concerning the trends in the heat of solution of H in the β-phase have also been obtained from the Compton profile measurements [5.241].

Mössbauer isomer shift (IS) measurements reveal a systematic decrease of the s contact electron density ϱ_s at the metal site with hydrogen uptake for binary compounds such as TaH_x, PdH_x, as well as for hydrides of intermetallic compounds $FeTiH_x$ [5.242]. For $FeTiH_x$ unlike for PdH_x, simple renormalization of the 4*s*-wavefunctions due to the lattice expansion accompanying hydrogen uptake did not appear to be sufficient to explain the observed decrease in the contact *s* density [5.242]. A microscopic calculation of ϱ_s at the nucleus was performed [5.243] to interpret the data of *Swartzendruber* et al. [5.242]. The calculated valence electron contribution $\varrho_s = 5.858$ for β-FeTiH is almost identical to the theoretical value obtained for pure iron metal by *Callaway* et al. [5.244] ($\varrho_s = 5.823$) and with the experimental result of *Shinohara* and *Fujioka* ($\varrho_s = 5.53 \pm 0.46$) [5.245]. Although it should be pointed out that the present calculation for β-FeTiH [5.243] and the calculation of *Callaway* et al. [5.244] for iron metal have been carried out using different techniques, we obtain nearly the same value of ϱ_s (Fe). This result is in very good agreement with the data of *Swarzendruber* et al. [5.242] since for the $FeTiH_{0.9}$ sample these workers did not observe any shift of the Mössbauer line compared with that for pure iron metal; indeed a negative IS ($-0.14\,\mathrm{mm\,s^{-1}}$) was observed from pure iron metal to pure FeTi and this shift disappears with hydrogenation when β-$FeTiH_{0.9}$ is formed [5.242].

$FeTiH_2$. The γ-phase has been shown to have a monoclinic structure [5.238] which can be regarded as a distortion of a tetragonal cell; the hydrogen atoms are ordered at octahedral interstices. In the electronic band structure calculation [5.6], the monoclinic distortion with an angle of about 97° was ignored and the lattice was treated as tetragonal.

A comparison of the DOS of $FeTiH_2$ plotted in Fig. 5.17 with the DOS of FeTiH reveals that on hydrogen uptake the metal-hydrogen related structure at low energies grows both in width and in intensity. Indeed, this structure which corresponds to two bands in FeTiH is formed from four bands in $FeTiH_2$ and thus accommodates twice as many electrons. The additional two bands found for $FeTiH_2$ result from the presence of two additional H atoms in the unit cell; the hydrogen states forming these low lying bands are also hybridized with the metal *d*-bands. The Fermi energy of $FeTiH_2$ falls above the valley of the DOS, as for FeTiH. For $FeTiH_2$ we obtain $N(E_F) = 1.68$ states of both spin per eV per $FeTiH_2$ molecule which corresponds to an unenhanced value of $\gamma = 1.93\,\mathrm{mJ\,mole^{-1}\,K^{-2}}$ [5.6].

The DOS analysis of $FeTiH_2$ into its angular momentum components *s*, *p*, and *d*, inside the MT hydrogen and metal spheres, in shown in Fig. 5.17. As for FeTiH, the structure at low energy is dominated by the H *s*-states and also by the metal *d*-states; a smaller contribution from the metal *s*- and *p*-states is also observed. It is interesting to note that the contribution of the Ti *d*-states to the low energy structure is much larger in $FeTiH_2$ than in FeTiH. This is due to the fact that in $FeTiH_2$ one of the four H atoms is located closer to two Ti atoms rather than to Fe atoms.

Susceptibility measurements indicate a further increase for $FeTiH_{1.9}$ to $\chi = (8.5 \pm 1) \times 10^{-6}$ emu/g at room temperature [5.233]. The valence electron

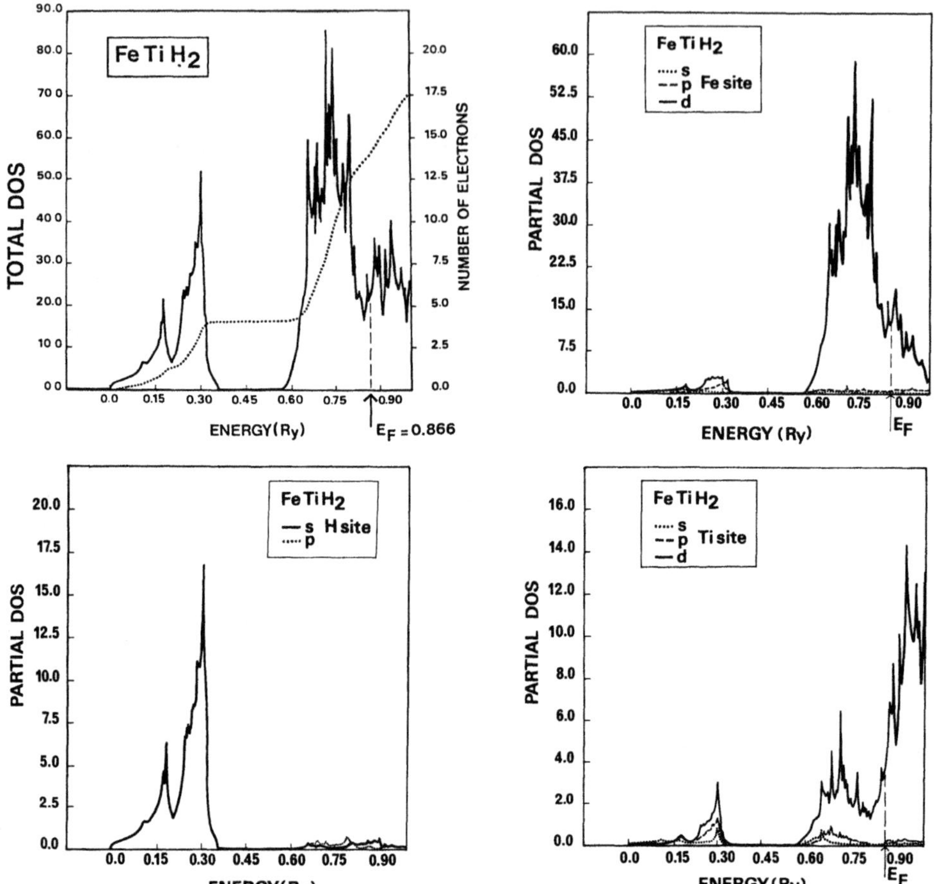

Fig. 5.17. The total DOS of $FeTiH_2$ and its partial wave analysis inside the Fe, Ti, and H muffin-tin spheres. Units are [states of both spins/Rydberg formula unit] [5.6]

contribution to ϱ_s decreases by 10.4% from β-FeTiH to γ-$FeTiH_2$. One can calculate the corresponding IS difference

$$\Delta\varepsilon = \alpha(\varrho_{\mathrm{FeTiH}} - \varrho_{\mathrm{FeTiH_2}}), \tag{5.9}$$

where the calibration constant α, which is related to the change in nuclear radius during the transition, has the value given by *Ingalls* [5.246] $\alpha = -0.37\,a_0^3\,\mathrm{mm\,s^{-1}}$. The theoretical result $\Delta\varepsilon = 0.23\,\mathrm{mm\,s^{-1}}$, is in good agreement with the experimental value [5.242].

$$\Delta\varepsilon = \alpha(\varrho_{\mathrm{FeTiH_{0.9}}} - \varrho_{\mathrm{FeTiH_{1.7}}}) = 0.27\,\mathrm{mm\,s^{-1}}. \tag{5.10}$$

The decrease in ϱ_s(Fe) with hydrogen uptake from FeTiH to $FeTiH_2$ cannot be ascribed to the pure effect of a 4*s*-wavefunction renormalization due to volume expansion since, on the assumption that this effect is proportional to $-\Delta$ volume/volume, it accounts for only half of the IS observed between the dilute α-phase of $FeTiH_x$ and γ-$FeTiH_{1.7}$. The theoretical results show that 93.6% of the decrease in ϱ_s(Fe) from FeTiH to $FeTiH_2$ is due to a decrease in the contribution of the low lying bands. Since, as discussed previously, the metal states which contribute to these bands are very much affected by the metal-hydrogen bonding we can conclude from the theoretical study [5.243] that the depletion of the 4*s* metal states at the iron site on hydrogen uptake is mainly due to the metal-hydrogen bonding.

Substoichiometric $FeTiH_x$ ($x<1$). The tight-binding CPA method mentioned in Sect. 5.2.1 has been applied to study substoichiometric hydrides of $FeTiH_{x<1.0}$ [5.227], using Slater-Koster (SK) fits to the band structures of FeTi [5.229] and $FeTiH_{1.0}$ [5.6]. The DOS and its site and angular momentum components studied for H concentrations $x=0.1$, $x=0.8$, and $x=0.9$ indicate a very nonrigid band behaviour. The low lying hydrogen induced states develop from previously occupied metal states while the additional electrons brought by the H atoms fill up previously unoccupied states pulled down below E_F. The readjustment of the Fermi level, is small and appears to be nearly independent of x, but the value of the DOS at E_F increases with x [5.227] in agreement with the heat capacity measurements [5.232]. Similar behaviour of the Fermi level position has been found by *Temmerman* and *Pindor* [5.42] from their KKR–CPA results on the Pd–Ag–H system.

One can say that most of the features of the bonding of H in the intermetallic compound FeTi do not differ fundamentally from the results knowns for binary TM hydrides. They are, however, more difficult to analyse due to the low symmetry of the crystal structure. Further work on the stability and on preferred site occupancy of H should be pursued.

5.4.3 Mg_2Ni and Isoelectronic Compounds and Their Hydrides

The hydrides Mg_2NiH_4, Mg_2CoH_5, and Mg_2FeH_6 are characterized by extremely high hydrogen content per unit volume and weight, and by very strong H-metal bonding together with nonmetallic appearance. One of the final goals of the investigations of these compounds therefore, is to learn how this H–metal bond could be weakened.

Mg_2Ni. Mg_2Ni has a hexagonal unit cell consisting of chains of Ni atoms separated by two layers of Mg atoms [5.247]. No theoretical investigation of the intermetallic compound Mg_2Ni is yet available, however electronic structure investigations of several AB_2–C 16 structure compounds by means of the tight-binding method have been performed [5.248, 249]. Photoelectron spectra of the valence band show that the Ni *d*-band moves away from E_F and that it becomes narrower and more symmetric than in elemental Ni. The intensity of the 6 eV satellite is smaller indicating a decrease of the number of *d*-holes [5.66]. Core-level spectra exhibit almost no shift ($\lesssim 0.2$ eV) for the Ni 2*p*-level and a shift of ≈ 0.4 eV to smaller binding energy for Mg 2*s*- and Mg 2*p*-levels as compared to the elemental metals [5.201, 84]. The measured magnetic susceptibility amounts to $\chi = (0.9 \pm 0.2) \times 10^{-6}$ emu/g at room temperature and increases below 80 K [5.250].

Crystal Structure of Mg_2NiH_4. Mg_2Ni absorbs hydrogen reversibly and forms an α-phase up to $H/M = 0.3$ and a β-phase which is based on the stoichiometric Mg_2NiH_4 composition, for higher hydrogen concentrations [5.251]. The formation of the β-phase is accompanied by a structural transformation of the metal lattice and by a substantial volume expansion ΔV of 4.80 Å^3 per H atom. A number of crystallographic determinations of the different phases of Mg_2NiH_x are available as discussed in Chap. 4. The interpretations of the data have been in some cases controversial.

Up to date, theoretical calculations on Mg_2NiH_4 have been undertaken only for the cubic high-temperature phase ($T > 245$ °C). X-ray diffraction [5.252] and neutron diffraction [5.253] data have shown unambiguously that in the cubic phase the metal atoms have an anti-fluorite-type arrangement in which the Ni atoms form an fcc lattice while the Mg atoms occupy the 1/4 (1, 1, 1) positions in the cubic cell. The location of the H atoms in this phase has been the object of discussions [5.253, 254].

For the interpretation of the crystallographic data and for the understanding of the electronic properties, it is interesting to keep in mind that Mg_2NiH_4 belongs to a family of isoelectronic ternary metal hydrides M'_2MH_x where M' is a divalent simple or rare earth metal (M' = Mg, Ca, Sr, Eu, Yb) and M a transition metal (M = Fe, Co, Ni, Rh, Ir, Ru, etc.). In this family, the maximum hydrogen content corresponds to $x = 6$, and one can notice that the maximum value of x which can be achieved for given elements M' and M is such that the total number of valence electrons is 18. Neutron diffraction analysis on $MgFeD_6$ samples [5.255] have been interpreted in terms of the K_2PtCl_6 structure in which the H atoms form octahedral cages around the TM atoms with positions close to (1/4, 0, 0). The positions of the H atoms have been shown to be consistent with Mössbauer, Raman and IR measurements [5.255]. This structure will hereafter be referred to as structure I. Another possible partial occupancy of H sites close to (1/4, 0, 0), hereafter referred to as structure II, has been obtained from neutron diffraction analysis on Mg_2NiD_4 [5.254] and appears to be consistent with NMR data [5.256]. However, most of the neutron diffraction data on the high temperature cubic phases of Mg_2NiD_4 [5.253] and Mg_2CoD_5 [5.257] are analyzed in terms of a disordered distribution of D atoms at the summits of octahedral cages around the TM atoms. These H locations are believed to result from the high temperature thermal average of orientationally square-planar [5.258] or tetrahedral [5.259] ions NiD_4^{-4} and square pyramidal complex ions CoD_5^{4-} observed respectively for Mg_2NiD_4 and Mg_2CoD_5 at room temperature.

Electronic Structure of Mg_2FeH_6. For clarity we shall first describe the theoretical APW band structure results of a compound corresponding to the maximum hydrogen capacity Mg_2FeH_6 in which the hydrogen atoms form octahedral cages around the transition metal atoms. The energy bands of Mg_2FeH_6 [5.260] are plotted along some high symmetry directions in Fig. 5.18a. In order to understand these results we should keep in mind the following essential features of the structural data. The *M–M* distances are large ($d_{Fe-Fe}=4.56$ Å in Mg_2FeH_6 while $d_{Fe-Fe}=2.48$ Å in bcc iron), therefore we expect a weak (*Md*)–(*Md*) interaction and narrow *Md*-bands. The *M*–H distances are short ($d_{Fe-H}=1.54$ Å in Mg_2FeH_6), and therefore strong (*Md*)–H interactions and thus large ligand field splittings are expected. The *M'*–H distances are much larger because the hydrogen atoms are located close to the centres of the faces of the *M'* cubes ($d_{Mg-H}=2.28$ Å in Mg_2FeH_6). We can thus expect the electronic properties of the compounds under study to be strongly dominated by the characteristic features of the octahedral transition metal complex MH_6 since much weaker *M'*–H and *M–M* interactions are expected. In order to illustrate this point we have reproduced in Fig. 5.18b the typical energy diagram expected for a free octahedral ML_6 complex where *M* is a transition element and the ligand *L* has occupied *s*- and empty *p*-orbitals. In the present case, since the octahedral complex is in a periodic array of metal atoms, we expect of course that the *Ms*- and *p*-states are partially filled and thus are located below the *Md*-orbitals. Nevertheless, we observe strong similarities between the bands and the schematic diagram plotted in Figs. 5.18a and b. At the Brillouin zone center Γ, we obtain in increasing order of energy (i) the Γ_1 state which is a hybridized (*M's*)–(*Ms*) state lowered by the H *s*-interactions (this state corresponds to the $1a_{1g}$ molecular orbital level), (ii) a doubly degenerate Γ_{12} state which is a bonding combination of H *s*-states and the Md_{z^2}- and $d_{x^2-y^2}$-orbitals whose lobes point towards the hydrogen sites (this state corresponds to the $1e_g$ molecular levels) and (iii) a triply degenerate H *s*-state which is non-bonding with respect to the transition metal atom (this state corresponds to the $1t_{1u}$ molecular degenerate levels). In the solid we thus obtain six low lying bands which result essentially from strong (*Md*)–(H*s*) and H–H interactions, and the corresponding states are hydridized with the *M sp*-states and the *M' sp*-states. A gap separates the first six bands from the next three-band complex which is formed by the narrow *M* $d(t_{2g})$-states. At Γ, the Γ'_{25} state corresponds to the molecular $1t_{2g}$ levels shown in Fig. 5.18b. The t_{2g} orbitals whose lobes are not pointing towards the hydrogen atoms are not affected by the H *s*-interaction. Another energy gap separates the *M* $d(t_{2g})$-states from the higher states in which we find the antibonding combination of the doubly degenerate d_{z^2} and $d_{x^2-y^2}$ orbitals with the H *s*-states.

The total DOS for Mg_2FeH_6 is plotted in Fig. 5.18c. It is characterized by a low lying structure due to strong *M*–H and H–H interactions with a weaker *M'* *sp*-hybridization. These states accommodate 12 electrons. The remaining six electrons of the compounds with 18 valence electrons fill the t_{2g} manifold of the split *M* *d*-states. The filled *M* *d*-bands are narrow and strongly peaked owing to

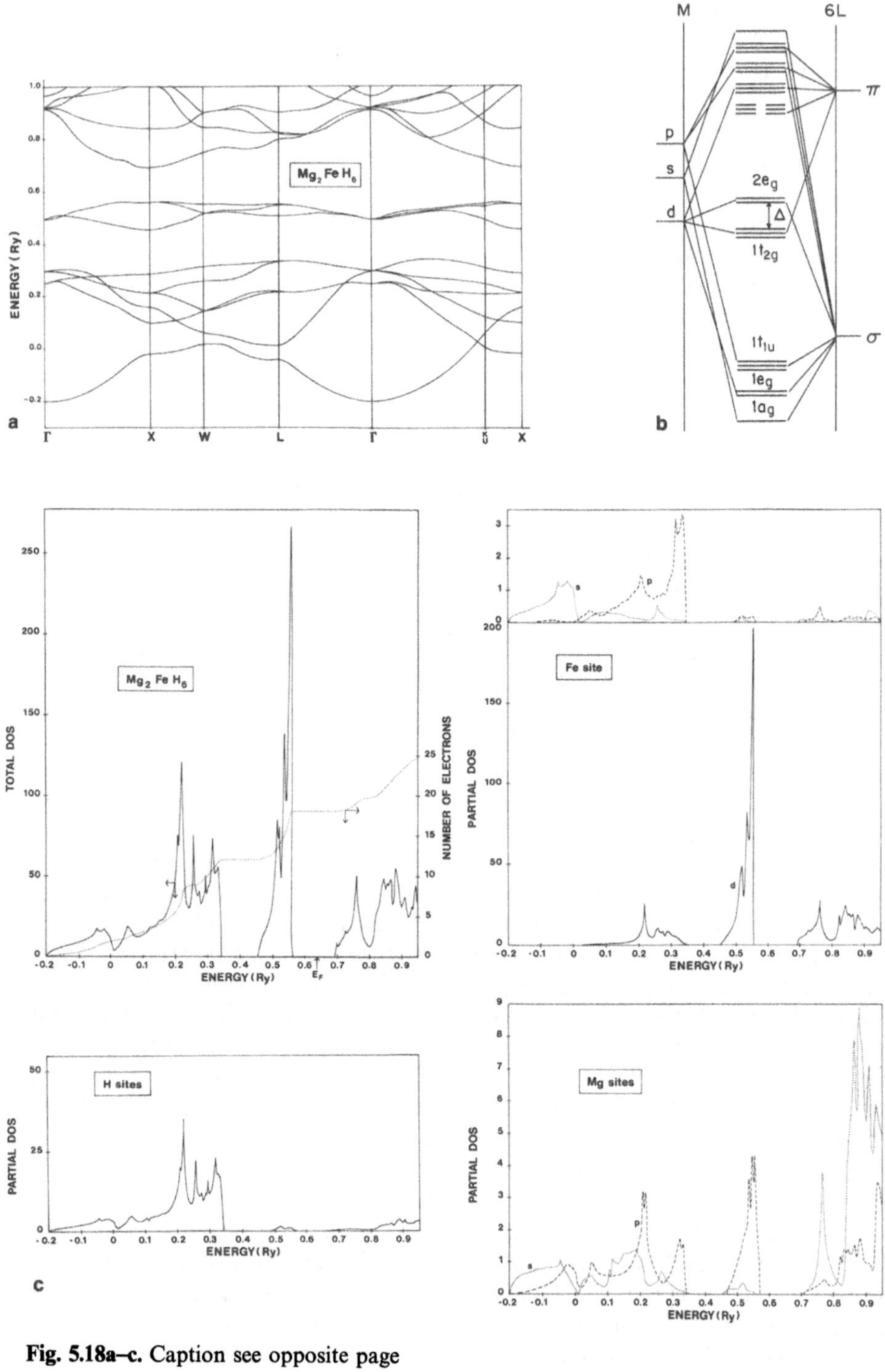

Fig. 5.18a–c. Caption see opposite page

the large M–M distances. An energy gap of 1.8 eV separates the M $d(t_{2g})$-states from the empty higher bands. The compound is thus found to be non-metallic. The origin of the gap between the filled non-bonding t_{2g} levels and the empty antibonding [M $d(e_g)$–H s]-states is due to the large ligand field splitting caused by the strong interactions between the transition metal elements and their octahedral environment of hydrogen atoms, the M–H distances being rather short.

The partial wave analysis of the wavefunctions was performed inside the MT spheres and the corresponding DOS for Mg_2FeH_6 are plotted in Fig. 5.18c. It is clear that the six low lying bands which form the structure at low energies consist mostly of H s-states and M d-states (which correspond to the e_g symmetry). The contribution of M s- and p-states also appears on the low energy side of this structure. The non-negligible contribution of the M' sp-states should also be noted. The second narrow filled structure of the DOS consists almost exclusively of M d-states of t_{2g} symmetry. It is interesting to note the contribution of the antibonding M d-states (of e_g symmetry) to the H s-states in the empty levels located above the energy gap. At higher energies we find mostly Mg p-states.

It would be very instructive to compare the present results with soft x-ray emission and absorption spectra. Some data of this type are available for Mg_2NiH_4 [5.261] and Mg_2CoH_5 [5.262] which belong to the same family, but such spectra have not been obtained for the compound under study here.

Electronic Structure of Mg_2NiH_4. As discussed previously the electronic structure calculation for Mg_2NiH_4 in its cubic phase was performed using structure I in which the H atoms partially occupy the summits of octahedral cages surrounding the Ni atoms. For our calculation we took into account the partial occupancy of the H sites by averaging the structure factor of the H atoms, this preserves the cubic symmetry around the Ni sites. The essential features are similar to those discussed for Mg_2FeH_6. Apart from small variations originating from the change in the lattice constants and atomic species the main difference is found in the relative position of the nine low-energy M–M, M–H and H–H bands. For example, at the BZ centre the triply degenerate H s-state is found in Mg_2NiH_4 to lie above the Γ_{12} and Γ'_{25} Ni–d derived bands while in Mg_2FeH_6 the H–s derived bands lie between the Γ_{12} and Γ'_{25} states. Indeed, with increasing hydrogen content, the TM–H interactions become stronger, the crystal field splitting of the TM d-bands increases and the M–H states gain in stability. As a consequence, unlike in Mg_2FeH_6 the narrow nonbonding Ni d–t_{2g} states are not separated by a gap from the low-energy metal-hydrogen bonding states.

Fig. 5.18. (a) The energy bands $E(\boldsymbol{k})$ of Mg_2FeH_6 along several high symmetry directions of the fcc BZ [5.260]. **(b)** Schematic energy level diagram of a octahedral ML_6 complex where M is a transition element and the ligand L has occupied s and empty p orbitals. **(c)** The total DOS of $MgFeH_6$ and its partial wave analysis inside the Fe, Mg, and H muffin-tin spheres. Units are [states of both spins/Rydberg formula unit] [5.260]

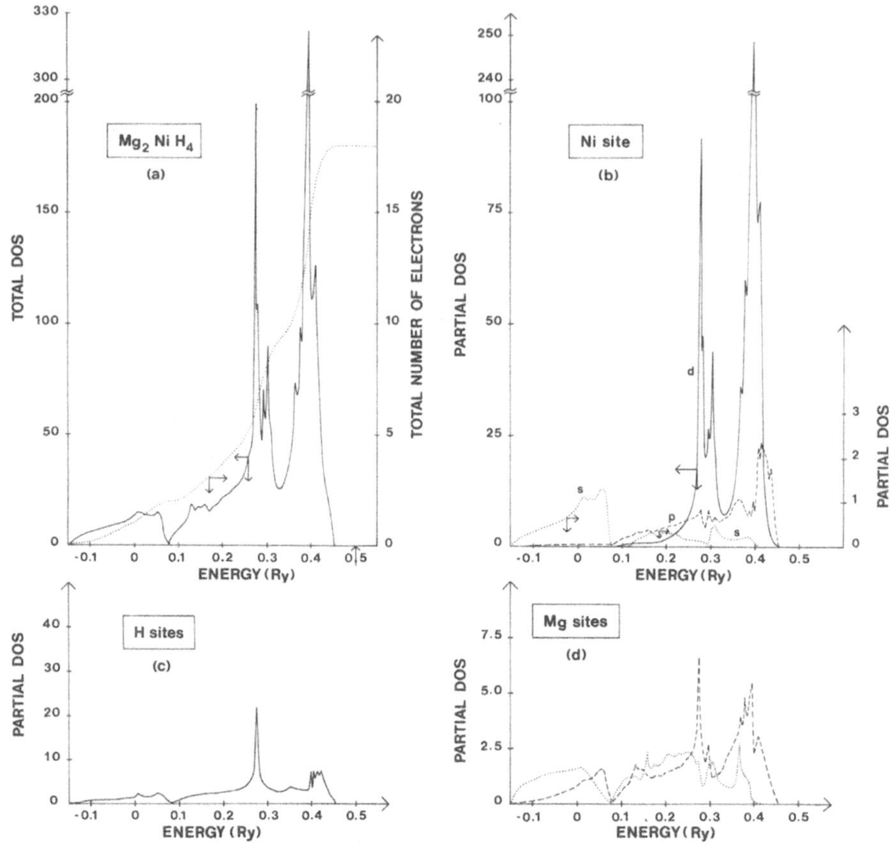

Fig. 5.19a–d. The total DOS of Mg_2NiH_4 and its partial wave analysis inside the Ni, H, and Mg sites. Units are [states of both spins/Rydberg formula unit] [5.261]

The total density of occupied states of Mg_2NiH_4 is plotted in Fig. 5.19. The structure of width 3 eV on the low-energy side is due to the H *s*-, Mg *s*-, and Ni *s*-interactions. This appears clearly from our partial DOS analysis performed inside the MT spheres which is also plotted in Fig. 5.19. The sharp structures in the DOS found at higher energies are due to the *M d*-states [bonding $d(e_g)$- and non-bonding $d(t_{2g})$-states] and to the H–H interactions. Since we have obtained nine bands of total width 8.1 eV separated by an energy gap from the higher states, the compound Mg_2NiH_4 with 18 valence electrons is found to be non-metallic. The width of the gap is 1.36 eV. As discussed previously for Mg_2FeH_6, the energy gap originates from a splitting of the transition metal *d*-bands due to the strong ligand field of the H atoms.

A band structure calculation for Mg_2NiH_4 was also performed using structure II in which the H atoms are located close to the (1/4, 0, 0) positions. Using structure II, the Ni–H distances are a factor of $2^{1/2}$ larger than those in structure I while the Mg–H distances are a factor of $2^{1/2}$ smaller. This leads to pronounced differences in the electronic structure, which can be briefly summarized as follows: (i) In structure II, the hydrogen atoms interact with delocalized *M* and *M'* *sp*-states, principally with the magnesium atoms. The interaction between the H *s*- and the Ni *d*-states is very weak, unlike the case of structure I. The bonding has an important ionic component because we observe a depletion of the interstitial charge in favour of the MT spheres. In contrast, in the structure I we have emphasized the strength of the (Ni *d*)–(H *s*) interactions. (ii) The Ni *d*-bands are not split by the ligand field because the Ni–H distances are a factor of $2^{1/2}$ larger than those in structure I. The filled Ni *d*-states are separated by a gap of 0.8 eV from the higher bands. The hydride is thus found to be non-metallic although the physical origin of the gap is different from that obtained with structure I. Ni core-level spectra of Mg_2NiH_4 measured by SXES [5.261] and XPS (Table 2, [5.84]), though suffering from oxygen contamination of the sample surface, both suggest a strong Ni–H interaction and favour structure I.

Electronic Structure of Mg_2CoH_5. The APW energy bands and DOS of the high temperature cubic phase Mg_2CoH_5 have been obtained [5.262]. They complete the study of the isoelectronic series Mg_2FeH_6 [5.260] and Mg_2NiH_4 [5.261, 263]. The ordering of the bands of Mg_2CoH_5 is found to be the same as that of Mg_2FeH_6. Thus for compounds with respectively 5 and 6 H atoms, unlike in Mg_2NiH_4, the triply degenerate H *s*-states at the BZ centre are located between the split transition metal $d(e_g)$- and $d(t_{2g})$-bands. The observed trend is towards a larger splitting of the TM *d*-bands and a stabilization of *M*–H and H–H induced states upon increasing hydrogen concentration. In Mg_2CoH_5, the narrow sharply peaked Co $d(t_{2g})$-states are separated by a small gap from the low energy metal–H and H–H induced states. Like Mg_2FeH_6, Mg_2CoH_5 is found to be non-metallic. A gap of 1.92 eV opens up between the filled Co $d(t_{2g})$-states and antibonding (Co–$d(e_g)$, H–*s*)-bands.

Several experimental and theoretical results appear to shed some light on the intriguing fact that the maximum hydrogen content in this family of compounds corresponds to 18 valence electrons (v.e.). The theoretical results indicate a filling of the low energy states by 18 v.e. and suggest that the filling of the antibonding (TM–$d(e_g)$, H–*s*)-states is energetically unfavourable owing to the existence of an energy gap. Mössbauer, Raman and IR data [5.255] have been obtained for Mg_2FeH_6 and have been shown to be consistent with the presence of octahedral low-spin iron (II) $(FeH_6)^{4-}$ ions. The calculation shows that the TM–$d(t_{2g})$-states are filled leaving the antibonding (TM–$d(e_g)$, H–*s*)-states empty. This picture confirms the experimental interpretation made in terms of a low spin d^6 configuration. It is however to be noted that the ionic picture proposed is certainly too schematic. We have emphasized in our partial-DOS analysis that the low lying bands are not just H-derived states. They contain also a strong TM (e_g) bonding contribution together with a Mg *sp*-component.

Despite the absence of precise experimental determinations of the energy gaps which is due to the difficulty of having powder samples, the compounds belonging to this family are known to be non-metallic from the preliminary resistivity data of *Moyer* and coworkers [5.264] and *Zolliker* et al. [5.257]. The conductivities of the Ca–Ir–H, Sr–Ir–H, Ca–Ru–H, and Sr–Ru–H systems range from 6.4×10^{-8} to 2.6×10^{-8} to $2.6 \times 10^{-7}\,\Omega^{-1}\,\mathrm{cm}^{-1}$; the conductivities of the Ca–Rh–H and Sr–Rh–H systems are also small but are nevertheless much higher than those of the other systems; they are of the order of $10^{-1}\,\Omega^{-1}\,\mathrm{cm}^{-1}$ [5.264]. Resistivity measurements using the four-point method yielded values of the order of 10^9–10^9 $\mu\Omega$ cm at room temperature for compacted powders of Mg_2CoH_5. The resistivity has been found to increase by a factor of 7 or 8 at 4 K [5.257]. Moreover, the hydrides do not have a metallic appearance, Mg_2NiH_4 and Mg_2FeH_6 are respectively reddish and olive-green powders. The low temperature phases of Mg_2NiH_4 exhibit a metal-like conductivity based on measurements with a conventional Volt–Ohm meter at room temperature [5.265], or a semiconducting behaviour with an activation energy of 0.05 eV according to other authors [2.266]. The theoretical results show also that the compounds are not metallic, the values of the energy gaps are 1.8 eV, 1.92 eV, and 1.36 eV for Mg_2FeH_6, Mg_2CoH_5, and Mg_2NiH_4 (using structure I) respectively. These values should however be modified by correlation effects which are expected to be important in these narrow *d*-band materials; smaller values of the exchange potential than in the strong X_α $\alpha = 1.0$ exchange used in the calculation should reduce the predicted energy gap.

Magnetic susceptibility measurements are difficult to analyse due to the presence of significant amounts of free Fe, Co, and Ni precipitates [5.255, 257, 250]. However, the results appear to be consistent with a weakly paramagnetic or possibly diamagnetic behaviour. A value of $\chi = (0.5 \pm 0.2) \times 10^{-6}$ emu/g was obtained for Mg_2NiH_4 [5.250]. Mössbauer data on Mg_2FeH_6 also suggest a diamagnetic low spin iron (II) complex [5.255].

As mentioned in Sect. 5.1.2, x-ray emission spectroscopy allows an experimental determination of the distributions around each type of atom of occupied electronic states having a symmetry consistent with the dipole selection rules. The densities of occupied Ni *s*-, *d*-; Co *s*-, *d*- and Mg *s*-, *p*-states in Mg_2NiH_4 [5.261] and Mg_2CoH_5 [5.262] have been investigated by a study of the Ni L_α, Co L_α and Mg K_β emission spectra at room temperature. After correction for a strong oxide contribution in the Mg spectra, the experimental results have been found to be in satisfactory agreement with the corresponding partial wave analysis of the theoretical DOS discussed previously, although these correspond to the high temperature hydride phases.

The experimentally determined occupied 3*d*-states have a FWHM of 2.35 ± 6.1 eV and 4.2 ± 0.2 eV respectively for Ni and Co compared to the theoretical values of 2.5 eV for Ni (using structure I) and 4 eV for Co. The two-peak structure obtained in the theoretical DOS in Fig. 5.19 are not resolved experimentally due to the large $2p_{3/2}$ core lifetime (1.3 eV for Co and 0.7 eV for

Ni) and to the experimental resolution. The experimental Ni d-spectrum does not agree with the theoretical results obtained with structure II which lead to much narrower Ni d-bands. The experimentally determined occupied Mg p-states have a FWHM of 3.8 ± 0.2 eV and an additional peak of lower intensity at 6.0 ± 0.3 eV and 6.5 ± 0.3 eV below the valence band edge respectively for Mg_2NiH_4 and Mg_2CoH_5. This result is in perfect agreement with the theoretical position of the Mg p-, H s-induced states.

An attempt to study the unoccupied states has been made for Mg_2CoH_5 by an analysis of Mg K absorption and by reabsorption of the Co L_α emission in the case of cobalt [5.262]. The presence of empty antibonding TM d-states above the energy gap predicted by theory [5.260, 263] has been confirmed. These states are characteristic of the K_2PtCl_6 type structure [5.260]. Moreover, the observed empty Mg p-states are found to be separated by a gap of 2.0 ± 0.2 eV from the valence band edge, a result which agrees with the calculated value 1.92 eV for Mg_2CoH_5.

Important features of the electronic properties of hydrides isoelectronic to Mg_2NiH_4 are now understood. However, further studies of fundamental properties such as electrical conductivity and also theoretical investigations of the low temperature phases are still needed. Due to the peculiarities of the crystal structures and to the presence of divalent metal atoms these ternary hydrides have a rather unusual electronic structure.

5.4.4 The AB_2 Laves Phase, and Related Compounds and Their Hydrides

Many cubic (C-15) and hexagonal (C-14) Laves phase compounds (AB_2) readily absorb hydrogen and form extended solid solutions or hydrides over a very wide range of stability and concentration. They not only show interesting hydrogen storage properties but are also fascinating because of their electronic, magnetic and superconducting properties (see Chap. 7).

The electronic structure of ZrV_2, $ZrFe_2$, and $ZrCo_2$ was calculated self-consistently using the APW method and the local-density form of exchange-correlation potential [5.267], that of ZrV_2 also by LMTO calculations [5.268]. The energy bands and the DOS of $ZrFe_2$ and $ZrCo_2$ are very similar. Their valence bands are dominated by occupied states of largely Fe d- and Co d-character and by empty states of largely Zr d-character. In ZrV_2 the Zr d-states contribute more bonding states below E_F than in $ZrFe_2$ and $ZrCo_2$. The Fermi levels in ZrV_2 (see Fig. 5.20) and $ZrFe_2$ fall in a peak in the DOS and $N(E_F)$ values are quite large.

Klein and *Pickett* [5.49] used the above results as the basis for a muffin-tin Green's function study of dilute hydrogen impurities in ZrV_2 and $ZrCo_2$. For the interstitial sites in ZrV_2, larger charge density near the proton is found than in the case of $ZrCo_2$. UV photoelectron spectra of ZrV_2 confirm that E_F is close to the top of the d-band, whereas in $ZrCr_2$, E_F is shifted to 0.6 eV above the d-band in a region of low DOS [5.269]. In a rigid-band-like filling of the calculated DOS of ZrV_2 by the additional valence electrons of $ZrCr_2$, E_F moves

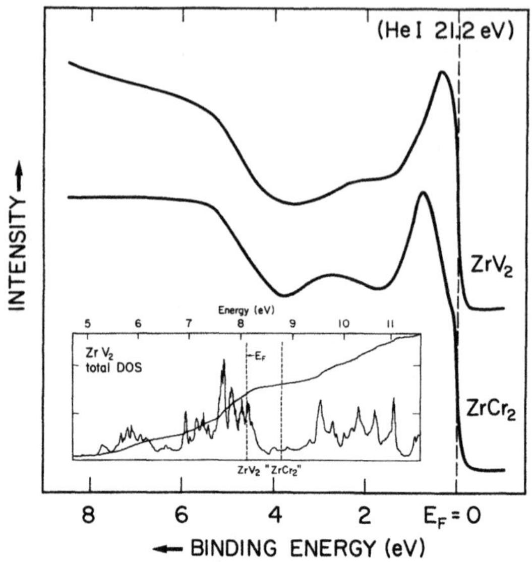

Fig. 5.20. Valence band photoelectron spectra of ZrV_2 and $ZrCr_2$, showing that the *d*-band of $ZrCr_2$ is shifted 0.6 eV away from E_F as compared with ZrV_2. *Inset:* Calculated total DOS of ZrV_2 with Fermi levels of ZrV_2 and "$ZrCr_2$". E_F of "ZrV_2" was determined by adding 4 electrons to E_F of ZrV_2, which corresponds to a shift of 0.65 eV, in agreement with the photoelectron spectra [5.267, 5.269]

up 0.65 eV, in good agreement with the UV spectra shown in Fig. 5.20. Differences of the heat of formation of binary transition metal hydrides were shown to be dominated by shifts of E_F. The measured shift of 0.6 eV accounts for a change of ΔH of $+60$ kJ/(mole H_2) from ZrV_2 to $ZrCr_2$ which compares surprisingly well with measured values of -78 kJ/(mole H_2) for ZrV_2 and -24 kJ/(mole H_2) for $ZrCr_2$, i.e. a change of 54 kJ/(mole H_2).

In ZrT_2 compounds (T = Mn, V, Cr) small core-level shifts of opposite sign were observed for Zr and T, respectively, as compared to the values of the elemental metals [5.7, 269, 270] (Table 5.2 and Fig. 5.21).

Hydrides of the hexagonal C-14 Laves phase $ZrMn_2$ were the first ternary hydrides studied by photoelectron spectroscopy [5.7]. A strong emission around 6.5 eV resulting from the hydrogen induced band and a 0.5 eV shift of the Zr 3*d* core-levels to larger binding energy together with some broadening of the Mn 2*p* core-level were observed (Fig. 5.21), in agreement with strong Zr–H-bonding. In disagreement with the XPS results, SXES spectra of $ZrMn_2$ showed neither a profile change nor any hydrogen-induced states on hydride formation [5.9]. As hydrogen induced features were observed in SXES spectra of more stable hydrides, we suppose that a rapid desorption of hydrogen from $ZrMn_2$ hydride was the reason for the discrepancy between the XPS and SXES results.

Valence band and core level spectra of the cubic Laves phase compounds ZrV_2 and $ZrCr_2$ and of their hydrides measured by XPS also are shown in Fig. 5.21. The intensity of the hydrogen induced peak at 6–7 eV below E_F varies with the hydrogen content of the compounds. Its position does not seem to correlate with the stability of these hydrides. The limited resolution, which is caused largely by the fast desorption, does not allow precise statements on other variations of the valence band. The core levels show interesting shifts for

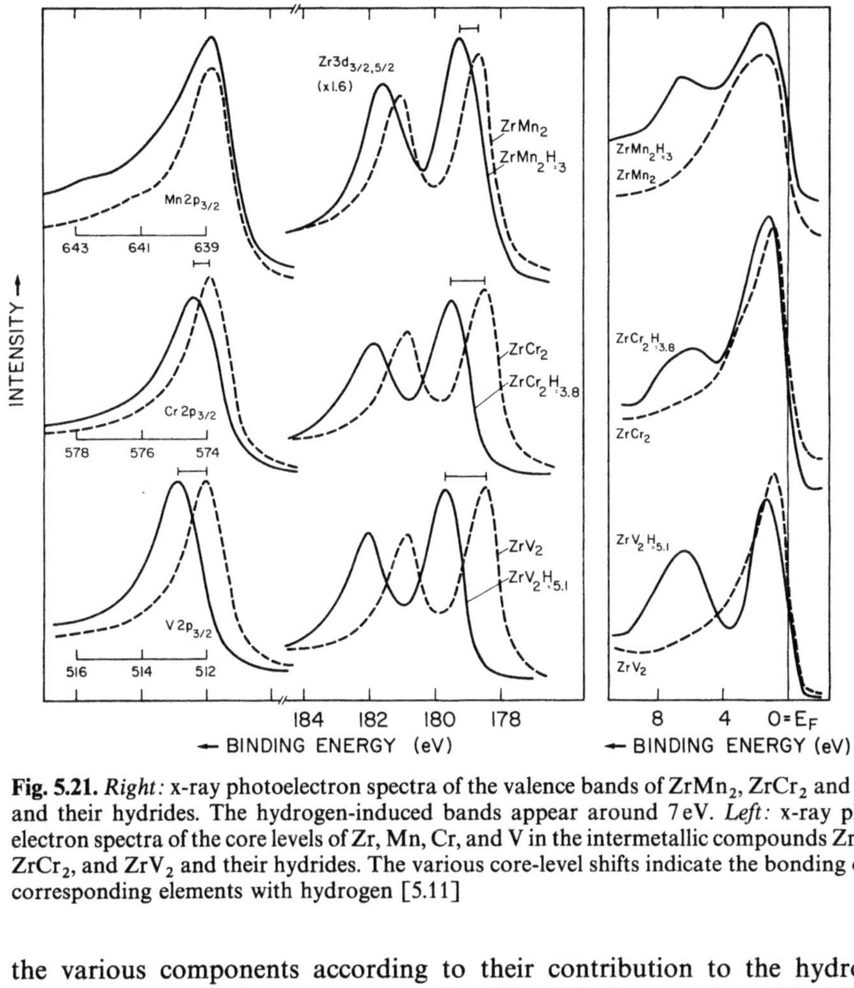

Fig. 5.21. *Right:* x-ray photoelectron spectra of the valence bands of $ZrMn_2$, $ZrCr_2$ and ZrV_2 and their hydrides. The hydrogen-induced bands appear around 7 eV. *Left:* x-ray photoelectron spectra of the core levels of Zr, Mn, Cr, and V in the intermetallic compounds $ZrMn_2$, $ZrCr_2$, and ZrV_2 and their hydrides. The various core-level shifts indicate the bonding of the corresponding elements with hydrogen [5.11]

the various components according to their contribution to the hydrogen bonding and the hydrogen concentration. The shifts increase from Mn (almost zero) through Cr and V to Zr (0.5–1.3 eV) and that of Zr from $ZrMn_2H_3$ (0.5 eV) through $ZrCr_2H_{3.8}$ to $ZrV_2H_{5.1}$ (1.3 eV) in the same sequence as the stability of the hydrides increases. It is remarkable that not only the core levels of the elements which form stable binary hydrides (Zr, V) but also those of Cr are shifted. The Zr core-level shift in $ZrV_2H_{5.1}$ (1.3 eV) is significant larger than in binary Zr hydrides. Attempts to correlate the core-level shifts to the heat of hydride formation [5.269] by analogy with the models which correlate the core-level shifts in binary alloys to the heat of alloy formation [5.202], yield a characteristic value for the shift of each component when normalized to equal hydrogen contents.

SXES studies of ZrV_2 deuterides show "hydrogen induced bands" in both the Zr L_β ($4d$–$2p$ transition) and V K_β ($3d$–$1s$ transition) spectra [5.271], revealing strong Zr–H and V–H bonding, in agreement with the XPS results. A

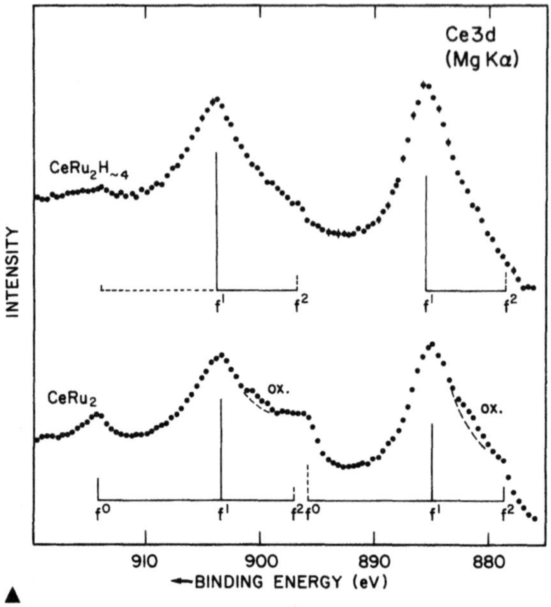

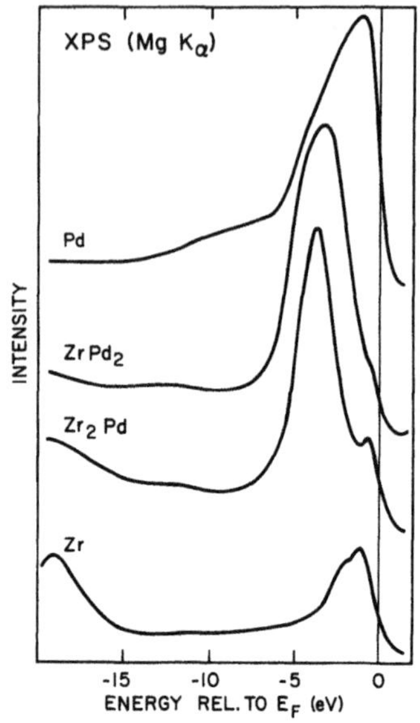

▲

Fig. 5.22. Ce $3d$ core level spectra (XPS) of $CeRu_2$ and its hydride with final state assignment. The intensity of the f° final state satellite is significantly weaker in the hydride, revealing a transition from nonmagnetic α-type Ce in $CeRu_2$ to magnetic γ-type Ce in the hydride, i.e. a hydrogen-induced change in the $4f$ localisation and hybridization [5.278]

Fig. 5.23. XPS valence band spectra of Pd, $ZrPd_2$, Zr_2Pd, and Zr. The Pd d-band becomes significantly narrower and shifts away from E_F with increasing Zr concentration. In $ZrPd_2$ the emission at E_F is very weak [5.15]

correlation between the position of the Zr–H bonding band as measured by SXES Zr L_β spectra in the hydrides $ZrVCuH_{2.8}$, $ZrVFeH_{3.7}$, $ZrVCrH_{2.8}$, and $ZrVCoH_{3.9}$ and the thermal stability of these hydrides has been discussed [5.272]. It is concluded that Zr d-states play the most important role in the formation of bonding states in this series of hydrides. V d-states also take part, but to a lesser extent, and Cr, Fe, Co and Cu are thought not to participate in the metal–H bonding. Susceptibility and specific heat measurements [5.273] of ZrV_2 and $ZrV_2H_{1.3}$ point to a decrease of $N(E_F)$ by a factor of 3 in the hydride, in qualitative agreement with NMR Knight shift measurements [5.274] and also with the calculated DOS of ZrV_2, which predicts a large decrease of $N(E_F)$ if E_F is shifted upwards.

The susceptibility of $TiCr_{1.8}$ is almost temperature independent (2.6×10^{-6} emu/g) and grows continuously with hydrogen content in the C-15 as well as C-14 modifications up to 4.3×10^{-6} emu/g for $TiCr_{1.8}H_{1.6}$ reflecting an increase of $N_d(E_F)$ [5.275]. Similarly, an increase of $N(E_F)$ upon hydride formation was also derived from susceptibility and specific heat of $Ti_{0.4}Mn_{0.6}$ [5.276]. TaV_2 exists in the ordered C-15 and in the random bcc modification and forms with hydrogen an extended solid solution or a dihydride phase, respectively. The magnetic susceptibility decreases linearly with increasing hydrogen content, however, the decrease in the case of the C-15 modification is about 4 times stronger [5.277].

An interesting hydrogen induced electronic effect was observed with XPS in the C-15 compound $CeRu_2$ [5.278]: the strong $4f$-conduction-band hybridization energy, which accounts for the nonmagnetic α-type Ce in the superconducting $CeRu_2$, is reduced upon hydride formation from 120 meV to $\approx$60 meV, accompanied by the reappearance of the magnetic moment of Ce. Hydrogen induced emission and core-level shifts are also seen (see Table 5.2 and Fig. 5.22).

It was shown [5.31] that major contributions to the heat of hydride formation are an exothermic term due to the lowering of occupied metal states and an endothermic term due to the upwards shift of E_F. If we assume that within an alloy system AB_x the endothermic term does not vary significantly with the alloy composition, the increase in E_F required to accommodate part of the additional electron determines the hydride stability. An illustrative example is the Zr–Pd system: Pd, Zr, and $PdZr_2$ form stable hydrides, whereas Pd_2Zr does not. Valence band photoelectron spectra show rather high DOS at E_F for Pd, Zr, and $PdZr_2$ (Fig. 5.23). For Pd_2Zr, however, the Pd $4d$-band is full and shifted away from E_F so that E_F falls into a region of low DOS. Accordingly, a large shift of E_F is required and the hydride is unstable [5.15].

5.4.5 Amorphous Alloys and Their Hydrides

Hydride forming amorphous or glassy alloys prepared by rapid quenching of the melt, by coevaporation on cooled substrates [5.280] or by solid-state reactions such as rapid interdiffusion or hydrogenation [5.281] could be particularly interesting for the study of the electronic properties for the following reasons: (i) the presence of effects due to the absence of long-range order, (ii) their possibly quite different behaviour from that of their crystalline counterparts if interstitial sites with quite different coordination numbers or much smaller H–H distance could be occupied, and (iii) the fact that they are experimentally more easily accessible by spectroscopic methods, since generally they do not disintegrate upon hydride formation.

Band-structure calculations of ordered and disordered crystalline alloys [5.194, 196, 197] and cluster calculations [5.282] are thought to yield a good approximation for amorphous alloys. In most calculations performed so far, at least one of the components is a transition metal.

Extensive studies of the electronic structure of amorphous alloys by photoemission and x-ray emission spectroscopy are available [5.203, 283]. A comparison of these spectra with the calculated DOS of the corresponding crystalline alloys suggests that the essential features of the electronic structure are very similar in amorphous alloys and corresponding ordered close-packed alloys.

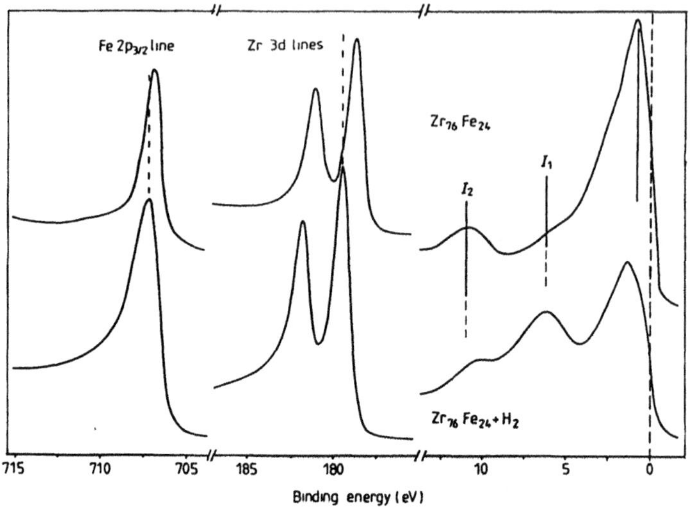

▲ **Fig. 5.24.** XPS valence band and core level spectra of amorphous $Zr_{76}Fe_{24}$ and its hydride [5.284]. Hydrogen-induced bonding states appear at 6 eV (I_1). The structure I_2, which has almost the same intensity for both samples, is possibly due to surface contamination. The Fe and Zr core levels of the hydride are shifted by 0.3 and 0.8 eV, respectively. The increased asymmetry of the Fe line points to a larger density of Fe states at E_F in the hydride

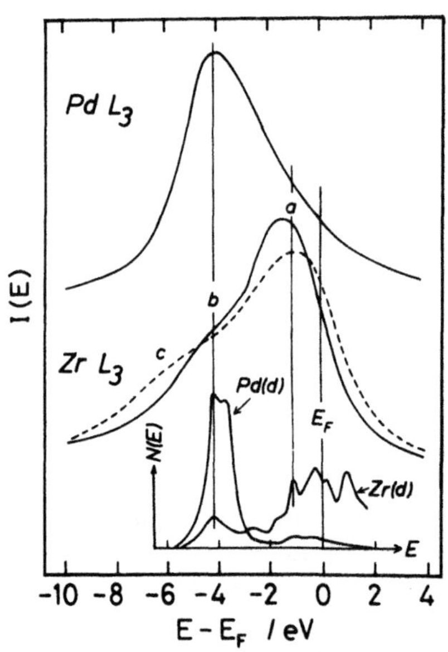

Fig. 5.25. Zr L_3 and Pd L_3 emission spectra of amorphous $Zr_{65}Pd_{35}$ and its hydride compared with calculated DOS for crystalline Zr_3Pd [5.10]

Relatively few hydrides of amorphous alloys have been studied so far [5.12]. No cluster calculations nor other electronic structure calculations for hydrides of amorphous or disordered alloys are available. Photoemission spectra of amorphous $Fe_{24}Zr_{76}H_{160}$ show [5.284] a broad (≈4 eV) hydrogen induced band at 6 eV which develops at the expense of the conduction band emission (Fig. 5.24). The Fe 2*p* and Zr 3*d* core levels are shifted by 0.3 and 0.8 eV, respectively, in the hydride. Furthermore the Fe 2*p* core level becomes rather asymmetric indicating an increased Fe DOS at E_F. In hydrided amorphous

$Ni_{26}Zr_{76}$ an increased asymmetry of the Ni $2p$ line and a core-level shift of both Ni (<0.2 eV) and Zr (0.7 eV) were also observed [5.285].

Soft x-ray emission investigations of amorphous $Pd_{35}Zr_{65}$ (Fig. 5.25) and $Ni_{33}Zr_{67}$ reveal [5.10] significant modification of the Zr $L_3(sd \rightarrow 2p_{2/3})$ spectra as a result of hydride formation (one hydrogen atom per metal atom). A low-energy shoulder similar to that observed in ZrH_x as a result of H–Zr bonding states is induced about 7 eV below the emission edge. In contrast to this, hydride formation has negligible effects on the L_3 spectra of palladium and nickel in these alloys, again indicating the strong H–Zr bonding. Similar behaviour was observed in crystalline alloys.

Apart from the small core-level shifts of Fe and Ni, which seem to be absent in the corresponding crystalline hydrides, photoemission and x-ray emission did not reveal significant differences between the electronic structure of hydrides of amorphous alloys and those of their crystalline counterparts. On the other hand, NMR experiments have suggested that the d-electron DOS at E_F is reduced in amorphous $PdZr_2$ hydride as compared to the hydride of the crystalline alloy [5.286]. Hydrogen absorption by amorphous Pd–Zr and Ni–Zr alloys has been shown, by measurements of the specific heat and electrical resistivity, to reduce both the DOS at E_F and also the superconducting transition temperature [5.287, 288].

5.5 Conclusions

One of the goals of the research efforts on properties of metal hydrides is – through a microscopic investigation of the bonding properties – to gain a better insight into the modification of their electronic properties and to establish the factors which control their stability and their hydrogen absorption capacity. One would also like to be able to modify these materials – e.g. by alloying – in order to make them suitable for the different applications.

A good qualitative, and in many examples even quantitative, agreement has been achieved between band-structure calculations for metal hydrides and measured spectra: the metal-hydrogen bonding interaction which leads to the observed additional low-energy structures in the densities of states as well as the modification of the valence bands upon hydrogen absorption are well explained by theoretical calculations. The drastic modifications of properties related to the Fermi energy observed with the hydride formation (electronic specific heat, the Pauli susceptibility, the Fermi surface geometry, the electrical resistivity and the superconducting properties), are generally also well accounted for by theoretical work. The early results concerning the metal-hydrogen bonding interactions [5.13] and the major contributions to the heat of hydride formation [5.31] have been confirmed by more recent work [5.32, 35, 36] and extended to other systems.

The sign of the core-level shift agrees with the sign of the calculated charge transfer. Variations of the line asymmetry can be described with variations of the partial DOS at E_F. The quantitative calculations of core-level shifts from first principles is rather difficult and not available up to now. The measured core-level shifts of binary and ternary hydrides point to a correlation with the thermochemical stability of the hydrides.

Johansson and *Mårtensson* [5.289] and Steiner and *Hüfner* [5.290] have developed a model based on the equivalent core or $Z+1$ approximation to correlate the thermochemical data and the measured core-level shift of a metallic element, *A*, upon the formation of an alloy A_xB. The model, which works so successfully that it was proposed that calorimetric measurements could be replaced by core-level shift measurements. However, it cannot be applied directly to correlate the core-level shifts measured upon hydride formation to corresponding thermodynamic data for the following reasons [5.190]: As long as the core-level shifts are not measured over a certain range of hydrogen concentration, further simplifying assumptions have to be made. For the hydrides of Ti, Zr, Nb, Pd, and Y a rather good agreement was obtained between measured core-level shifts and those calculated from thermochemical data. The calculations predict core-level shifts of 0.26, 0.14, and 1.0 eV for the hydrides $VH_{0.5}$, HfH_2, and $TaH_{0.8}$, respectively. For hydrides of rare earths, however, the same approach fails, apparently because the core excitation of rare earths cannot be described by an adiabatic process. An extension of core-level shifts of ternary hydrides is possible, but requires their concentration dependence, which is not yet available, or further assumptions and free parameters.

Since the publication of the *Hydrogen in Metals* volumes [5.13] theoretical calculations have gained in numerical accuracy in the density of states and Fermi energy properties. Self-consistent treatments of the crystal potential are now included and in some cases, wave functions have been obtained in order to predict the values of several observables. Accurate total energy calculations performed on a series of binary hydrides [5.35, 46] give a good account of the trends in the experimental values.

Theoretical work on stoichiometric and non-stoichiometric intermetallic hydrides with complex crystal structure is progressing [5.6, 208, 227, 260–262]. The origin of the lower stability of the technologically important material Mg_2NiH_4 as compared to MgH_2 appears clearly from the theoretical results which have emphasized the role of the Ni–H bonds in these systems as opposed to the rather ionic Mg–H interactions in MgH_2. Theoretical results have also provided an explanation for the electrical conductivity and for the electronic origin of the maximum *H* absorption capacity in the isoelectronic series Mg_2FeH_6, Mg_2CoH_5, and Mg_2NiH_4. Total energy calculations are however still lacking for ternary hydrides and most of the intense activity on the prediction of heats of formation (see Chap. 6) relies on semi-empirical predictions based on physical concepts of the bonding obtained from theoretical calculations on binary systems. No quantitative account of the maximum hydrogen absorption capacity of intermetallic hydrides has yet been given, and to date most of the discussions rely on arguments such as the number of empty metal *d*-states above the Fermi energy of the metallic matrix available for *M*–H bonding [5.224]. We would like however to point out that the effective medium theory developed by *Norskov* and coworkers [5.36] could provide some insight. The existence of an optimum host electron density at the interstitital H sites, which appears to be necessary for hydrogen dissolution, could provide an explanation for the preferential H-site occupancy and the maximum H content in complex systems.

Not all of the goals mentioned at the beginning of this section have yet been achieved; theoretical efforts need to be continued in the field of bonding properties and total energies of complex intermetallic hydrides. Experimental developments of in situ implanation of protons in thin films of FeTi, $LaNi_5$ etc. could enable the spectroscopic studies not possible to date.

References

5.1 W.M. Müller, J.P. Blackledge, G.G. Libowitz: *Metal Hydrides* (Academic Press, New York 1968)

5.2 J. Friedel: Phil. Mag. **43**, 153 (1952); Ber. Bunsenges. Phys. Chemie **76**, 828 (1972)

5.3 N.N. Engel, J.E. Johnston: In *L'Hydrogène dans les Métaux*, Congrès International Paris, ed. by P. Bastien, Science et Industrie, Paris (1972)

5.4 A.C. Switendick: *Hydrogen in Metals I*, Topics Appl. Phys., Vol. 28, ed. by G. Alefeld, J. Völkl (Springer, Berlin, Heidelberg 1978)

5.5 J.H. Weaver, D.J. Peterman, D.T. Peterson: In *Electronic Structure and Properties of Hydrogen in Metals*, ed. by P. Jena, C.B. Satterthwaite, NATO Conference Series VI, Vol. 6 (Plenum, New York 1983) p. 207

5.6 M. Gupta: J. Phys. F **12**, 57 (1982); J. Less-Common Met. **88**, 221 (1982); **101**, 35 (1984)

5.7 L. Schlapbach: Phys. Lett. A **91**, 303 (1982)

5.8 K. Tanaka, N. Hamasaka, M. Yasuda, Y. Fukai: Proc. Jimis 2, *Hydrogen in Metals*, Suppl. Trans. Japan Institute of Metals **21**, 65 (1980); Y. Sato, K. Tanaka, M. Yasuda: J. Less-Common Met. **88**, 251 (1982)

5.9 N. Nishimiya, A. Suzuki, T. Sekine, S. Ono, E. Asada: J. Less-Common Met. **88**, 263 (1982)

5.10 K. Tanaka, M. Higatani, K. Kai, K. Suzuki: J. Less-Common Met. **88**, 317 (1982)

5.11 L. Schlapbach, J. Osterwalder, T. Riesterer: J. Less-Common Met. **103**, 295 (1984)

5.12 H. Bernas, A. Traverse, C. Janot: *Amorphous Metals and Semiconductors*, ed. by P. Haasen, R. I. Jaffee (Pergamon, Oxford 1985) p. 435; A. Traverse, H. Bernas: J. Less-Common Met., **129**, 1 (1987)

5.13 G. Alefeld, J. Völkl (eds.): *Hydrogen in Metals I, II*, Topics Appl. Phys., Vols. 28, 29 (Springer, Berlin, Heidelberg 1978)

5.14 T. Sohmura, F.E. Fujita: J. Phys. F **10**, 743 (1980)

5.15 L. Schlapbach, S. Büchler, T. Riesterer, S. Hüfner: J. Less-Common Met. **130**, 301 (1987) and references therein

5.16 H. Hohenberg, W. Kohn: Phys. Rev. **136**, B 864 (1964); W. Kohn, L.J. Sham: Phys. Rev. **140**, A 1133 (1965)

5.17 J.M. Ziman: Solid State Phys. **26**, 1 (1971); M.L. Cohen, V. Heine, D. Weaire: Solid State Phys. **24**, 1 (1970); W.A. Harrison: *Electronic Structure and the Properties of Solids* (Freeman, San Francisco 1980)

5.18 O.K. Andersen: Phys. Rev. B **12**, 3060 (1975); *The Electronic Structure of Complex System*, ed. by P. Phariseau, W.M. Temmerman, NATO ASI Series, Vol. 113 (Plenum, New York 1984) p. 11

5.19 E.S. Alekseev, N.I. Kulikov, A.F. Tatarchenko: J. Less-Common. Met. **130**, 261 (1987)
5.20 A.C. Switendick: Ber. Bunsenges. Phys. Chem. **76**, 535 (1972)
5.21 G. Lehmann, M. Taut: Phys. Status Solidi (b) **54**, 469 (1972); O. Jepsen, O.K. Andersen: Solid State Commun. **9**, 1763 (1971)
5.22 F.M. Mueller, J.W. Garland, M.H. Cohen, K.H. Bennemann: Ann. Phys. (NY) **67**, 19 (1971)
5.23 J.C. Slater, G.F. Koster: Phys. Rev. **94**, 1448 (1954)
5.24 D.A. Papaconstantopoulos: *Handbook of the Band Structure of Elemental Solids* (Plenum, New York 1985)
5.25 J. Friedel: *Physics of Metals*, ed. by J.M. Ziman, Vol. 1 (Cambridge University Press, London 1978)
5.26 D.G. Pettifor: J. Phys. F **7**, 613 (1977); Phys. Rev. Lett. **42**, 846 (1979)
5.27 F. Gautier, J. van der Rest, F. Brouers: J. Phys. F **5**, 1884 (1975)
5.28 L.H. Bennett (ed.): *Theory of Alloy Phase Formation*, AIME, N.Y. (1980) see contributions by C.M. Varma, R.E. Watson, L.H. Bennett; C.M. Varma: Solid State Commun. **31**, 295 (1979); R.E. Watson, L.H. Bennett; Phys. Rev. Lett. **43**, 1130 (1979)
5.29 V.L. Moruzzi, J.F. Janak, A.R. Williams: *Calculated Electronic Properties of Metals* (Pergamon, New York 1978)
5.30 A.R. Williams, C.D. Gelatt, V.L. Moruzzi: Phys. Rev. B **25**, 6509 (1982)
5.31 C.D. Gelatt, Jr., H. Ehrenreich, J.A. Weiss: Phys. Rev. B **17**, 1940 (1978)
5.32 A.R. Williams, J. Kübler, C.D. Gelatt, Jr.: Phys. Rev. B **19**, 6094 (1979)
5.33 M. Methfessel, J. Kübler: J. Phys. F **12**, 141 (1982)
5.34 K.-M. Ho, H.-J. Tao, X.-Y. Zhu: Phys. Rev. Lett. **53**, 1586 (1984)
5.35 A.C. Switendick: J. Less-Common Met., **130**, 249 (1987)
5.36 J.K. Nørskov: Phys. Rev. B **26**, 2875 (1982); **28**, 1138 (1983);
J.K. Nørskov, F. Besenbacher: J. Less-Common Met. **130**, 475 (1987);
P. Nordlander, J.K. Nørskov, F. Besenbacher: J. Phys. F **16**, 1161 (1986)
5.37 D.G. Westlake: J. Less-Common Met. **103**, 203 (1984)
5.38 R. Griessen, A. Driessen: Phys. Rev. B **30**, 4372 (1984); J. Less-Common Met. **103**, 245 (1984)
5.39 M. Cyrot, F. Cyrot-Lackmann: J. Phys. F **6**, 2257 (1976); A. Pasturel, P. Hicter, F. Cyrot-Lackmann: Solid State Commun. **48**, 561 (1983)
5.40 A.R. Miedema: J. Less-Common Met. **32**, 117 (1973); K.H.J. Buschow, P.C.P. Bouten, A.R. Miedema: Rep. Prog. Phys. **45**, 937 (1982)
5.41 M.S. Daw, M.I. Baskes: Phys. Rev. Lett. **50**, 1285 (1983); S.M. Foiles, M.S. Daw: J. Vac. Sci. Technol. A **3**, 1565 (1985)
5.42 W.M. Temmerman, A.J. Pindor: J. Phys. F **13**, 1869 (1983)
5.43 A. Bansil, R. Prasad, L. Schwarz: *Electronic Structure and Properties of Hydrogen in Metals*, ed. by P. Jena, C.B. Satterthwaite, NATO Conf. Series VI, Vol. 6 (Plenum, New York 1983) p. 249
5.44 P. Soven: Phys. Rev. **156**, 809 (1967)
5.45 J.S. Faulkner: Phys. Rev. B **13**, 2391 (1976)
5.46 D.A. Papaconstantopoulos, P.M. Laufer, A.C. Switendick: *Hydrogen in Disordered and Amorphous Solids*, ed. by G. Bambakidis, R.C. Bowman, NATO ASI Series, Vol. 136 (Plenum, New York 1986) p. 139; D.A. Papaconstantopoulos, P.M., Laufer: J. Less-Common Met., **130**, 229 (1987)
5.47 M. Manninen, N. Nieminen: J. Phys. F **9**, 1333 (1979); F. Perrot, M. Rasolt: Phys. Rev. B **23**, 6534 (1981); M. Manninen, J.M. Puska, R.M. Nieminen, P. Jena: Phys. Rev. B **30**, 1065 (1984)
5.48 M. Gupta: *The Electronic Structure of Complex Systems*, ed. by P. Phariseau, W.M. Temmerman, NATO ASI Series, Vol. 113 (Plenum, New York 1984) p. 243
5.49 B.M. Klein, W.E. Pickett: J. Less-Common Met. **88**, 231 (1982)
5.50 R.P. Gupta: J. Less-Common Met. **88**, 299 (1982)
5.51 R.P. Messmer, C.L. Briant: Acta Metall. **30**, 457 (1982); C.L. Briant, R.P. Messmer: Acta Metall. **32**, 2043 (1984)
5.52 J.L. Whitten, T.A. Pakkanen: Phys. Rev. B **21**, 4357 (1980)

5.53 M.A. Kahn, J.C. Parlebas, C. Demangeat: *Electronic Structure and Properties of Hydrogen in Metals*, ed. by P. Jena, C.B. Satterthwaite, NATO Conf. Series VI, Vol. 6 (Plenum, New York 1983) p. 265

5.54 W.L. McMillan: Phys. Rev. **167**, 331 (1968)

5.55 G.D. Gaspari, B.L. Gyorffy: Phys. Rev. Lett. **29**, 801 (1972)

5.56 L. Schlapbach: J. Phys. F **10**, 2477 (1980); L. Schlapbach, C. Pina-Perez, T. Siegrist: Solid State Commun. **41**, 135 (1982)

5.57 R.M. Cotts: In *Hydrogen in Metals I*, ed. by G. Alefeld, J. Völkl, Topics Appl. Phys., Vol. 28 (Springer, Berlin, Heidelberg 1978) p. 227

5.58 R.G. Barnes: *Handbook of the Physics and Chemistry of Rare Earth*, ed. by K.A. Gschneidner, L. Eyring, Vol. 2 (North-Holland, Amsterdam 1979) p. 387

5.59 A. Schenk: *Muon Spin Rotation Spectroscopy* (Adam Hilger, Bristol 1985)

5.60 F.N. Gygax, A. Hintermann, W. Rüegg, A. Schenck, W. Studer, A.J. van der Wal: J. Less-Common Met. **101**, 97 (1984)

5.61 B. Lengeler: *De Haas van Alphen Studies of The Electronic Structure of the Noble Metals and their Dilute Alloys*, Springer Tracts Mod. Phys., Vol. 82 (Springer, Berlin, Heidelberg 1978) p. 1

5.62 R. Griessen, L.M. Huisman: In *Electronic Structure and Properties of Hydrogen in Metals*, ed. by P. Jena, C.B. Satterthwaite, NATO Conf. Series VI, Vol. 6 (Plenum, New York 1983) p. 235

5.63 R. Lässer, B. Lengeler: Phys. Rev. B **18**, 637 (1978)

5.64 M. Hasegawa, S. Koike, H. Asano, M. Hirabayashi, S. Nuguchi: Proceedings Jimis 2, *Hydrogen in Metals*, Suppl. Trans. Japan Institute of Metals **21**, 61 (1980)

5.65 A. Ostrasz, E. Debowska: J. Less-Common Met. **130**, 307 (1987); E. Debowska, B. Rozenfeld: Z. Physik, Chemie NF **145**, 71 (1985)

5.66 J.C. Fuggle, F.U. Hillebrecht, R. Zeller, Z. Zolnierek, P.A. Bennett, Ch. Freiburg: Phys. Rev. B **27**, 2145 (1982)

5.67 R.E. Watson, M.L. Perlman: In *Structure and Bonding*, ed. by J.D. Dunitz (Springer, Berlin, Heidelberg 1975) p. 83
M. Cardona, L. Ley (eds.): *Photoemission in Solids I, II*, Topics Appl. Phys., Vols. 26, 77 (Springer, Berlin, Heidelberg 1978, 1979)

5.68 W. Speier, J.C. Fuggle, R. Zeller, B. Ackermann, K. Szot, F.U. Hillebrecht, M. Campagna: Phys. Rev. B **30**, 6921 (1984)

5.69 C.F. Hague, R.H. Fairlie, W.M. Temmerman, B.L. Gyorffy, P. Oelhafen, H.J. Güntherodt: J. Phys. F **11**, L 95 (1981); D.S. Urch: In *Electron Spectroscopy*, ed. by C.R. Brundle, A.D. Baker, Vol. 3 (Academic, New York 1979) p. 1

5.70 J. Osterwalder: Z. Physik B **61**, 113 (1985)

5.71 B. Lengeler, R. Zeller: J. Less-Common Met. **103**, 337 (1984)

5.72 J.H. Weaver, R. Rosei, D.T. Peterson: Phys. Rev. B **19**, 4855 (1979)

5.73 D.A. Papaconstantopoulos, M. Klein: Phys. Rev. Lett. **39**, 574 (1977); D.A. Papaconstantopoulos, B.M. Klein, E.N. Economou, L.L. Boyer: Phys. Rev. B **17**. 141 (1978); Phys. Rev. B **18**, 2784 (1978)

5.74 M. Gupta, A.J. Freeman: Phys. Rev. B **17**, 3029 (1978)

5.75 C.T. Chan, S.G. Louie: Phys. Rev. B **27**, 3325 (1983)

5.76 D.E. Eastman, J.K. Cashion, A.C. Switendick: Phys. Rev. Lett. **27**, 35 (1971)

5.77 L. Schlapbach, J.P. Burger: J. de Phys. **43**, L 273 (1982); and in *Electronic Structure and Properties of Hydrogen in Metals*, ed. by P. Jena, C.B. Satterthwaite, NATO Conf. Series VI, Vol. 6 (Plenum, New York 1983) p. 229

5.78 T. Riesterer, J. Osterwalder, L. Schlapbach: Phys. Rev. B **32**, 8405 (1985)

5.79 P.A. Bennett, J.C. Fuggle: Phys. Rev. B **26**, 6030 (1982)

5.80 E. Belin, L. Schlapbach, M. Gupta: J. Phys. F **13**, L 193 (1983)

5.81 S. Sinha, S. Badrinarayanan, A.P.B. Sinha: J. Phys. F **16**, L 229 (1986)

5.82 A. Traverse, T. Kachnowski, L. Thomé, H. Bernas: Phys. Rev. B **22**, 4355 (1980)

5.83 M. Gupta: *Metal Hydrides*, ed. by G. Bambakidis, NATO ASI Series, Vol. 76 (Plenum, New York 1981) p. 255; M. Gupta, J.P. Burger: J. Phys. F **10**, 2649 (1980)

5.84 L. Schlapbach, T. Riesterer, J. Osterwalder, S. Büchler: unpublished
5.85 D. Horn-Chandesris: Thèse d'Etat, Université Paris-Sud Orsay, unpublished (1983) No. 2708
5.86 O. Blaschko: Phys. Rev. B **29**, 5187 (1984)
5.87 W. Wampler, B. Lengeler: Phys. Rev. B **15**, 4614 (1977)
5.88 H.L.M. Bakker, R. Griessen, N.J. Koeman, P. Albers, G.H. Sicking: J. Phys. F **16**, 721 (1986)
5.89 C.L. Wiley, G. Cinader, F.Y. Fradin: Bull. Am. Phys. Soc. **21**, 404 (1976)
5.90 B. Stritzker, H. Wühl: In [5.13] Vol. II, p. 243 and references therein
5.91 M. Gupta: *Electronic Structure and Properties of Hydrogen in Metals*, ed. by P. Jena, C.B. Satterthwaite, NATO Conf. Series IV, Vol. 6 (Plenum, New York 1983) p. 321; Phys. Rev. B **24**, 7099 (1981)
5.92 D.S. McLachlan, I. Papadopoulos, T.B. Doyle: J. Physique Coll. **6**, 430 (1978)
5.93 J. Eckert, C.F. Majkzrak, L. Passell, W.B. Daniels: Phys. Rev. B **29**, 3700 (1984)
5.94 D.A. Papaconstantopoulos: *Metal Hydrides*, ed. by G. Bambakidis, NATO ASI Series, Vol. 76 (Plenum, New York 1981) p. 215
5.95 F. Ducastelle, R. Caudron, P. Costa: J. de Phys. **31**, 57 (1970)
5.96 M. Gupta: Phys. Rev. B **25**, 1027 (1982); Solid State Commun. **29**, **47** (1979)
5.97 N.I. Kulikov, V.N. Borzunov, D.A. Zvonkov: Phys. Status Solidi (b) **86**, 83 (1978); N.I. Kulikov: Phys. Status Solidi (b) **91**, 753 (1979)
5.98 A. Fujimori, N. Tsuda: J. Less-Common Met. **88**, 269 (1982); Solid State Commun. **41**, 491 (1982)
5.99 D.A. Papaconstantopoulos, A.C. Switendick: J. Less-Common Met. **103**, 317 (1984)
5.100 K. Bohmhammel, G. Wolf, G. Gross, H. Mägde: J. Low Temp. Phys. **43**, 521 (1981)
5.101 C. Korn: Phys. Rev. B **28**, 95 (1983); C. Korn, S. Goren: J. Less-Common Met. **104**, 113 (1984)
5.102 R.C. Bowman, B.D. Craft, J.S. Cantrell, E.L. Venturini: Phys. Rev. B **31**, 5604 (1985); R.C. Bowman, Jr., E.L. Venturini, B.D. Craft, A. Attah, D.B. Sullenger: Phys. Rev. B **27**, 1474 (1983)
5.103 K. Gesi, Y. Takagi, T. Takeuchi: J. Phys. Soc. Japan **18**, 306 (1965)
5.104 K. Kundasamy, N.A. Surplice: J. Phys. D **18**, 1377 (1985)
5.105 J.H. Weaver, D.J. Peterman, D.T. Peterson, A. Franciosi: Phys. Rev. B **23**, 1692 (1981)
5.106 B.W. Veal, D.J. Lam, D.G. Westlake: Phys. Rev. B **19**, 2856 (1979)
5.107 B.C. Lamartine, T.W. Haas, J.S. Solomon: Appl. Surf. Sci. **4**, 537 (1980)
5.108 K. Tanaka, N. Hamasaka, M. Yasuda, Y. Fukai: Solid State Commun. **30**, 173 (1979)
5.109 T.A. Sasaki, Y. Baba: Phys. Rev. B **31**, 791 (1985)
5.110 R.L. Tapping: J. Nucl. Materials **107**, 151 (1982)
5.111 D.A. Papaconstantopoulos: Proc. 17th Conf. on Low Temperature Physics Part I, (North Holland, Amsterdam 1984) p. 129
5.112 T. Springer: In [5.13] Vol. I, p. 75, and references therein
5.113 P.W. Birkel, T.G. Berlincourt: Phys. Rev. B **2**, 4807 (1970)
5.114 B. Stritzker, J.D. Meyer: Z. Physik B **38**, 77 (1980)
5.115 A.C. Switendick: J. Less-Common Met. **103**, 309 (1984)
5.116 E.L. Venturini: In [5.35]
5.117 T. Schober, H. Wenzl: In [5.13], Vol. II, p. 11
5.118 D.J. Peterman, D.K. Misemer, J.H. Weaver, D.T. Peterson: Phys. Rev. B **27**, 799 (1983)
5.119 Y. Fukai, S. Kazama, K. Tanaka, M. Matsumoto: Solid State Commun. **19**, 507 (1976)
5.120 V.V. Nemoshkalenko, M.M. Antonova, M.M. Kindrat, V.P. Krivitskii, N.N. Shevchenko: Inorg. Mat. **19**, 675 (1983)
5.121 L.M. Brown, A.P. Stephens: Acta Metall. **33**, 827 (1985)
5.122 K. Tanaka, C. Sugiura, S. Nakai, Y. Ohno: Japan. J. Appl. Phys. **20**, 41 (1981)
5.123 N. Stump, G. Alefeld, T. Tocchetti: Solid State Commun. **19**, 805 (1976)
5.124 R. Sen Gupta, S. Chatterjee: J. Phys. F **12**, 1923 (1982)
5.125 D. Ohlendorf, E. Wicke: J. Phys. Chem. Solids **40**, 721 (1979)
5.126 B. Nowak, M. Minier: J. Phys. F **14**, 1291 (1984)

5.127 G.G. Libowitz, A.J. Mealand: *Handbook on the Physics and Chemistry of Rare Earth*, ed. by K.A. Gschneidner, Jr., L. Eyring, Vol. 3 (North Holland, Amsterdam 1979) p. 299
5.128 R.R. Arons: Landolt-Börnstein, Group III, Vol. 12c, *Magnetic and other Properties of Oxides and Related Compounds*, Chap. 6.3 "Rare Earth Hydrides" (Springer, Berlin, Heidelberg 1982) p. 372
5.129 P. Knappe, H. Müller: J. Less-Common Met. **95**, 323 (1983); Z. anorg. allg. Chemie **487**, 63 (1982)
5.130 E. Kaldis, M. Tellefsen, R. Bischof: J. Less-Common Met., **130**, 242 (1986)
5.131 J.J. Didisheim, K. Yvon, P. Fischer, W. Hälg, L. Schlapbach: Phys. Lett. **78** A, 111 (1980)
5.132 T. Ito, B.J. Beaudry, K.A. Geschneidner, Jr.: J. Less-Common Met. **88**, 425 (1982); T. Ito, B.J. Beaudry, K.A. Gschneidner, Jr., T. Takeshita: Phys. Rev. B **27**, 2830 (1983)
5.133 P. Klavins, R.N. Shelton, R.G. Barnes, B.J. Beaudry: Phys. Rev. B **29**, 5349 (1984)
5.134 I.O. Bashkin, M.E. Kost, E.G. Ponyatovskii: Phys. Status Solidi (a) **83**, 461 (1984)
5.135 J. Schefer, P. Fischer, W. Hälg, J. Osterwalder, L. Schlapbach, J.D. Jorgensen: J. Phys. C **17**, 1575 (1984)
5.136 M. Drulis, Z. Bieganski: Phys. Status Solidi (a) **53**, 277 (1979); M. Drulis, B. Stalinski: Phys. Status Solidi (a) **69**, 385 (1982)
5.137 J.H. Weaver, D.T. Peterson, R.L. Benbow: Phys. Rev. B **20**, 5301 (1979)
5.138 D.J. Peterman, J.H. Weaver, D.T. Peterson: Phys. Rev. B **23**, 3903 (1981)
5.139 L. Schlapbach, H.R. Scherrer: Solid State Commun. **41**, 893 (1982)
5.140 L. Schlapbach, J. Osterwalder: Solid State Commun. **42**, 271 (1982); L. Schlapbach, J. Osterwalder, H.C. Siegmann: J. Less-Common Met. **88**, 291 (1982)
5.141 A. Fujimori, N. Tsuda: Phys. Status Solidi (b) **114**, K 139 (1982)
5.142 D.J. Peterman, J.H. Weaver, M. Croft, D.T. Peterson: Phys. Rev. **27**, 808 (1983)
5.143 A. Fujimori, J. Osterwalder: J. Phys. C **17**, 2869 (1984)
5.144 A. Fujimori, L. Schlapbach: J. Phys. C **17**, 341 (1984)
5.145 J.H. Weaver, D.T. Peterson, R.A. Butera, A. Fujimori: Phys. Rev. B **32**, 3562 (1985)
5.146 A.C. Switendick: J. Less-Common Met. **74**, 199 (1980)
5.147 M. Gupta: Solid State Commun. **27**, 1355 (1978)
5.148 M. Gupta, J.P. Burger: Phys. Rev. B **22**, 6074 (1980)
5.149 D.K. Misemer, B.N. Harmon: Phys. Rev. B **26**, 5634 (1982)
5.150 N.I. Kulikov: J. Less-Common Met. **88**, 307 (1982); N.I. Kulikov, A.D. Zvonkov: Z. Phys. Chem. **117**, 113 (1979)
5.151 D.J. Peterman, B.N. Harmon: Phys. Rev. B **20**, 5313 (1979); D.J. Peterman, B.N. Harmon, J. Marchiando, J.H. Weaver: Phys. Rev. B **19**, 4867 (1979)
5.152 A. Fujimori, F. Minami, T. Tsuda: Phys. Rev. B **22**, 3573 (1980); A Fujimori, N. Tsuda: J. Phys. C **14**, 1427 (1981)
5.153 N.I. Kulikov: J. Less-Common Met. **107**, 111 (1985)
5.154 R.A. Alikhanov, V.I. Buzin, N.I. Kulikov, M.E. Kost, W. Sikora, L.S. Smirnov: J. Less-Common Met. **101**, 291 (1984)
5.155 R.G. Barnes, B.J. Beaudry, R.B. Creel, D.R. Torgeson, D.G. de Groot: Solid State Commun. **36**, 105 (1980)
5.156 A. Fujimori: Phys. Rev. B **27**, 3992 (1983)
5.157 H. Shaked, J. Faber, Jr., M.H. Mueller, D.G. Westlake: Phys. Rev. B **16**, 340 (1977)
5.158 M. Gupta: Solid State Commun. **50**, 439 (1984)
5.159 P. Vorderwisch, S. Hautecler, W.D. Teuchert: Solid State Commun. **25**, 213 (1978)
5.160 C. Stassis, T. Gould, O.D. McMasters, K.A. Gschneidner, Jr.: Phys. Rev. B **19**, 5746 (1979)
5.161 D.L. Johnson, D.K. Finnemore: Phys. Rev. **158**, 376 (1967); Z. Bieganski, D. Gonzales Alvarez, F.W. Klaaijsen: Physics **37**, 153 (1967)
5.162 G.G. Libowitz: Ber. Bunsenges., Phys. Chem. **76**, 873 (1972)
5.163 J.N. Daou: C.R. Acad Sc. **250**, 3165 (1960); J.N. Daou, J. Bonnet: C.R. Acad. Sc. **261**, 1675 (1975)
5.164 A. Fujimori, M. Ishii, N. Tsuda: Phys. Status Solidi (b) **99**, 673 (1980); A. Fujimori, N. Tsuda: J. Phys. C **14**, L 69 (1981)

5.165 D.R. Torgeson, L.T. Lu, T.T. Phua, R.G. Barnes, D.T. Peterson, E.F.W. Seymour: J. Less-Common Met. **104**, 79 (1984)
5.166 D. Zamir, R.G. Barnes, N. Salibi, R.M. Cotts, T.T. Phua, D.R. Torgeson, D.T. Peterson: Phys. Rev. B **29**, 61 (1984)
5.167 A. Raizman, D. Zamir, R.M. Cotts: Phys. Rev. B **31**, 3384 (1985)
5.168 L. Schlapbach, H.R. Ott, E. Felder, H. Rudigier, P. Thiry, J. Bonnet, Y. Petroff, J.P. Burger: J. Less-Common Met. **130**, 239 (1987); L. Schlapbach, J.B. Burger, P. Thiry, J. Bonnet, Y. Petroff: Phys. Rev. Lett. **57**, 2219 (1986)
5.169 O. Blaschko, G. Krexner, J.N. Daou, P. Vajda: Phys. Rev. Lett. **55**, 2876 (1985) I.S. Anderson, J.J. Rush, T. Udovic, J.M. Rowe: Phys. Rev. Lett. **57**, 2822 (1986)
5.170 P. Vajda, J.N. Daou, J.P. Burger, A. Lucasson: Phys. Rev. B **31**, 6900 (1985)
5.171 W.E. Pickett, A.J. Freeman, D.D. Koelling: Phys. Rev. B **23**, 1266 (1981)
5.172 L. Schlapbach: J. Less-Common Met. **111**, 291 (1985)
5.173 O.J. Zogal, B. Stalinski: Proc. Jimis 2, *Hydrogen in Metals*, Suppl. Trans. Japan. Inst. of Metals **21**, 77 (1980)
5.174 R. Bischof, E. Kaldis, I. Lacis: J. Less-Common Met. **84**, 117 (1983)
5.175 P. Fischer, J. Schefer, K. Tichy, R. Bischof, E. Kaldis: J. Less-Common Met. **94**, 151 (1983)
5.176 J. Osterwalder, H.R. Ott, L. Schlapbach, J. Schefer, P. Fischer: J. Less-Common Met. **94**, 129 (1983)
5.177 R.R. Arons, A. Heidel, F.R. De Boer: Solid State Commun. **51**, 159 (1984)
5.178 R.R. Arons, A. Heidel, J.W. Cable: J. Less-Common Met., **130**, 205 (1987)
5.179 R. Bischof, E. Kaldis, P. Wachter: J. Mag. Magn. Mater. **31–34**, 255 (1983)
5.180 J.C. Fuggle, N. Mårtensson: J. Electron Spectr. Rel. Phenomena **21**, 275 (1980)
5.181 J.H. Weaver, M. Gupta, D.T. Peterson: Solid State Commun. **51**, 805 (1984)
5.182 J. Felsteiner, M. Heilper, I. Gerter, A.C. Tanner, R. Opher: Phys. Rev. B **23**, 5156 (1981)
5.183 G. Krasko: In Proc. Int. Symp. on Metal-Hydrogen Systems, ed. by T.N. Veziroglu, Z. Naturforsch. **36a**, 1146 (1981) (Pergamon, Oxford 1982)
5.184 J.W. Ward: Chap. 1 in *Handbook on the Physics and Chemistry of the Actinides*, ed. by A.J. Freeman, C. Keller (Elsevier, Amsterdam 1985)
5.185 A.C. Switendick: J. Less-Common Met. **88**, 257 (1982)
5.186 J.W. Ward, L.E. Cox, J.L. Smith, G.R. Stewart, J.H. Wood: J. de Phys. Colloque C 4, **40**, 15 (1979)
5.187 J.C. Fernandes, M.A. Continentino, A.P. Guimarzes: Solid State Commun. **55**, 1011 (1985)
5.188 J.H. Weaver, J.A. Knapp, D.E. Eastman, D.T. Peterson, C.B. Satterthwaite: Phys. Rev. Lett. **39**, 639 (1977)
5.189 J.F. Miller, R.H. Caton, C.B. Satterthwaite: Phys. Rev. B **14**, 2795 (1976)
5.190 T. Riesterer: Submitted to Z. Physik B
5.191 L. Schlapbach: Solid State Commun. **38**, 117 (1981)
5.192 E. Jensen, D.M. Wieliczka: Phys. Rev. B **30**, 7340 (1984)
5.193 B.D. Padalia, W.C. Lang, P.R. Norris, L.M. Watson, D.J. Fabian: Proc. R. Soc. London A **354**, 269 (1977)
5.194 V.L. Moruzzi, P. Oelhafen, A.R. Williams, R. Lapka, H.J. Güntherodt, J. Kübler: Phys. Rev. B **27**, 2049 (1983)
5.195 V.L. Moruzzi, P. Oelhafen, A.R. Williams: Phys. Rev. B **27**, 7194 (1983)
5.196 P. Weinberger, J. Staunton, B.L. Gyorffy: J. Phys. F **12**, 2229 (1982)
5.197 A. Bansil, M. Pessa: Physica Scripta T **4**, 52 (1983)
5.198 M.O. Robbins, L.M. Falicov: Phys. Rev. B **29**, 1333 (1984)
5.198a E.Z. da Silva, P. Strange, W.M. Temmerman, B.L. Gyorffy: Phys. Rev. B **36**, 3015 (1987)
5.199 G.M. Stocks, H.W. Winter: *The Electronic Structure of Complex Systems*, ed. by P. Phariseau, W.M. Temmerman, NATO ASI Series B: Physics, Vol. 113 (Plenum, New York 1983) p. 463

5.200 J. Hafner: Helv. Phys. Acta **56**, 257 (1983)
5.201 F.U. Hillebrecht, J.C. Fuggle, P.A. Bennet, Z. Zolnierek, Ch. Freiburg: Phys. Rev. B **27**, 2179 (1983)
5.202 N. Mårtensson, R. Nyholm, H. Calén, J. Hedman, B. Johansson: Phys. Rev. B **24**, 1725 (1981)
5.203 P. Oelhafen: In *Glassy Metals* II, Topics Appl. Phys., Vol. 53, ed. by H. Beck, H.-J. Güntherodt (Springer, Berlin, Heidelberg 1983) p. 283
5.204 R. Lapka, P. Oelhafen, U.M. Gubler, H.-J. Güntherodt: Phys. Rev. B **31**, 7734 (1985)
5.205 A. Amamou: Solid State Commun. **33**, 1029 (1980)
5.206 A. Percheron-Guégan, C. Lartigue, J.C. Achard: J. Less-Common Met. **109**, 287 (1985); C. Lartigue, A. Percheron-Guégan, J.C. Achard, J.L. Soubeyroux: J. Less-Common Met. **113**, 127 (1985)
5.207 S.K. Malik, F. Arlinghaus, W.E. Wallace: Phys. Rev. B **25**, 6488 (1982)
5.208 M. Gupta: J. Less-Common Met., **130**, 219 (1987)
5.209 S. Cabus, K. Gloos, U. Gottwick, S. Horn, M. Klemm, J. Kübler, F. Steglich, R.D. Parks: Solid State Commun. **51**, 909 (1984)
5.210 S. Nasu, H.H. Neumann, N. Marzouk, R.S. Craig, W.E. Wallace: J. Phys. Chem. Solids **32**, 2779 (1971)
5.211 T. Takeshita, G. Dublon, O.D. McMasters, K.A. Gschneidner, Jr.: In *The Rare Earth in Modern Science and Technology*, ed. by G.J. McCarthy, J.J. Rhyne, H.B. Silber (Plenum, New York 1980) p. 563; T. Takeshita, K.A. Gschneidner, Jr., D.K. Thome, O.D. McMaster: Phys. Rev. B **21**, 5636 (1980)
5.212 D. Ohlendorf, H.E. Flotow: J. Chem. Phys. **73**, 2973 (1980)
5.213 J.H. Weaver, A. Franciosi, D.J. Peterman, T. Takeshita, K.A. Gschneidner, Jr.: J. Less-Common Met. **86**, 195 (1982)
5.214 T.K. Hatwar, D.R. Chopra: J. of Electron Spectrosc. Relat. Phenom. **35**, 77 (1985)
5.215 P. Thompson, J.J. Reilly, L.M. Corliss, J.M. Hastings, R. Hempelmann: J. Phys. F **16**, 675 (1986)
5.216 Y. Chung, T. Takeshita, O.D. McMasters, K.A. Gschneidner, Jr.: J. Less-Common Met. **74**, 217 (1980)
5.217 J. Palleau, G. Chouteau: J. de Phys. Lett. **41**, L 227 (1980)
5.218 K.H.J. Buschow: In *Handbook of Physics and Chemistry of Rare Earth*, ed. by K.A. Gschneidner, L. Eyring, Vol. 6 (North Holland, Amsterdam 1984)
5.219 F.N. Gygax, A. Hintermann, W. Ruegg, A. Schenck, W. Studer, L. Schlapbach, F. Stucki: In *Recent Developments in Condensed Matter Physics*, Vol. 2, ed. by J.T. De Vreese et al. (Plenum, New York 1981) p. 1
5.220 N. Marzouk, K.S. Craig, W.E. Wallace: J. Phys. Chem. Sol. **34**, 15 (1973)
5.221 G.Y. Adachi, K.I. Niki, J. Shiokawa: J. Less-Common Met. **81**, 345 (1981)
5.222 J.W. Larsen, M.L. Fuller, B.R. Livesay: Proceedings of the Second Miami International Conference on Alternative Energy Sources, Miami Beach, Florida (1979)
5.223 W.M. Walsh, Jr., L.W. Rupp, Jr., P.H. Schmidt, L.D. Longinotti: AIP Conf. Proc. **29**, 686 (1975)
5.224 W.E. Wallace, F. Pourarian: J. Phys. Chem. **86**, 4958 (1982); W.E. Wallace: J. Less-Common Met. **88**, 141 (1982)
5.225 K.A. Gschneidner, Jr., T. Takeshita, Y. Chung, O.D. McMasters: J. Phys. F: Met. Phys. **12**, L 1 (1982)
5.226 T. Takeshita, K.A. Gschneidner, Jr., J.F. Lakner: J. Less-Common Met. **78**, 43 (1981)
5.227 D.A. Papaconstantopoulos, A.C. Switendick: J. Less-Common Met. **88**, 273 (1983); Phys. Rev. B **32**, 1289 (1985)
5.228 J. Yamashita, S. Asano: Progress of Theoretical Physics **48**, 2119 (1972)
5.229 D.A. Papaconstantopoulos: Phys. Rev. B **11**, 4801 (1975)
5.230 G. Schadler, P. Weinberger: J. Phys. F **16**, 27 (1986)
5.231 G.N. Kamm: Phys. Rev. B **12**, 3013 (1975)
5.232 R. Hempelmann, D. Ohlendorf, E. Wicke: *Hydrids for Energy Storage*, ed. by A.F. Andresen, A.J. Maeland (Pergamon, Oxford 1978) p. 407

5.233 L. Schlapbach, T. Riesterer: Appl. Phys. A **32**, 169 (1983)
5.234 J.H. Weaver, D.T. Peterson: Phys. Rev. B **22**, 3624 (1980)
5.235 N. Motta, M. De Crescenzi, A. Balzarotti: Phys. Rev. B **27**, 4712 (1983)
5.236 L. Schlapbach, A. Seiler, F. Stucki, P. Zürcher, P. Fischer, J. Schefer: Z. Physik Chemie N.F., **117**, 205 (1979)
5.237 J.M. Welter, G. Arnold, H. Wenzl: J. Phys. F **13**, 1773 (1983)
4.238 P. Thomspon, P.A. Pick, F. Reidinger, L.M. Corliss, J.M. Hastings, J.J. Reilly: J. Phys. F **8**, L 75 (1978)
5.239 P. Fischer, W. Hälg, L. Schlapbach, F. Stucki, A.F. Andresen: Mater. Res. Bull. **13**, 931 (1978)
5.240 G. Hilscher, G. Wiesinger, R. Hempelmann: J. Phys. F **11**, 2161 (1981)
5.241 R. Lässer, B. Lengeler, G. Arnold: Phys. Rev. B **22**, 663 (1980)
5.242 L.J. Swartzendruber, L.H. Bennett, R.E. Watson: J. Phys. F **6**, L 331 (1976)
5.243 M. Gupta: Solid State Commun. **42**, 501 (1982)
5.244 J. Callaway, R.A. Tawil, C.S. Wang: Phys. Lett. A **46**, 161 (1973)
5.245 T. Shinohara, M. Fujika: Phys. Rev. B **7**, 37 (1973)
5.246 R. Ingalls: Phys. Rev. **155**, 157 (1967)
5.247 K. Schubert, K. Anderko: Z. Metallkde. **42**, 321 (1951)
5.248 M. Cyrot, M. Lavagna: J. de Phys. **40**, 763 (1979)
5.249 R. Visnov, F. Ducastelle, G. Treglia: J. Phys. F **12**, 441 (1982)
5.250 F. Stucki, L. Schlapbach: J. Less-Common Met. **74**, 143 (1980)
5.251 J.J. Reilly, R.H. Wiswall: Inorg. Chem. **18**, 2254 (1968)
5.252 Z. Gavra, M.H. Mintz, G. Kimmel, Z. Hadari: Inorg. Chem. **18**, 3595 (1979); J. Genossar, P.S. Rudman: J. Phys. Chem. Solids **42**, 611 (1981); S. Ono, H. Hayakawa, A. Suzuki, K. Nomura, N. Nishimiya, T. Tabata: J. Less-Common Met. **88**, 63 (1982)
5.253 K. Yvon, J. Schefer, F. Stucki: Inorg. Chem. **20**, 2776 (1981); D. Noreus, L.G. Olsson: J. Chem. Phys. **78**, 2419 (1983)
5.254 B. Darriet, J.L. Soubeyroux, M. Pezat, D. Fruchart: J. Less-Common Met. **103**, 153 (1984)
5.255 J.J. Didisheim, P. Zolliker, K. Yvon, P. Fischer, J. Schefer, M. Gubelmann, A.F. Williams: Inorg. Chem. **23**, 1953 (1984)
5.256 J.P. Darnaudery, M. Pezat, B. Darriet, P. Hagenmuller: Mater. Res. Bull. **16**, 1237 (1981)
5.257 P. Zolliker, K. Yvon, P. Fischer, J. Schefer: Inorg. Chem. **24**, 4177 (1985)
5.258 P. Zolliker, K. Yvon, P. Fischer, J. Schefer: J. Less-Common Met. **129**, 211 (1987)
5.259 P. Lindberg, D. Noreus, M.R.A. Blomberg, P.E.M. Siegbahn: J. Chem. Phys. **85**, 4530 (1986)
5.260 M. Gupta: J. Less-Common Met. **103**, 325 (1984); E. Orgaz, M. Gupta: J. Less-Common Met. **130**, 293 (1987)
5.261 M. Gupta, E. Belin, L. Schlapbach: J. Less-Common Met. **103**, 389 (1984)
5.262 E. Belin, M. Gupta, P. Zolliker, K. Yvon: J. Less-Common Met., **130**, 267 (1987)
5.263 M. Gupta: Unpublished
5.264 R.O. Moyer, Jr., C. Stanitzki, J. Tanoka, M.I. Kay, K. Kleinberg: J. Solid State Chem. **3**, 541 (1971); J.S. Thompson, R.O. Moyer, Jr., R. Lindsay: Inorg. Chem. **14**, 1866 (1975); R. Lindsay, R.O. Moyer, Jr., J.S. Thompson, D. Kuhn: Inorg. Chem. **15**, 3050 (1976)
5.265 Z. Gavra, G. Kimmel, Y. Gefen, M.H. Mintz: J. Appl. Phys. **57**, 4548 (1985)
5.266 D. Noreus, K. Jansson, M. Nygren: Z. Physik. Chemie N.F. **146**, 191 (1985)
5.267 B.M. Klein, W.E. Pickett, D.A. Papaconstantopoulos, L.L. Boyer: Phys. Rev. B **27**, 6721 (1983)
5.268 T. Jalborg, A.J. Freeman: Phys. Rev. B **22**, 2332 (1980)
5.269 T. Riesterer, P. Kofel, J. Osterwalder, L. Schlapbach: J. Less-Common Met. **101**, 221 (1984); T. Riesterer: Z. Physik B **66**, 441 (1987)
5.270 I. Jacob, M. Polak: Mater. Res. Bull. **16**, 1311 (1981)
5.271 Y.M. Yarmoshenko, V.N. Kozhanov, V.M. Cherkashenko, E.Z. Kurmaev, E.P. Romanov: Solid State Commun. **55**, 19 (1985)

5.272 S.G. Porutsky, E.A. Zhurakovsky: J. Less-Common Met. **120**, 273 (1986)
5.273 C. Geibel, W. Goldacker, H. Keiber, V. Oestreich, H. Rietschel, H. Wühl: Phys. Rev. B **30**, 6363 (1984)
5.274 M. Peretz, J. Barak, D. Zamir: Phys. Rev. B **23**, 103 (1981)
5.275 J.F. Lynch, J.R. Johnson, R.C. Bowman, Jr.: In *Electronic Structure and Properties of Hydrogen in Metals*, ed. by P. Jena, C.B. Satterthwaite, NATO Conf. Series VI, Vol. 6 (Plenum, New York 1983) p. 437
5.276 R. Hempelmann, E. Wicke, G. Hilscher, W. Wiesinger: Ber. Bunsenges. Phys. Chem. **87**, 48 (1983)
5.277 J.F. Lynch, R. Lindsay, R.O. Moyer, Jr.: Solid State Commun. **41**, 9 (1982)
5.278 J. Osterwalder, T. Riesterer, L. Schlapbach, F. Vaillant, D. Fruchart: Phys. Rev. B **31**, 8311 (1985)
5.279 G. Teisseron, P. Vulliet, L. Schlapbach: J. Less-Common Met., **130**, 163 (1987)
5.280 H.-J. Güntherodt, H. Beck (eds.): *Glassy Metals I, II*, Topics Appl. Phys., Vols. 46, 53 (Springer, Berlin, Heidelberg 1981, 1983)
5.281 K. Samwer: In *Hydrogen in Disordered and Amorphous Solids*, ed. by G. Bambakidis, R.C. Bowman, NATO ASI Series, Series B: Physics, Vol. 136, 173 (1986); J. Less-Common Met., to be published
5.282 R.H. Fairlie, W.M. Temmerman, B.L. Gyorffy: J. Phys. F **12**, 1641 (1982)
5.283 C.F. Hague, J.M. Mariot, E. Belin, E. Beauprez: In *Amorphous Metals and Non-Equilibrium Processing*, ed. by M. von Allmen (Les Editions de Physique, Paris 1984) p. 331; J.M. Mariot, C.F. Hague, P. Oelhafen, H.-J. Güntherodt: J. Phys. F **16**, 1197 (1986)
5.284 S.M. Fries, H.-G. Wagener, S.J. Campell, U. Gonser, N. Blaes, P. Steiner: J. Phys. F **15**, 1179 (1985)
5.285 X. Yu, L. Schlapbach: Phys. Rev. B (1988) in press
5.286 R.C. Bowman, Jr., W.L. Johnson, A.J. Maeland, W.K. Rhim: Phys. Lett. **94** A, 181 (1983)
5.287 M. Kulik, G. von Minnigerode, K. Samwer: Z. Physik B **60**, 357 (1985)
5.288 U. Mizutani, S. Ohta, T. Matsuda: J. Phys. Soc. Japan **54**, 3406 (1985)
5.289 B. Johansson, N. Mårtensson: Phys. Rev. B **21**, 4427 (1980)
5.290 P. Steiner, S. Hüfner: Acta Metall. **29**, 1885 (1981)

6. Heat of Formation Models

Ronald Griessen and Thomas Riesterer

With 23 Figures and 3 Tables

First principles calculations of the cohesive properties of metal-hydrogen systems are still inaccurate and time consuming. There is thus presently a need for simple empirical models. Empirical and semi-empirical models for the heat of formation and heat of solution of binary and ternary hydrides are reviewed. Many of the proposed empirical correlations are in general based on too small a number of metal-hydrogen systems and are therefore often only useful (and valid) for restricted classes of materials.

In this chapter it is shown that semi-empirical models for the heat of alloy formation have to satisfy several general relations. The two semi-empirical models proposed so far for the heat of formation and heat of solution of metal hydrides (models by Miedema, Griessen, and Driessen) are shown to follow these relations. The semi-empirical band structure model of Griessen and Driessen is more reliable than the cellular model of Miedema although it involves a much smaller number of fit parameters. The cellular model suffers furthermore from the fact that it does not allow phase separation. Both models show however that the energetics of the metal-hydrogen interaction depend both on geometric and on electronic factors.

The chapter is completed with the derivation of the thermodynamic quantities relevant for hydride stability and with a set of tables containing experimental data for the heats of solution and the heats of formation of binary and ternary hydrides and calculated values for the heats of formation of ternary transition metal hydrides.

6.1 Introduction

One of the fundamental characteristics of a metal–hydrogen system is the heat of formation ΔH, i.e. the amount of heat absorbed or liberated during hydrogen absorbtion by the metallic host. ΔH determines essentially the stabilitity of metal hydrides because it is directly related to the dissociation pressure p via the van 't Hoff relation

$$\frac{1}{2}\ln p=\frac{\Delta H}{RT}-\frac{\Delta S}{R}, \tag{6.1}$$

where the entropy of formation ΔS is approximately constant for all metal–hydrogen systems as it arises mainly from the entropy loss of gaseous hydrogen during hydrogen uptake by the metal.

Despite the importance of ΔH, for example for technical applications, first principles calculations of the heat of formation of metal hydrides are still lacking for ternary hydrides $A_{y_A}B_{y_B}H_x$ because the theory of cohesive properties of multicomponent systems is inherently complicated and a priori calculations are still very time consuming. Until now ΔH has been calculated for selected binary hydrides MH_x only [6.1–6]. Almost no calculations are available for ΔH in ternary metal hydrides although the band structure of some of them has been calculated (e.g. $Ti_{1-y}V_yH_x$ [6.7], $TiFeH_2$ and $TiPdH_2$ [6.2], $Pd_{1-y}Ag_yH_x$ and $Pd_{1-y}Rh_yH_x$ [6.8], $FeTiH_x$ [6.9, 10], $Pd_{1-y}Ag_yH_x$ [6.11], Ca_2RuH_6 and Mg_2FeH_6 [6.12], and Mg_2NiH_4 [6.13]).

The difficulty in determining ΔH from first principles has stimulated a search for empirical correlations between ΔH and physical parameters such as the unit cell volume [6.14], interstitial site size [6.15, 16], elastic modulii [6.17–20], Debye temperatures [6.21] and electronic specific heat [6.17, 22, 23]. These empirical correlations are useful for the optimization of metal–hydrogen systems for specific technical applications. In general, however, they do not provide us with more insight into the physics of hydride formation and are difficult to generalize.

This is probably the reason why semi-empirical models were developed to predict ΔH for large classes of metal hydrides. The best known example of such a model is due to Miedema and his collaborators [6.24–29]. In this model each metal is characterized by two parameters (essentially the work function and the electron density at the boundary of the Wigner-Seitz cell of the elemental metal). The heat of formation is given as a universal function of these characteristic parameters. For the application of this model to ternary hydrides one uses the rule of reversed stability which relates, in a straightforward manner, the heat of formation of the hydride of an intermetallic compound to the heats of formation of the intermetallic compound itself and of the binary hydrides of the metals constituting the compound. The rule of reversed stability is found to work well for intermetallics made of a minority metal with high hydrogen affinity and a majority metal with a low hydrogen affinity (e.g. $LaNi_5$, YCo_5, but not FeTi). For other types of compounds it generally leads to too negative values for ΔH. Several workers [6.30–32] have tried to give a microscopic basis to the model of Miedema, but without too much success.

Recently the problem of the heat of formation of hydrides has been tackled along two different lines, which as shall be discussed later, have some points in common.

First, *Nørskov* [6.3] proposed a relatively simple scheme based on the effective medium theory to calculate the embedding energy of an atom in the host metal. The host metal is replaced by a homogeneous electron gas of a density equal to that seen by the impurity atom. Nørskov's scheme is especially useful for rare gases in metals because their core levels are so far from any of the

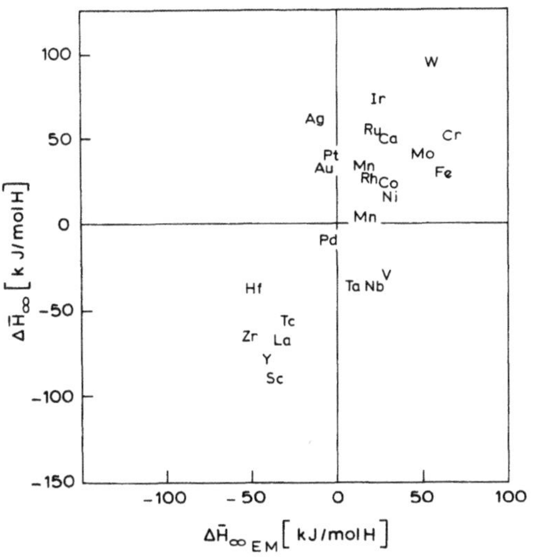

Fig. 6.1. Comparison of experimental heats of solution at infinite dilution for binary metal hydrides with values calculated by *Nørskov* by means of the effective medium theory [6.3]

host metal levels that covalent effects can be neglected. For hydrogen in metals this is however not the case and covalent effects must be explicitly taken into account. Using an appropriate expression for these effects (derived from the atomic sphere approximation formalism of *Andersen* et al. [6.33, 34]) he obtained ΔH values which reproduce the general trends in the experimental data. However, as shown in Fig. 6.1 the theoretical values for ΔH at infinite dilution may be off by a factor 2 or 3 or even have the wrong sign (for Ag, Pt, Au, V, Nb, and Ta). Although considerably simpler than first principles calculations Nørskov's scheme is at present not appropriate for calculations of a large number of more complicated hydrides as is required for example in the tailoring of intermetallics for hydrogen storage or other specific applications.

Second, *Griessen* and *Driessen* [6.35–38] proposed a semi-empirical model in which each metal is characterized by one parameter (essentially the difference between the Fermi energy and the energy of the lowest conduction band of the host metal). This model reproduces well the heat of formation of binary hydrides and, when generalized to ternary hydrides, predicts ΔH values in good agreement with existing experimental data. Due to its simplicity this model requires only very modest computing time. It can thus be used effectively in a screening of attractive materials for specific technical applications.

The purpose of this chapter is to review some of these recent developments and to provide the reader with up-to-date information on the heat of formation of metal–hydrogen systems. It is by no means meant to be comprehensive and the interested reader is referred to the excellent reviews of *Buschow* et al. [6.29], *Westlake* [6.39], *Flanagan* and *Oates* [6.40], *Wenzl* [6.41] and *Carter* and *Carter* [6.42].

This article is organized as follows. In Sect. 6.2 we introduce the thermodynamic quantities relevant to the stability of metal hydrides. In Sect. 6.3 several empirical correlations between the heats of formation of metal hydrides and characteristic physical parameters of the host metal are indicated and their applicability is briefly discussed. The two semi-empirical models mentioned above are described in Sects. 6.4.2 and 6.4.3. The *local* semi-empirical band structure model proposed very recently by one of the authors is briefly presented in Sect. 6.4.4. Conclusions are given in Sect. 6.5. Tables of the heats of formation of binary and ternary metal–hydrides are given in Sects. 6.6.1 and 6.6.2. Theoretical values for the heats of formation of ternary hydrides obtained by means of the Griessen-Driessen model are tabulated in Sect. 6.6.3.

6.2 Thermodynamics of Hydrogen Absorption in Metals

Hydrogen may in principle be dissolved in any metal to form a (dilute) homogeneous solution. Hydrogen is then in thermodynamic equilibrium with *one* metal hydride phase MH_x and consequently

$$\tfrac{1}{2}\mu_{H_2}(p,T)=\mu_H(p,T,x)\,, \tag{6.2}$$

where μ_{H_2} is the chemical potential of pure molecular hydrogen (which is in the gaseous state at pressures lower than 53 kbar [6.43] at room temperature) and μ_H is the chemical potential of atomic hydrogen in solution in the metal. By differentiating (6.2) with respect to $1/T$ at constant concentration $x\equiv[\mathrm{H}]/[M]$ we obtain

$$\begin{aligned}&\frac{1}{2}\frac{\partial(\mu_{H_2}/T)}{\partial(1/T)}\bigg|_p+\frac{1}{2}\frac{\partial(\mu_{H_2}/T)}{\partial p}\bigg|_T\frac{\partial p}{\partial(1/T)}\bigg|_x\\&\quad=\frac{\partial(\mu_H/T)}{\partial(1/T)}\bigg|_{p,x}+\frac{\partial(\mu_H/T)}{\partial p}\bigg|_T\frac{\partial p}{\partial(1/T)}\bigg|_x\,.\end{aligned} \tag{6.3}$$

Using the general relations

$$\frac{\partial\mu}{\partial p}\bigg|_{n,T}=\frac{\partial^2 G}{\partial p\,\partial n}\bigg|_T=\frac{\partial V}{\partial n}\bigg|_{p,T}\equiv\bar{V} \tag{6.4}$$

and

$$\frac{\partial(\mu/T)}{\partial(1/T)}\bigg|_{p,n}=\frac{\partial H}{\partial n}\bigg|_{p,T}\equiv\bar{H} \tag{6.5}$$

we can rewrite (6.3) in the following form

$$\bar{H}_H-\frac{1}{2}\bar{H}_{H_2}=\left(\bar{V}_H-\frac{1}{2}\bar{V}_{H_2}\right)\frac{\partial p}{\partial\ln T}\bigg|_x\,. \tag{6.6}$$

In (6.4–6) G is the Gibbs free energy of the total system, n the number of moles, $\bar{H}$ is the partial molar enthalpy and $\bar{V}$ the partial molar volume. The quantity $\Delta\bar{H} \equiv \bar{H}_{\mathrm{H}} - \frac{1}{2}\bar{H}_{\mathrm{H}_2}$ is the partial molar heat of solution of hydrogen in a metal per gram-atom of hydrogen.

There is some experimental evidence that $\bar{V}_{\mathrm{H}}$, the partial molar volume of hydrogen in a metal, is independent of pressure and temperature [6.44–47]. It is therefore possible to determine $\Delta\bar{H}$ from a plot of p versus $\ln T$ once the pressure and temperature dependence of the enthalpy $\bar{H}_{\mathrm{H}_2}$ and $\bar{V}_{\mathrm{H}_2}$ of pure hydrogen are known. Values of the thermodynamic properties in the temperature range 100 K to 1000 K at pressures up to 1 Mbar are given by *Hemmes* et al. [6.48]. Values for $\bar{V}_{\mathrm{H}}$ are given in the review articles of *Peisl* [6.44] and *Westlake* [6.49]. *Bouten* and *Miedema* [6.28] and *Griessen* and *Feenstra* [6.38] apply their semi-empirical models to the calculation of the partial molar volume.

At low pressures ($p < 100$ bar) the partial molar volume of hydrogen gas is much larger than $\bar{V}_{\mathrm{H}}$ (which is typically 1.7 cm^3/mol H) and H_2 behaves approximatly as an ideal gas, i.e. $p\bar{V}_{\mathrm{H}_2} \simeq RT$. Equation (6.6) then simplifies to

$$\bar{H}_{\mathrm{H}} - \frac{1}{2}\bar{H}^0_{\mathrm{H}_2} \simeq \frac{R}{2} \left.\frac{\partial \ln p}{\partial(1/T)}\right|_x . \tag{6.7}$$

Since $\bar{H}_{\mathrm{H}_2}$ is nearly independent of pressure in this regime, we have replaced $\bar{H}_{\mathrm{H}_2}$ by its value at the standard pressure of one atmosphere.

As indicated in the Tables of Sect. 6.6 the absorption of hydrogen in a metal can be exothermic or endothermic. For the early transition-metal hydrides, for example one finds large negative values for $\Delta\bar{H}$ (exothermic). For the late transition metals as well as for the majority of the simple metals $\Delta\bar{H}$ is positive (endothermic). In contrast to exothermic systems, the metals with $\Delta\bar{H} > 0$ absorb only small quantities of hydrogen at moderate pressures. For such systems one often carries out experiments at constant pressure and determines the temperature dependence of the solubility of hydrogen in a given metal. From these data one can derive the partial heat of solution in the following way.

At constant pressure (6.2) leads to

$$-\left.\frac{\partial(\mu_{\mathrm{H}}/T)}{\partial x}\right|_{p,T} \left.\frac{\partial x}{\partial(1/T)}\right|_p = \bar{H}_{\mathrm{H}} - \frac{1}{2}\bar{H}_{\mathrm{H}_2} . \tag{6.8}$$

The first derivative in (6.8) can be written as

$$\frac{1}{T}\left.\frac{\partial \bar{H}_{\mathrm{H}}}{\partial x}\right|_{p,T} - \left.\frac{\partial \bar{S}_{\mathrm{H}}}{\partial x}\right|_{p,T} . \tag{6.9}$$

In the limit of infinite dilution $(\partial\bar{H}_{\mathrm{H}}/\partial x)|_{p,T}$ tends towards a constant. The entropy $\bar{S}_{\mathrm{H}}$ however, diverges because of the contribution $R\ln[x/(1-x)]$ of the mixing entropy term. We then have for $x \to 0$

$$\left.\frac{\partial(\mu_H/T)}{\partial x}\right|_{p,T} \to \frac{R}{x} \quad \text{so that,} \tag{6.10}$$

$$\Delta \bar{H} = -R \left.\frac{\partial \ln x}{\partial(1/T)}\right|_p . \tag{6.11}$$

The partial molar heat of solution may thus be obtained from a $\ln x$ versus inverse temperature plot.

Until now we have only considered the situation of a homogeneously hydrogenated metal in equilibrium with pure hydrogen. However, many metal–hydrogen systems exhibit miscibility gaps for which two metal-hydride phases, say α and β, coexist and are in equilibrium with the surrounding hydrogen[1]. Below a critical temperature the pressure-composition isotherms have a plateau for hydrogen concentrations between x_α and x_β. The conditions for thermodynamic equilibrium between MH_{x_α}, MH_{x_β}, and H_2 are

$$\mu_H^\alpha(p, T, x_\alpha) = \mu_H^\beta(p, T, x_\beta) = \tfrac{1}{2}\mu_{H_2}(p, T) \quad \text{and} \tag{6.12}$$

$$\mu_M^\alpha(p, T, x_\alpha) = \mu_M^\beta(p, T, x_\beta) . \tag{6.13}$$

The atomic fractions x_α and x_β are independent variables. The corresponding Gibbs-Duhem equations are, at constant p and T,

$$x_\alpha \frac{\partial \mu_H^\alpha}{\partial x_\alpha} + \frac{\partial \mu_M^\alpha}{\partial x_\alpha} = 0 \quad \text{and} \tag{6.14}$$

$$x_\beta \frac{\partial \mu_H^\beta}{\partial x_\beta} + \frac{\partial \mu_M^\beta}{\partial x_\beta} = 0 . \tag{6.15}$$

Consider now the following derivative taken along the coexistence line defined by the equilibrium conditions (6.12, 13) given above,

$$\frac{d(\mu_j^\nu/T)}{d(1/T)} = \frac{\partial(\mu_j^\nu/T)}{\partial(1/T)} + \frac{\partial(\mu_j^\nu/T)}{\partial p}\frac{dp}{\partial(1/T)} + \frac{\partial(\mu_j^\nu/T)}{\partial x_\nu}\frac{dx_\nu}{d(1/T)} . \tag{6.16}$$

Using the standard relations (6.4, 5) we obtain

$$\frac{d(\mu_j^\nu/T)}{d(1/T)} = \bar{H}_j^\nu + \frac{\bar{V}_j^\nu}{T}\frac{dp}{d(1/T)} + \frac{\partial(\mu_j^\nu/T)}{\partial x_\nu}\frac{dx_\nu}{d(1/T)} , \tag{6.17}$$

where $\nu = \alpha$ or β and $j = H$ or M.

---

1 The labeling of the phases α and β does not refer to specific metal-hydride phases but indicates any coexisting phases.

Along the coexistence line we have, as a direct consequence of (6.12, 13)

$$\frac{d(\mu_H^\alpha/T)}{d(1/T)} = \frac{d(\mu_H^\beta/T)}{d(1/T)} = \frac{1}{2}\frac{d(\mu_{H_2}/T)}{d(1/T)} \tag{6.18}$$

and

$$\frac{d(\mu_M^\alpha/T)}{d(1/T)} = \frac{d(\mu_M^\beta/T)}{d(1/T)}. \tag{6.19}$$

Multiplying (6.17) for $v=\alpha$ and $j=\mathrm{H}$ by x_α and adding it to (6.17) for $v=\alpha$ and $j=M$ we obtain

$$x_\alpha \frac{d(\mu_H^\alpha/T)}{d(1/T)} + \frac{d(\mu_M^\alpha/T)}{d(1/T)} = H_\alpha + \frac{V_\alpha}{T}\frac{dp}{d(1/T)} \tag{6.20}$$

with

$$H_\alpha \equiv x_\alpha \bar{H}_H^\alpha + \bar{H}_M^\alpha \tag{6.21}$$

and

$$V_\alpha \equiv x_\alpha \bar{V}_H^\alpha + \bar{V}_M^\alpha . \tag{6.22}$$

For the β phase we have similarly

$$x_\beta \frac{d(\mu_H^\beta/T)}{d(1/T)} + \frac{d(\mu_M^\beta/T)}{d(1/T)} = H_\beta + \frac{V_\beta}{T}\frac{dp}{d(1/T)}. \tag{6.23}$$

Subtraction of (6.20) from (6.23) leads, in combination with (6.18) to

$$\frac{dp}{Td(1/T)} = \frac{H_\beta - H_\alpha - \frac{1}{2}(x_\beta - x_\alpha)\bar{H}_{H_2}}{\frac{1}{2}(x_\beta - x_\alpha)\bar{V}_{H_2} - (V_\beta - V_\alpha)}. \tag{6.24}$$

This is a general expression which is also valid under high pressure conditions. It can be simplified by using the fact that $\bar{V}_H^v$ depends only weakly on the hydrogen concentrations and that $\bar{V}_M^\beta \simeq \bar{V}_M^\alpha$. We then obtain

$$\Delta H_{\alpha\to\beta} \simeq \left(\bar{V}_H - \frac{1}{2}\bar{V}_{H_2}\right)\frac{dp}{d\ln T} \tag{6.25}$$

where $\Delta H_{\alpha\to\beta}$ is the enthalpy of formation (expressed per mol H) of the hydride $M\mathrm{H}_{x_\beta}$ from the hydrogen saturated metal solid solution of composition $M\mathrm{H}_{x_\alpha}$, i.e.

$$\Delta H_{\alpha\to\beta} = \frac{H_\beta - H_\alpha}{x_\beta - x_\alpha} - \frac{1}{2}\bar{H}_{H_2} \tag{6.26}$$

or equivalently $\Delta H_{\alpha\to\beta}$ is the heat of the reaction of the process

$$MH_{x_\alpha}+\tfrac{1}{2}(x_\beta-x_\alpha)H_2\to MH_{x_\beta}. \tag{6.27}$$

To gain some insight into the various contributions to the right-hand side of (6.26) it is useful to consider a simple lattice-gas model for which the chemical potential of hydrogen in a metal is given by an expression of the type

$$\mu_H^{LG}=RT\ln\left(\frac{x}{1-x}\right)+A+ax+\bar{V}_H p+\frac{3}{2}R\theta_E+3RT\ln(1-e^{-\theta_E/T}), \tag{6.28}$$

where A, a, and $\bar{V}_H$ and the Einstein temperature θ_E of the lattice-gas vibrations (optical phonons) are treated as constant parameters. We then have

$$\bar{H}_H^{LG}=A+ax+\bar{V}_H p+\frac{3}{2}R\theta_E+3R\theta_E\frac{1}{e^{\theta_E/T}-1} \tag{6.29}$$

and

$$\bar{S}_H^{LG}=-R\ln\frac{x}{1-x}-3R\ln(1-e^{-\theta_E/T})+\frac{3R\theta_E}{T}\frac{1}{e^{\theta_E/T}-1} \tag{6.30}$$

and from the Gibbs-Duhem equations (6.14, 15)

$$\bar{H}_M^{LG}=-\frac{ax^2}{2}+\bar{H}_M^{\infty} \tag{6.31}$$

$$\bar{S}_M^{LG}=-R\ln(1-x)+\bar{S}_M^{\infty}, \tag{6.32}$$

where $\bar{H}_M^{\infty}$ and $\bar{S}_M^{\infty}$ are the molar enthalpy and entropy of the pure host metal. When introduced into the expression (6.26) for $\Delta H_{\alpha\to\beta}$ these relations lead to

$$\Delta H_{\alpha\to\beta}=A+\frac{a}{2}(x_\beta+x_\alpha)+\bar{V}_H p+\frac{3}{2}R\theta_E+3R\theta_E\frac{1}{e^{\theta_E/T}-1}-\frac{1}{2}\bar{H}_{H_2}. \tag{6.33}$$

Since in this model $x_\alpha+x_\beta=1$, we have that $\Delta H_{\alpha\to\beta}=\bar{H}_H^{LG}(x=1/2)$. The slope of a $\ln p$ versus $1/T$ plot for $x=1/2$ is thus continuous at the critical point.

In the low pressure regime $\bar{H}_{H_2}$ is independent of pressure and varies approximately linearly with temperature between 200 and 1000 K with a slope $\partial\bar{H}_{H_2}/\partial T=14.5$ J/K mol H_2. This temperature variation is exactly compensated by the optical phonon term at the temperature $T_{comp}=0.38\,\theta_E$. As metal hydrides have typically $\theta_E\simeq 1000$ K this implies that the temperature dependence of the last two terms in (6.33) cancel each other around room temperature. For the simple lattice-gas model considered here one thus expects the $\ln p$ versus $1/T$ plots to be linear over a large interval of temperature. For a detailed discussion of this point the reader is referred to the articles by *Flanagan* [6.50], *Flanagan* and *Lynch* [6.51], *Wicke* and *Blaurock* [6.52], and *Wicke* and *Brodowsky* [6.53].

To conclude this section we consider once more the case of endothermic metal–hydrogen systems. Recently, several late transition metals have been loaded with hydrogen at pressures up to 70 kbar [6.47]. Unfortunately in many cases only one isotherm has been measured so that relation (6.6) cannot be used for the evaluation of the heat of hydride formation. An estimate of $\Delta\bar{H}$ or $\Delta H_{\alpha\to\beta}$ may however be obtained in the following way.

For a single phase metal hydride the equilibrium condition (6.2) leads directly to

$$\Delta\bar{H} = T\bar{S}_{\mathrm{H}} - \tfrac{1}{2}T\bar{S}_{\mathrm{H}_2}. \tag{6.34}$$

If only half of the available sites of a given type are occupied then the mixing entropy term in $\bar{S}_{\mathrm{H}}$ vanishes. The magnitude of $\bar{S}_{\mathrm{H}}$ is therefore generally small and $\Delta\bar{H}$ is readily evaluated from the value of $\bar{S}_{\mathrm{H}_2}$ at a given pressure and temperature [6.48, 54].

Similarly for a two-phase metal hydride the equilibrium conditions (6.12, 13) lead to

$$\Delta H_{\alpha\to\beta} = T\Delta S_{\alpha\to\beta} \quad \text{with} \tag{6.35}$$

$$\Delta S_{\alpha\to\beta} = \frac{S_\beta - S_\alpha}{x_\beta - x_\alpha} - \frac{1}{2}\bar{S}_{\mathrm{H}_2} \quad \text{and} \tag{6.36}$$

$$S_i = x_i\bar{S}^i_{\mathrm{H}} + \bar{S}^i_M. \tag{6.37}$$

Within the simple lattice-gas model given above the first term on the right-hand side of (6.36) vanishes identically so that

$$\Delta H_{\alpha\to\beta} = -\tfrac{1}{2}T\bar{S}_{\mathrm{H}_2}. \tag{6.38}$$

At low pressures

$$\bar{S}_{\mathrm{H}_2} = \bar{S}^0_{\mathrm{H}_2} - R\ln p \tag{6.39}$$

and (6.38) reduces to

$$\frac{1}{2}\ln p = \frac{\Delta H_{\alpha\to\beta}}{RT} - \frac{S^0_{\mathrm{H}_2}}{2R}, \tag{6.40}$$

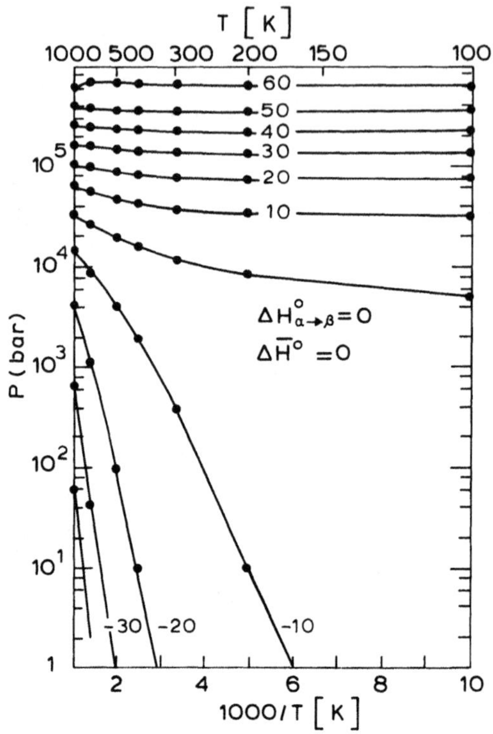

Fig. 6.2. Variation of plateau pressure with temperature for a representative metal hydride according to relation (6.43) with $\bar{V}_H = 1.7\ cm^3/mol\ H$ and $\theta_E = 850\ K$. For each line the standard heat of formation $\Delta H^0_{\alpha\to\beta}$ is given in kJ/mol H. The values of the chemical potential of pure hydrogen used in (6.43) are from *Hemmes* et al. [6.48]. As (6.34) for $\bar{S}_H = 0$ reduces to the same form as (6.38) the curves shown in this figure are also valid for a single-phase metal hydride in which only half of the sites of a given type are occupied. The numbers indicated correspond then to the standard heat of solution $\Delta\bar{H}^0$

an expression which has often been used (see for example [6.29]) to get estimates of the heat of formation of metal hydrides with $S^0_{H_2} = 130.8\ J/K\ mol\ H_2$.

One common feature of (6.34) (with $\bar{S}_H = 0$) and (6.38) is that $\Delta\bar{H}$ or $\Delta H_{\alpha\to\beta}$ are always negative. This seems at first sight to be in contradiction with solubility measurements in many simple metals and late transition metals (except Pd and Mn) where positive heats of solution (reaction) are found (see first column in Table 6.6.1). This apparent discrepancy is easily resolved by noting that $\bar{S}_H = 0$ in (6.34) is not possible at infinite dilution. The discrepancy seems however to persist for concentrated hydrides of Cr, Co, Fe, Mo, and Rh for which slightly positive $\Delta H_{\alpha\to\beta}$ have been found (see second column in Table 6.6.1). The explanation is that $\Delta\bar{H}$ and $\Delta H_{\alpha\to\beta}$ refer to equilibrium states at a given pressure (often in the kbar range) and temperature well above room temperature. In order to reduce the enthalpies $\Delta\bar{H}$ or $\Delta H_{\alpha\to\beta}$ to a chosen standard state we use the fact that

$$\bar{H}_H - \bar{H}^0_H = \int_{p_0}^{p} \bar{V}_H \left(1 - \left.\frac{\partial \ln \bar{V}_H}{\partial \ln T}\right|_p\right) dp + \int \bar{c}^H_p \, dT, \tag{6.41}$$

where $\bar{c}^H_p$ is the partial molar specific heat (of hydrogen in a metal) at constant pressure. It has only been measured for a limited number of metal hydrides. A direct evaluation of (6.41) is thus not possible in general. The specific heat $\bar{c}_p^{\mathrm{H}}$ can, however, be estimated by means of a simple Einstein model for the optical phonons in a metal hydride. Furthermore one expects, on the basis of existing experimental data [6.44–47], that $\bar{V}_{\mathrm{H}}$ depends only weakly on pressure and temperature. With these approximations we find that the standard enthalpy ΔH^0 (at a suitable reference state with pressure p_0 and temperature T_0) is related to the enthalpy ΔH (at p and T) by means of [see (6.33)]

$$\Delta H^0 = \Delta H - \bar{V}_{\mathrm{H}}(p-p_0) - 3R\theta_{\mathrm{E}}[n_{\mathrm{BE}}(T) - n_{\mathrm{BE}}(T_0)] + \tfrac{1}{2}(\bar{H}_{\mathrm{H}_2} - \bar{H}^0_{\mathrm{H}_2}), \qquad (6.42)$$

where n_{BE} is the Bose-Einstein distribution function

$$n_{\mathrm{BE}} = \frac{1}{\mathrm{e}^{\theta_{\mathrm{E}}/T} - 1}.$$

The Einstein temperature $\theta_{\mathrm{E}} = \hbar\omega_{\mathrm{E}}/k_{\mathrm{B}}$ appropriate for the sites occupied by hydrogen (for octahedral sites θ_{E} is typically 600 K while for tetrahedral sites $\theta_{\mathrm{E}} \approx 1100$ K) is assumed to be volume independent in the derivation of (6.42).

Within the approximations used to simplify (6.24) to (6.25), the relation (6.42) is valid for both $\Delta\bar{H}$ and $\Delta H_{\alpha\to\beta}$. A useful form of (6.42) can be obtained by combining it with (6.38). We then have

$$\Delta H^0_{\alpha\to\beta} = \tfrac{1}{2}\mu_{\mathrm{H}_2} - \bar{V}_{\mathrm{H}}(p-p_0) - 3R\theta_{\mathrm{E}}[n_{\mathrm{BE}}(T) - n_{\mathrm{BE}}(T_0)] - \tfrac{1}{2}\bar{H}^0_{\mathrm{H}_2}. \qquad (6.43)$$

At $p_0 = 1$ atm and $T_0 = 298$ K, $\bar{H}^0_{\mathrm{H}_2} = 8.45$ kJ/mol H_2. For a quick estimate of the standard heat of formation $\Delta H^0_{\alpha\to\beta}$ from limited high-pressure data we indicate in Fig. 6.2 how $\ln p$ varies with $1/T$ at constant $\Delta H^0_{\alpha\to\beta}$ for a representative hydride (with $\bar{V}_{\mathrm{H}} = 1.7\ \mathrm{cm}^3$/mol H, $\theta_{\mathrm{E}} = 850$ K) according to (6.43). In the low-pressure regime ($p \leqq 1$ kbar) (6.43) reduces to the well-known van 't Hoff relation. At higher pressures significant deviations are observed.

With expressions of the type of (6.43) *Driessen* et al. [6.54] determined $\Delta H^0_{\alpha\to\beta}$ for metal hydrides synthesized under high pressures. Typical values are $\Delta H^0_{\alpha\to\beta} \simeq +5.4$ kJ/mol H for Mo, +10 for Rh, $+(15\pm2)$ for Co, $-(3\pm2)$ for Ni, $-(8\pm1)$ (desorption) and +2.3 (absorption) for Cr, $+(10\pm1.5)$ for Fe and $-(8\pm1)$ for Mn.

A compilation of enthalpies of solution at infinite dilution $\Delta\bar{H}_\infty$ and heats of formation $\Delta H_{\alpha\to\beta}$ is given in Sect. 6.6.1 for binary metal hydrides and in Sect. 6.6.2 for selected ternary metal hydrides. Note that in general the literature values are not available for standard state conditions. In order to demonstrate the trends in enthalpies from element to element $\Delta\bar{H}_\infty$ and $\Delta H_{\alpha\to\beta}$ are indicated in Figs. 6.3 and 4 as a function of position of the host metal in the periodic table. Roughly speaking the metals on the left of the group VI metals react exothermically with hydrogen. The other metals (with the exeption of Pd and Ni) react endothermically or weakly exothermically with hydrogen.

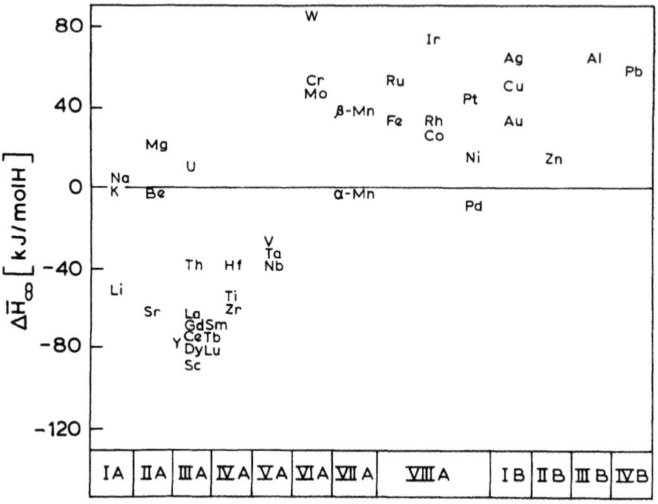

Fig. 6.3. Heat of solution at infinite dilution for binary metal hydrides. For references see Table 6.6.1. Note that in general the literature values are not available for standard state conditions; the values indicated here are thus not necessarily *standard* heats of solution

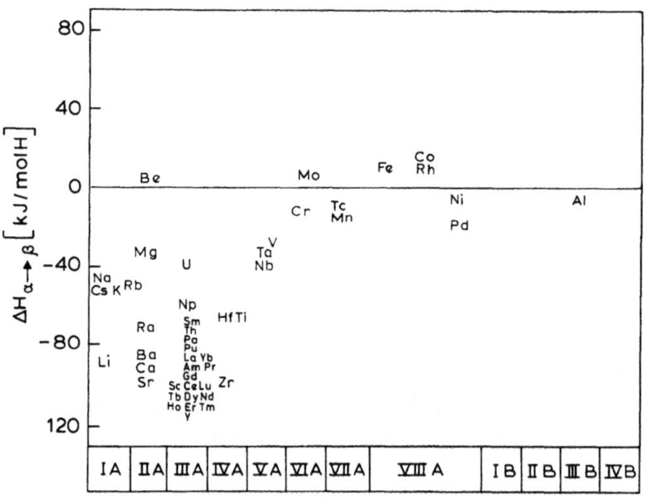

Fig. 6.4. Heat of formation of concentrated metal hydrides. For references see Table 6.6.1. Same comment as for Fig. 6.3

For all binary hydrides investigated at infinite dilution and at higher concentrations the enthalpy of solution decreases with increasing hydrogen content. As we shall see this general behaviour is in many cases related to the lattice expansion accompanying hydrogen absorption in a metal.

6.3 Empirical Correlations

In order to rationalize trends in hydride stability and site occupation, several empirical correlations have been proposed between the enthalpy of hydride formation and physical parameters of the host metal. Among these parameters the unit cell volume or the radius of interstitial holes played an important role. *Lundin* et al. [6.15, 16] showed for example that the free energy of formation of various AB_5 hydrides decreases linearly with increasing tetrahedral hole radius when the B atoms are partially substituted by larger atoms. Such relations are most useful for the tailoring of materials with specific characteristics but they do not shed light on the actual interaction of hydrogen with metals. It is however intuitively clear that larger interstitial holes will tend to stabilize a hydride (see "Volume effects" in Sect. 6.4.3).

Since the interstitial hole size varies in most cases in proportion to the unit cell volume, the stability of hydrides has often been related to the cell (molar) volume of the host metal. A well-documented example is that of the rare-earth-Ni_5 and rare-earth-Co_5 compounds. As shown in Fig. 6.5, the logarithm of the plateau pressure at constant temperature varies linearly with the volume of the unit cell (except for $CeCo_5$). The simple correlation suggested by the data in Fig. 6.5 is however not confirmed by the data in Fig. 6.6. As follows from the high-pressure measurements of *Takeshita* et al. [6.55], YNi_5 and $LaPt_5$ have similar heats of hydride formation although their unit cell volumes are very different. This shows that other effects such as band filling also influence the stability of metal hydrides.

Yei and *McLellan* [6.22] suggest that there is a correlation between $\Delta\bar{H}_\infty$ and γ/V where γ is the coefficient of the linear electronic specific heat and V the molar volume of the metal. As shown in Fig. 6.7 such a correlation (which was proposed by comparing the heats of hydrogen solution of four transition metals and two noble metals only) is not confirmed by the data presently available for binary hydrides.

Various authors have also tried to find a correlation between the heat of solution of binary hydrides and some of the cohesive properties of the host metal. On the basis of continuum elasticity theory *Welch* and *Pick* [6.20] derived the following expression for $\Delta\bar{H}_\infty$

$$\Delta\bar{H}_\infty = \tfrac{1}{2}D - E_B + E_D\,, \tag{6.44}$$

where D is the dissociation energy of the H_2 molecule (218 kJ/mol H), E_B is the energy gained by forming electronic bonding states and E_D is the distortion energy due to the size mismatch between hydrogen and the interstitial hole. The later is approximately given by

$$E_D = \frac{(G/V)}{a + b(G/V)}\,, \tag{6.45}$$

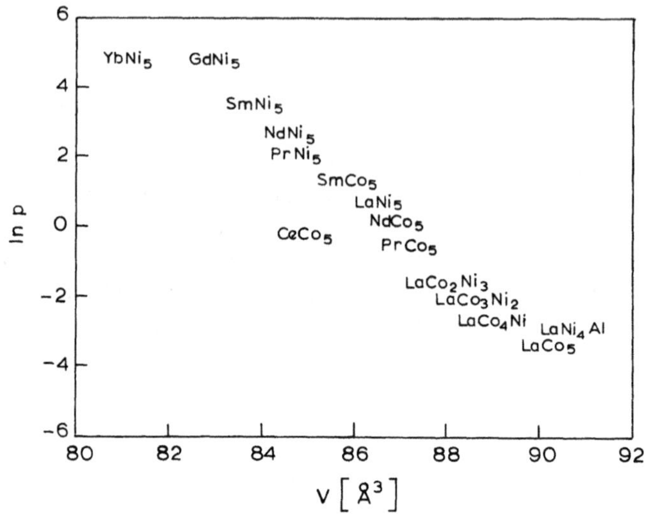

Fig. 6.5. Variation of plateau pressures at constant temperature with unit cell volume for AB_5 intermetallic compounds [6.14, 115]

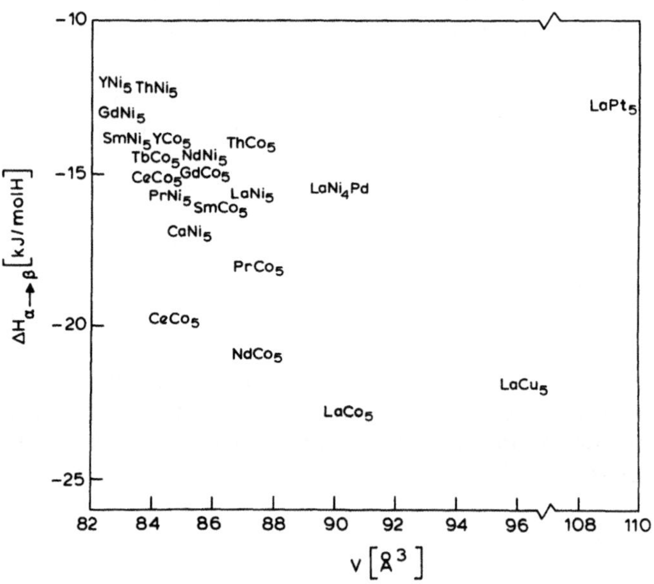

Fig. 6.6. Heat of formation of ternary AB_5H_x hydrides versus unit cell volume. For references see Table 6.6.2

where G is the shear modulus of the host metal and V the molar volume. By fitting the parameters a, b, and E_B Welch and Pick were able to reproduce nicely the heats of solution for binary transition metal hydrides. The fact that a correlation was also found [6.18, 19] between $\Delta\bar{H}_\infty$ and the volume-reduced

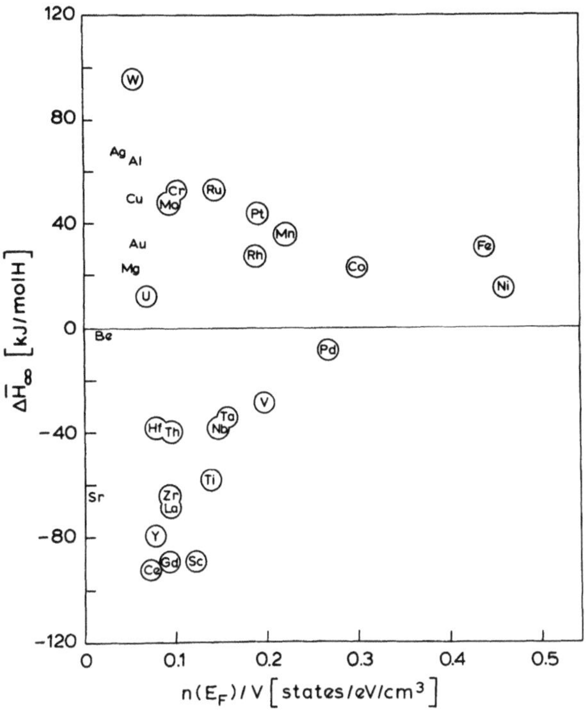

Fig. 6.7. Heat of solution at infinite dilution for binary metal hydrides as a function of the volume-reduced electronic specific heat coefficient. The transition metals, actinides and rare earths are indicated within a circle

bulk modulus B/V suggests, however, that the volume-reduced shear modulus is not really a fundamental characteristic of the ability of a metal to absorb hydrogen. This conclusion is confirmed by the observation that relations (6.44, 45) fail badly in predicting the heats of solution for noble and simple metals. As shown in Fig. 6.8 most of these metals have a low shear modulus and a relatively large molar volume. This results in small values for the volume-reduced shear modulus G/V. None of these metals however has a particularly strong affinity for hydrogen. Another prediction of the relations (6.44, 45) which is not confirmed by experiment concerns liquid metals. For a liquid $G=0$ and, consequently, one expects that $\Delta\bar{H}_\infty=(1/2)D-E_B=-82$ kJ/mol H. However, as can be seen in Table 6.6.1, the heat of solution of hydrogen in liquid metals is to within ≈ 15 kJ/mol H the same as that of the corresponding metal in the solid state. For liquid Co $\Delta\bar{H}_\infty=32$, for Fe $\Delta\bar{H}_\infty=33$, for Mn $\Delta\bar{H}_\infty=30$, for Ni $\Delta\bar{H}_\infty=24$, for Nb $\Delta\bar{H}_\infty=-31$, and for Ti $\Delta\bar{H}_\infty=-47$ kJ/mol H. This casts some doubt on the soundness of the physical basis for the model of Welch and Pick.

Villars [6.56] has shown that most of the tabulated properties of the elements show systematic trends across the periodic table. Since ΔH also varies systematically across the periodic table this opens up a wide variety of potential empirical correlations between ΔH and a simple function of tabulated properties.

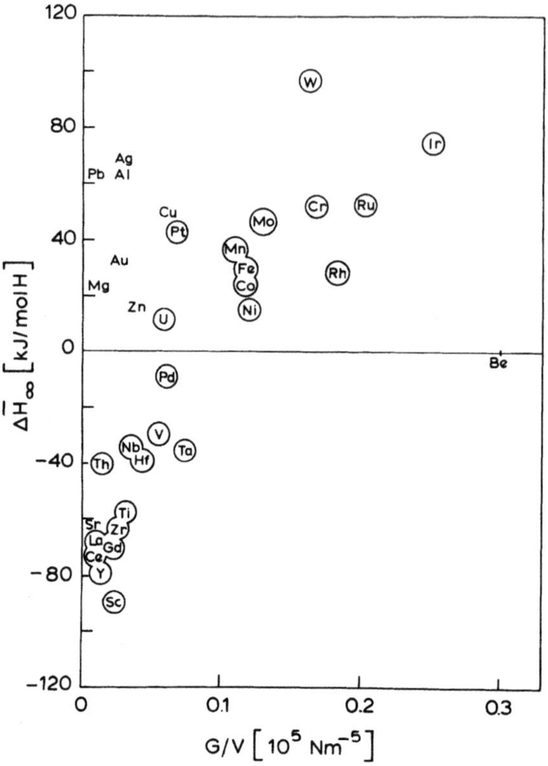

Fig. 6.8. Heat of solution at infinite dilution for binary metal hydrides as a function of the volume reduced shear modulus. The transition metals, actinides and rare earths are indicated within a circle

Finally, it is worth noting that correlations between the heat of hydride formation and some physical parameter of an alloy are only useful if this parameter can be predicted from a model or from theoretical calculations. For example, if a correlation between ΔH and, say the compressibility of the host compound really did exist, then ΔH could easily be predicted for new (not yet investigated) alloys as it is well known that for transition metal alloys the compressibility is rather accurately given by the arithmetic mean of the compressibilities of their constituent metals. If this is not the case, empirical correlations will only be of marginal use in the search for new metal hydrides.

6.4 Semi-empirical Models

As mentioned in the introductory section, two semi-empirical models for the heat of formation of binary and ternary hydrides have been proposed so far: (i) the "atomic" cell model of *Miedema* et al. [6.24–29] and (ii) the "band structure" model of *Griessen* and *Driessen* [6.35–38]. These two models will be dealt with at the end of this section. First, however, we shall consider semi-empirical models from a more general point of view and derive expressions which must be satisfied by any semi-empirical model for the heat of formation of alloys.

6.4.1 General Properties of Semi-empirical Models

For historical reasons we shall consider in this section primarily the *integral* enthalpy of formation $\Delta\tilde{H}$ which represents the total heat involved in the preparation of one mol formula unit of a given compound (or alloy). The heat of formation ΔH given per mol H is related to $\Delta\tilde{H}$ by means of

$$\Delta H(MH_x) \equiv \frac{\Delta\tilde{H}(MH_x)}{x}. \tag{6.46}$$

The heat of solution $\Delta\bar{H}$ for the absorption of hydrogen in MH_x is connected to $\Delta\tilde{H}$ by the relation

$$\Delta\bar{H}(MH_x) \equiv \frac{\partial\Delta\tilde{H}(MH_x)}{\partial x}. \tag{6.47}$$

If the integral heat of formation varies linearly with hydrogen concentration then $\Delta H = \Delta\bar{H}$. For a metal–hydrogen system such as PdH_x, for which a highly nonlinear behaviour of $\Delta\tilde{H}$ is observed at concentrations $x > 0.3$, ΔH and $\Delta\bar{H}$ are markedly different from each other [6.53, 57, 58].

a) Binary Alloys

As an introduction to the problem of the heat of formation of hydrides we consider first the simple case of intermetallic binary alloys $A_{1-y}B_y$. The purpose of a semi-empirical model is to express the integral heat of formation of an alloy $A_{1-y}B_y$ in terms of characteristic parameters A_i and B_i (with $i = 1, \ldots, n$) of the constituent metals. Mathematically this means that

$$\Delta\tilde{H}(A_{1-y}B_y) = f(\boldsymbol{A}, \boldsymbol{B}, y), \tag{6.48}$$

e.g. that the integral heat of formation $\Delta\tilde{H}$ is a function of the vectors $\boldsymbol{A} = (A_1, \ldots, A_n)$ and $\boldsymbol{B} = (B_1, \ldots, B_n)$ which characterize the metals A and B, and of the fraction y of B atoms in A. The function f has some general properties which derive directly from the facts

$$\Delta\tilde{H}(A_{1-y}B_y) = \Delta\tilde{H}(B_yA_{1-y}), \tag{6.49}$$

$$\Delta\tilde{H}(A_{1-y}A_y) = 0 \quad \text{for all } y, \tag{6.50}$$

$$\Delta\tilde{H}(A_1B_0) = 0, \tag{6.51}$$

$$\Delta\tilde{H}(A_0B_1) = 0. \tag{6.52}$$

From (6.49) it follows that

$$f(\boldsymbol{A}, \boldsymbol{B}, y) = f(\boldsymbol{B}, \boldsymbol{A}, 1-y); \tag{6.53}$$

from (6.50) that

$$f(\boldsymbol{A},\boldsymbol{A},y)=0; \tag{6.54}$$

from (6.51) that

$$f(\boldsymbol{A},\boldsymbol{B},0)=0; \tag{6.55}$$

and from (6.52) that

$$f(\boldsymbol{A},\boldsymbol{B},1)=0. \tag{6.56}$$

The general conditions (6.53–56) can be satisfied by writing the function f in the form

$$\begin{aligned} f(\boldsymbol{A},\boldsymbol{B},y)=&\,y(1-y)\sum_{i=1}^{n}\sum_{j=1}^{\infty}(A_i-B_i)^{2j}a_{ij}\\ &+y(1-y)\sum_{\substack{i=1\\l=1}}^{n}\sum_{\substack{k=0\\m=0}}^{\infty}(A_i-B_i)^{2k+1}(A_l-B_l)^{2m+1}a_{ilkm}, \end{aligned} \tag{6.57}$$

where the coefficients a_{ij} and a_{ilkm} are such that simultaneous permutation of $\boldsymbol{A}$ with $\boldsymbol{B}$, and y with $1-y$ leave them unchanged (i.e. these coefficients must obey the symmetry condition given in (6.53). Simple examples of such functions are

$$a=h(y(1-y)) \tag{6.58}$$

$$a=h((1-y)A_i+yB_i) \tag{6.59}$$

$$a=h_1(h_2(A_i)h_3(B_j)+h_3(A_i)h_2(B_j)), \tag{6.60}$$

where h, h_1, h_2, and h_3 are arbitrary functions.

In (6.57) we have not included more complicated terms such as

$$(A_i-B_i)^{2j}(A_l-B_l)^{2j'} \quad \text{or}$$

$$(A_i-B_i)^{2k+1}(A_l-B_l)^{2m+1}(A_{i'}-B_{i'})^{2k'+1}(A_{l'}-B_{l'})^{2m'+1}$$

since in this work we shall mainly be concerned with scalar $[A=(A)]$ and two-dimensional $[\boldsymbol{A}=(A_1,A_2)]$ semi-empirical models.

In a scalar model each metal is characterized by one parameter only. Equation (6.57) reduces then simply to

$$f(A,B,y)=y(1-y)\sum_{j=1}^{\infty}(A-B)^{2j}a_j(A,B,y). \tag{6.61}$$

For transition metals, for example, it is reasonable to choose the number Z of d-electrons as characteristic parameter. Equation (6.61) can then be viewed as an expansion in $\Delta Z = Z_A - Z_B$. It is a generalization of an expression proposed by *Williams* et al. [6.30] (see also [6.6]). Within a tight-binding model for the density of states of transition metal alloys, *Pettifor* [6.31] has indeed derived an expression of the form (6.61) for $\Delta\tilde{H}$ of AB compounds where both A and B belong to the same transition metal series. He obtained

$$\Delta\tilde{H}(AB) = a(\bar{Z})(\Delta Z)^2 \quad \text{with} \tag{6.62}$$

$$\bar{Z} = \tfrac{1}{2}(Z_A + Z_B) \quad \text{and} \tag{6.63}$$

$$a(\bar{Z}) = \frac{1}{8}\left[1 - \frac{3}{50}\bar{Z}(10 - \bar{Z})\right] - \frac{W}{24}(5 - \bar{Z})\frac{d\ln V}{dZ}\bigg|_{\bar{Z}}, \tag{6.64}$$

where the last term represents the contribution to $\Delta\tilde{H}$ of volume changes during alloying (W is the width of the d-band in eV). The numerical parameters give the heat of formation in units of eV per atom. Pettifor found that expression (6.62) was in reasonable agreement with experimental values. This indicates that in the general expression (6.61) the coefficients a_j decrease rapidly with increasing j so that the lowest order terms are dominant.

In a two-dimensional vectorial model each metal is characterized by two parameters. To lowest order in the differences of these parameters we obtain from (6.57) the following expression

$$\begin{aligned} f(\boldsymbol{A}, \boldsymbol{B}, y) = y(1-y)\,[&(A_1 - B_1)^2 a_1(\boldsymbol{A}, \boldsymbol{B}, y) \\ &+ (A_2 - B_2)^2 a_2(\boldsymbol{A}, \boldsymbol{B}, y) \\ &+ (A_1 - B_1)(A_2 - B_2) a_{12}(\boldsymbol{A}, \boldsymbol{B}, y)]\,. \end{aligned} \tag{6.65}$$

As we shall see later, Miedema's model resembles expression (6.65) in the case of alloys consisting of two metals of the same type, for example, two transition metals. Even for this special category of alloys there are however essential differences: a term of the form $(A_1 - B_1)(A_2 - B_2)$ does not appear in the semi-empirical model of Miedema. Furthermore the coefficients a_1 and a_2 depend here only on A_1, A_2, B_1, B_2, and y while in Miedema's model they depend on additional parameters such as the molar volume, the valence and the type (transition or non-transition) of the metals A and B.

b) Binary Metal Hydrides

One possible approach to modelling the binary metal hydrides is based on the assumption that AH_x can be treated as a *normal* alloy after gaseous hydrogen has been brought to a suitable condensed state. We then have

$$\Delta\tilde{H}(AH_x) = \Delta\tilde{H}(AH_x^*) + x\Delta H_{\text{trans}}, \tag{6.66}$$

where ΔH_{trans} is the heat of transformation of gaseous H_2 into condensed (metallic) hydrogen H^*, and $\Delta\tilde{H}$ is the integral heat of formation of the alloy

AH_x made of A metal and condensed (metallic) hydrogen. To make contact with the treatment of binary intermetallic alloys given above we express $\Delta\tilde{H}(AH_x^*)$ in terms of $\Delta\tilde{H}(A_{1-y}H_y^*)$, i.e.

$$\Delta\tilde{H}(A_{1-y}H_y^*)=\frac{1}{1+x}\Delta\tilde{H}(AH_x^*) \tag{6.67}$$

with $y=x/(1+x)$. We can then use the results obtained for binary alloys and write

$$\Delta\tilde{H}(A_{1-y}H_y^*)=f(\boldsymbol{A},\boldsymbol{H},y), \tag{6.68}$$

where the function f is now the same as in (6.48).

Within the framework of a scalar model one obtains from (6.61) and (6.68) to lowest order the result

$$\Delta\tilde{H}(AH_x)=\frac{x}{1+x}b(A-H)^2+x\Delta H_{\text{trans}} \tag{6.69}$$

and for a two-dimensional model from (6.65)

$$\Delta\tilde{H}(AH_x)=\frac{x}{1+x}[a_1(A_1-H_1)^2+a_2(A_2-H_2)^2 +a_{12}(A_1-H_1)(A_2-H_2)]+x\Delta H_{\text{trans}}. \tag{6.70}$$

This relation (with $a_{12}=0$) has a form similar to that of *Bouten* and *Miedema* [6.28] for the integral heat of formation of binary transition metal hydrides.

Another way to set up a semi-empirical model for binary metal hydrides is simply to write

$$\Delta\tilde{H}(AH_x)=f(\boldsymbol{A},x). \tag{6.71}$$

The function f is not subject to any symmetry condition. We know only that

$$f(\boldsymbol{A},0)=0 \tag{6.72}$$

which implies that

$$f(\boldsymbol{A},x)=x\phi(\boldsymbol{A},x). \tag{6.73}$$

For a scalar model we then have

$$\Delta\tilde{H}(AH_x)=x\sum_{i=0}^{\infty}a_i(x)A^i. \tag{6.74}$$

In contrast to the case of binary intermetallic alloys all powers of A may appear. In the simplest model one writes

$$\Delta\tilde{H}(AH_x)=x(c_1+c_2A)\,. \tag{6.75}$$

For the partial molar heat of solution $\Delta\bar{H}$ this implies that

$$\Delta\bar{H}=\alpha A+\beta\,, \tag{6.76}$$

where α and β are constants. *Griessen* and *Driessen* [6.35–38] found that such a simple form did indeed give a very satisfactory description of the heats of formation of binary hydrides when A represents the difference between the energy of the Fermi level and the energy of the lowest conduction band of the host metal. Remarkably enough the parameters α and β were even found to be the same for transition metals, simple metals, rare earths and actinides.

c) Ternary Metal Hydrides

In analogy with the treatment of binary alloys one could indicate general properties for the heat of formation of ternary alloys $A_{y_A}B_{y_B}C_{y_C}$. In the context of this chapter we shall however restrict ourselves to the important class of ternary metal hydrides $A_{1-y}B_yH_x$. In the spirit of a semi-empirical model, the integral heat of hydride formation can be written as

$$\Delta\tilde{H}(A_{1-y}B_yH_x)=f(\boldsymbol{A},\boldsymbol{B},y,x)\,. \tag{6.77}$$

In contrast to the function f used for binary alloys $f(\boldsymbol{A},\boldsymbol{A},y,x)\neq 0$ since

$$f(\boldsymbol{A},\boldsymbol{A},y,x)=\Delta\tilde{H}(AH_x) \tag{6.78}$$

and similarly

$$f(\boldsymbol{B},\boldsymbol{B},y,x)=\Delta\tilde{H}(BH_x)\,. \tag{6.79}$$

Furthermore the function $f(\boldsymbol{A},\boldsymbol{B},y,x)$ must also satisfy the following symmetry relation

$$f(\boldsymbol{A},\boldsymbol{B},y,x)=f(\boldsymbol{B},\boldsymbol{A},1-y,x) \tag{6.80}$$

for all values of y and x. Since the function $f(\boldsymbol{A},\boldsymbol{B},y,x)$ does not have the same properties as the function $f(\boldsymbol{A},\boldsymbol{B},y)$ used for binary intermetallic alloys, the results given in (6.57, 61, 65) cannot be used. It is, however, possible to define a new function

$$\begin{aligned}g(\boldsymbol{A},\boldsymbol{B},y,x)&\equiv f(\boldsymbol{A},\boldsymbol{B},y,x)-(1-y)f(\boldsymbol{A},\boldsymbol{A},y,x)-yf(\boldsymbol{B},\boldsymbol{B},1-y,x)\\&\equiv\Delta\tilde{H}(A_{1-y}B_yH_x)-(1-y)\Delta\tilde{H}(AH_x)-y\Delta\tilde{H}(BH_x)\end{aligned} \tag{6.81}$$

so that

$$g(\boldsymbol{A},\boldsymbol{A},y,x)=0 \tag{6.82}$$

$$g(\boldsymbol{A},\boldsymbol{B},0,x)=0 \tag{6.83}$$

$$g(\boldsymbol{A},\boldsymbol{B},1,x)=0 \tag{6.84}$$

and also

$$g(\boldsymbol{A},\boldsymbol{B},y,x)=g(\boldsymbol{B},\boldsymbol{A},1-y,x)\,. \tag{6.85}$$

This means that the function $g(\boldsymbol{A},\boldsymbol{B},y,x)$ can be expressed in the same way as the function $f(\boldsymbol{A},\boldsymbol{B},y)$ in (6.57) with the only difference that the coefficients a_{ij}, a_{ilkm}, depend now also on the hydrogen concentration x. As the integral heat of hydride formation $\Delta\tilde{H}(A_{1-y}B_yH_x)$ vanishes for $x=0$, all these coefficients must be proportional to x and we can write $a_j=x\alpha_j$.

In a scalar model

$$g(\boldsymbol{A},\boldsymbol{B},y,x)=xy(1-y)\sum_{j=1}^{\infty}(A-B)^{2j}\alpha_j(A,B,y,x) \tag{6.86}$$

and in the special case where both in (6.61) and (6.86) the dominant term is that with $j=1$ we can express the function g in terms of the heat of formation of the host alloy, i.e.

$$g(A,B,y,x)=x\Delta\tilde{H}(A_{1-y}B_y)\frac{\alpha_1(A,B,y,x)}{a_1(A,B,y,x)}\,. \tag{6.87}$$

Combining (6.77–79, 81, 87) we obtain finally

$$\Delta\tilde{H}(A_{1-y}B_yH_x)\simeq(1-y)\Delta\tilde{H}(AH_x)+y\Delta\tilde{H}(BH_x)+kx\Delta\tilde{H}(A_{1-y}B_y) \tag{6.88}$$

with $k=\alpha_1/a_1$. As we shall see later, this relation resembles the so-called *rule of reversed stability* first proposed by *van Mal* et al. [6.26] for the description of the hydrogen absorption in various $LaNi_4M$ (with $M=$ Pd, Ag, Cu, Co, Fe, or Cr) compounds.

In a two-dimensional model one obtains to lowest order in the differences of characteristic parameters

$$g(\boldsymbol{A},\boldsymbol{B},y,x)=xy(1-y)\,[\alpha_1(A_1-B_1)^2+\alpha_2(A_2-B_2)^2 \\ +\alpha_{12}(A_1-B_1)(A_2-B_2)]\,. \tag{6.89}$$

Although this resembles the functional dependence for $\Delta\tilde{H}(A_{1-y}B_y)$ [see (6.65)] it is not expected that (6.88) holds for this two-dimensional model (except of course if $\alpha_i/a_i=k$ for $i=1$, 2, 12, and x and y constant in (6.65)).

Quite often one of the components of an alloy $A_{1-y}B_y$, say component B, is substituted by another metal C. We obtain then *quarternary* metal hydrides of the form $A_{1-y}B_{y(1-z)}C_{yz}H_x$. For a *given* value of y and for a *given* metal A one can write

$$\Delta\tilde{H}(A_{1-y}B_{y(1-z)}C_{yz}H_x)=h(\boldsymbol{B},\boldsymbol{C},z,x). \tag{6.90}$$

Proceeding as for the ternary metal hydrides one defines a function

$$g(\boldsymbol{B},\boldsymbol{C},z,x)=h(\boldsymbol{B},\boldsymbol{C},z,x)-(1-z)h(\boldsymbol{B},\boldsymbol{B},z,x)-zh(\boldsymbol{C},\boldsymbol{C},1-z,x) \tag{6.91}$$

with the properties (6.82–85) (with the variables $\boldsymbol{A}, \boldsymbol{B}, y, x$ replaced by $\boldsymbol{B}, \boldsymbol{C}, z, x$).

For a scalar model,

$$g(B,C,z,x)=K(y)x\Delta\tilde{H}(B_{1-z}C_z). \tag{6.92}$$

The function $K(y)$ vanishes identically for $y=0$. For $y=1$, the relations (6.91, 92) lead to

$$\Delta\tilde{H}(B_{1-z}C_zH_x)=(1-z)\Delta\tilde{H}(BH_x)+z\Delta\tilde{H}(CH_x)+K(y=1)x\Delta\tilde{H}(B_{1-z}C_z). \tag{6.93}$$

Now assuming for simplicity that $K(y)=yK(y=1)$ we obtain

$$\begin{aligned}\Delta\tilde{H}(A_{1-y}B_{y(1-z)}C_{yz}H_x)=&(1-z)\Delta\tilde{H}(A_{1-y}B_yH_x)\\&+z\Delta\tilde{H}(A_{1-y}C_yH_x)+y\Delta\tilde{H}(B_{1-z}C_zH_x)\\&-y(1-z)\Delta\tilde{H}(BH_x)-yz\Delta\tilde{H}(CH_x).\end{aligned} \tag{6.94}$$

This expression shows that it is certainly not correct to estimate the heat of formation of a quarternary metal hydride $A_{1-y}B_{y(1-z)}C_{yz}H_x$ from a weighted average of the $\Delta\tilde{H}$ of the ternary hydrides $A_{1-y}B_yH_x$ and $A_{1-y}C_yH_x$.

d) Local or Site-Dependent Enthalpies

The basic weakness of the semi-empirical models considered above is that they do not include structural information. There is no distinction between the various sites which can be occupied by hydrogen. Even for the simplest class of metal hydrides, the binary hydrides (e.g. PdH_x, NbH_x), the semi-empirical models do not make a distinction between octahedral and tetrahedral sites. In NbH_x, for example, the difference in energy between these sites is not considerable [6.45, 59–62] but for PdH_x where only octahedral sites are occupied the energy of the tetrahedral sites might well be 50 to 100 kJ/mol H higher than that of the octahedral sites. This clearly indicates that site-dependent properties (e.g. local electron density, geometry, size) also play a role.

In a pure metal larger interstitial sites will tend to be occupied first by hydrogen. In compounds and alloys the local chemical composition around each site (i.e. in a first approximation the ratio of A to B atoms in the immediate neighbourhood of the site) is a further factor which determines the heat of hydride formation. To circumvent this weakness one can attribute to each site a *local* heat of hydride formation. From a microscopic point of view this is a reasonable assumption as it is well-known that the electronic structure of transition metals is essentially determined by the overlap of nearest-neighbour wave functions. A general formulation of such *local* semi-empirical models has not been given yet. Attempts have, however, been made to rationalize the observed site occupancies in the hydrides of intermetallic compounds.

In a series of articles *Westlake* [6.39, 49] proposed a purely geometric model which is based on a minimum interstitial hole radius of 0.40 Å for occupation by hydrogen, a minimum hydrogen–hydrogen distance of 2.10 Å and the assumption that larger interstitial sites are occupied first. This model predicts correctly the site occupancy by hydrogen in Friauf-Laves phases with the cubic C15 structures such as $ZrCr_2$, ZrV_2, $ZrFe_2$, $TiCr_2$, HfV_2, $ZrTi_2$, and TaV_2, Friauf-Laves phases with the hexagonal C14 structure as $ErMn_2$ and $ZrMn_2$, Haucke compounds such as $LaNi_5$, $LaNi_4Al$, and $LaNi_4Mn$ [6.63], A_6B_{23} compounds [6.64, 65], ZrNi [6.66, 67] and the hydride $V_2H(D)$ [6.68, 69]. Recently it has also been applied to FeTi [6.70] and Be-based compounds [6.71]. Especially interesting are the hydrides of Ti_2Ni [6.72] in which it seems that H also occupies large tetrahedral sites coordinated by four Ni atoms. This is considered to be clear evidence that hydrogen site preference depends more heavily on interstitial hole size than on the affinity of the coordinating atoms for hydrogen. In all other cases however, the geometric model predicts the same site occupancy as a model based on the hydrogen affinity of the metallic nearest neighbours since in these systems the larger interstitial sites are usually those coordinated by the higher number of atoms of the metal with the greater affinity for hydrogen.

Another model which was proposed to explain the site occupancy by hydrogen is the so-called *imaginary binary hydrides model*. This model, originally proposed by *Jacob* and *Shaltiel* [6.73, 74], has been refined in succesive steps. A good description may be found in [6.75] and [6.76]. The basic idea of the model is that the affinity for hydrogen of a given site in an intermetallic compound A_mB_n may be characterized by an effective *local* enthalpy of formation defined as

$$\Delta H'_{a/b}(A_mB_nH_x) \equiv \Delta H\left(A_mH_{\frac{xa}{a+b}}\right) + \Delta H\left(B_nH_{\frac{xa}{a+b}}\right), \tag{6.95}$$

where x is the number of hydrogen atoms per formula unit A_mB_n occupying an interstitial site coordinated by a A atoms and b B atoms. By means of a semi-empirical model it is possible to calculate the two heats of formation of the binary hydrides on the right-hand side of (6.95). At $x=0$ the effective local heats of formation of the various sites differ markedly from each other (in ZrV_2 this difference is for example typically 40 kJ/mol H). It is then normally argued that around room temperature only the site with the lowest $\Delta H'$ will be occupied because of the strong energy dependence of the Boltzmann factor [6.75, 77].

Within the model of Miedema one finds that $\Delta H'_{a/b}$ is a monotonically increasing function of x. For some sites (for example 2 Zr + 2 V tetrahedral sites in ZrV_2) this increase is however much faster than for other sites (e.g. 1 Zr + 3 V site). It is therefore possible that different sites with different hydrogen occupancy are characterized by the same $\Delta H'$ at a certain concentration x. The simultaneous occupation of these sites is then expected to occur. This is in qualitative agreement with experimental structural data on the hydrides (or deuterides) of intermetallic compounds [6.76]. Recently *Riesterer* [6.78] argued that the affinity of a site for hydrogen should be characterized by the heat of solution $\Delta\bar{H}'_{a/b}$ instead of the heat of formation [see (6.46, 47)]. For $ZrV_2D_{1.5}$, $ZrV_2D_{2.6}$, $ZrV_2D_{4.5}$, and $ZrCr_3D_3$ his results agree well with experimental data. However for $ZrNiD_3$, $ZrTi_{1.9}D_{3.8}$, and $FeTiD_{1.9}$ his approach leads to the same wrong predictions as those derived from (6.95).

In an attempt to predict quantitatively the occupancy of various sites in intermetallics, *Jacob* et al. [6.76] assumed that the number $x_{a/b}$ of hydrogen atoms in an a/b-site is given by a Boltzmann distribution so that

$$\frac{x_{a/b}}{x}=\frac{\exp(-\Delta H'_{a/b}/kT)}{\sum\limits_{\text{sites}}\exp(-\Delta H'_{a/b}/kT)}, \tag{6.96}$$

where x is the total number of hydrogens per formula unit. They obtained site occupancies in remarkably good agreement with structural data for FeTiH, $TiCr_2H_{0.15}$, $ZrV_2H_{1.5-4.9}$, $ZrCr_2H_{0.15-4}$, $ZrMn_2H_3$, and $LaNi_5H_{0.15-6}$. This agreement is in fact rather surprising since it follows from *Lacher*'s statistical treatment [6.79] of hydrogen absorption (see also the works of *Kirchheim* [6.80, 81] and *Griessen* [6.82] on this point) that the Fermi-Dirac distribution must be used instead of Boltzmann's distribution function. Furthermore (6.96) does not take into account that the number of sites per formula unit is different for each type of site. As pointed out by Jacob et al. it is however possible that the errors introduced by their simplifying assumptions are to a certain extent self-compensating.

The problem of the local enthalpies has been tackled recently by *Griessen* and *Driessen* [6.37]. In their work on the pressure–composition isotherms of disordered alloys they showed that the semi-empirical band-structure model could be used to estimate the heat of formation for hydrogen absorption at a given site. This model takes into account both the hydrogen affinity of the coordinating atoms and their interatomic separation in the alloy. It might be interesting to apply this model (or an improved version of it) to the problem of site occupancy in intermetallics.

6.4.2 The Atomic-Cell Model

In a series of articles, *Miedema* and coworkers [6.24, 27] have developed a semi-empirical model for the intgral enthalpy of formation of intermetallic compounds. We give here a brief summary of this model and describe its applications to binary and ternary metal hydrides [6.25, 26, 28, 29].

a) Intermetallic Compounds and Alloys

The basis of Miedema's model is that the Wigner-Seitz concept of atomic cells is still meaningful in alloys or compounds. This means that in an ordered compound, say AB, the atomic cells of A and B are assumed to resemble strongly the atomic cells of the pure metals A and B. There are, however, modifications in the boundary conditions when cells are brought into contact with each other. According to Miedema et al. the changes in boundary conditions lead to two energy effects. The first one is related to the mismatch of the electronic density n_{ws} of metal A and B at the boundary of their respective Wigner-Seitz cells. The second effect is related to the difference in the chemical potential ϕ^* of electrons in metals A and B. The integral enthalpy of formation of a compound $A_{y_A}B_{y_B}$ (with $y_A+y_B=1$) is then

$$\Delta\tilde{H}(A_{y_A}B_{y_B})=fgP\left[-(\Delta\phi^*)^2+\frac{Q}{P}(\Delta n_{ws}^{1/3})^2-\frac{R}{P}\right] \tag{6.97}$$

with

$$f=y_A^s y_B^s[1+8(y_A^s y_B^s)^2] \tag{6.98}$$

$$\Delta\phi^*=\phi_A^*-\phi_B^* \tag{6.99}$$

$$\Delta n_{ws}^{1/3}=(n_{ws}^A)^{1/3}-(n_{ws}^B)^{1/3} \tag{6.100}$$

$$g=\frac{y_A(V_A^{al})^{2/3}+y_B(V_B^{al})^{2/3}}{\langle n_{ws}^{-1/3}\rangle} \tag{6.101}$$

$$\langle n_{ws}^{-1/3}\rangle=\tfrac{1}{2}[(n_{ws}^A)^{-1/3}+(n_{ws}^B)^{-1/3}] \tag{6.102}$$

$$y_i^s=\frac{y_i(V_i^{al})^{2/3}}{y_A(V_A^{al})^{2/3}+y_B(V_B^{al})^{2/3}} \tag{6.103}$$

$$(V_i^{al})^{2/3}=V_i^{2/3}\left[1+a\frac{f}{y_i^s}(\phi_i^*-\phi_j^*)\right], \tag{6.104}$$

where i and j represent A or B. The parameters y_i^s represent effective surface concentrations. If both atoms A and B have the same molar volume ($V_A=V_B$) then $y_i^s=y_i$. The function f is an empirical function derived from a fit to experimental data on the heat of formation of ordered binary compounds. Since

V_i^{al} depends both on y_i^{s} and the function f, the equations (6.98, 103, 104) must be solved iteratively, except, of course, if volume effects can be neglected i.e. when the parameter a is negligible.

The numerical values for ϕ^* and n_{ws} are given in [6.84]. The parameters P, Q, R, and a depend on the type of metals involved in an alloy. For the alkali metals $a=0.14$, for the divalent metals $a=0.10$, for the noble metals and trivalent metals $a=0.07$ and $a=0.04$ for the other elements. The so-called hybridization term R/P depends in a complicated way on the position of the metals A and B in the periodic table. For alloys of two transition metals, however $R/P=0$. For tables of values of these parameters the reader is referred to [6.83, 84].

Except for the cases where $\Delta\tilde{H}(A_{y_A}B_{y_B})$ is close to zero, the model predicts values which are only 10–20% different from the experimental values for the heat of formation of ordered compounds of two transition metals. The error is somewhat larger for alloys of a transition metal with a non-transition metal. Despite this remarkable accuracy, Miedema's model has been criticized by *Williams*, *Gelatt*, and *Moruzzi* [6.30, 32] and by *Pettifor* [6.31, 85]. These authors argue basically that the success of an empirical model is not necessarily a confirmation of the validity of the underlying physical concepts which have been used to set up the model.

In the context of our discussion of the general properties of semi-empirical models, it should be noted that Miedema's model is in fact a six-dimensional model as each metal is characterized by ϕ^*, n_{ws}, V, a, R', P'. This high dimensionality is not immediately apparent because the model is not formulated in a consistent vectorial form. In fact it could be written as follows

$$f(\boldsymbol{A},\boldsymbol{B},y)=y(1-y)F(y,a_i,V_i,n_{\mathrm{ws}}^i,\phi_i^*,P_i')$$
$$\times\{(\phi_A^*-\phi_B^*)^2+k_1[(n_{\mathrm{ws}}^A)^{1/3}-(n_{\mathrm{ws}}^B)^{1/3}]^2+k_2(R_A'-R_B')^2\}\,. \quad (6.105)$$

In this form the hybridization term $k_2(R_A'-R_B')^2$ (in Miedema's notation R/P) is explicitly shown to vanish for two metals of the same type, i.e. for two metals with $R_A'=R_B'$. It is also apparent that no elastic energy term due to a size mismatch, i.e. a term proportional to $(V_A-V_B)^2$ is included in Miedema's model.

b) Binary Metal Hydrides

The atomic-cell model has been extended to the case of binary transition metal hydrides by *Bouten* and *Miedema* [6.28] by using the procedure described in Sect. 6.4.1. They treat a binary metal hydride as a "normal" alloy of a metal with metallic hydrogen and include in the expression for $\Delta\tilde{H}$ the enthalpy $\Delta H_{\mathrm{trans}}$ for the transformation at room temperature of gaseous hydrogen into "metallic" hydrogen. The parameters used to characterize metallic hydrogen are $\phi_{\mathrm{H}}^*=5.2$ eV, $n_{\mathrm{ws}}^{1/3}=1.5$ d.u., $V_{\mathrm{H}}=1.7\ \mathrm{cm}^3/\mathrm{mol}$, $R/P=3.9$, $P=12.3$, $Q/P=9.4$, and $\Delta H_{\mathrm{trans}}=100$ kJ/mol H. With these parameters Bouten and Miedema obtained

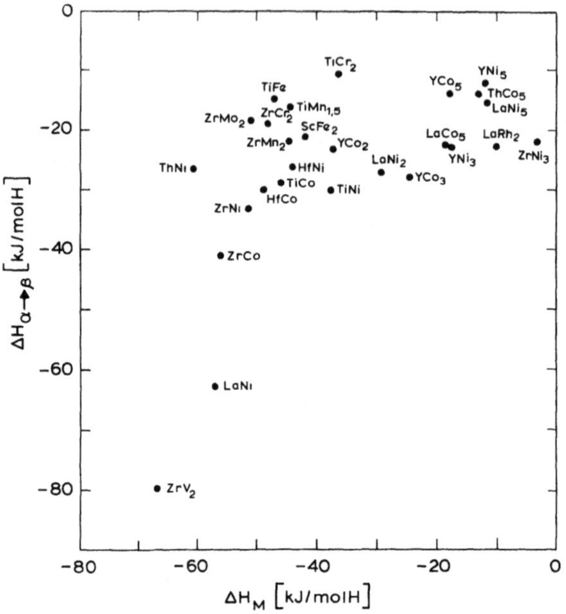

Fig. 6.9. Comparison of experimental heats of formation for ternary metal hydrides with values calculated by means of *Miedema*'s semi-empirical model [6.29]. References for experimental data are given in Table 6.6.2

a relatively good agreement between calculated and experimental values for concentrated hydrides (see Fig. 6.9).

In the limit of low hydrogen concentrations the heat of solution is given by

$$\Delta\bar{H} = \frac{V_{\mathrm{H}}^{2/3}}{\langle n_{\mathrm{ws}}^{-1/3}\rangle}(1+a\Delta\phi^*)\left[1-2x\left(1+\frac{V_{\mathrm{H}}^{2/3}}{V_M^{2/3}}(1+2a\Delta\phi^*)\right)\right]$$
$$\times P\left[-(\Delta\phi^*)^2+\frac{Q}{P}(\Delta n_{\mathrm{ws}}^{1/3})^2-\frac{R}{P}\right]+\Delta H_{\mathrm{trans}}, \tag{6.106}$$

where $x = y_{\mathrm{H}}/y_M$, $\Delta\phi^* = \phi_{\mathrm{H}}^* - \phi_M^*$, and $\Delta H_{\mathrm{trans}} = 100$ kJ/mol H. The heats of solution at infinite dilution $\Delta\bar{H}_\infty$ calculated by means of this equation (with $x=0$) are compared to experimental values in Fig. 6.10. One sees that in general the trend within the periodic table is correctly reproduced although the calculated values may be out by as much as a factor of two.

A difficulty in the application of Miedema's model to binary hydrides is that the model predicts that $d\Delta\bar{H}/dx > 0$ since

$$\frac{d\Delta\bar{H}}{dx} = 2(\Delta H_{\mathrm{trans}} - \Delta\bar{H}_\infty)\left[1+\frac{V_{\mathrm{H}}^{2/3}}{V_M^{2/3}}(1+2a\Delta\phi^*)\right] \tag{6.107}$$

and $|\Delta\bar{H}_\infty| < \Delta H_{\mathrm{trans}}$ for practically all elements. This conclusion is not confirmed by experiment since for most metal hydrides $\Delta\bar{H}$ decreases with increasing hydrogen concentration (see Table 6.6.1). This is probably due to the absence of an elastic energy term in (6.97) and (6.105).

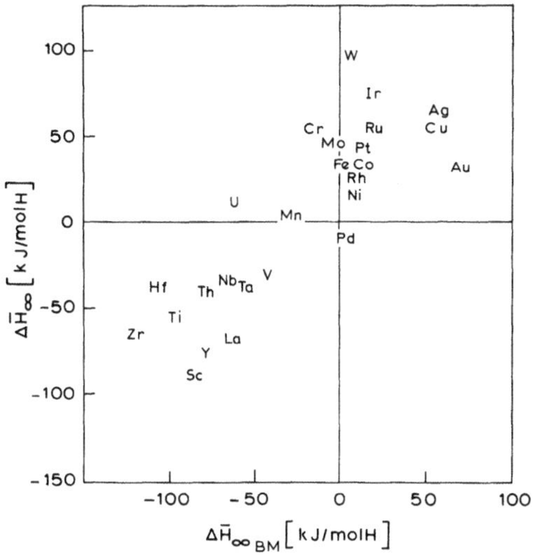

Fig. 6.10. Comparison of experimental heats of solution at infinite dilution for binary metal hydrides with the values predicted by the cellular semi-empirical model of *Miedema* and coworkers [6.28, 29]. References for the experimental data are given in Table 6.6.1

c) Ternary Metal Hydrides

On the basis of their work on hydrogen absorption in $LaNi_5$ and related compounds, *van Mal* et al. [6.26, 86] suggested that the integral heat of hydride formation can be calculated from

$$\Delta\tilde{H}(AB_n\mathrm{H}_{2m}) = \Delta\tilde{H}(A\mathrm{H}_m) + \Delta\tilde{H}(B_n\mathrm{H}_m) - \Delta\tilde{H}(AB_n) \tag{6.108}$$

when n is much larger than unity. The transition metal A is assumed to have a stronger affinity for hydrogen than the transition metal B. As a consequence hydrogen is attracted by the A atoms in the AB_n compound and isolates the A atoms from the surrounding B atoms. The breaking of A–B bonds is accounted for by the last term in (6.108). In compounds AB_n with $n<5$, however, not all A–B bonds are broken up. In successive steps *Miedema* et al. [6.28] generalized (6.108) to the case of AB_n compounds with n as small as one and proposed that

$$\Delta\tilde{H}(AB_n\mathrm{H}_{x+y}) = \Delta\tilde{H}(A\mathrm{H}_x) + \Delta\tilde{H}(B_n\mathrm{H}_y) - (1-F)\Delta\tilde{H}(AB_n)\,. \tag{6.109}$$

The factor $(1-F)$ represents the extent to which A–B contacts between atomic cells are broken by hydrogen. In AB_5 compounds it is empirically found that 90% of the A–B contacts are broken while only 40% are broken in AB compounds. Values of F for various ternary hydrides are tabulated in [6.29] for selected values of x, y, and n. Unfortunately no analytical dependence of F on x, y, and n has been proposed by the authors. It is, for example, quite evident that for small values of $x+y$ the factor $(1-F)$ must tend towards zero as all the other terms in (6.109) vanish for $x=y=0$.

It is worth noting that the so-called *rule of reversed stability* given in (6.109) resembles the relation (6.88) derived from a general scalar model. Using the notation of *Miedema* [6.29] (6.88) may be written as

$$\Delta\tilde{H}(AB_n\mathrm{H}_{x+y})=\Delta\tilde{H}\left(A\mathrm{H}_{\frac{x+y}{n+1}}\right)+\Delta\tilde{H}\left(B_n\mathrm{H}_{\frac{(x+y)n}{n+1}}\right)+K\frac{x+y}{n+1}\Delta\tilde{H}(AB_n)\,. \tag{6.110}$$

The main difference from Miedema's expression (6.109) is that here hydrogen is shared in the same proportions between A and B atoms, while in (6.109) y/n is always smaller than x. This basic difference probably arises from the fact that some structural information is implicitly contained in the atomic-cell model while no such ingredients are included in the "general" semi-empirical models discussed in Sect. 6.4.1.

Recently *Shilov* et al. [6.87] proposed a relation for the heats of formation of ternary hydrides based essentially on parameters derived from Miedema's model. However, because the present authors were not able to reproduce the numerical results of Shilov et al. using the parameter values from the cited references, it is presently difficult to assess the validity of their astonishingly accurate approach and its applicability to systems not considered by these authors.

d) Quaternary Metal Hydrides

The cellular approach has been used by *Pasturel* et al. [6.88–90] to calculate the integral heat of hydride formation of $AB_{n-s}C_s\mathrm{H}_{2m}$. In order to derive a relation for quaternary hydrides they consider first the case of ternary hydrides and write

$$\Delta\tilde{H}(AB_n\mathrm{H}_{2m})=\frac{m}{3}\Delta\tilde{H}(A\mathrm{H}_3)+\frac{m}{3}\Delta\tilde{H}(B_n\mathrm{H}_3)-\frac{m}{3}\Delta\tilde{H}(AB_n) \tag{6.111}$$

in agreement with their assumptions that for $2m<6$ some A–B bonds coexist with both A–H and B–H bonds and that the A and B atoms form binary hydrides of the form $A\mathrm{H}_3$ and $B_n\mathrm{H}_3$.

For the quaternary hydride they find that

$$\Delta\tilde{H}(AB_{n-s}C_s\mathrm{H}_{2m})=\frac{m}{3}\Delta\tilde{H}(A\mathrm{H}_3)+\frac{m}{3}\frac{(n-s)}{n}\Delta\tilde{H}(B_n\mathrm{H}_3)$$
$$+\frac{m}{3}\frac{s}{n}\Delta\tilde{H}(C_n\mathrm{H}_3)-\frac{m}{3}\Delta\tilde{H}(AB_{n-s}C_s)+\Delta\tilde{H}_1\,. \tag{6.112}$$

The term $\Delta\tilde{H}_1$ is a mixing term due to the substitution of B atoms by C atoms. It is given by

$$\Delta\tilde{H}_1=\Delta\tilde{H}(AB_{n-s}C_s)-\frac{n-s}{n}\Delta\tilde{H}(AB_n)-\frac{s}{n}\Delta\tilde{H}(AC_n)\,. \tag{6.113}$$

Although (6.112, 113) have been shown to nicely reproduce the experimental results for $LaNi_5$-type compounds, they erroneously predict that for $m=0$, the hydriding heat of reaction is equal to $\Delta\tilde{H}_1$ instead of vanishing. One can thus conclude that (6.112, 113) are probably only valid for $2 \leqq 2m < 6$.

6.4.3 The Band-Structure Model

a) Binary Metal Hydrides

In a search of a correlation between the standard heat of formation of binary metal hydrides and the smallest possible number of band-structure parameters of the host metal *Griessen* and *Driessen* [6.35] found that the experimental values for $\Delta H_{\alpha\to\beta}$ (see Fig. 6.11) derived from plateau dissociation pressures (see Sect. 6.2) or from calorimetric measurements in concentrated hydrides could be reasonably well reproduced by the following simple relation

$$\Delta H^0_{\alpha\to\beta} = \frac{n_s}{2}(\alpha\Delta E + \beta) \quad \text{with} \tag{6.114}$$

$$\Delta E \equiv E_F - E_s \quad \text{and} \tag{6.115}$$

$$\alpha = 29.62\ \text{kJ/eV mol H} \tag{6.116}$$

$$\beta = -135.0\ \text{kJ/mol H}\,. \tag{6.117}$$

For the standard heat of solution $\Delta\bar{H}^0_\infty$ at infinite dilution one finds a similar relation (see Fig. 6.12)

$$\Delta\bar{H}^0_\infty = \frac{n_s}{2}(\alpha_\infty \Delta E + \beta_\infty) \quad \text{with} \tag{6.118}$$

$$\alpha_\infty = 29.62\ \text{kJ/eV mol H} \tag{6.119}$$

$$\beta_\infty = -122.0\ \text{kJ/mol H}\,. \tag{6.120}$$

In (6.114, 115, 118) E_F represents the Fermi energy and n_s the number of electrons per atom in the lowest s-like conduction band of the host metal. E_s is the energy for which the integrated density of states of the host metal is equal to $n_s/2$ electrons per atom. The relative positions of E_F and E_s are indicated for several representative metals in Fig. 6.13. Except for the alkali metals (for which $n_s=1$) we have $n_s \simeq 2$ and E_s corresponds approximately to the centre of the lowest conduction band since even for transition metals this band has a predominantly s-character with respect to the interstitial sites occupied by the hydrogen atoms [6.1, 91, 92].

A remarkable feature of the correlations shown in Figs. 6.11 and 6.12 is that irrespective of the "type" of host metal (noble, simple or transition metals, rare-earths or actinides) the heats of formation of their hydrides are reasonably well

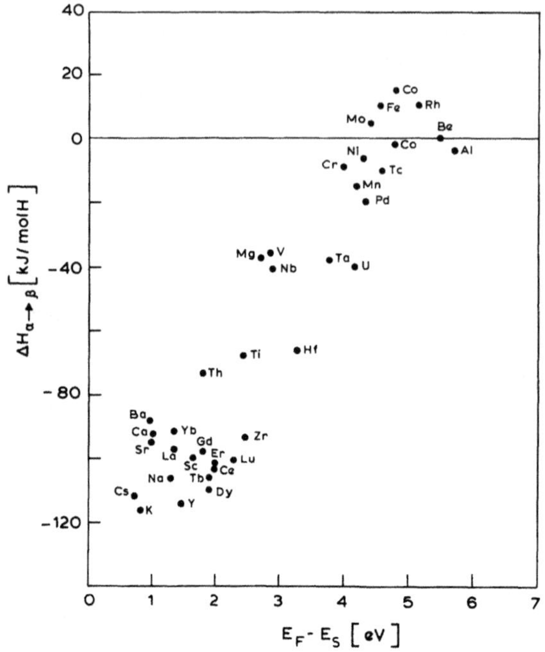

Fig. 6.11. Correlation between experimental heats of formation and the characteristic band structure energy $E_F - E_s$. E_F is the Fermi energy and E_s is essentially the centre of the lowest *s*-band (see also Fig. 6.13) [6.35]. For the alkali metals $n_s = 1$. Their ΔH values have been multiplied by 2 to correlate them with the other metals for which $n_s = 2$

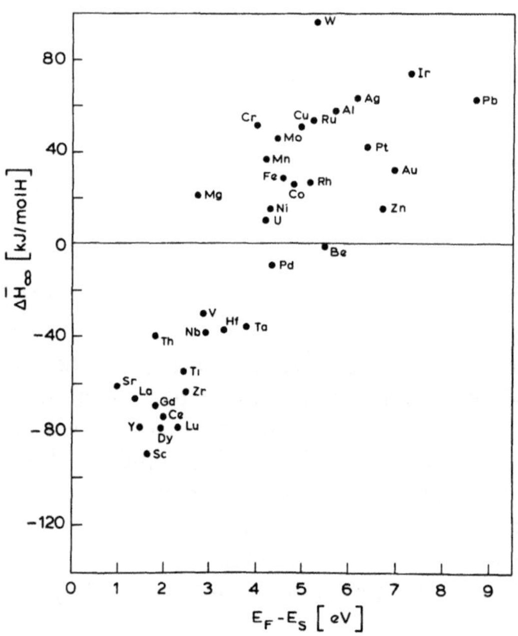

Fig. 6.12. Correlation between experimental heats of solution at infinite dilution and the characteristic band structure energy $E_F - E_s$ [6.35]

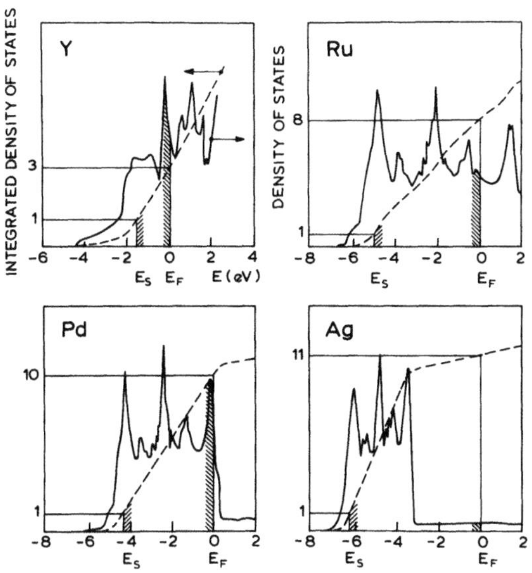

Fig. 6.13. Relative position of E_F and E_s for representative metals. E_s is defined as the energy for which the integrated electronic density of states is equal to 1 state per atom [6.116]

described by the same relation (6.114). Furthermore the large difference in hydriding properties of similar metals such as Be and Mg or Pd and Pt is correctly described.

Until now no first-principles derivation of (6.114) or (6.118) has been given. In the following we shall, however, show that the effective medium theory (including covalent effects) of *Nørskov* [6.3] can be cast in the form of (6.118). In this theory the standard heat of solution at infinite hydrogen dilution is given by

$$\Delta\bar{H}^0_\infty = \Delta E^{\rm hom}_{\rm eff}(\bar{n}_0) + \Delta E^{\rm hyb} - E_{\rm H_2}, \tag{6.121}$$

where $E_{\rm H_2} = -2.4$ eV is the binding energy in the H_2 molecule (expressed per atom). The effective embedding energy consists of two terms

$$\Delta E^{\rm hom}_{\rm eff}(\bar{n}_0) = \Delta E^{\rm hom}(\bar{n}_0) - \alpha_{\rm tot}\bar{n}_0, \tag{6.122}$$

where $\Delta E^{\rm hom}$ is the embedding energy of a hydrogen atom into the free electron gas of the host metal which in the simplest effective medium approximation is treated as a homogeneous electron gas of appropriate density $\bar{n}_0$. The second term $\alpha_{\rm tot}\bar{n}_0$ is a correction to the zeroth-order (in electron density variation) energy $\Delta E^{\rm hom}(\bar{n}_0)$. The second term in (6.121), $\Delta E^{\rm hyb}$ is due to the hybridization of the hydrogen 1*s*-like state with the host metal *d*-states. Within the atomic-sphere approximation [6.33, 34]

$$\Delta E^{\rm hyb} = -40(1-f)\frac{\Delta_d\Delta_s}{C_d - V^0}\sum_{\boldsymbol{R}}\frac{1}{R^6}, \tag{6.123}$$

where Δ_d (for the host metal *d*-states) and Δ_s (for the hydrogen *s*-state) are functions of the potentials and of the radii of the metal and the hydrogen atomic spheres, respectively. C_d is the centre of the d-band and V^0 is the value of the effective potential of the host metal at the hydrogen site. In most cases $C_d - V^0$ is much larger than the d-band width. The summation is over all lattice sites, the hydrogen at an interstitial site being taken as origin. The parameter f is the degree of filling of the d-bands which have an s-character at the hydrogen site. It is equal to 0.15 for Sc and increases up to 0.85 for Ni. For noble metals, $f=1$ because of their full d-band [6.3].

To simplify the notation we introduce an effective band width W_{eff} defined as

$$W_{\text{eff}} = 40 \frac{\Delta_d \Delta_s}{C_d - V^0} \sum_R \frac{1}{R^6} \tag{6.124}$$

and rearrange the terms in (6.123) so as to obtain

$$\Delta \bar{H}_\infty = [E_{\text{eff}}^{\text{hom}}(\bar{n}_0) - W_{\text{eff}}] + f W_{\text{eff}} - E_{\text{H}_2} . \tag{6.125}$$

The expression in square brackets has the interesting property of being almost constant for all transition metals, the largest value being -2.97 eV for La and the smallest -4.24 eV for Pt. The average value is -3.51 eV.

The second term on the other hand varies strongly across the $3d$-transition metals and is therefore naturally associated with the parameter ΔE of the semi-empirical band structure model. As shown in Fig. 6.14 there is an approximately linear relation between $f W_{\text{eff}}$ and ΔE.

Finally from (6.125) and Fig. 6.14 it follows that

$$\Delta \bar{H}_\infty^0 \simeq 27(E_{\text{F}} - E_s) - 110 \tag{6.126}$$

if E_{F} and E_s are expressed in eV and $\Delta \bar{H}_\infty^0$ in kJ/mol H. The numerical values are remarkably close to the values of the parameters α_∞ and β_∞ given in (6.119, 120). For transition metals it is thus possible to give a justification of the semi-empirical model of Griessen and Driessen. There remains the interesting question of why this semi-empirical model also works well for non-transition metals.

The validity of the underlying physical basis of the Griessen-Driessen model can also be tested by looking at implications of (6.114) or (6.118).

One implication, for example, concerns the expansion of the metal lattice during hydrogen absorption or, equivalently, the standard partial molar volume $\bar{V}_{\text{H}}$ of hydrogen in a metal. From relation (6.4) and the definition of $\Delta \bar{H} = \bar{H}_{\text{H}} - \frac{1}{2}\bar{H}_{\text{H}_2}$ we obtain

$$\bar{V}_{\text{H}} = -\frac{1}{B} \frac{\partial \Delta \bar{H}}{\partial \ln V}\bigg|_{T,x} + T \frac{\partial \bar{V}_{\text{H}}}{\partial T}\bigg|_{p,x} , \tag{6.127}$$

where B is the bulk modulus of the metal hydride which is usually approximately equal to that of the host metal [6.93, 94]. As $\bar{V}_{\text{H}}$ is experimentally found

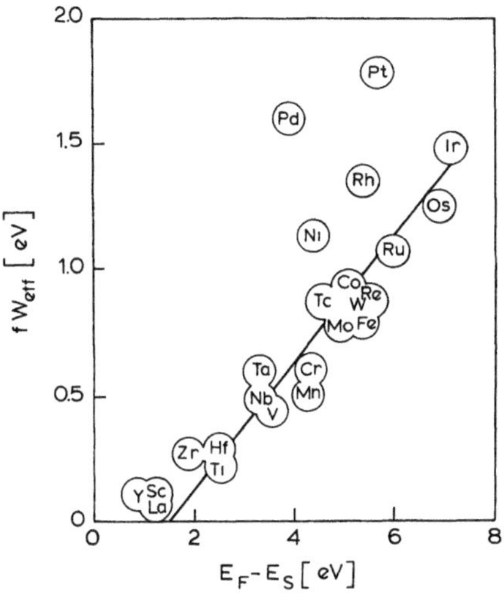

Fig. 6.14. Relation between the parameter f $W_{\rm eff}$ in *Nørskov*'s theory [6.3] and the characteristic band structure energy $E_{\rm F}-E_s$ [6.35]

to be only weakly temperature dependent, the last term in (6.127) may safely be neglected. One then finds if $\Delta\bar{H}\simeq\Delta H^0_{\alpha\to\beta}$ that

$$\bar{V}_{\rm H}=\frac{1}{B}\left[\left(\frac{n}{3}-\frac{\partial\ln\alpha}{\partial\ln V}\right)\Delta H^0_{\alpha\to\beta}-\beta\left(\frac{n}{3}-\frac{\partial\ln\alpha}{\partial\ln V}+\frac{\partial\ln\beta}{\partial\ln V}\right)\right] \qquad (6.128)$$

if $E_{\rm F}-E_s$ is assumed to vary with the host metal volume as $V^{-n/3}$. The exponent n depends on the type of metal under consideration. For simple metals the nearly-free-electron model predicts that $n\simeq 2$. For transition metals band-structure calculations lead to values close to $n=5$ [6.95, 96]. If for simplicity the parameters α and β are taken to be independent of volume then the partial molar volume is given by

$$\bar{V}_{\rm H}\approx\frac{1}{B}\frac{n}{3}(\Delta H^0_{\alpha\to\beta}-\beta). \qquad (6.129)$$

For transition metal hydrides such as VH_x and PdH_x (6.129) leads to $\bar{V}_{\rm H}\approx 1.0\,{\rm cm^3/mol\ H}$ for the former and $\bar{V}_{\rm H}\approx 1.1\,{\rm cm^3/mol\ H}$ for the latter in semi-quantitative agreement with the experimental results $\bar{V}_{\rm H}=1.42\,{\rm cm^3/mol\ H}$ for VH_x and $\bar{V}_{\rm H}=1.69\,{\rm cm^3/mol\ H}$ for PdH_x. This is quite remarkable as the partial molar volume $\bar{V}_{\rm H}$ is a *derivative* of the enthalpy. In other words the empirical correlation given in (6.114, 118) combined with theoretical results concerning the volume dependence of electronic band widths predicts $\Delta\bar{H}$ as well as $\Delta\bar{H}/\partial\ln V$.[2] This strongly suggests that the physical basis

2 Miedema's model [6.28] also leads to good predictions for the volume dilation of metals during hydrogen absorption. This is, however, not surprising, as $\bar{V}_{\rm H}=1.7\,{\rm cm^3/mol\ H}$ is introduced as a fit parameter in this model.

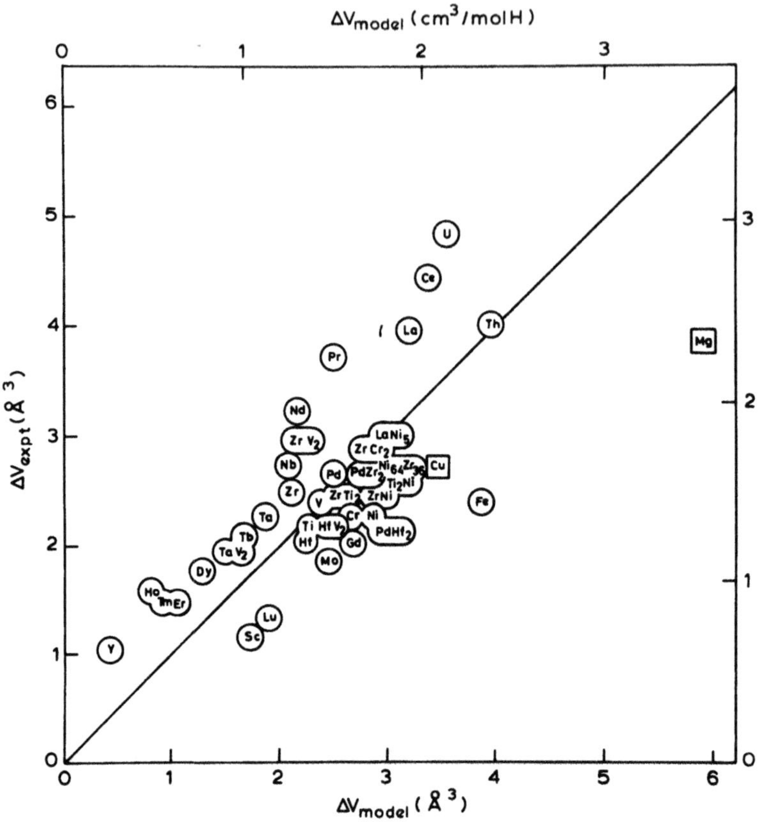

Fig. 6.15. Partial molar volume for hydrogen in various metals and intermetallic compounds compared to the predictions of the band-structure semi-empirical model [6.35, 38]

of (6.114, 118) is sound. As shown by *Griessen* and *Feenstra* [6.38] the agreement between calculated and measured $\bar{V}_H$ may be improved by allowing for volume-dependent parameters α and β. From a fit to existing data they obtain

$$\bar{V}_H = \frac{1}{B}(2.69\,\Delta H^0_{\alpha\to\beta} + 326) \tag{6.130}$$

for transition metal hydrides and

$$\bar{V}_H = \frac{1}{B}(1.67\,\Delta H^0_{\alpha\to\beta} + 189) \tag{6.131}$$

for simple metal hydrides when $\bar{V}_H$ is expressed in cm^3/mol H and $\Delta H^0_{\alpha\to\beta}$ in kJ/mol H. A comparison between $\bar{V}_H$ values calculated by means of (6.130, 131) and experimental values is shown in Fig. 6.15.

b) Ternary Metal Hydrides

The most interesting use of the empirical linear relation between the standard heat of formation ΔH and the characteristic band-structure energy parameter ΔE is to predict ΔH for ternary hydrides $A_{y_A}B_{y_B}H_x$. As band-structure calculations are available only for a very limited number of alloys or compounds Griessen and Driessen used the model of *Cyrot* and *Cyrot-Lackmann* [6.97] to evaluate ΔE for intermetallics of two d-band metals. The basic idea is that within the tight-binding approximation the exact density of states $n(E; y_A, y_B)$ calculated by means of the coherent-potential approximation formalism for an alloy $A_{y_A}B_{y_B}$ has the same moments

$$\mu^\nu \equiv \int E^\nu n(E; y_A, y_B) dE \tag{6.132}$$

for $\nu=0$, 1, and 2 (and even $\nu=3$ in the case of two metals with equal bandwidths) as the "simple" density-of-states (DOS) $n^*(E)$ given by

$$n^*(E; y_A, y_B) = \frac{y_A n'_A(E) + y_B n'_B(E)}{y_A + y_B}, \tag{6.133}$$

where $n'_i(E)$ is a scaled DOS function of metal i. The various steps involved in this scaling are shown schematically in Fig. 6.16 for the special case of an alloy AB ($y_A = y_B = 1$) when the d-band width of metal A is larger than that of metal B, i.e. $W_A > W_B$.

In the first step the widths of the d-bands of both metals are set equal to their weighted average

$$W^* = \frac{y_A W_A + y_B W_B}{y_A + y_B}. \tag{6.134}$$

Since the number of available Bloch states contained in the d-band is constant, a change in band width implies a change in height of the DOS curve so that

$$n'_i\left(\frac{E' - \varepsilon_i}{W^*}\right) = \frac{W_i}{W^*} n_i \left(\frac{E - \varepsilon_i}{W_i}\right) \quad \text{with} \tag{6.135}$$

$$E' - \varepsilon_i = (E - \varepsilon_i)\frac{W^*}{W_i}. \tag{6.136}$$

In (6.135, 136) ε_i is the centre of the DOS curve of metal i ($i = A$ or B) measured with respect to the vacuum zero. For the individual metals

$$\varepsilon_i = (E_F - E_s)_i - \frac{W_i}{2} + \phi_i, \tag{6.137}$$

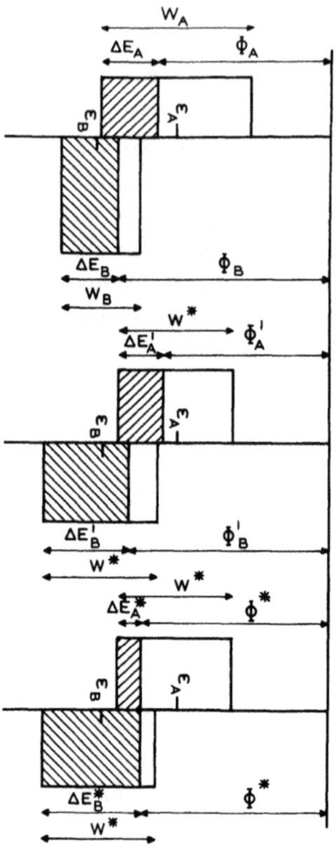

Fig. 6.16. The various steps involved in the construction of the appropriate density-of-states function $n^*(E)$ for the special case of an alloy AB [6.35, 97]

where ϕ_i is the work function of metal i. After having brought the DOS curves $n_i(E)$ to a common width W^* the band energy parameters $\Delta E_i'$ and work functions ϕ_i' are given by

$$\Delta E_i' = \Delta E_i \frac{W^*}{W_i} \quad \text{and} \tag{6.138}$$

$$\phi_i' = \phi_i + \Delta E_i - \Delta E_i' + \tfrac{1}{2}(W^* - W_i) \tag{6.139}$$

since ε_A and ε_B are assumed to be constant. In both (6.138) and (6.139) $\Delta E_i \equiv (E_F - E_s)_i$ and $\Delta E_i' \equiv (E_F - E_s)_i'$. On an absolute energy scale the Fermi energies E_{F_A}' and E_{F_B}' are different. This leads to an electron transfer until the Fermi energy is constant throughout the alloy.

In the second step of the construction of $n^*(E; y_A, y_B)$ one thus equilibrates the Fermi energies by requiring to first order that

$$(\phi^* - \phi_A') y_A n_A'(E_{F_A}') = (\phi_B' - \phi^*) y_B n_B'(E_{F_B}'), \tag{6.140}$$

where E_{F_i}' is the scaled Fermi energy of metal i obtained by setting $E = E_{F_i}$ in (6.136) and ϕ^* is the work function of the alloy $A_{y_A}B_{y_B}$ under consideration. As seen from (6.137, 139, 140) ϕ^* depends on the bandwidths W_i, the DOS $n_i(E_F)$ and the work functions ϕ_i of the pure metals A and B and, of course, on the concentration y_A and y_B of A and B atoms.

Since according to the result of *Cyrot* and *Cyrot-Lackmann* [6.97], a good approximate DOS $n^*(E; y_A, y_B)$ can be obtained by taking a weighted average of the scaled DOS, $n_i'(E)$, of the pure constituents one expects that for the heat of formation of the hydride

$$\Delta H^0_{\alpha\to\beta}(A_{y_A}B_{y_B}H_x)=\alpha\left(\frac{y_A\Delta E_A^*+y_B\Delta E_B^*}{y_A+y_B}\right)+\beta \tag{6.141}$$

with

$$\Delta E_i^*=\Delta E_i'+\phi_i'-\phi^*. \tag{6.142}$$

As indicated in Fig. 6.16, ΔE_i^* corresponds to the band-structure energy parameter of the scaled DOS of metal i after allowing a transfer of electrons to bring E_{F_i} to a common level.

Values for $\Delta H^0_{\alpha\to\beta}$ were obtained from (6.114–117) by using experimental values for ϕ_i, calculated DOS curves for the band widths W_i and the optimized ΔE values given in [Ref. 6.35, Table II]. These values are such that when inserted into (6.114) the experimental values for the heats of formation of the binary hydrides are reproduced [using the values for α and β given in (6.116, 117)]. In most cases the optimized ΔE differ by no more than a few tenths of an eV from the values derived from band-structure calculations.

As shown in Fig. 6.17 the values for $\Delta H^0_{\alpha\to\beta}$ calculated by means of (6.141) are in remarkably good agreement with experimental values. The average mean-square deviation is more than twice as small as that in Fig. 6.10 which shows the predictions of the cellular model of Miedema and co-workers which involves many more fit parameters. From a practical point of view this means that the predictive power of the Griessen-Driessen model is sufficiently high for a meaningful overall search of potential hydrogen storage alloys. Calculated heats of hydride formation are given in Sect. 6.6.3 for AB_5, AB_2, AB, A_2B, and A_5B intermetallics (A is Sc, Ti, V, Y, Zr, Nb, La, Hf, Ta, Pd or Ni and B is a transition metal).

It is important to point out to the user of these tables that the results listed correspond to a situation where each hydrogen atom is surrounded by a fraction $y_A/(y_A+y_B)$ of A atoms and a fraction $y_B/(y_A+y_B)$ of B atoms. This is an implicit assumption made in writing (6.141). This is in general not true for intermetallic compounds (in AB_2 Laves phases, for example, there are three types of interstitial sites A_2B_2, AB_3, and B_4). Furthermore in disordered alloys there is a whole distribution of sites. In *dilute* disordered alloys most of the interstitial sites are coordinated by atoms of the most abundant metal. The heats of formation given in Tables 6.6.3a–e correspond then approximately to hydrogen concentrations which are high enough so that most hydrogens

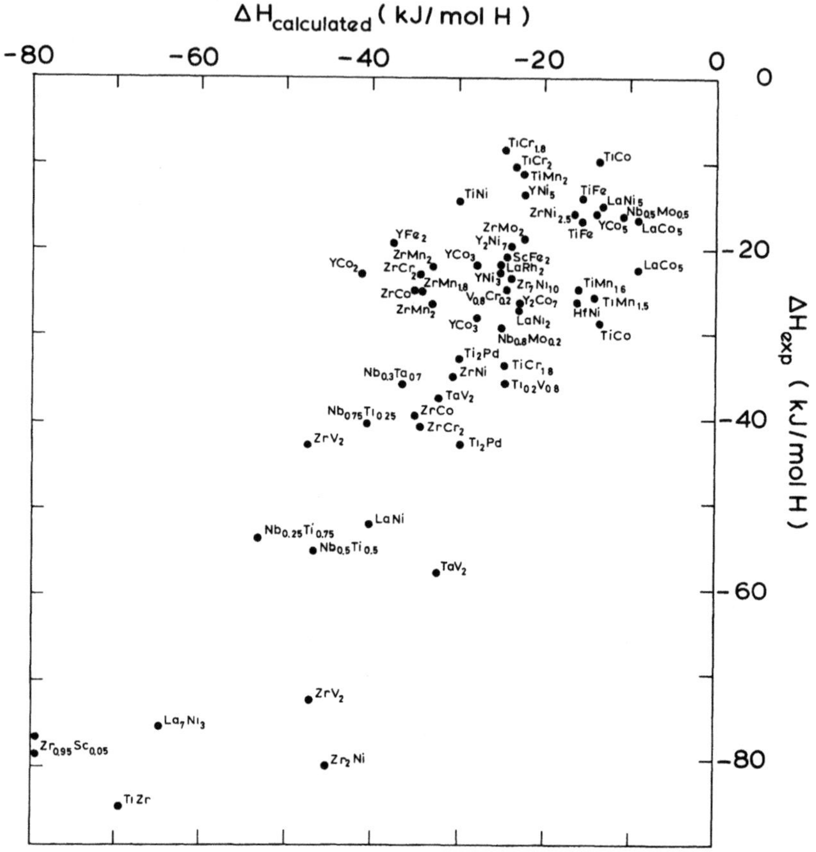

Fig. 6.17. Comparison of experimental heats of formation with values predicted from the band-structure semi-empirical model of *Griessen* and *Driessen* [6.35]

occupy sites of composition $A_{y_A}B_{y_B}$. For example, for a A_5BH_x disordered hydride the tabulated values correspond roughly to $x \simeq 4$ if one assumes that the interstitial sites are coordinated by 6 nearest-neighbour atoms.

Another factor which has been neglected is the volume change due to alloying of a metal A with a metal B. This point will be treated in some detail in the next section.

To conclude this section it is worth considering a few special cases of the Griessen-Driessen model in the light of our discussion of general semi-empirical models. The simplest model is the square electron density-of-states model for which n_A and n_B are constant in the energy interval $[\varepsilon_i-(W_i/2),\, \varepsilon_i+(W_i/2)]$ and vanish everywhere else. For this model of a d-band metal

$$n_A W_A = n_B W_B = 10 \qquad (6.143)$$

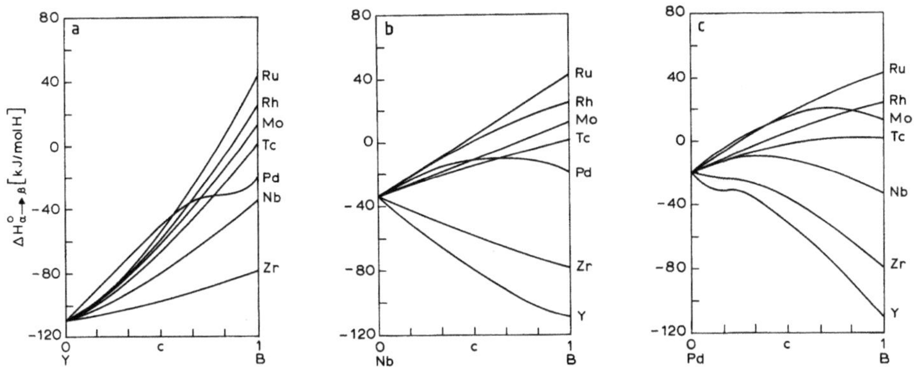

Fig. 6.18a–c. Heats of formation of the hydrides of Y-, Nb-, and Pd-based alloys calculated by means of (6.141) with the assumption that each hydrogen is surrounded by a fraction $1-y$ of Y, Nb or Pd atoms. Volume effects are not included

and the integral heat of formation is given by

$$\Delta\tilde{H}^0(A_{1-y}B_yH_x) = (1-y)\Delta\tilde{H}^0(AH_x) + y\Delta\tilde{H}^0(BH_x) + \alpha x(Z_A - Z_B)(W_B - W_A)y(1-y)\,. \tag{6.144}$$

This expression has the same form as the general relation (6.89) [see also (6.81)] except that it does not involve squared terms such as $(Z_A-Z_B)^2$ or $(W_B-W_A)^2$. In this simple model the two relevant parameters which characterize a metal are Z, the number of d-electrons and W, the width of the d-band. It is quite remarkable that ε, the position of the centre of the d-band, does not appear in (6.144). This is due to the fact that when an electron flow is allowed from one metal to the other (see Fig. 6.16) to bring the Fermi energies of both metals to a common value, the total energy remains constant. This, however, is a pecularity of this simple model as can easily be shown by considering another density-of-states model where both metals A and B have the same d-band width, i.e. $W_A = W_B$ but *different* density of states at the Fermi energy. Then

$$\Delta\tilde{H}^0(A_{1-y}B_yH_x) = (1-y)\Delta\tilde{H}^0(AH_x) + y\Delta\tilde{H}^0(BH_x) - \alpha xy(1-y)(\phi_A-\phi_B)(n_A-n_B)\frac{1}{(1-y)n_A+yn_B}\,. \tag{6.145}$$

This expression is consistent with the general form (6.89) derived in Sect. 6.4.1 for a two-dimensional model, each metal being characterized by the work function ϕ_i and the value of the density of states n_i at E_{F_i}. As discussed by Griessen and Driessen this expression explains qualitatively why alloys of Pd and an early transition metal have a tendency to be poorer hydrogen absorbers than pure Pd. This is confirmed by the results shown in Fig. 6.18 which have been obtained from expression (6.141). For Pd-based alloys (as well as for Pt-

and Ni-based alloys) the $\Delta\bar{H}^0$ versus y plots do indeed exhibit a large positive concavity. Conversely for transition metals of neighbouring columns in the periodic table $\phi_A \simeq \phi_B$ and $n_A \simeq n_B$ (except for Cr, Mo, and W based alloys) and the nonlinear term in $\Delta\bar{H}^0$ is negligible.

c) Volume Effects

Due to the importance attached to correlations between heat of formation and host metal volume or interstitial site size we consider here in some detail another implication of (6.114) [or (6.118)].

By combining (6.127) and (6.130) we obtain for transition metals a volume dependence of the heat of formation given by

$$\left.\frac{\partial \Delta H^0_{\alpha\to\beta}}{\partial \ln V}\right|_{T,x} = -2.69\Delta H^0_{\alpha\to\beta} - 326\,. \qquad (6.146)$$

This means that, as expected, a dilation of the host lattice induces a lowering of the heat of formation. This prediction cannot easily be checked by direct mechanical compression of the lattice because of the very high pressures which would be required for a sizeable volume change. A much easier way is to alloy a metal A to a metal B with a smaller affinity for hydrogen. At low hydrogen concentrations the sites coordinated exclusively by A atoms will thus be occupied first and the heat of solution determined in the dilute regime will correspond to the $\Delta\bar{H}$ of the pure host alloy at a volume V_{alloy} which may be larger or smaller than V_A depending on whether B atoms act as dilation or contraction centres in the A matrix.

For Pd based alloys the data compiled in Fig. 6.19 imply that

$$\frac{\partial \Delta\bar{H}^0_\infty}{\partial \ln V} = -223\ \text{kJ/mol H}$$

in reasonable agreement with the value of -299 kJ/mol H obtained by assuming that (6.146) is also valid at infinite dilution. For Nb alloys the data of *Machlin* [6.98] lead to

$$\frac{\partial \Delta\bar{H}^0_\infty}{\partial \ln V} = -197\ \text{kJ/mol H}$$

and the model gives a value of -232 kJ/mol H.

Another very interesting class of materials are the hydrides of the $LaNi_5$ type compounds and some substituted quaternary hydrides such as $LaNi_{4-y}Al_yH_x$. As shown in Fig. 6.20, $LaNi_5H_x$ is stabilized when Ni is replaced by a larger atom. The slope of the straight line fitted to the experimental data is

$$\frac{\partial \Delta H^0_{\alpha\to\beta}}{\partial \ln V} = -231\ \text{kJ/mol H}\,,$$

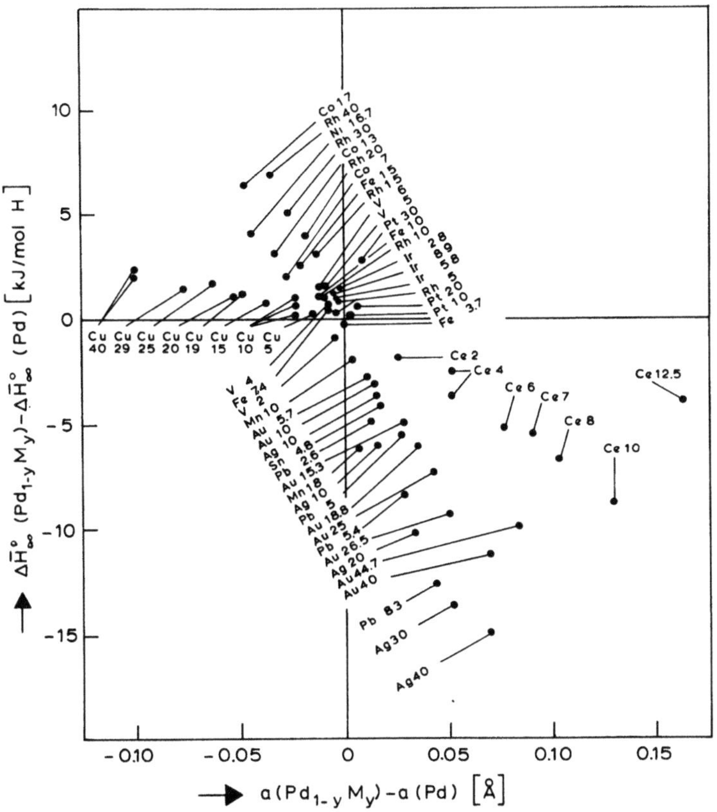

Fig. 6.19. Heat of solution at infinite dilution for Pd-based disordered alloys as a function of lattice spacing. The numbers indicate the amount (in at.%) of metal alloyed to palladium [6.98, 117]

which is in agreement with the theoretical value of −272 kJ/mol H. As pointed out by *Percheron-Guégan* [6.90] this behaviour of substituted $LaNi_5$ hydrides cannot be explained by means of Miedema's rule of reversed stability since the intermetallic compounds $LaNi_4M$ (M=Cu, Fe, Ni, and Mn) exhibit similar heats of alloy formation while their heats of hydride formation are different especially for Mn. Furthermore, $LaNi_4Al$ forms a very stable hydride although the intermetallic itself is not unstable.

Further, somewhat indirect evidence for the importance of volume effects in the heat of formation may be found in the isoelectronic alloys such as $Nb_{1-y}V_y$. The heat of solution (formation) of the binary hydrides VH_x is almost the same as that of NbH_x. However it is found that the critical point of the Nb–H phase diagram is substantially lowered by the addition of vanadium. As shown by *Griessen* and *Driessen* [6.37] this critical-point lowering may easily be understood by taking into account that vanadium clusters in a niobium-rich

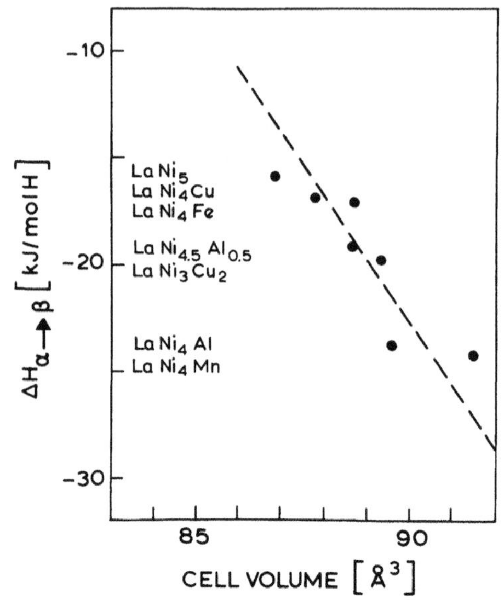

Fig. 6.20. Heat of formation of $LaNi_{5-y}M_yH_x$ as a function of unit cell volume. The dashed line corresponds to the prediction of the band structure semi-empirical model [6.35, 38]

alloy have a strong hydrogen affinity because the V–V separation in such a cluster is larger than in pure vanadium. In other words, V acts as a trapping centre in Nb in agreement with the work of *Futran* et al. [6.99] and *Pick* and *Welch* [6.100]. In non-isoelectronic alloys volume effects play in general a secondary role. In $Nb_{1-y}Mo_y$, *Fenzl* and *Peisl* [6.101] found for example a decrease of 123 K in T_c for $y=0.05$. Such a large variation can be semiquantitatively explained by the large difference in hydrogen affinity of Mo and Nb [6.37].

Another manifestation of the dependence of ΔH on interatomic spacing is the preference of hydrogen (or its isotopes) for occupying specific sites of the metallic host lattice. In fcc metals such as Ni, Pd, and Cu, hydrogen occupies the octahedral interstitial sites while in the group V bcc metals V, Nb, and Ta it occupies tetrahedral sites (at low concentrations). These site preferences can be explained by means of the geometric model of *Westlake* [6.63]. In Pd, for example, the tetrahedral hole radius is $r_h \simeq 0.37$ Å. It remains unoccupied because r_h is smaller than the empirical critical hole radius of 0.40 Å. The octahedral site on the other hand is easily occupied (according to this criterion) as $r_h \simeq 0.64$ Å. In our opinion the existence of a sharp criterion for the minimum hole size must reflect a strong dependence of the metal–hydrogen interaction potential on interatomic spacing.

6.4.4 The Local Band-Structure Model

Very recently *Griessen* [6.102] proposed a new model, the so-called local band-structure model in which the heat of hydrogen solution at infinite dilution $\Delta\bar{H}_\infty$ in a transition metal is given by

$$\Delta\bar{H}_{\infty} = a\Delta E W^{1/2} \sum_j R_j^{-4} + b \quad \text{with} \tag{6.147}$$

$$a = 18.6\,(\text{kJ/mol H})\,\text{Å}^4\,\text{eV}^{-3/2} \quad \text{and} \tag{6.148}$$

$$b = -90\ \text{kJ/mol H} \tag{6.149}$$

if ΔE and the d-band width W are given in eV and the distance R_j between a given hydrogen and the j-nearest neighbour host metal atom is expressed in Å. As in the band-structure model of *Griessen* and *Driessen*, $\Delta E \equiv E_F - E_s$ [see (6.114)]. This model incorporates both nonlocal effects (through the electronic band energies ΔE and W) and local information on the site actually occupied by a hydrogen atom (through the purely geometric term $\sum_j R_j^{-4}$, the summation being over all host metal atoms in the first nearest neighbour shell of a given hydrogen atom).

As shown in Fig. 6.21 the correlation between the experimental values for $\Delta\bar{H}_{\infty}$ and the parameter $\Delta E W^{1/2} \sum_j R_j^{-4}$ is excellent. The discrepancies for tungsten and iridium are not serious as the experimental uncertainties in the $\Delta\bar{H}_{\infty}$ of strongly endothermal metal–hydrogen systems may be considerable. The remarkable agreement between calculated ($\Delta\bar{H}_{\infty}^{\text{mod}}$) and experimental ($\Delta\bar{H}_{\infty}^{\text{exp}}$) values for the heats of solution is also clearly indicated in Fig. 6.22.

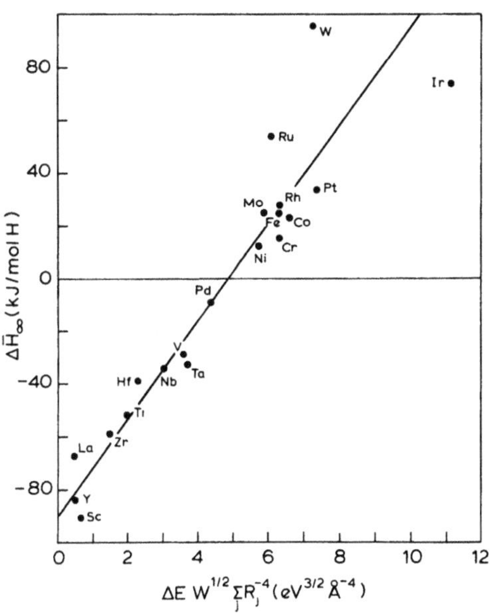

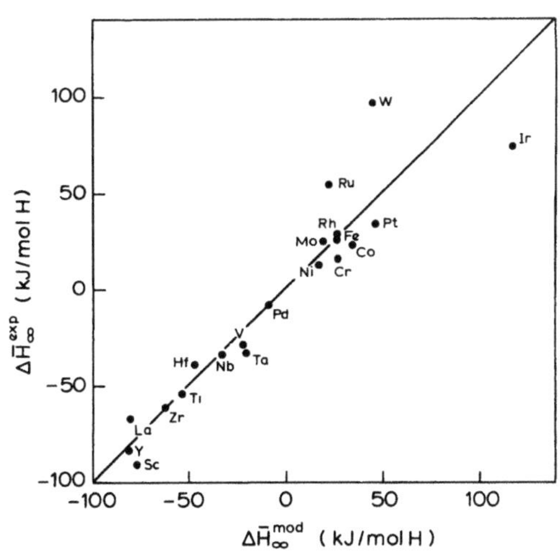

▲

Fig. 6.21. Correlation between experimental values of the enthalpy of hydrogen solution at infinite dilution and the local band-structure parameter $\Delta E W^{1/2} \sum_j R_j^{-4}$

Fig. 6.22. Comparison between experimental $\Delta\bar{H}_{\infty}^{\text{exp}}$ and values calculated by means of the local band-structure model (6.147–149)

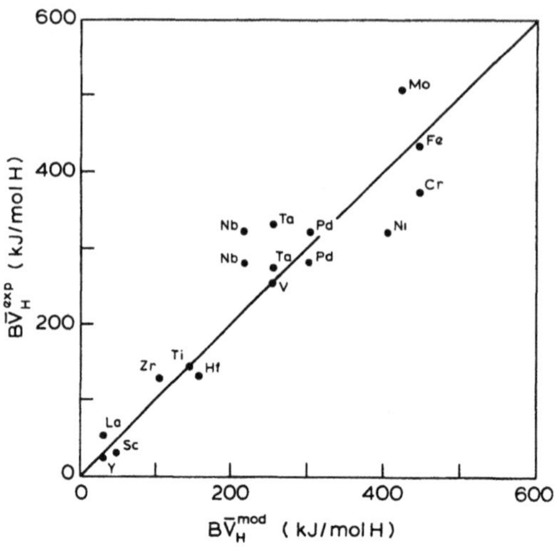

Fig. 6.23. Comparison between experimental $B\bar{V}_H^{exp}$ and values calculated by means of the local band-structure model (6.150 or 151). For ferro- or antiferromagnetic metals the values used for the bulk modulus B are those calculated by *Moruzzi* et al. [6.103] (see also [6.38])

The model also predicts the correct volume dependence of $\Delta\bar{H}_\infty$ and can therefore be used to evaluate the molar volume of hydrogen in a transition metal. This is easily seen by using (6.127) and the fact that for a transition metal both ΔE and W scale as $V^{-5/3}$ so that

$$B\bar{V}_H = -\frac{\partial \Delta\bar{H}_\infty}{\partial \ln V} = 71.2\Delta E W^{1/2} \sum_j R_j^{-4} \tag{6.150}$$

or in terms of $\Delta\bar{H}_\infty$,

$$B\bar{V}_H = 3.83\Delta\bar{H}_\infty + 345\,. \tag{6.151}$$

A comparison between experimental values ($B\bar{V}_H^{exp}$) and calculated values ($B\bar{V}_H^{mod}$) is shown in Fig. 6.23. The simple relation (6.147) thus well reproduces both $\Delta\bar{H}_\infty$ *and* its volume dependence without requiring additional fit parameters as was the case in Miedema's model and in the band-structure model of Griessen and Driessen.

The tendency of hydrogen to occupy certain sites in a given crystal structure follows directly from (6.147) by noting that ΔE and W are by definition site independent so that the heats of solution for hydrogen in two different types of sites (say type I and type II) are related to each other by

$$\frac{\Delta H_\infty^{\mathrm{I}} + 90}{\Delta H_\infty^{\mathrm{II}} + 90} = \frac{\sum_{\text{site I}} R_j^{-4}}{\sum_{\text{site II}} R_j^{-4}}\,. \tag{6.152}$$

This simple relation correctly predicts that octahedral sites are occupied in palladium, tetrahedral sites in group V metals and that in the early transition metals dual occupancy (of the octahedral and tetrahedral sites of the hcp structure) is possible.

Other applications of the model, for example, to calculate the binding energy of hydrogen to a vacancy or the heat of solution of hydrogen in intermetallic compounds are discussed in [6.102, 104].

6.5 Conclusions

On the basis of the work reviewed in this chapter it can be concluded that:

1) The heat of solution (formation) of a metal–hydrogen system depends *both* on lattice spacing and electronic structure. For recent articles on this subject the reader is referred to [6.105–109]. Only in special cases (such as hydrogen in dilute alloys and in isoelectronic alloys) are lattice spacing effects dominant. It is therefore not expected that empirical correlations involving only cell volume or hole size will hold for a wide class of materials.

2) Empirical correlations between ΔH and a physical parameter (for example, shear modulus, compressibility, electronic specific heat, atomic volume, interstitial hole volume, etc.) have in general been proposed on the basis of too small a number of metal hydrogen systems. For some of them it is shown that the proposed correlations do not even hold for the binary hydrides investigated up to now.

3) For restricted classes of materials empirical correlations may be useful for the tailoring of metal hydrides with precisely defined characteristics (a given plateau pressure at a chosen temperature, for example) [6.110, 111].

4) For binary hydrides the trends in ΔH through long series of the periodic system are well reproduced by first-principles calculations. For a fast and accurate evaluation of ΔH however, one still requires semi-empirical models.

5) For an overall search for metal hydrides with specific properties semi-empirical models are most valuable. For ternary metal hydrides the band-structure semi-empirical model of Griessen and Driessen appears to be more reliable than the cellular model of Miedema and co-workers (although it involves far less fit-parameters than Miedema's model). One should keep in mind however that Miedema's model was designed to describe a much wider class of alloys than just the metal hydrides.

6) The constancy of the volume dilation upon hydrogen absorption observed in many transition-metal–hydrogen systems is related to the general behaviour of the volume dependence of their electronic *d*-band width.

7) The two models proposed so far to explain the site preference of hydrogen in intermetallics cannot be used for an evaluation of the actual heat of solution (formation) associated with the various interstitial sites. The imaginary-binary-hydrides model of Jacob and Shaltiel is based on an ad hoc definition of local enthalpies and Westlake's geometrical model does not involve any thermodynamic parameters at all.

8) The local band-structure model recently proposed by Griessen incorporates both site-dependent effects and electronic band-structure effects. It can advantageously be used as long as sufficiently accurate first-principles calculations are not available.

6.6 Tables

Table 6.6.1. Heat of solution and formation of binary metal hydrides. $\Delta\bar{H}_\infty$ is the heat of solution at infinite dilution and ΔH is the heat of formation of concentrated hydrides (indicated in parenthesis). Both enthalpies are given in kJ/mol H. For most metals the values for $\Delta\bar{H}_\infty$ and ΔH are for the reaction of hydrogen with the solid metal (*s*) unless specified explicity by (*l*) for the liquid state

Element	$\Delta\bar{H}_\infty$ [kJ/mol H]	ΔH [kJ/mol H]	Ref.
Al	+63		[6.118]
(*l*)	+59		[6.118]
(*l*)	+59 (1000–2000 K)		[6.119, 120]
	+26 (298 K)	−7/+3 ($AlH_{0.5}$) (298 K)	[6.54]
		−4 (AlH_3) calor (298 K)	[6.121]
Ag	+63 (823–1234 K)		[6.120]
(*l*)	+68 (1234–1923 K)		[6.120]
Am		−95 (AmH_2) (750–1150 K)	[6.122]
Au	+32 (966–1373 K)		[6.120]
B		−4 ($B_{10}H_{14}$)	[6.119]
Ba		−88 (BaH_2) (743–823 K)	[6.119]
		−86 (BaH_2) calor (298 K)	[6.119]
Be	− 2		[6.118]
		~0 (BeH_2) estim (298 K)	[6.119]
Bi			
Ca		−92 (CaH_2) (873–1053 K)	[6.119]
		−85 (CaH_2) (1053–1173 K)	[6.119]
		−94 (CaH_2) calor (298 K)	[6.119]
Cd			
Ce	−74 (800–1100 K)		[6.123]
	−73 (824–1173 K)	−109 (CeH_2) (824–1173 K)	[6.125]
		−102 (CeH_2) (870–1270 K)	[6.120]
		−103 (CeH_2)	[6.124]
Co	+26 (923–1768 K)		[6.120]
	+21 (298 K)	+15 ($CoH_{0.5}$) (298 K)	[6.54]
(*l*)	+32 (1768–2073 K)		[6.120]
Cr	+52 (1000–1620 K)	−6 (CrH) (298–423 K)	[6.120]
	+28/−2 (298 K)	−8/+2 ($CrH_{0.5}$) (298 K)	[6.54]
Cs (*l*)		−56 (CsH) (518–651 K)	[6.119]
Cu	+46 (770–1356 K)		[6.120]
	+55		[6.125]
	+49		[6.118]
	+43 (1356–1823 K)		[6.120]
Dy	−79 (830–1250 K)	−104 (DyH_2) (830–1250 K)	[6.120]
		−110 (DyH_2)	[6.124]
Er		−112 (ErH_2)	[6.124]
Eu			
Fe (α)	+29 (280–1184K)		[6.120]
(γ)	+28 (1184–1667 K)		[6.120]
(δ)	+29 (1667–1811 K)		[6.120]
(*l*)	+33 (1811–2093 K)		[6.120]
	+21 (298 K)	+10 ($FeH_{0.5}$) (298 K)	[6.54]

Table 6.6.1 (continued)

Element	$\Delta\bar{H}_\infty$ [kJ/mol H]	ΔH [kJ/mol H]	Ref.
Ga			
Gd	−62 (823–1173 K)	−103 (GdH_2) (823–1173 K)	[6.125]
	−76 (830–1250 K)	−101 (GdH_2) (830–1250 K)	[6.120]
		− 98 (GdH_2)	[6.124]
Ge	(+221) (1073–1183 K)		[6.126]
(*l*)	+14 (1770–1979 K)		[6.120]
Hf (α)	−38	−66 (HfH_2)	[6.118]
Hg			
Ho		−113 (HoH_2)	[6.124]
In			
Ir	+74 (1666–1853 K)		[6.127]
K	~0		[6.118]
(*l*)		−56 (KH) (561–688 K)	[6.119]
		−58 (KH) calor (298 K)	[6.119]
La	−67 (900–1050 K)		[6.118]
		−97 (LaH_2)	[6.124]
		−104 (LaH_2) (600–1150 K)	[6.120]
Li (*l*)	−51 (980–1180 K)		[6.120]
(*l*)		−97 (LiH) (<967 K)	[6.128]
		−72 (LiH) (>967 K)	[6.128]
		−90 (LiH) calor (298 K)	[6.119]
Lu	−79(830–1250 K)	−102 (LuH_2) (830–1250 K)	[6.120]
		−104 (LuH_2)	[6.124]
		−102 (LuH_2) (1098–1223 K)	[6.129]
Mg	+21 (370–940 K)		[6.120]
(*l*)	+27 (970–1070 K)		[6.120]
		−37 (MgH_2) (713–833 K)	[6.119]
Mn (α)	(−2) (413–898 K)		[6.120]
(β)	+37 (1050–1330 K)		[6.120]
(*l*)	+30 (1623–1718 K)		[6.120]
	+1 (298 K)	−8 ($MnH_{0.5}$) (298 K)	[6.54]
Mo	+46(973–1773 K)		[6.120]
	+34/+15 (298 K)	+5 ($MoH_{0.5}$) (298 K)	[6.54]
Na (*l*)	+2 (380–670 K)	−53 (NaH) (380–670 K)	[6.120]
		−56 (NaH) calor (298 K)	[6.119]
Nb	−38 (275–2275 K)		[6.120]
	−33 (275–353 K)		[6.120]
	−35 (625–944 K)	−44 ($NbH_{0.5}$) (625–944 K)	[6.130]
		−38 ($NbH_{0.5}$)	[6.41]
(*l*)	−31		[6.118]
Nd	−50 (823–1173 K)	−103 (NdH_2) (823–1173 K)	[6.125]
		−106 (NdH_2)	[6.124]
Ni	+16 (623–1673 K)		[6.120]
	+12 (473–873 K)		[6.131]
	+17 (1000–1516 K)		[6.127]
	+10 (298 K)	−3 ($NiH_{0.5}$) (298 K)	[6.54]
(*l*)	+24 (1763–1973 K)		[6.120]
Np		−73 (NpH_{2+x}) (808–883 K)	[6.132]
(β)		−56 (NpH_{2+x}) (743 K)	[6.198]
(γ)		−61 (NpH_{2+x}) (875 K)	[6.198]

Table 6.6.1 (continued)

Element	$\Delta\bar{H}_\infty$ [kJ/mol H]	ΔH [kJ/mol H]	Ref.
Os			
Pa		(−36/−48) ($PaH_{1.3}$)	[6.133]
Pb	+62 (298–600 K)		[6.123]
(*l*)	+58 (600–1973 K)		[6.123]
Pd	−10 (273–420 K)		[6.118]
		−20 ($PdH_{0.6}$)	[6.53,118]
Pm			
Pr	−68 (823–1173 K)	−106 (PrH_2) (823–1173 K)	[6.125]
		−104 (PrH_2)	[6.124]
Pt	+42 (575–1670 K)		[6.120]
	+27 (883–1466 K)		[6.22]
Pu		−77 (PuH_2)	[6.122]
		−86 (PuH_2) calor (298 K)	[6.134]
Ra		(−72) (RaH_2)	[6.135]
Rb (*l*)		−54 (RbH) (519–623 K)	[6.119]
Re			
Rh	+27 (1131–1797 K)		[6.127]
		+10 ($RhH_{0.5}$) (298 K)	[6.54]
Ru	+54 (1275–1776 K)		[6.127]
Sb			
Sc	−90	−100 (ScH_2)	[6.118]
Se	(+39/+33)		[6.135]
Si	~+180		[6.118]
(*l*)	+110		[6.118]
Sm	−70 (775–925 K)	−98 (SmH_2) (775–1100 K)	[6.120]
		−100 (SmH_2)	[6.136]
Sn (*l*)	+125 (1273–1573 K)		[6.123]
Sr	−61	−92 (SrH_2)	[6.137]
		−100 (SrH_2) (...–1273 K)	[6.119]
		−88 (SrH_2) calor (298 K)	[6.119]
Ta	−36 (313–1213 K)		[6.120]
	−32 (623–904 K)		[6.130]
		−38 ($TaH_{0.5}$) (573–973 K)	[6.119]
Tb	−78 (830–1250 K)	−104 (TbH_2) (830–1250 K)	[6.120]
		−106 (TbH_2)	[6.124]
Tc		−12 ($TcH_{0.5}$) (298 K)	[6.54]
Te	(−84)		[6.135]
Th	−40 (573–1073 K)		[6.123]
		−73 (ThH_2)	[6.119]
Ti (α)	−52 (573–1073 K)		[6.138]
(β)	−58 (1173–1423 K)		[6.138]
(*l*)	−47 (1928–2073 K)		[6.123]
		−67 (TiH_2) (723 K)	[6.123]
		−68 (TiH_2) calor (298 K)	[6.119]
Tl			
Tm		−112 (TmH_2)	[6.124]
U (α)	+7 (<941 K)		[6.123]
(β)	+17 (941–1048 K)		[6.123]
(γ)	+4 (1048–1405 K)		[6.123]
(γ)	+8		[6.139]

Table 6.6.1 (continued)

Element	$\Delta\bar{H}_\infty$ [kJ/mol H]	ΔH [kJ/mol H]	Ref.
U (continued)			
(l)	+11 (>1405 K)		[6.123]
(α)		−42 (UH_3) (580–933 K)	[6.120]
(β)		−39 (UH_3) (933–1047 K)	[6.120]
(γ)		−42 (UH_3) (1023–1320 K)	[6.120]
V	−26 (253–373 K)	−35 ($VH_{0.5}$) (253–373 K)	[6.140]
	−33 (293–328 K)	−42 ($VH_{0.5}$) (283–413 K)	[6.120]
		−36 ($Vh_{0.5}$)	[6.41]
	−29(519–827 K)	−35 ($VH_{0.5}$) (519–827 K)	[6.130]
W	+96 (1100–3070 K)		[6.120]
Y	−79		[6.118]
		−114 (YH_2) (800–1200 K)	[6.120]
		−107 (YH_2) (1200–1600 K)	[6.120]
		−93 (YH_2) (>1173 K)	[6.124]
Yb		−91 (YbH_2)	[6.124]
Zn	(+15) (473–673 K)		[6.123]
(l)	+20		[6.141]
Zr (α)	−64 (773–1073 K)		[6.139]
(α)	−52 (700–1170 K)		[6.120]
(β)	−64 (1073–1223 K)		[6.120]
(l)	−51 (2200–2760 K)		[6.120]
(α)		−94 (ZrH_2)	[6.118]
(β)		−106 (ZrH_2)	[6.118]
		−82 (ZrH_2) calor (298 K)	[6.119]

Table 6.6.2. Heat of formation of selected ternary metal hydrides. In most cases ΔH is determined from the dissociation pressure of plateaus in the pressure–composition isotherms. As the coexisting phases may be different for each metal hydride we use the symbol ΔH instead of $\Delta H_{\alpha \to \beta}$. Heats of formation which have been determined calorimetrically are indicated by "calor". Values for ΔH estimated by means of relation (6.40) are indicated by "estim". Values of uncertain origin are given in parenthesis. Intermetallics which do not form hydrides are given with the comment "no hydride" and those which probably decompose during hydrogen uptake are indicated by "decomp?"

Hydride		ΔH [kJ/mol H]	Ref.
$AlHfH_1$		−21 estim	[6.142]
$AlTh_2H_4$		(<−49)	[6.29]
$AlTi_3H_{\approx 0}$		−53	[6.143]
$Be_5HfH_{0.2}$		(−24)	[6.144]
$Be_2HfH_{1.1}$		(<−27)	[6.144]
Be_2TiH_3		(−19−−13)	[6.144]
$Be_5ZrH_{0.2}$		(−24)	[6.144]
$Be_2ZrH_{1.5}$		(<−27)	[6.144]
$CaNi_2H_{3.4}$	(decomp?)	−42 calor	[6.145]
$CaNi_3H_{4.5}$		−29 calor	[6.145]
$CaNi_5H_{4.5}$		−19 calor	[6.145]
$CaNi_5H_4$		−17 calor	[6.146]
$CaNi_5H_1$		−22 calor	[6.147]

Table 6.6.2 (continued)

Hydride		ΔH [kJ/mol H]	Ref.
$CaNi_5H_1$		−22	[6.148]
$CaNi_5H_{4.5}$		−16	[6.149]
$CeCo_2H_{4.1}$		(< −36)	[6.150]
$CeCo_3H_{3.8}$		−23 estim	[6.150]
$CeCo_3H_3$		−18	[6.120]
$Ce_2Co_7H_7$		−22	[6.151]
$CeCo_5H_{2.5}$		−20	[6.152]
$CeCo_5H_3$		−19	[6.86]
$CeCo_5H_{2.5}$		−15	[6.120]
$CeMg_{12}H_2$		−36 estim	[6.153]
$CeNi_2H_{3.9}$		(< −36)	[6.150]
$CeNi_3H_{4.2}$		−22 calor	[6.145]
$Ce_2Ni_7H_{4.2}$		−23 estim	[6.150]
$CeNi_5H_6$		−18 estim	[6.150]
$CeNi_5H_6$		−7	[6.15]
$Co_7Dy_2H_{2.6}$		−23	[6.151]
$Co_3DyH_{1.0}$		−24	[6.154]
$Co_7Er_2H_{...}$		−20	[6.151]
$Co_3ErH_{1.1}$		−22	[6.155]
$Co_7Gd_2H_{2.6}$		−29	[6.151]
Co_3GdH_2		−27	[6.155]
$Co_2GdH_{4.5}$		−27	[6.73]
$Co_5GdH_{2.2}$		−15 estim	[6.152]
$CoHfH_{3.2}$		−30 estim	[6.142]
$CoHf_2H_{3.8}$		(< −36)	[6.142]
$Co_7Ho_2H_1$		−21	[6.151]
Co_3HoH_1		−24	[6.155]
$Co_5LaH_{3.4}$		−23	[6.152]
Co_3LuH_1		−18	[6.156]
$Co_5NdH_{2.7}$		−21	[6.86]
$Co_7Nd_2H_{2.7}$		−36	[6.151]
Co_3NdH_2		−32	[6.155]
$Co_5PrH_{2.9}$		−17	[6.157]
$Co_5PrH_{2.9}$		−19	[6.152]
$Co_7Pr_2H_{2.5}$		−32	[6.151]
$Co_7Pr_2H_{2.5}$		−17	[6.157]
Co_3PrH_4		−27	[6.157]
Co_2PrH_4	(decomp?)	(< −33)	[6.157]
$Co_5SmH_{2.5}$		−16	[6.152]
$Co_5TbH_{...}$		−14	[6.152]
$Co_7Tb_2H_{2.7}$		−24	[6.151]
Co_3TbH_1		−25	[6.155]
$Co_{17}Th_2$	no hydride at 313 K, 40 bar		[6.86]
$Co_5ThH_{2.6}$		−14 estim	[6.152]
$Co_7Th_2H_{3.5}$		−22 estim	[6.86]
$CoThH_4$		(< −24)	[6.86]
$Co_3Th_7H_{29}$		(< −24)	[6.86]
$CoTiH_{1.4}$		−29	[6.158]
$CoTiH_{1.5}$		−31	[6.159]
$CoTiH_{0.9}$		−27	[6.160]

Table 6.6.2 (continued)

Hydride		ΔH [kJ/mol H]	Ref.
$CoTiH_{1.5}$		−24 calor	[6.145]
$Co_3TmH_{1.3}$		−22	[6.161]
$Co_5YH_{2.8}$		−14	[6.120]
$Co_7Y_2H_{1.5}$		−27 estim	[6.150]
$Co_3YH_{1.0}$		−28	[6.155]
$Co_2YH_{3.7}$		−23 estim	[6.150]
$CoZrH_{0.65}$		−41	[6.162, 142]
$CoZr_2H_{4.8}$	(decomp)		[6.142]
$CrNi_2H_{\approx 0}$		+29 (573–773 K)	[6.120]
		+53 (773–1 073 K)	[6.120]
$Cr_{1.9}TiH_{2.5}$	(C14 struct)	−13	[6.163]
$Cr_{1.8}TiH_{\approx 0}$	(C15 struct)	−35	[6.164]
$Cr_{1.8}TiH_{0.1}$	(C15 struct)	−22	[6.164]
$Cr_{1.8}TiH_{0.2}$	(C15 struct)	−12	[6.164]
$Cr_{1.8}TiH_{0.9}$	(C15 struct)	−9	[6.164]
$Cr_{1.7}TiH_{2.6}$		−10	[6.165]
$Cr_2ZrH_{\approx 0}$		−23	[6.166]
$Cr_2ZrH_{\dots}$		−19 calor	[6.167]
$Cu_5LaH_{2.5}$		−22	[6.87]
Cu_2Th	no hydride (?)		[6.144]
$CuTiH_1$		−38	[6.168]
$CuTiH_{\approx 0}$	(decomp. for $x > 0.2$)	−38	[6.169]
$CuTi_2H_{\approx 0}$	(decomp. for $x > 0.2$)	−33	[6.169]
$CuZr_2H_1$	($T > 1\,000$ K disprop?)	−60	[6.170]
$CuZr_2H_{0.2}$	(decomp. at higher conc.?)	−74	[6.171]
$DyFe_2H_2$		−29	[6.172]
$DyFe_3H_2$		−24	[6.172]
$DyRu_2H_{3.1}$		−28 calor	[6.145]
$ErFe_2H_{1.5}$		−29	[6.173]
$ErFe_3H_{2.8}$		−22	[6.172]
$Er_6Fe_{23}H_{12}$		−25	[6.174]
$ErMn_2H_{4.9}$		(< −27)	[6.153]
$ErNiH_{3.1}$		−52 calor	[6.175]
$ErRu_2H_{\dots}$		−23	[6.87]
$EuNi_5H_{5.5}$	(ex. prob. more stable $EuNi_5H_{\approx 1.2}$)	−13	[6.176]
$Fe_3GdH_{1.5}$		−25	[6.172]
$Fe_2GdH_{2.7}$	(decomp. at first desorption)	(−15)	[6.73, 173]
Fe_2Hf	no hydride		[6.29]
$FeHf_2H_{3.1}$		(< −36)	[6.142]
$Fe_3HoH_{1.5}$		−22	[6.172]
$Fe_{23}Ho_6H_{12}$		−26	[6.174]
$Fe_{23}Lu_6H_8$		−20	[6.174]
$FePd_3H_{\sim 0.1}$		−15 estim	[6.197]
Fe_2ScH_2		−15	[6.196]
$Fe_2ScH_{\dots}$		−18	[6.87]
$Fe_3TbH_{1.5}$		−24	[6.172]
$Fe_5ThH_{2.4}$	no plateau at 313 K		[6.177]
$Fe_7Th_2H_{6.1}$		−22 estim	[6.177]
Fe_2Ti	no hydride		[6.178]
$FeTiH_{\approx 0}$		+10	[6.179]

Table 6.6.2 (continued)

Hydride		ΔH [kJ/mol H]	Ref.
$FeTiH_1$		−14	[6.165]
$FeTiH_{1.6}$		−17	[6.165]
$FeTiH_{1.6}$		−12 calor	[6.179]
$Fe_2TmH_{1.6}$		−29	[6.180]
$FeU_6H_{3.2}$		−32 estim	[6.29]
$Fe_{23}Y_6H_{21.5}$		(<−36)	[6.150]
$Fe_3YH_{4.8}$		(<−36)	[6.150]
Fe_2YH_2	(decomp. at first desorption)		[6.173]
$Ga_2PrH_{\approx 0.3}$			[6.144]
$GdMn_2H_{2.7}$		(−44)	[6.73]
$GdNi_5H_{2.9}$		−13 estim	[6.86]
$GdNi_2H_3$		−45	[6.73]
$GdRh_2H_{2.8}$		−25	[6.73]
$GdRu_2H_{2.7}$		−30	[6.73]
$HfNiH_{3.2}$		−26 estim	[6.142]
$Hf_2PdH_{1.9}$			[6.181]
HfPd	no hydride		[6.181]
$Hf_2PtH_{0.9}$			[6.29]
HfPt	no hydride		[6.29]
$Hf_2RhH_{2.2}$			[6.144]
HfRh	no hydride		[6.29]
Ir_2La	no hydride		[6.29]
La_3NiH_2		−111	[6.182]
$La_3NiH_{8.8}$	(probably decomp.)		[6.183]
$La_7Ni_3H_{\ldots}$	(decomp.)		[6.184]
$LaNiH_{3.8}$		−52 calor	[6.184]
$LaNiH_{1.3}$		−88	[6.182]
$LaNiH_4$		−63 calor	[6.175]
$LaNi_2H_2$		−27	[6.184]
$LaNi_5H_{5.5}$		−16	[6.86]
$LaNi_5H_{5.5}$		−15	[6.186]
$LaNi_5H_{5.5}$		−16 calor	[6.146]
$LaNi_5H_{5.5}$		−15 calor	[6.167, 185]
$LaPtH_{2.8}$			[6.153]
$LaPt_2$	no hydride (?)		[6.29]
$LaPt_5H_{1.2}$		−13 estim	[6.55]
$LaRh_2H_{1.4}$		(−22)	[6.73]
$LaRu_2H_{4.5}$		very stable	[6.73]
$LiPdH_{0.8}$		−38	[6.187]
$LiPtH_{0.7}$		−67	[6.187]
$LuNiH_{2.9}$		−50 calor	[6.175]
$LuPdH_{\ldots}$		>−14 estim	[6.175]
$Mg_2NiH_{0.06}$		−13	[6.188]
Mg_2NiH_4		−32	[6.165]
$Mn_2ScH_{\ldots}$		−32	[6.87]
$Mn_{1.5}TiH_{2.4}$		−16	[6.102]
$Mn_{2.8}ZrH_2$		−18	[6.190]
$Mn_{2.8}ZrH_{\ldots}$		−16 calor	[6.167]
$Mn_2ZrH_{3.6}$		−22	[6.189]
$Mn_2ZrH_{\ldots}$		−20 calor	[6.167]

Table 6.6.2 (continued)

Hydride		ΔH [kJ/mol H]	Ref.
$Mo_2ZrH_{\approx 0}$		−19	[6.166]
$NdNi_5H_4$		−14	[6.15]
Ni_5PrH_6		−15	[6.15/86]
$Ni_7Pr_2H_{...}$			[6.151]
$Ni_5SmH_{1.6}$		−13 estim	[6.86]
$Ni_5SmH_{1.6}$		−15 estim	[6.191]
$NiThH_{3.5}$		−26 estim	[6.177]
$NiTiH_{0.9}$		−30	[6.160]
$NiTi_2H_1$		−31 estim	[6.192]
Ni_5YH_1		−12 estim	[6.55]
$Ni_7Y_2H_3$		−20 estim	[6.150]
Ni_3YH_4		−23 estim	[6.150]
Ni_5YbH_3		−13 estim	[6.191]
$Ni_2YbH_{3.1}$		−26 calor	[6.145]
$NiYbH_{2.7}$		−73 calor	[6.175]
$Ni_5Zr_2H_{1.7}$		−22	[6.193]
$Ni_{10}Zr_7H_{15}$		−24	[6.193]
$NiZrH_{2.8}$		−33 estim	[6.142]
$NiZr_2H_{4.5}$	(decomp.)	−81 estim	[6.142]
$NiZr_2H_{0.35}$	(decomp. at higher conc.?)	−92	[6.171]
PTi_3H_0		−43	[6.194]
$PdTi_2H_{1.5}$		−46	[6.169]
Pd_3U	no hydride		[6.117]
$PdYbH_{2.7}$		−33 calor	[6.175]
Pd_2Zr	no hydride		[6.181]
$PdZr_2H_{2.7}$			[6.181]
$SbTi_3H_{\approx 0}$		−55	[6.194]
$SnTi_3H_{\approx 0}$		−51	[6.143]
$TaV_2H_{\approx 0}$	(C15 struct.)	−58	[6.195]
$V_2ZrH_{\approx 0}$		−77	[6.166]
$V_2ZrH_{1.5}$		−80	[6.166]

Table 6.6.3. Heat of formation of some ternary metal hydrides $A_mB_nH_x$ calculated with the band-structure semi-empirical model [6.35, 36]. Volume effects are not included and it is explicity assumed that hydrogen is surrounded by a fraction $m/(m+n)$ of A atoms and a fraction $n/(m+n)$ of B atoms

Part a) Heats of formation of A_5BH_x in kJ/mol H
Part b) Heats of formation of A_2BH_x in kJ/mol H
Part c) Heats of formation of ABH_x in kJ/mol H
Part d) Heats of formation of AB_2H_x in kJ/mol H
Part e) Heats of formation of AB_5H_x in kJ/mol H

Table 6.6.3a. Calculated heats of formation of A_5BH_x

B	Heat of formation [kJ/mol H]										
	$A\equiv$Sc	$A\equiv$Ti	$A\equiv$V	$A\equiv$Y	$A\equiv$Zr	$A\equiv$Nb	$A\equiv$La	$A\equiv$Hf	$A\equiv$Ta	$A\equiv$Pd	$A\equiv$Ni
Sc	−98.0	−67.0	−43.5	−107.8	−82.3	−46.8	−103.9	−68.4	−49.1	−12.0	− 4.1
Ti	−92.8	−59.4	−34.9	−104.5	−75.5	−38.5	− 99.3	−62.1	−40.8	− 8.6	0.7
V	−89.4	−54.6	−29.8	−103.0	−71.6	−33.5	− 96.5	−57.3	−36.0	− 8.3	2.2
Cr	−84.0	−50.3	−25.3	− 98.1	−68.2	−29.2	− 92.1	−51.9	−31.7	−10.7	0.5
Mn	−84.5	−49.4	−25.0	−101.0	−68.0	−28.8	− 93.1	−50.9	−31.5	−13.4	− 1.9
Fe	−85.2	−44.3	−19.3	−109.1	−67.5	−23.7	− 95.6	−49.3	−26.7	−10.0	2.6
Co	−80.4	−42.4	−18.0	−103.2	−65.1	−22.4	− 90.9	−46.2	−25.3	−13.3	− 0.4
Ni	−74.3	−41.3	−17.4	− 87.0	−61.5	−21.6	− 82.5	−43.8	−24.8	−17.0	− 4.7
Y	−99.5	−69.7	−46.9	−109.8	−85.6	−50.5	−106.4	−70.6	−52.8	−30.3	−22.4
Zr	−94.2	−62.4	−38.6	−106.2	−78.7	−42.4	−101.3	−64.3	−44.7	−23.1	−13.3
Nb	−89.4	−55.3	−30.5	−102.9	−72.2	−34.3	− 96.7	−57.6	−36.7	−11.6	− 0.9
Mo	−82.5	−48.1	−22.4	− 96.8	−66.1	−26.4	− 90.6	−49.9	−28.9	− 5.0	6.3
Tc	−83.3	−48.5	−23.9	− 99.5	−67.1	−27.8	− 91.9	−50.1	−30.4	−11.9	− 0.5
Ru	−78.6	−42.1	−16.8	− 98.1	−62.9	−21.2	− 88.1	−45.4	−23.8	− 5.5	6.9
Rh	−77.6	−41.4	−16.9	− 98.2	−62.8	−21.2	− 87.6	−44.4	−23.9	−10.8	1.7
Pd	−75.9	−43.0	−19.6	− 88.5	−63.3	−23.6	− 84.1	−44.2	−26.6	−19.5	− 7.4
La	−99.0	−68.3	−45.3	−109.4	−83.9	−48.9	−105.4	−70.6	−51.1	−21.3	−13.3
Hf	−92.7	−60.6	−36.0	−103.5	−77.0	−39.5	− 99.7	−60.9	−41.9	−16.2	− 6.6
Ta	−89.8	−55.7	−31.1	−103.4	−72.6	−34.8	− 97.1	−58.1	−37.2	−11.6	− 1.1
W	−80.5	−46.8	−20.7	− 93.4	−64.2	−24.7	− 88.6	−47.4	−27.0	− 1.9	9.3
Re	−82.2	−45.4	−20.0	−101.4	−66.2	−24.5	− 91.3	−50.0	−27.0	− 6.3	6.1
Os	−80.0	−40.7	−14.8	−102.9	−63.6	−19.6	− 90.2	−46.4	−22.2	− 2.4	11.2
Ir	−76.9	−37.4	−11.7	−102.0	−61.8	−16.8	− 87.8	−43.8	−19.4	− 3.3	11.2
Pt	−68.6	−36.9	−12.8	− 81.6	−55.4	−17.0	− 76.8	−38.6	−19.4	− 8.4	4.5

Table 6.6.3b. Calculated heats of formation of A_2BH_x

B	Heat of formation [kJ/mol H]										
	$A \equiv$ Sc	$A \equiv$ Ti	$A \equiv$ V	$A \equiv$ Y	$A \equiv$ Zr	$A \equiv$ Nb	$A \equiv$ La	$A \equiv$ Hf	$A \equiv$ Ta	$A \equiv$ Pd	$A \equiv$ Ni
Sc	−98.0	−74.1	−56.3	−105.8	−85.0	−58.6	−102.6	−75.2	−60.4	−19.4	−13.1
Ti	−87.1	−59.4	−39.9	−97.2	−72.4	−42.7	−92.4	−62.4	−44.5	−11.0	−3.5
V	−79.4	−49.8	−29.8	−92.0	−64.1	−32.7	−85.7	−52.7	−34.8	−6.0	2.5
Cr	−69.6	−41.2	−20.8	−84.1	−57.1	−24.3	−77.9	−42.8	−26.2	−3.7	4.3
Mn	−70.4	−39.8	−20.6	−87.4	−56.8	−23.7	−79.0	−41.6	−26.1	−9.2	0.2
Fe	−67.7	−29.7	−9.6	−91.1	−51.9	−13.6	−77.9	−35.7	−16.3	−1.9	8.6
Co	−62.2	−27.1	−7.7	−87.7	−50.3	−12.1	−73.6	−32.6	−14.5	−7.2	3.5
Ni	−51.7	−27.0	−8.3	−65.6	−45.3	−12.5	−60.6	−29.1	−15.4	−11.4	−4.7
Y	−101.2	−79.5	−63.2	−109.8	−91.9	−65.9	−107.3	−79.7	−67.6	−36.4	−30.7
Zr	−90.6	−65.4	−47.2	−101.8	−78.7	−50.2	−97.1	−67.5	−52.0	−26.1	−19.0
Nb	−80.0	−51.1	−31.3	−92.6	−65.4	−34.3	−86.6	−53.7	−36.2	−9.5	−1.2
Mo	−66.1	−36.5	−15.1	−80.7	−52.5	−18.6	−74.3	−38.3	−20.5	6.9	15.3
Tc	−68.0	−37.8	−18.2	−84.8	−54.8	−21.4	−76.7	−39.5	−23.7	−6.1	2.6
Ru	−57.8	−24.8	−4.1	−79.1	−45.1	−8.2	−67.9	−29.1	−10.4	7.0	17.4
Rh	−57.3	−24.5	−5.1	−81.0	−46.3	−9.3	−68.3	−28.9	−11.6	−2.7	7.7
Pd	−54.9	−29.7	−12.0	−68.5	−48.9	−15.8	−63.8	−32.5	−18.5	−19.5	−9.9
La	−100.1	−76.8	−59.3	−108.9	−88.8	−62.5	−105.4	−79.1	−64.2	−29.2	−23.0
Hf	−87.2	−61.5	−41.9	−96.7	−74.9	−44.5	−93.5	−60.9	−46.4	−14.7	−7.6
Ta	−80.8	−51.9	−32.3	−93.5	−66.2	−35.2	−87.3	−54.6	−37.2	−10.0	−1.6
W	−62.6	−33.9	−11.6	−75.4	−49.1	−15.0	−70.9	−33.8	−16.8	−13.9	22.0
Re	−64.4	−31.3	−10.3	−84.7	−51.2	−14.6	−73.7	−37.1	−16.6	4.5	14.9
Os	−57.5	−21.0	0.6	−81.2	−43.4	−4.0	−68.0	−27.7	−6.2	13.2	25.4
Ir	−52.5	−15.2	6.2	−80.1	−40.4	1.0	−64.3	−23.5	−1.2	12.6	26.2
Pt	−40.4	−15.3	3.4	−55.4	−33.2	−0.8	−49.4	−17.7	−2.4	2.2	13.0

Table 6.6.3c. Calculated heats of formation of ABH_x

B	Heat of formation [kJ/mol H]										
	$A \equiv$ Sc	$A \equiv$ Ti	$A \equiv$ V	$A \equiv$ Y	$A \equiv$ Zr	$A \equiv$ Nb	$A \equiv$ La	$A \equiv$ Hf	$A \equiv$ Ta	$A \equiv$ Pd	$A \equiv$ Ni
Sc	−98.0	−80.8	−68.4	−103.8	−86.9	−69.7	−101.3	−81.4	−71.0	−35.5	−30.7
Ti	−80.8	−59.4	−44.9	−88.7	−69.2	−46.9	−84.9	−62.1	−48.2	−20.0	−14.5
V	−68.4	−44.9	−29.8	−78.4	−56.2	−32.0	−73.4	−47.5	−33.5	−7.7	−2.2
Cr	−54.9	−32.2	−16.6	−67.9	−45.5	−19.4	−62.5	−33.6	−20.8	1.0	6.6
Mn	−55.6	−30.8	−16.6	−70.5	−45.0	−19.1	−63.3	−32.6	−21.0	−6.9	0.3
Fe	−47.0	−15.6	−0.3	−66.1	−34.2	−3.7	−55.7	−21.1	−5.9	5.4	13.7
Co	−43.4	−13.6	1.0	−66.5	−34.6	−3.1	−54.0	−19.8	−5.1	−1.3	7.2
Ni	−30.7	−14.5	−2.2	−46.3	−30.7	−6.0	−40.3	−16.4	−7.7	−12.3	−4.7
Y	−103.0	−88.7	−78.4	−109.8	−97.0	−80.0	−108.2	−88.2	−81.3	−50.5	−46.3
Zr	−87.3	−68.5	−55.6	−96.7	−78.7	−57.9	−92.8	−70.6	−59.1	−36.1	−30.7
Nb	−69.7	−46.9	−32.0	−80.0	−58.3	−34.3	−75.2	−49.3	−35.7	−11.5	−6.0
Mo	−48.8	−24.8	−7.8	−61.5	−37.9	−10.7	−56.2	−26.2	−12.1	15.5	22.1
Tc	−51.8	−27.4	−12.7	−66.7	−41.7	−15.3	−59.8	−29.0	−17.1	−1.7	4.2
Ru	−35.3	−7.6	8.3	−54.4	−25.5	4.7	−44.6	−11.9	3.0	18.1	26.5
Rh	−36.8	−9.1	5.5	−59.3	−29.2	1.5	−47.5	−14.3	−0.3	5.0	13.2
Pd	−35.5	−20.0	−7.7	−50.5	−36.1	−11.5	−45.0	−21.6	−13.6	−19.5	−12.3
La	−101.3	−84.9	−73.4	−108.2	−93.2	−75.2	−105.4	−86.7	−76.3	−45.0	−40.3
Hf	−81.4	−48.2	−47.5	−89.4	−71.9	−49.3	−86.7	−60.9	−50.7	−21.6	−16.4
Ta	−71.0	−62.1	−33.5	−81.3	−59.5	−35.7	−76.3	−50.7	−37.2	−13.6	−7.7
W	−43.9	−20.7	−2.5	−55.2	−33.2	−5.3	−51.9	−19.8	−6.6	27.0	32.9
Re	−44.6	−17.1	−0.9	−62.3	−34.2	−4.6	−52.9	−22.8	−6.1	12.9	21.7
Os	−31.1	−0.4	16.4	−50.9	−19.4	12.5	−40.3	−6.4	10.7	27.8	38.1
Ir	−24.8	7.1	23.9	−49.9	−15.3	19.1	−35.7	−0.9	17.4	28.5	40.1
Pt	−14.1	4.7	18.2	−31.7	−12.9	13.9	−23.9	0.3	13.0	12.1	20.8

Table 6.6.3d. Calculated heats of formation of AB_2H_x

B	Heat of formation [kJ/mol H]										
	$A \equiv$ Sc	$A \equiv$ Ti	$A \equiv$ V	$A \equiv$ Y	$A \equiv$ Zr	$A \equiv$ Nb	$A \equiv$ La	$A \equiv$ Hf	$A \equiv$ Ta	$A \equiv$ Pd	$A \equiv$ Ni
Sc	−98.0	−87.1	−79.4	−101.8	−90.6	−80.0	−100.1	−87.2	−80.8	−54.9	−51.7
Ti	−74.1	−59.4	−49.8	−79.5	−65.4	−51.1	−76.8	−61.5	−51.9	−29.7	−27.0
V	−56.3	−39.9	−29.8	−63.2	−47.2	−31.3	−59.8	−41.9	−32.3	−12.0	−8.3
Cr	−39.5	−23.3	−12.6	−49.5	−33.2	−14.8	−45.7	−24.5	−15.6	2.8	6.7
Mn	−40.2	−22.4	−13.0	−51.2	−32.9	−14.8	−46.1	−24.0	−16.2	−6.0	−1.5
Fe	−24.3	−1.8	8.4	−37.5	−15.2	6.0	−30.5	−6.0	4.4	12.2	18.0
Co	−24.0	−1.8	7.9	−41.5	−18.3	4.6	−32.3	−7.5	3.2	4.6	10.5
Ni	−13.1	−3.5	2.5	−30.7	−19.0	−1.2	−23.0	−7.6	−1.6	−9.9	−4.7
Y	−105.0	−97.2	−92.0	−109.8	−101.8	−92.6	−108.9	−96.1	−93.5	−68.5	−65.6
Zr	−84.1	−71.7	−63.7	−91.2	−78.7	−65.2	−88.2	−73.4	−66.0	−48.9	−45.3
Nb	−58.6	−42.7	−32.7	−65.9	−50.2	−34.3	−62.5	−44.5	−35.2	−15.8	−12.5
Mo	−30.1	−12.7	−0.6	−39.4	−22.2	−2.8	−36.0	−13.6	−3.7	20.1	25.4
Tc	−34.9	−17.3	−7.5	−46.0	−27.9	−9.4	−41.4	−18.7	−10.7	1.2	4.7
Ru	−11.1	9.4	20.3	−25.3	−4.2	17.5	−18.4	5.8	16.3	27.8	34.1
Rh	−16.3	4.5	14.3	−34.0	−11.6	11.0	−25.0	−0.6	9.7	12.2	18.0
Pd	−19.4	−11.0	−6.0	−36.4	−26.1	−9.5	−29.2	−14.7	−10.0	−19.5	−11.4
La	−102.6	−92.4	−85.7	−107.3	−97.1	−86.6	−105.4	−93.5	−87.3	−63.8	−60.6
Hf	−75.2	−62.4	−52.7	−81.1	−67.5	−53.7	−79.1	−60.9	−54.6	−32.5	−29.1
Ta	−60.4	−44.5	−34.8	−67.6	−52.0	−36.2	−64.2	−46.4	−37.2	−18.5	−15.4
W	−23.9	−7.0	6.7	−32.4	−16.0	4.5	−30.7	−5.5	3.7	35.7	40.5
Re	−22.9	−2.8	8.4	−35.7	−15.5	5.5	−29.2	−7.3	4.5	19.3	26.1
Os	−1.4	21.1	32.7	−15.1	7.3	29.8	−8.1	16.7	28.5	41.4	49.1
Ir	5.9	29.7	41.3	−11.7	12.7	37.5	−2.5	23.3	36.3	44.2	53.1
Pt	8.3	21.5	30.1	−11.7	4.1	25.5	−1.9	14.3	25.5	15.4	23.3

Table 6.6.3e. Calculated heats of formation of AB_5H_x

B	Heat of formation [kJ/mol H]										
	$A \equiv$ Sc	$A \equiv$ Ti	$A \equiv$ V	$A \equiv$ Y	$A \equiv$ Zr	$A \equiv$ Nb	$A \equiv$ La	$A \equiv$ Hf	$A \equiv$ Ta	$A \equiv$ Pd	$A \equiv$ Ni
Sc	−98.0	−92.8	−89.4	−99.9	−94.2	−89.4	−99.0	−92.7	−89.8	−75.9	−74.3
Ti	−76.0	−59.4	−54.6	−69.7	−62.4	−55.3	−68.3	−60.6	−55.7	−43.0	−41.3
V	−43.5	−34.9	−29.8	−47.0	−38.6	−30.5	−45.3	−36.0	−31.1	−19.6	−17.4
Cr	−23.4	−14.6	−9.0	−28.9	−20.1	−10.3	−27.1	−15.3	−10.7	0.8	3.3
Mn	−24.4	−14.6	−10.0	−30.1	−20.4	−11.0	−27.5	−15.7	−11.7	−6.1	−4.2
Fe	−0.2	11.7	16.8	−6.9	4.6	15.5	−3.5	9.4	14.7	18.7	21.7
Co	−4.2	8.1	13.0	−13.7	−1.3	11.0	−8.9	4.5	10.3	10.4	13.5
Ni	−4.1	0.7	2.2	−22.4	−13.3	−0.9	−13.3	−6.6	−1.1	−7.4	−4.7
Y	−107.2	−104.5	−103.0	−109.8	−106.2	−102.9	−109.4	−103.3	−103.4	−88.5	−87.0
Zr	−81.3	−75.1	−71.5	−85.1	−78.7	−72.2	−83.6	−76.1	−72.5	−53.3	−61.5
Nb	−46.8	−38.5	−33.5	−50.5	−42.4	−34.3	−48.9	−39.5	−34.8	23.6	−21.6
Mo	−9.6	−0.1	6.4	−14.5	−5.3	5.1	−13.1	−0.5	4.7	19.8	23.5
Tc	−17.2	−7.6	−2.6	−23.1	−13.4	−3.8	−20.7	−8.5	−4.4	2.3	4.0
Ru	14.8	26.2	31.8	7.3	18.6	30.2	10.7	24.0	29.6	36.0	39.6
Rh	4.3	16.1	21.0	−5.7	6.4	18.9	−0.8	12.5	18.3	18.9	22.0
Pd	−12.0	−8.6	−8.3	−30.3	−23.1	−11.6	−21.3	−16.2	−11.6	−19.5	−17.0
La	−103.9	−99.3	−96.5	−106.4	−101.3	−96.7	−105.4	−99.7	−97.1	−84.1	−82.5
Hf	−68.4	−62.1	−57.3	−71.4	−64.3	−57.6	−70.6	−60.9	−58.1	−44.2	−43.8
Ta	−49.1	−40.8	−36.0	−52.8	−44.7	−36.7	−51.1	−41.9	−37.2	−26.6	−24.8
W	−1.7	7.7	15.8	−7.5	3.0	14.6	−6.1	9.4	14.2	37.2	41.3
Re	0.7	11.6	17.4	−6.0	4.7	15.8	−2.8	9.1	15.3	23.7	27.7
Os	31.3	43.4	49.4	24.4	36.0	47.8	27.6	41.1	47.1	54.2	58.6
Ir	39.3	52.4	58.5	30.2	43.0	56.3	34.7	48.8	55.6	59.8	64.8
Pt	22.0	30.5	34.7	2.5	14.5	29.9	12.8	20.6	30.6	24.6	29.0

Acknowledgements. We are grateful to Dr. A. Driessen, R. Feenstra, D. G. de Groot, and S. Büchler for valuble discussions and their critical and careful reading of the manuscript. One of the authors (T.R.) would like to acknowledge financial support from the National Energy Research Foundation (NEFF).

References

6.1 C.D. Gelatt, H. Ehrenreich, J.A. Weiss: Phys. Rev. **B 17**, 1940 (1978)
6.2 C.D. Gelatt: "Calculated Heats of Formation of Metal and Metal Alloy Hydrides", in *Hydrides for Energy Storage*, ed. by A.F. Andresen, J.A. Maeland (Pergamon, Oxford 1978) p. 193
6.3 J.K. Nørskov: Phys. Rev. **B 26**, 2875 (1982); **B 28**, 1138 (1983)
6.4 M. Methfessel, J. Kübler: J. Phys. F **12**, 141 (1982)
6.5 M. Gupta, L. Schlapbach: Chap. 5 this volume and references therein
6.6 A.R. Williams, J. Kübler, C.D. Gelatt: Phys. Rev. **B 19**, 6094 (1979)
6.7 A.C. Switendick: "The Changes in Electronic Structure on Hydrogen Alloying and Hydride Formation" in *Topics Appl. Phys.*, Vol. 28 (Springer, Berlin, Heidelberg 1978) p. 101
6.8 D.A. Papaconstantopoulos, E.N. Economou, B.M. Klein, L.L. Boyer: Phys. Rev. **B 20**, 177 (1979)
6.9 M. Gupta: Phys. Lett. **88**A, 469 (1982); J. Less-Common Met. **88**, 221 (1982)
6.10 D.A. Papaconstantopoulos, A.C. Switendick: J. Less-Common Met. **88**, 273 (1982)
6.11 W.M. Temmermann, A.J. Pindor: J. Phys. F **13**, 1869 (1983)
6.12 M. Gupta: J. Less-Common Met. **103**, 325 (1984)
6.13 M. Gupta, E. Belin, L. Schlapbach: J. Less-Common Met. **103**, 389 (1984)
6.14 M.H. Mendelsohn, D.M. Gruen, A.E. Dwight: Nature **269**, 45 (1977)
6.15 C.E. Lundin, F.E. Lynch, C.B. Magee: J. Less-Common Met. **56**, 19 (1977)
6.16 C.B. Magee, J. Liu, C.E. Lundin: J. Less-Common Met. **78**, 119 (1981)
6.17 T. Takeshita, K.A. Gschneidner, D.K. Thome, O.D. McMasters: Phys. Rev. **B 21**, 5636 (1980)
6.18 H.M. Lee: J. Mater. Sci. **13**, 1374 (1978)
6.19 H.M. Lee: J. Mater. Sci. **14**, 1002 (1979)
6.20 D.O. Welch, M.A. Pick: Phys. Lett. **99**A, 183 (1983)
6.21 I. Jacob, A. Wolf, M.H. Mintz: Solid State Commun. **40**, 877 (1981)
6.22 Wei-Ming Yei, R.B. McLellan: J. Less-Common Met. **64**, P11 (1979)
6.23 M. Peretz, D. Zamir, D. Shaltiel, J. Shinar: Z. Phys. Chem. NF **117**, 221 (1979)
6.24 A.R. Miedema: J. Less-Common Met. **32**, 117 (1973)
6.25 A.R. Miedema, K.H.J. Buschow, H.H. van Mal: J. Less-Common Met. **49**, 463 (1976)
6.26 H.H. van Mal, K.H.J. Buschow, A.R. Miedema: J. Less-Common Met. **35**, 65 (1974)
6.27 A.R. Miedema, F.R. de Boer, R. Boom: Physica **103**B, 67 (1981)
6.28 P.C.P. Bouten, A.R. Miedema: J. Less-Common Met. **71**, 147 (1980)
6.29 K.H.J. Buschow, P.C.P. Bouten, A.R. Miedema: Rep. Prog. Phys. **45**, 937 (1982)
6.30 A.R. Williams, C.D. Gelatt, V.L. Moruzzi: Phys. Rev. Lett. **44**, 429 (1980)
6.31 D.G. Pettifor: Phys. Rev. Lett. **42**, 846 (1979)
6.32 A.R. Williams, C.D. Gelatt, V.L. Moruzzi: Phys. Rev. **B 25**, 6509 (1982)
6.33 A.R. Mackintosh, O.K. Andersen: "The Electronic Structure of Transition Metals" in *Electrons at the Fermi Surface*, ed. by M. Springford (Cambridge Univ. Press, London 1980) p. 149
6.34 O.K. Andersen, H.L. Skriver, H. Nohl, B. Johansson: Pure Appl. Chem. **52**, 93 (1979)
6.35 R. Griessen, A. Driessen: Phys. Rev. **B 30**, 4372 (1984)
6.36 R. Griessen, A. Driessen, D.G. de Groot: J. Less-Common Met. **103**, 235 (1984)
6.37 R. Griessen, A. Driessen: J. Less-Common Met. **103**, 245 (1984)

6.38 R. Griessen, R. Feenstra: J. Phys. F**15**, 1013 (1985)
6.39 D.G. Westlake: "Stabilities, Stoichiometries and Site Occupancies in Hydrides of Intermetallic Compounds" in *Electronic Structure and Properties of Hydrogen in Metals*, ed. by P. Jena, C.B. Satterthwaite (Plenum, New York 1983) p. 85; J. Less-Common Met. **91**, 1 (1983)
6.40 W.A. Oates, T.B. Flanagan: Prog. Solid St. Chem. **13**, 193
6.41 H. Wenzl: Int. Met. Rev. **27**, 140 (1982)
6.42 G.C. Carter, F.L. Carter: "Metal Hydrides for Hydrogen Storage: A Review of Theoretical and Experimental Research and Critically Compiled Data" in *Metal–Hydrogen Systems*, ed. by T.N. Veziroglu (Pergamon, Oxford 1981) p. 503
6.43 V. Diatschenko, C.W. Chu: Science **212**, 1393 (1982)
6.44 H. Peisl: "Lattice Strains due to Hydrogen in Metals" in *Topics Appl. Phys.*, Vol. 28 (Springer, Berlin, Heidelberg 1978) p. 53
6.45 Y. Fukai: Jpn. J. Appl. Phys. **22**, 207 (1983)
6.46 A. Fukizawa, Y. Fukai: J. Phys. Soc. Jpn. **52**, 2102 (1983)
6.47 E.G. Ponyatowskii, V.E. Antonov, I.T. Belash: Usp. Fiz. Nauk **137**, 663 (1982) [English transl.: Phys. Usp. **25**, 596 (1982)]
6.48 H. Hemmes, A. Driessen, R. Griessen: J. Phys. C **19**, 3571 (1986)
6.49 D.G. Westlake: J. Less-Common Met. **90**, 251 (1983)
6.50 T.B. Flanagan: J. Less-Common Met. **63**, 209 (1979)
6.51 T.B. Flanagan, J.F. Lynch: J. Phys. Chem. **79**, 444 (1975)
6.52 E. Wicke, J. Blaurock: Ber. Bunsengesell. Phys. Chem. **85**, 1091 (1981)
6.53 E. Wicke, H. Brodowsky: "Hydrogen in Palladium and Palladium Alloys" in *Topics Appl. Phys.*, Vol. 29 (Springer, Berlin, Heidelberg 1978) p. 73
6.54 A. Driessen, H. Hemmes, R. Griessen: Z. Phys. Chem. NF **143**, 145 (1985)
6.55 T. Takeshita, K.A. Gschneidner, J.F. Lakner: J. Less-Common Met. **78**, P43 (1981)
6.56 P. Villars: J. Less-Common Met. **92**, 215 (1983)
6.57 T. Kuji, W.A. Oates, B.S. Bowerman, T.B. Flanagan: J. Phys. F **13**, 1785 (1983)
6.58 R. Feenstra, R. Griessen, D.G. de Groot: J. Phys. F **16**, 1933 (1986)
6.59 H. Sugimoto, Y. Fukai: Phys. Rev. B **22**, 670 (1980)
6.60 H. Sugimoto, Y. Fukai: J. Phys. Soc. Jpn. **50**, 3709 (1981)
6.61 H. Sugimoto, Y. Fukai: J. Phys. Soc. Jpn. **51**, 2554 (1982)
6.62 Y. Fukai: J. Less-Common Met. **101**, 1 (1984)
6.63 D.G. Westlake: J. Less-Common Met. **91**, 275 (1983)
6.64 D.G. Westlake: Scripta Met. **16**, 1049 (1982)
6.65 D.G. Westlake: J. Mater. Sci. **18**, 605 (1983)
6.66 D.G. Westlake: J. Less-Common Met. **75**, 177 (1980)
6.67 D.G. Westlake: J. Less-Common Met. **88**, 17 (1982)
6.68 D.G. Westlake: Mat. Lett. **1**, 107 (1982)
6.69 D.G. Westlake: J. Chem. Phys. **79**, 4532 (1983)
6.70 D.G. Westlake: J. Mater. Sci. **19**, 316 (1984)
6.71 D.G. Westlake: Mat. Res. Bull. **18**, 1409 (1983)
6.72 D.G. Westlake: J. Less-Common Met. **105**, 69 (1985)
6.73 I. Jacob, D. Shaltiel: J. Less-Common Met. **65**, 117 (1979)
6.74 J. Shinar, I. Jacob, D. Davidov, D. Shaltiel: "Hydrogen Sorption Properties in Binary and Pseudobinary Intermetallic Compounds" in *Hydrides for Energy Storage*, ed. by A.F. Andresen, A.J. Maeland (Pergamon, Oxford 1978) p. 337
6.75 J.J. Didisheim: "Structure et Stabilité des Hydrures métalliques ternaire ZrV_2H_x et $ZrMn_2H_x$", Thesis No. 2013, University of Geneva, Switzerland, 1981
6.76 I. Jacob, J.M. Bloch, D. Shaltiel, D. Davidov: Solid State Commun. **35**, 155 (1980)
6.77 J.J. Didisheim, K. Yvon, D. Shaltiel, P. Fischer: Solid State Commun. **31**, 47 (1979)
6.78 T. Riesterer: J. Less-Common Met. **103**, 219 (1984)
6.79 J.R. Lacher: Proc. R. Soc. (London) A **161**, 525 (1937)
6.80 R. Kirchheim, F. Sommer, G. Schluckebier: Acta Metall. **30**, 1059 (1982)
6.81 R. Kirchheim: Acta Metall. **30**, 1069 (1982); Acta Metall. **29**, 845 (1981)

6.82 R. Griessen: Phys. Rev. B **27**, 2069 (1983)
6.83 P.C.P. Bouten, A.R. Miedema: J. Less-Common Met. **65**, 217 (1979)
6.84 A.K. Niessen, F.R. de Boer, R. Boom, P.F. de Châtel, W.C.M. Mattens, A.R. Miedema: CALPHAD **7**, 51 (1983)
6.85 D.G. Pettifor: "Electron Theory of Metals" in *Physical Metallurgy Part I*, ed. by R.W. Cahn, P. Haasen (North-Holland, Amsterdam 1983) p. 73
6.86 H.H. van Mal: Philips Res. Repts. Suppl. **1** (1976)
6.87 A.L. Shilov, M.E. Kost, N.T. Kuznetsov: J. Less-Common Met. **105**, 221 (1985)
6.88 A. Pasturel, C. Chatillon-Colinet, A. Percheron-Guégan, J.C. Achard: J. Less-Common Met. **84**, 73 (1982)
6.89 A. Pasturel, F. Liautaud, C. Colinet, C. Allibert, A. Percheron-Guégan, J.C. Achard: J. Less-Common Met. **96**, 93 (1984)
6.90 A. Percheron-Guégan, C. Lartigue, J.C. Achard: J. Less-Common Met. **109**, 287 (1985)
6.91 R. Griessen, L.M. Huisman: "Electronic Structure of Metal Hydrides and Deuterides from De Haas-Van Alphen Measurements" in *Electronic Structure and Properties of Hydrogen in Metals*, ed. by P. Jena, C.B. Satterthwaite (Plenum, New York 1983) p. 235
6.92 H.L.M. Bakker, R. Feenstra, R. Griessen, L.M. Huisman, W.J. Venema: Phys. Rev. B **26**, 5321 (1982)
6.93 B.M. Geerken, R. Griessen, L.M. Huisman, E. Walker: Phys. Rev. B **26**, 1637 (1982)
6.94 D.K. Hsu, R.G. Leisure: Phys. Rev. B **20**, 1339 (1979)
6.95 W. Joss, R. Griessen, E. Fawcett: "Electron States and Fermi Surfaces of Homogeneously Strained Metallic Elements" in *Landolt-Börnstein New Series, Metals*: Vol. 12: Phonon States, Electron States and Fermi Surfaces, Subvol. b; ed. by K.H. Hellwege, J.L. Olsen (Springer, Berlin, Heidelberg 1983) p. 1
6.96 V. Heine: "Electronic Structure of Metals" in *The Physics of Metals I. Electrons*, ed. by J.M. Ziman (Cambridge Univ. Press, Cambridge 1969) p. 1
6.97 M. Cyrot, F. Cyrot-Lackmann: J. Phys. F **6**, 2257 (1976)
6.98 E.S. Machlin: J. Less-Common Met. **64**, 1 (1979)
6.99 M. Futran, S.G. Coats, C.K. Hall: J. Chem. Phys. **77**, 6223 (1982)
6.100 M.A. Pick, D.O. Welch: Z. Phys. Chem. NF **114**, 37 (1979)
6.101 W. Fenzl, J. Peisl: Phys. Rev. Lett. **54**, 2064 (1985)
6.102 R. Griessen: Phys. Rev. B (1988)
6.103 V.L. Moruzzi, J.F. Janak, A.R. Williams: *Calculated Electronic Properties of Metals* (Pergamon, New York 1978)
6.104 E. Salomons, R. Griessen: to be published
6.105 R. Griessen: "Hydrogen in Disordered Solids: Model and Calculations" in *Hydrogen in Disordered and Amorphous Solids*, ed. by G. Bambakidis, R.C. Bowman Jr., NATO ASI Series B: Physics Vol. 136 (Plenum, New York 1986) p. 153
6.106 R. Feenstra, D.G. de Groot, J.H. Rector, E. Salomons, R. Griessen: J. Phys. F **16**, 1953 (1986)
6.107 R. Feenstra, R. Griessen, D.G. de Groot: J. Phys. F **16**, 1933 (1986)
6.108 R. Feenstra, B.M. Geerken, D.G. de Groot, R. Griessen: J. Phys. F **16**, 1965 (1986)
6.109 R. Feenstra, R. Griessen: Solid State Commun. **59**, 905 (1986)
6.110 J.J. Reilly: Z. Phys. Chem. NF **117**, 155 (1979)
6.111 R. Wiswall: "Hydrogen Storage in Metals" in *Topics Appl. Phys.*, Vol. 29 (Springer, Berlin, Heidelberg 1978) p. 201
6.112 M.J. Puska, R.M. Nieminnen: Phys. Rev. B **29**, 5382 (1984); **32**, 1353 (1985)
6.113 M.J. Puska, R.M. Nieminnen: Physica **127** B, 417 (1984)
6.114 K.M. Ho, H.J. Tao, X.Y. Zhu: Phys. Rev. Lett. **53**, 1586 (1984)
6.115 G. Busch, L. Schlapbach, A. Seiler: "The plateau pressure of $RENi_5$ and $RECo_5$ hydrides" in *Hydrides for Energy Storage*, ed. by A.F. Andresen, J.A. Maeland (Pergamon, Oxford 1978) p. 293
6.116 V.L. Moruzzi, J.F. Janak, A.R. Williams: *Calculated Electronic Properties of Metals* (Pergamon, New York 1978)

6.117 R. Feenstra, D.G. de Groot, R. Griessen, J.P. Burger, A. Menowsky: J. Less-Common Met. **130**, 375 (1987)
6.118 E. Fromm, G. Hörz: Intern. Metals Rev. **25**, 269 (1980)
6.119 W.M. Mueller, J.P. Blackledge, G.G. Libowitz: *Metal Hydrides* (Academic, New York, London 1968)
6.120 H. Behrens, G. Ebel (eds.): "Gases and Carbon in Metals", in *Physics Data* Series 5 (Fachinformationszentrum Energie, Physik, Mathematik, Karlsruhe)
6.121 G.C. Sinke, L.C. Walker, F.L. Oetting, D.R. Stull: J. Chem. Phys. **47**, 2759 (1967)
6.122 J.W. Ward, J. Less-Common Met. **93**, 279 (1983)
6.123 E. Fromm, E. Gebhardt (eds.): *Gase und Kohlenstoff in Metallen* (Springer, Berlin, Heidelberg 1976)
6.124 G.G. Libowitz, A.J. Maeland: "Hydrides" in *Handbook on the Physics and Chemistry of Rare Earths Vol. 3*, ed. by K.A. Gschneidner, L. Eyring (North-Holland, Amsterdam 1979) p. 299
6.125 H.M. Lee: Metallurgical Trans. **7**A, 431 (1976)
6.126 R.C. Frank, J.E. Thomas: J. Phys. Chem. Solids **16**, 144 (1960)
6.127 R.B. McLellan, W.A. Oates: Acta Metall. **21**, 181 (1973)
6.128 E. Veleckis: J. Nucl. Mater. **79**, 20 (1979)
6.129 P.R. Subramanian, J.F. Smith: J. Less-Common Met. **87**, 205 (1982)
6.130 E. Veleckis, R.K. Edwards: J. Phys. Chem. **73**, 683 (1969)
6.131 C. Papastaikoudis, B. Lengeler, W. Jäger: J. Phys. F **13**, 2257 (1983) and refs. therein
6.132 M.H. Mintz, Z. Hadari, M. Bixon: J. Less-Common Met. **48**, 183 (1976)
6.133 J.M. Haschke, J.W. Ward, W. Bartscher: J. Less-Common Met. **107**, 159 (1985)
6.134 C.M. Smith, A.E. Hodges, J.M. Haschke, F.L. Oetting: J. Chem. Thermodyn. **14**, 115 (1982)
6.135 N.A. Galaktionowa: *Hydrogen–Metal Systems Databook* (Ordentlich, Holon Israel 1980)
6.136 H. Uchida, Y.C. Huang, M. Tada, K. Fujita: Ber. Bunsenges. phys. Chemie **114**, 51 (1979)
6.137 D.T. Peterson, S.O. Nelson: J. Less-Common Met. **72**, 251 (1980)
6.138 M. Nagasaka, T. Yamashina: J. Less-Common Met. **45**, 53 (1976)
6.139 W.J. Arnoult, R.B. Mc Lellan: Acta Metall. **21**, 1397 (1973)
6.140 K. Papathanassopoulos, H. Wenzl: J. Phys. F **12**, 1369 (1982)
6.141 V.G. Mogilatenko, D.F. Chernega, K.I. Vashchenko: Sov. Non-ferrous Metals Res. **9**, 346 (1981)
6.142 R.M. van Essen, K.H.J. Buschow: J. Less-Common Met. **64**, 277 (1979)
6.143 P.S. Rudmann, J.J. Reilly, R.H. Wiswall: Ber. Bunsenges. phys. Chemie **82**, 611 (1978)
6.144 A.J. Maeland: J. Less-Common Met. **89**, 173 (1983)
6.145 H. Oesterreicher, K. Ensslen, E. Bucher: Appl. Phys. **22**, 303 (1980)
6.146 J.J. Murray, M.L. Post, J.B. Taylor: J. Less-Common Met. **80**, 201 (1981)
6.147 J.J. Murray, M.L. Post, J.B. Taylor: J. Less-Common Met. **90**, 65 (1983)
6.148 G.D. Sandrock, J.J. Murray, M.L. Post, J.B. Taylor: Mater. Res. Bull. **17**, 887 (1982)
6.149 G.D. Sandrock: "Development of Low Cost Nickel-Rare Earth Hydrides for Hydrogen Storage" in *Hydrogen Energy Systems*, ed. by T.N. Veziroglu, W. Seifritz (Pergamon, Oxford 1978) p. 1625
6.150 R.H. van Essen, K.H.J. Buschow: J. Less-Common Met. **70**, 189 (1980)
6.151 A. Goudy, W.E. Wallace, R.S. Craig, T. Takeshita: "Thermodynamics and Kinetics of Hydrogen Absorption in R_2Co_7, RCo_3, and RFe_3 Compounds" in *Adv. in Chemistry Series, Vol. 167* (American Chem. Soc., Washington D.C. 1978) p. 312
6.152 F.A. Kuijpers: Philips Res. Repts. Suppl. 1973 No. 2
6.153 K.H.J. Buschow: "Hydrogen Absorption in Intermetallic Compounds" in *Handbook on the Physics and Chemistry of Rare Earth, Vol. 6*, ed. by K.A. Gschneidner, L. Eyring (North-Holland, Amsterdam 1984) p. 1
6.154 H.A. Kierstead: J. Less-Common Met. **73**, 61 (1980)

6.155 H.A. Kierstead: "Thermodynamics and Structural Properties of the Rare Earth-Co_3 Hydrides" in *The Electronic Structure and Properties of Hydrogen in Metals*, ed. by P. Jena, C.B. Satterthwaite (Plenum, New York 1983) p. 103
6.156 H.A. Kierstead: J. Less-Common Met. **96**, 133 (1984)
6.157 J. Clinton, H. Bittner, H. Oesterreicher: J. Less-Common Met. **41**, 187 (1975)
6.158 Y. Osumi, H. Susuki, A. Kato, K. Oguro, M. Nakone: J. Less-Common Met. **74**, 271 (1980)
6.159 M. Someno, M. Arita, R. Kinaka, Y. Ichinose: Suppl. Trans. Jap. Inst. Metals **21**, 325 (1980)
6.160 R. Burch, N.B. Mason: J. Chem. Soc. Faraday Trans. I **75**, 561 (1979)
6.161 H.A. Kierstead: J. Less-Common Met. **86**, L5 (1982)
6.162 S.J.C. Irvine, I.R. Harris: "The Effect of Induced Disorder on the Hydrogenation Behaviour of the Phase ZrCo" in *Hydrides for Energy Storage*, ed. by A.F. Andresen, A.J. Maeland (Pergamon, Oxford 1978) p. 431
6.163 J.R. Johnson: J. Less-Common Met. **73**, 345 (1980)
6.164 J.F. Lynch, J.R. Johnson, J.J. Reilly: Z. Phys. Chemie NF **117**, 229 (1979)
6.165 J.J. Reilly: "Synthesis and Properties of Useful Metal Hydrides" in *Hydrides for Energy Storage*, ed. by A.F. Andresen, A.J. Maeland (Pergamon, Oxford 1978) p. 301
6.166 A. Pebler, E.A. Gulbransen: Trans. Met. Soc. AIME **239**, 1593 (1976)
6.167 A.T. Pedziwiatr, R.S. Craig, W.E. Wallace, F. Pourarian: J. Solid State Chem. **46**, 336 (1983)
6.168 A.J. Maeland: "Preparation and Properties of TiCuH" in *Adv. in Chemistry Series, Vol. 167* (American Chem. Soc., Washington D.C. 1978) p. 302
6.169 R. Kadel, A. Weiss: Ber. Bunsenges. phys. Chemie **82**, 1290 (1978), extracted from Figs. 11 and 13
6.170 R. Kadel, A. Weiss: J. Less-Common Met. **65**, 89 (1979)
6.171 A. Pebler, E.A. Gulbransen: Electrochem. Technology **4**, 211 (1966)
6.172 C.A. Bechmann, A. Goudy, T. Takeshita, W.E. Wallace, R.S. Craig: Inorg. Chem. **15**, 2184 (1976)
6.173 H.A. Kierstead: J. Less-Common Met. **86**, L1 (1982)
6.174 H.K. Smith, W.E. Wallace, R.S. Craig: J. Less-Common Met. **94**, 89 (1983)
6.175 K. Ensslen, E. Bucher, H. Oesterreicher: J. Less-Common Met. **92**, 343 (1983)
6.176 Z. Gavra, J.J. Murray, L.D. Calvert, J.B. Taylor: J. Less-Common Met. **105**, 291 (1985)
6.177 K.H.J. Buschow, H.H. van Mal, A.R. Miedema: J. Less-Common Met. **42**, 163 (1975)
6.178 J.J. Reilly, R.H. Wiswall: Inorg. Chem. **13**, 218 (1974)
6.179 H. Wenzl: J. Less-Common Met. **74**, 351 (1980)
6.180 H.A. Kierstead: J. Less-Common Met. **85**, 213 (1982)
6.181 A.J. Maeland, J.J. Libowitz: J. Less-Common Met. **74**, 295 (1980)
6.182 D.H.W. Carstens: J. Less-Common Met. **61**, 253 (1978)
6.183 V.I. Mikhewa, M.E. Kost, A.L. Shilov: Russ. J. Inorg. Chem. **23**, 657 (1978)
6.184 G. Busch, L. Schlapbach, Th. von Waldkirch: "Hydrides of Rare Earth-Nickel Compounds: Structure and Formation Enthalpies" in *Hydrides for Energy Storage*, ed. by A.F. Andresen, A.J. Maeland (Pergamon, Oxford 1978) p. 287
6.185 B.S. Bowerman, C.A. Wulff, T.B. Flanagan: Z. Phys. Chemie NF **116**, 197 (1979)
6.186 D. Ohlendorf, H.E. Flotow: J. Less-Common Met. **73**, 25 (1980)
6.187 B. Nachen, W. Bronger: J. Less-Common Met. **52**, 323 (1977)
6.188 S. Ono, K. Imanari, K. Nomura, H. Sato: "Hydrogen Solubility in Mg_2Ni" in *Hydrogen Energy Progress V*, ed. by T.N. Veziroglu, J.B. Taylor (Pergamon, New York 1984) p. 1405
6.189 Y. Machida, T. Yamadaya, M. Asanuma: "Hydride Formation of C14-Type Ti Alloys" in *Hydrides for Energy Storage*, ed. by A.F. Andresen, A.J. Maeland (Pergamon, Oxford 1978) p. 329
6.190 F. Pourarian, V.K. Shinha, W.E. Wallace, A.T. Pedziwiatr, R.S. Craig: "Hydrogenation Entropies of the $ZrMn_{2+y}$ System" in *The Electronic Structure and Properties of Hydrogen in Metals*, ed. by P. Jena, C.B. Satterthwaite (Plenum, New York 1983) p. 385

6.191 J.L. Anderson, T.C. Wallace, A.L. Bowman, C.L. Radosewich, M.L. Courtney, Los Alamos Rept. LA-5320-MS (1973)
6.192 H. Buchner, M.A. Gutjahr, K.-D. Beccu, H. Säufferer: Z. Metallkde. **63**, 497 (1972)
6.193 F.H.M. Spit, J.W. Drijver, W.C. Turkenburg, S. Radelaar: "Thermodynamics and Kinetics of Hydrogen Absorption in Amorphous NiZr-Alloys" in *Metal Hydrides*, ed. by G. Bambakidis (Plenum, New York, London 1981) p. 345
6.194 K.V.S. Rama Rao, M. Mrowietz, A. Weiss: Ber. Bunsenges. phys. Chemie **86**, 1135 (1982)
6.195 J.F. Lynch: J. Phys. Chem. Solids **42**, 411 (1980)
6.196 K.N. Semenenko, R.A. Sirotina, A.P. Savchenkova, V.V. Burnasheva, M.V. Lototskii, E.E. Fokina, S.L. Troitskaya, V.N. Fokin: J. Less-Common Met. **106**, 349 (1985)
6.197 T.B. Flanagan, S. Majchrzak, B. Baranowski: Philos. Mag. **25**, 257 (1972)
6.198 J.W. Ward, W. Bartscher, J. Rebizant: J. Less-Common Met. **130**, 431 (1987)

Additional References

Baranowski, B., Hochheimer, H.D., Strössner, K., Hönle, W.: "High Pressure X-ray Investigation of AlH_3 and Al at Room Temperature", J. Less-Common Met. **113**, 341 (1985)
Fukai, Y.: "Atomistic and Electronic Approaches to Hydrogen in Metals", Cryst. Latt. Def. and Amorph. Mat. **11**, 85 (1985)
Nordlander, P., Nørskov, J.K., Besenbacher, F.: "Trends in Hydrogen Heats of Solution and Vacancy Trapping Energies in Transition Metals", J. Phys. F **16**, 1161 (1986)
Riesterer, T.: "Electronic Structure and Bonding in Metal Hydrides, Studied with Photoelectron Spectroscopy", Z. Phys. B **66**, 441 (1987)
Salomon, E., Griessen, R., de Groot, D.G.: "Surface Tension and Subsurface Sites of Metallic Nanocrystals determined from H Absorption", Europhys. Lett. (1988)
Watanabe, K., Fukai, Y.: "Calorimetric Studies of the Behaviour of Hydrogen in Vanadium and Vanadium Alloys", J. Phys. Soc. Japan **54**, 3415 (1985)

7. Magnetic Properties, Mössbauer Effect and Superconductivity

Gerfried Wiesinger and Günter Hilscher

With 20 Figures

Investigations of the bulk magnetic properties, of Mössbauer spectroscopy- and superconductivity studies on hydrides of intermetallic compounds, predominately of those containing a rare earth metal, are reviewed. The chapter begins with some introductory remarks dealing with concepts of magnetism and of Mössbauer spectroscopy. Experimental results covering both fields are then discussed with reference to the compound's stoichiometry. Although strictly speaking they do not belong to the category of "intermetallic compounds", hydrides of ternary oxides (oxygen stabilized intermetallics) and of amorphous alloys are covered too. The present review closes with a section concerning the hydrogen induced change of the superconducting properties of crystalline compounds and amorphous alloys.

7.1 Introduction and Scope

When studying the bulk magnetic properties and the hyperfine interactions by magnetisation- and susceptibility measurements and by Mössbauer spectroscopy respectively, one has two experimental techniques (macroscopic and microscopic respectively) which complement one another and form an ideal combination to elucidate a great many physical and chemical properties of intermetallic hydrides such as the chemical nature of hydrogen and its location, the formation and the stability of various structural phases and the influence of hydrogen absorption on electronic, magnetic, and superconducting properties of the host lattice. Moreover, from studying the hydrides, conclusions may frequently be drawn upon properties of the parent system, particularly in case of the electronic structure and the magnetic interaction mechanism. Thus such studies have attracted considerable attention not only in terms of basic research but also from the point of view of applications.

Several review papers describing magnetism and/or the Mössbauer effect in hydrides already have appeared in the last decade, among which the two Mössbauer review articles by *Wortmann* [7.1] and by *Wagner* and *Wortmann* [7.2] covered predominantely binary metal-hydrogen systems. Selected rare earth (*R*)–transition metal hydrides have been considered in the papers by *Wallace* [7.3–8], where emphasis was laid on the bulk magnetic properties and hydrogen was assumed to act generally as an electron acceptor (anionic model). *Shenoy* et al. [7.9, 10] reviewed Mössbauer results obtained in particular from hydrogen-storage materials. In his paper *Cohen* [7.11] stressed Mössbauer studies performed on rare-earth nuclei. In several other reviews the present topic has only been touched upon, e.g. those of *Oesterreicher* [7.12], *Cohen* and *Wernick* [7.13], *Kuijpers* [7.14], and *Buschow* [7.15]. In the late seventies *Buschow* and co-workers [7.16–18] made another attempt to find, apart from Wallace's suggestions, a general interpretation of the hydrogen induced changes of the magnetic properties in *R*–transition metal compounds. Besides the alteration in the interatomic distances – especially important in case of the Mn compounds – the importance of differences in the electronegativity for Fe, Co, and Ni compounds has been emphasized. Finally two comprehensive reviews from *Buschow* et al. [7.19] and *Buschow* [7.20] have appeared, where alongside formation of ternary hydrides, their thermodynamic and their chemical properties, magnetic features and Mössbauer spectroscopy are also included; however, non-*R* metal compounds are not considered. Attempts have been made to interpret several properties of the hydrides in terms of Miedema's atomic cell model.

As can be seen from the reference list, the number of papers on this subject has continued to grow in the last few years, and thus it seems justified to review the topic again. Moreover the present article is arranged in a way differing from the previous reviews and furthermore considers new materials with outstanding physical properties ($R_2Fe_{14}B$-type compounds and amorphous alloys). We have deliberately made no separation between results obtained from magnetic and from Mössbauer measurements since the intentions as well as the outcome of many investigations often prevent an unambiguous assignment to a certain topic. The chapter dealing with superconductivity remained rather short since this subject, although of great importance in binary metal-hydrogen systems (see e.g. the review by *Stritzker* and *Wühl* [7.21], has been hardly investigated for ternary systems.

The present chapter begins with an introductory section discussing some relevant physical concepts in magnetism with emphasis laid on recent developments in the theory of 3*d*-magnetism. Subsequently a few remarks about basics of Mössbauer spectroscopy are made, followed by the review of the experimental work in these fields, where we have tried to stress those kinds of materials which have received less attention in previous reviews.

7.2 Theoretical Background

7.2.1 Theoretical Concepts in Magnetism

Metallic magnetism covers a wide range of phenomena which are intimately correlated with the electronic structure but also with the metallurgy. The latter appears to be particularly important for intermetallics and for the binary and ternary metal hydrides. Although no general prediction of the effect of hydrogen absorption on the magnetic properties can be given, attempts to understand the magnetic properties, however, have led to a deeper insight into the fundamental structure and mechanisms of many different metal-hydrogen systems. This understanding is often still far from being complete since even the high-temperature magnetism of the well-known elements such as Fe, Co, and Ni remains a controversial topic [7.22–24].

In contrast to the magnetism of the 3*d*-metals which reflect the old controversy of localized versus itinerant magnetism, the magnetic properties of the rare earth (*R*) elements are successfully described in terms of the RKKY theory; because of the localized nature of the 4*f*-electrons there is no overlap between 4*f*-wavefunctions on different lattice sites and the magnetic coupling proceeds indirectly via the spatially non-uniform polarization of the conduction electrons.

The rare earths *R* form hydrides RH_x with two stoichiometries $x=2$ and x approaching or equalling 3 (see e.g. [7.4, 25] and references therein). Electrical resistivity measurements revealed that metallic conductivity disappears when x approaches 3. According to *Switendick* [7.26] hydrogen forms in the trihydrides a low lying *s*-band with the capacity to hold six valence electrons supplied by one *R* and three H.

Since this low lying band is filled with electrons – the RH_3 are semiconductors – and due to the lack of conduction electrons the RKKY interaction cannot be transmitted. This accounts for the suppression of the magnetic interactions which is generally observed. The discussion as to whether hydrogen is anionic (H^-) or protonic (H^+) in binary hydrides, has been settled by *Wallace* and *Mader* [7.27]. By combining susceptibility measurements with crystal-field calculations in PrH_2 they could demonstrate the validity of the anionic model. This means that H, with its low-lying *s*-band, acts as an acceptor in the rare-earth series; this is further supported by heat capacity and inelastic neutron scattering studies [7.28–31].

The preceding discussion demonstrates that magnetic studies of the rare earth hydrides are exceedingly valuable and confirms the validity of the RKKY interaction in the binary rare-earth hydrides which is suppressed upon hydrogen absorption.

Although the general characteristics of the rare earths can be regarded as having been explained in terms of a semiempirical model based in fact on bandstructure calculations of *Switendick* [7.26], CeH_x may serve as an example to illustrate that the details still pose unsolved physical problems [7.32–34]. Very recently *Schlapbach* [7.34] reported on the electronic specific heat coefficient γ of $CeH_{2.6}$ ($\gamma=110$ mJ/mole K^2) to be more than about one order of magnitude larger than that of γ-Ce ($\gamma=10$ mJ/mole K^2). This indicates that in $CeH_{2.6}$ the 4*f*-band is located close to the Fermi energy. On the other hand enormous specific heat values are obtained for high effective mass *f*-electron materials such as $CeCu_2Si_2$, UBe_{13}, UPt_3, $CeCu_6$, $NpBe_{13}$, U_2Zn_{17}, and UCd_{11} which furthermore exhibit maxima in the resistivity at low temperatures and large values of the magnetic susceptibility; for obvious reasons these are called heavy-fermion systems [7.35, 36]. By analogy with these compounds it was suggested by *Schlapbach* et al. [7.34] that $CeH_{2.6}$, although ordered at low temperatures should also exhibit heavy-fermion-like behaviour. From crystal field calculations *Osterwalder* et al. [7.32] and *Schefer* et al. [7.33] could furthermore clarify the striking changes of the magnetic interactions between the Ce^{3+} ions which occur in these hydrides at low temperatures as function of x: no ordering is observed for $CeH_{2.93}$ down to 1.3 K, $CeH_{2.43}$ orders ferromagnetically ($T_c = 4.2$ K) and $CeH_{3.00}$ orders antiferromagnerically ($T_N = 1.9$ K).

In contrast to the binary 4*f*-hydrides, for the 3*d* metals and their intermetallics no similar simple and convincing conclusions can be drawn about changes of magnetism in terms of a protonic or an anionic behaviour. The only general statement which can be made is that changes in the magnetic properties in the Fe compounds occur in the opposite direction to those in the Ni and Co compounds. Particulary in Sc-, Y-, Hf-, and R-Fe intermetallics, the Fe moment is found to be increased on hydrogen absorption while changes in the magnetic ordering temperatures can be obtained in either direction [7.19, 20].

As already mentioned, magnetism in the 3*d* elements and their intermetallics remains a controversial topic and is by no means a solved problem. The reason for this controversy is the absence of a general agreement upon the microscopic nature of the magnetic state above and below the Curie temperature. Two opposite standpoints have so far been used to explain the magnetic order as a function of temperature. In a Heisenberg model the description is in terms of localized moments and the magnetisation disappears at T_c because of disorder in the local moments due to thermal fluctuations. Nevertheless the absolute value of the magnetic moments remains almost constant. The usual result of a mean field theory leads within this type of model to the well-known Brillouin function which in its argument contains the ratio of the magnetic energy to the thermal energy. Thus a rise in temperature reduces the magnetisation due to the thermal disorder of the magnetic moments.

In the Stoner-Wohlfarth itinerant-electron model the magnetic moment and the magnetisation vector are determined by the unpaired electrons of the exchange-split spin-up and spin-down bands; this appears to be a realistic description of the magnetism in metals at 0 K. A given material becomes magnetic if the Stoner criterion [$N(E_F)I \geqq 1$] is fulfilled [with $N(E_F)$ the density of states at the Fermi level and I the intra-atomic exchange or Hubbard exchange energy]. In this model the thermal excitations of electron-hole pairs reduce the exchange splitting and drive the transition towards the paramagnetic state. Consequently the magnetisation vanishes only if the absolute value of the magnetic moment goes to zero which only happens if the exchange splitting is zero. Unfortunately this yields Curie temperatures which are 5–10 times larger than those observed experimentally. The controversy concerning the localized versus the itinerant electron model has been settled recently and all current theories are now based on the latter model. It is obvious that the order parameter in the magnetic state has to be related to the magnetisation vector.

On the other hand, the question of how the magnetisation arises from the spins of mobile electrons and of which forces are responsible for its fluctuations are only partially understood. An improved theory based on the Stoner-Wohlfarth model (namely a spin-polarized band theory) must allow for the existence of a magnetic polarisation whose direction varies from one unit cell to the other. Then the global magnetization vanishes at the Curie temperature not because the entities, the absolute values of the magnetic moments which may be called local moments, are zero, but because they point in random directions. For a very recent formulation of a first-principles theory of ferromagnetic phase transitions and the electronic structure of metallic ferromagnets we refer to *Gyorffy* et al. [7.23] and *Staunton* et al. [7.37].

The phenomena of 3*d* magnetism is also strongly connected with the width and structure of the 3*d*-band and the density of states at the Fermi level. In a semiempirical model *Griessen* and *Feenstra* [7.38] recently related the heat of formation of a metal hydride to the difference between the Fermi energy and the centre of the *s*-band of the host metal. They found that the product of the bulk modulus and the lattice expansion due to hydrogen absorption depends on the heat of formation of the hydride in a linear way. The bandwidth W is, according to *Heine* [7.39], proportional to a^{-5} (where a is the interatomic distance). The simple band model states that narrow bands with high $N(E_F)$ favour the occurrence of magnetism. When the hydrogen induced volume increase is considered together with magnetovolume data of the parent compound at least a first estimate of changes in the magnetic ordering temperature may be obtained. If the electronic structure is only slightly affected, T_c may be extrapolated from its pressure dependence. Such a correlation was proposed by *Buschow* and *Sherwood* [7.17] for crystalline, and by *Coey* et al. [7.40] for amorphous Y-Fe compounds and by *Hilscher* et al. [7.41] for the pseudobinaries Ti(Fe, Co).

Realistic electronic band structure calculations for the parent intermetallics and their hydrides are therefore necessary to gain a deeper insight into the basic physical behaviour of these systems. Otherwise semiempirical models have to be used in order to interpret the effect of hydrogen upon magnetism.

Besides $LaNi_5$, TiFe and the pseudobinaries Ti(Fe, Co) also serve as relevant examples of intermetallic compounds in which hydrogen absorption affects not only the bulk 3*d* magnetism but also the metallurgy, giving rise to a decomposition at the surface. For TiFe this leads to the formation of Ti and Fe clusters at the surface accompanied by a growing number of antistructure atoms in the bulk. These additional contributions to the global magnetisation further complicate the analysis of the data actually measured. The change of the magnetic properties upon hydrogenation may thus be of different origins which need not necessarily be correlated with altered bulk or intrinsic magnetic properties of the material.

Intermetallic compounds of 3*d* metals (Mn, Fe, Co, Ni) with rare earth elements exhibit a large variety of interesting magnetic properties. A large number of hydrogen absorption studies have also been performed on such compounds. The magnetic properties of these intermetallics have been reviewed by *Buschow* [7.42, 43], *Kirchmayr* and *Poldy* [7.44] and *Wallace* [7.45]. These properties are of interest for two major reasons: Firstly their study helps to shed light on some of the fundamental principles of magnetism such as the indirect exchange or RKKY interaction, crystal field effects, valence instabilities, magnetoelastic properties, the occurrence of superconductivity and magnetic ordering in the same compound. Secondly these properties are of technical interest due to their potential application in the production of permanent magnets such as the RCo_5 and the Nd–Fe–B based materials. Furthermore, several of those compounds have been proved to be suitable for hydrogen storage purposes.

In *R*–3*d* intermetallics and their hydrides, where both the *R* and the 3*d* element are magnetic we can distinguish between three main types of magnetic interactions which are quite different in nature: the magnetic interaction (i) between the localized 4*f* moments, (ii) between the more itinerant 3*d* moments and (iii) between the 3*d* and the 4*f* moments. Generally it is observed [7.43, 44] that the magnetic interaction in *R*–3*d* intermetallics, where both the *R* and the 3*d* element are magnetic, decreases in the following sequence: 3*d*–3*d*, 4*f*–3*d*, 4*f*–4*f*.

Depending upon the composition and the type of combination of these elements hydrogen absorption usually leads to a weakening of the magnetic coupling between the 4*f* moment and the 3*d* moment and to substantial changes in the 3*d* transition metal moment. In Co and Ni compounds hydrogen absorption usually results in a reduction of the 3*d* moment, while the opposite is observed for *R*-Fe compounds. The weakening of the 4*f*–3*d* exchange interaction upon hydrogen absorption may be explained by a reduced overlap of the 3*d* electron wave functions with the 5*d*-like ones which are polarized by the 4*f* moment and thus transmit the interaction to the 3*d* electrons. Hydrogen reaction can furthermore lead to concentration fluctuations of H atoms over a few atomic distances so that the electron concentration may differ from one *R* site to the other and therefore changes of the magnetic coupling strength can be expected. Moreover resistivity measurements [7.25, 46] show that the residual resistivity of the hydrides is higher than that of the parent intermetallics which indicates that the mean free path of the conduction electrons is reduced. This can lead to a damping of the RKKY conduction electron polarization which tends to decrease the magnetic coupling strength.

Apart from the pure 4*f* elements and their binary hydrides, compounds where the transition metal moment is zero also allow the 4*f*–4*f* interaction to be studied directly. As a first approximation one would expect that changes in magnetic properties can be explained by analogy with the pure *R* elements and their hydrides, in terms of the anionic model, by a reduction of the conduction electron concentration which consequently reduces the RKKY interaction. This is obviously the case for $GdRu_2$ and $GdRh_2$, where the Curie temperatures of 83 and 73 K are reduced upon hydrogen absorption to 65 and 35 K respectively [7.47]. However the opposite is found for $GdCu_2$ and GdAg, where, after hydrogen uptake, the paramagnetic Curie temperatures change from 7 to 57 K and -57 to $+50$ K (i.e. from antiferro- to ferromagnetism) [7.48]. According to *Buschow* [7.20] the latter compounds may serve as an example in which altered conduction electron polarization can cause changes in both magnitude and sign of the magnetic coupling due to the oscillatory character of the RKKY interaction.

If magnetism in *R*-intermetallics is dominated by the 4*f* moments the concept of charge transfer, namely that *R* atoms donate electrons to the H atoms seems by analogy with binary *R*-hydrides to be a reasonable explanation for changes in magnetism. This is in fact confirmed by numerous isomer shift results obtained from Mössbauer experiments on rare earth nuclei.

In the case where 3*d* magnetism is dominant in *R*–3*d* compounds, hydrogen absorption leads to a reduction of the 3*d* moment in Ni and Co based intermetallics, but to an enhancement of the Fe moment. For Mn-intermetallics both changes from para- to ferromagnetism and in the opposite direction are observed.

As already mentioned, the 3*d*-moment formation from itinerant electrons is a controversial topic and the introduction of hydrogen makes the situation even less clear. *Wallace* [7.4, 45] explained the controversial changes of the 3*d* moment in Fe, Co, and Ni compounds as well as the occurrence or breakdown of the magnetic order in Mn-compounds in terms of a depopulation of the exchange split 3*d* bands as a consequence of a sizeable charge transfer from the 3*d* metal to the H atoms. From various Mössbauer experiments [7.20, 47, 48] and according to reasonable arguments by *Buschow* [7.15, 20, 32] such a large charge transfer appears to be rather unlikely to explain the substantial changes of the 3*d* moment. For binary 3*d* and 4*d* transition metal hydrides and their intermetallics *Wicke* [7.49] proposed a screened-proton model combined with the concept of *d*-band filling. In terms of these models recent results of UV and x-ray photoelectron spectroscopy can also be interpreted [7.50, 51]. Since band structure calculations for these intermetallics and their hydrides are not available, somewhat speculative semiempirical models are used. In view of the Mössbauer results on *R*-nuclei and the above cited experiments, we agree with *Buschow* [7.15, 20] that hydrogen exists as a screened proton in these *R*-intermetallic hydrides with a screening charge larger than one. The excess screening charge is due to charge transfer from *R*-atoms to hydrogen rather than from the 3*d* atoms.

7.2.2 Physical Concepts of Mössbauer Spectroscopy

a) Introductory Remarks

Only some brief remarks will be given here. For more detailed discussions we refer the reader to the numerous textbooks dealing with Mössbauer spectroscopy, among which two most recent ones are included in the reference list [7.52, 53]. The reason for observing the Mössbauer effect is based on the fact that in case of medium γ-energies ($\lesssim 100$ keV) a certain probability exists for the recoil-free emission and absorption of γ-rays. Because of the range of energies involved, such a process can only occur if the Mössbauer nucleus is bound in a solid. Since transitions between nuclear levels are observed by Mössbauer spectroscopy, this technique is sensitive to the influences of the electronic environment of the probe nucleus which may cause a perturbation to these levels (electron-nucleus- or hyperfine interactions). When γ-rays are emitted or absorbed under the above conditions they suffer no energy loss, and the resulting lines exhibit the natural width. Thus a sufficiently high resolution can be obtained to monitor the minute hyperfine interactions by periodically changing the Doppler velocity in the range of mm/s up to cm/s corresponding to energy changes of the order of a few μeV or even less.

Three of these hyperfine interactions are known: (i) An electrostatic one (Coulomb-like) between the nuclear charge and the electronic charge at the nucleus, causing a shift of the nuclear levels. Therefore we commonly observe deviations from the theoretical resonance frequency (isomer shift). (ii) The interaction between the nuclear quarupole moment and the electric field gradient (EFG) at the nuclear site results in a partial (quadrupole-) splitting of the nuclear levels. (iii) The degeneracy of the nuclear states is lifted completely, when the nuclear magnetic moments experience a magnetic field (nuclear Zeeman effect). Consequently in the last two cases we observe a splitting of the γ-ray pattern, i.e. multi-line Mössbauer spectra are obtained. Further parameters which can be derived from the analysis of a spectrum are the width of the Mössbauer line which can yield information about diffusion, or relaxation processes and the Mössbauer-Lamb (f) factor representing the fraction of recoil-free γ-rays emitted or absorbed, thus giving insight into the dynamic behaviour of the lattice.

b) Experimental Techniques

Since there exist numerous books on Mössbauer spectroscopy which include a discussion of experimental techniques we will only touch upon this topic by mentioning those problems which arise specifically in the study of hydrides of intermetallic compounds. In principle, two different kinds of experimental set up are employed: (i) the sample is prepared as a source, the spectrum is scanned by using a single line absorber (source experiment); (ii) a single line source is used in combination with the sample positioned as absorber. In contrast to binary metal–hydrogen systems [7.2], no source experiment has yet been reported for the case of ternary hydrides. When spectra are to be recorded by applying the latter technique in transmission geometry, reasonable measuring times (some hours up to a few days) are only obtained when the Mössbauer isotope is present in the absorber in an abundance of at least a few percent. This value may, however, be considerably lowered, if enriched isotopes are used.

Mössbauer absorbers are chiefly prepared by melting the intermetallic host compound and by subsequently loading the sample with hydrogen gas. An exception is found in case of hydriding amorphous alloys, where electrolytical charging is preferred. In order to keep the amount of dissolved hydrogen in the sample constant during the measuring time, sealing the hydride is highly recommended (e.g. with SO_2) (see Volume II, Chap. 2), which particularly holds for Mössbauer studies at or above room temperature.

Among the vast number of intermetallic compounds which have been studied, attention has been focussed on those materials which have proved suitable for hydrogen storage purposes ($LaNi_5$, TiFe, ...). Consequently ^{57}Fe or a rare-earth isotope (^{151}Eu, ^{155}Gd, ^{161}Dy, ^{166}Er, ^{169}Tm) are frequently mentioned in the studies under consideration. ^{181}Ta, which is widely used to investigate binary hydrides [7.1, 2], is of no relevance in the present case.

Because of the γ-energies involved, the absorber thickness has to be in the range of about 20 μm (^{57}Fe, 14.4 keV) up to 1 mm (^{166}Er, 81 keV). For studying the parent compounds, powder absorbers are generally prepared by grinding the material. After hydrogen absorption, however, the solid material has usually already converted into a fine powder. It can thus be directly used as a proper absorber material preventing further damage to the sample.

c) Hyperfine Interactions

The isomer shift contains simultaneous information about the difference in the nuclear radii (δR) of excited and ground state respectively and the difference in the charge at the nuclear site in absorber and source respectively. Because δR can be of either sign (e.g. <0 for ^{57}Fe, ^{155}Gd and >0 for ^{119}Sn) a shift of the resonance line to more positive (negative) energies corresponds in the former case to a smaller (larger) electron density, whereas the opposite is obtained in the latter case. The isomer shift depends upon valence state and chemical bonding. It can be directly influenced by altering the population of the s-like conduction electrons or indirectly via a shielding of the s-electrons by those with d-character. *Walker* et al. [7.54] made an attempt to calibrate the observed ^{57}Fe isomer shifts in (ionic) iron compounds and iron alloys in terms of the total s-electron density. This method may be of some relevance when the problem of hydrogen induced charge transfer is considered. *Buschow* [7.20], however, reported to have obtained unreasonably large values for the transferred charge when interpreting the change in isomer shift in terms of the Walker data.

Wagner and *Wortmann* [7.2] had already pointed out the problem of asigning the total change in isomer shift after hydrogen absorption to different origins. If the isomer shift S is considered as a function of hydrogen concentration n and volume v, the overall change after charging can be expressed as

$$\frac{dS}{dn} = \left(\frac{\partial S}{\partial \ln v}\right)_n \frac{d\ln v}{dn} + \left(\frac{\partial S}{\partial n}\right)_v .$$

dS/dn is measured by Mössbauer spectroscopy. $(\partial S/\partial \ln v)_n$ displays the pure volume dependence of S and can be obtained from high pressure data. The term

$d \ln v/dn$ represents the hydrogen induced rise in volume and may be evaluated from x-ray diffraction data. The change at constant volume, $(\partial S/\partial n)_v$, describing the actual effect of hydrogen on the electronic structure of the Mössbauer atom, can thus be readily calculated. It turns out that this term is rather small in magnitude, i.e. of the order of $+0.1$ mm/s. An appreciable $3d$ charge transfer from Fe to H which has frequently been proposed in order to explain the hydrogen induced changes in the magnetic moment (see e.g. [7.7]) can by no means account for it, since this would in fact lead to the wrong sign for dS.

Very recently *Gupta* [7.55] interpreted the isomer shift of TiFe hydride by applying the APW method and found excellent agreement with the experimental data. She proposed a depletion of the Fe4s states in order to form a low-lying metal-hydrogen band which, for the case of binary metal-hydrogen systems, has already been suggested by *Switendick* [7.26]. This finding is in complete accordance with XPS data recently reported by *Schlapbach* [7.56] for $ZrMn_2H_3$. He was able to verify experimentally that this hydrogen induced band lies 6.5 eV below the Fermi level. The shift of the Zr $3d$ levels towards higher energies can be regarded as a further indication of a charge transfer from Zr to H. The Fe $3d$ states seem to be hardly affected by hydrogen absorption.

Buschow [7.20] questioned the justification of making up the hydrogen induced change in isomer shift from an electronic and a volume term. He suggested instead that the model of *Miedema* and *van der Woude* [7.57] should be used, where the differences in electronegativity and in the electron density at the atomic cell boundary have to be considered. An attempt was made to explain the isomer shift data for $Th_7Fe_3H_x$, but no firm conclusion has been reached as yet.

The electric quadrupole interaction occurs if both a nuclear quadrupole moment and a nonzero electric field gradient (EFG) exist at the nucleus. As a consequence, the nuclear state can split into several sublevels, leading to the so-called quadrupole splitting of the Mössbauer line. Commonly one assumes the EFG to arise from two sources – from a lattice and an electronic contribution. The former is due to charges on lattice sites surrounding the Mössbauer atom in a non-cubic symmetry; the latter is attributed to an anisotropic distribution of the conduction electrons of the Mössbauer atom. In the hydrides the influence of interstitial H atoms on the EFG is far too small to be completely resolved in a Mössbauer spectrum. At most a line broadening or an asymmetric line shape will be observed.

A nucleus with a spin quantum number $I>0$ exhibits a nuclear moment which interacts with the magnetic field at the nuclear site (effective or hyperfine field). This commonly yields a splitting of the nuclear states into $2I+1$ sublevels, leading to multi-line hyperfine patterns. Because the presence of a magnetic field is required for this interaction, it will usually only be observed in magnetically ordered solids. In paramagnetic systems a magnetic hyperfine splitting can only occur if an external field of several T is applied to the sample or when the relaxation times of the atomic spins are sufficiently long compared with the nuclear precession time. The magnetic field acting on the nucleus is found to lie in the range between several T (α-Fe: $B_{eff} = 33$ T at room temperature) and a few hundred T in case of rare earths. B_{eff} consists of different terms which originate from the Fermi contact interaction, from polarized $d(f)$ electrons, from dipolar contributions of the atomic moments and if present also from a contribution due to the external field. It should be mentioned that the three hyperfine interactions just described can only in a few fortunate cases be observed separately. The simultaneous occurrence of at least two of them is more likely.

From the discussion of the hyperfine interactions it is obvious that a given phase in the sample under investigation should exhibit specific hyperfine parameters and consequently can be identified by applying Mössbauer spectroscopy. Quantative phase analysis using this technique should, however, be performed with great caution, since the f-factors in the individual phases will in general differ from each other and even are completely unknown in some cases. This holds especially in the case of high γ-energies. Although the assumption that the area under a given hyperfine pattern is proportional to the number of Mössbauer atoms in the corresponding phase may be applied frequently, clustering of the Mössbauer atoms can again lead to an error in the evaluation of the phase concentration.

Further properties which can be investigated by analyzing a Mössbauer spectrum, e.g. hydrogen diffusion or hydrogen induced change of the lattice dynamics are beyond the scope of the present chapter and are considered in Chap. 3 of Volume II of this series.

7.3 Magnetism and Mössbauer Spectroscopy–Experimental Results and Discussion

7.3.1 General Remarks

The organisation of the following section deviates from that used in earlier reviews. Since up to now only rare earth–transition metal hydrides have been considered a subdivision into Mn, Fe, Co, and Ni compounds seemed meaningful. In the present article, however, Zr and Ti compounds, amorphous alloys and even oxides (Ti_2FeO_z) are also included and thus the arrangement following the stoichiometry may be more favourable. Within a certain composition the results are again reported in the conventional sequence.

Since in case of magnetism and Mössbauer effect the difference between hydrogen and deuterium is insignificant, no distinction has been made between those two isotopes. This is furthermore justified by the fact that exclusively deuterides have been used in neutron diffraction studies because the large incoherent scattering cross section of hydrogen prevents the recording of reasonable neutron spectra in case of the hydrides.

7.3.2 Transition-Metal-Rich *R*-TM Compounds

a) R_2TM_{17}

This stoichiometry occurs principally in two possible crystal structures, the rhombohedral Th_2Zn_{17} (space group $R\bar{3}m$, No. 166) and the hexagonal Th_2Ni_{17} structure (space group $P6_3/mmc$, No. 194) (see e.g. [7.44]). In case of TM = Fe most compounds show both structure types, the trigonal predominantely for the light, and the hexagonal for the heavy lanthanides. Altogether four inequivalent TM lattice sites are present for both types which for R_2Fe_{17} leads to rather complex ^{57}Fe Mössbauer patterns with up to eight subspectra, the particular number depending on the direction of the easy axis of magnetisation [7.58, 59].

Up to now only one study on this kind of composition has been reported: *Zukrowski* et al. [7.60] investigated some Fe-rich $Dy_2(Fe_{1-y}Al_y)_{17}$-hydrides using ^{161}Dy as well as ^{57}Fe Mössbauer spectroscopy. For the host compounds only the Th_2Ni_{17} type of structure was obtained. The sample with $y = 0.18$, however, could be converted to the Th_2Zn_{17} type after hydrogen uptake. The storage capacity was found to be rather limited: between 4 and 5 H atoms/f.u. could be dissolved. While upon hydrogenation isomer shift and quadrupole splitting had increased in both cases, only the Fe-hyperfine field showed a remarkable increase, the hyperfine field at the Dy nuclei remaining essentially unaffected.

b) RTM_5

In this stoichiometry, except for $ThFe_5$, only Co and Ni compounds exist, crystallizing in the $CaCu_5$ type of structure (space group $P6/mmm$, No. 191). RMn_5 compounds do not form at all.

TM = Fe, $ThFe_5$

Gubbens and *van der Kraan* [7.61] examined the host compound by means of Mössbauer spectroscopy, x-ray diffraction and neutron depolarisation and concluded that $ThFe_5$ has a basal plane magnetic anisotropy of a complex ferrimagnetic character. Studying the hydrides *Gubbens* et al. [7.62] obtained a maximum hydrogen uptake corresponding to the composition $ThFe_5H_{1.7}$, leading to a rise in volume of about 6%. Both magnetisation and Curie temperature were found to have only slightly increased after hydrogenation. As in case of the parent compound the hydride exhibited a sharp hyperfine pattern, however, the easy axis of magnetization was concluded to have turned from a direction between the *a*- and the *b*-axis to a direction along the *a*-axis. The increase in isomer shift after charging has been explained by *Gubbens* et al. on the basis of a modified Miedema–van der Woude model [7.57].

TM = Co

Only compounds containing light rare earths up to Gd have been examined so far. The reason lies in the outstanding permanent magnetic properties of these

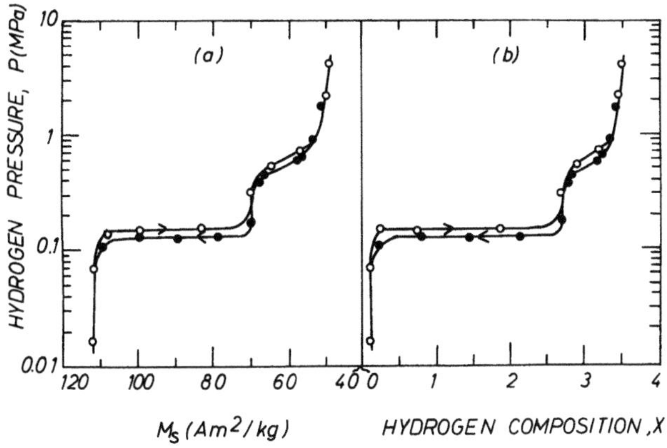

Fig. 7.1. Hydrogen pressure vs saturation magnetisation *(left)* and hydrogen pressure vs composition *(right)* in absorption (○) and desorption (●) process for $NdCo_5H_x$ at $T = 303.2$ K [7.64]

ferromagnetic materials which are lost in the case of the heavy rare earths, where ferrimagnetic coupling takes place between the *R* and the Co moments. The ferromagnetic behaviour of the light RCo_5's is preserved after charging with hydrogen; however, a weakening of the *R*-Co and the Co–Co exchange interaction is observed [7.63–65]. Co as well as *R* moments were found by *Kuijpers* [7.14] to decrease in a nonlinear way with increasing hydrogen content H_x. An extrapolation yielded a value of $x \sim 4.5$ for the loss of ferromagnetism. By studying $LaCo_5$ doped with 10% Gd with the ^{155}Gd Mössbauer effect *Bauminger* et al. [7.66] could confirm the presence of different hydride phases already reported by *Kuijpers* and *Loopstra* [7.63]. While isomer shift and electric field gradient changed monotonically with the amount of absorbed hydrogen, no such variation was obtained for the Gd hyperfine field.

Since the magnetisation strongly depends upon the hydrogen concentration in the sample, pressure–magnetisation isotherms yield similar information about the presence of different phases and about phase transformations as pressure–composition isotherms. Considering this, *Yamaguchi* et al. [7.64] developed a vibrating-sample magnetometer connected directly to a pressure–composition measuring system. This technique proved to be especially useful for studying the magnetisation of less stable hydrides as a function of applied field and hydrogen pressure. The authors could demonstrate the versatility of this method by examining the different phases (α, β^{I}, β^{II}, γ) in the system $NdCo_5H_x$ [7.64], $GdCo_5H_x$ and YCo_5H_x [7.65] and $PrCo_5H_x$ and $LaCo_5H_x$ [7.67]. The peculiar phases have been identified by the stepwise changes in magnetisation observed in the magnetisation–temperature isobars which are displayed in Fig. 7.1.

For $R = Y$ and $R = Gd$, ^{59}Co NMR spin-echo measurements have been carried out [7.68, 69]. Some discrepancies occurred in assigning the various resonance lines to the two Co sites. Probably the hydrogen atoms are located preferentially in the Co-only layers, causing a greater effect on the Co(II) signal than on that orginating, from the Co(I) sites.

Magnetic relaxation studies on hydrided (deuterated) light RCo_5's have been performed by *Herbst* and *Kronmüller* [7.70, 71] who measured the magnetic after-effect of the initial susceptibility. They concluded that the concentration of hydrogen (deuterium) in the planes containing R atoms, by far exceeds that in the Co-only planes. This result is in disagreement with the analysis of the ^{59}Co NMR by *Yamaguchi* et al. [7.68]. However, considering the weakly resolved spectra and the large discrepancies in the resonance frequences with the investigation of *Figiel* [7.69], we tend to favour the interpretation of Herbst and Kronmüller.

TM = Ni

Emphasis has so far been laid on pure $LaNi_5$ and on $LaNi_5$-containing materials. Because of their outstanding hydrogen storage properties they have received considerable attention particularly for technical application. With this kind of compound we meet a typical example in which bulk and surface effects occur simultaneously. In certain cases this has not quite been realized and has sometimes led to apparently inconsistant results.

A large variety of techniques has been applied in order to elucidate the complex hydrogen absorption mechanism in this kind of material. These can be roughly divided into surface sensitive methods (photo emission and related spectroscopies, and to some extent ESR) and experiments in which only the bulk properties can be studied (magnetic measurements, x-ray diffraction, transmission Mössbauer spectroscopy). Careful susceptibility studies, however, can also be a useful tool for identifying ferro- or at least superparamagnetic clusters at the surface if the results obtained from freshly cut samples in high vacuum at variable temperature are compared with those from samples exposed to some reactive atmosphere. Such disintegration has indeed been observed on uncharged $LaNi_5$ if oxygen or water was present [7.72–76]. When loaded with hydrogen under medium pressures ($\lesssim 10$ bar) the disintegration continues which, however, is not the case if the hydrogen pressure exceeds this value substantially [7.75]. This might be the reason for the unusual interpretation of *Palleau* and *Chouteau* [7.77].

The results from the various magnetic measurements can be summarized as follows: although by applying spin-polarized energy band calculations *Malik* et al. [7.78] predicted $LaNi_5$ to be weakly ferromagnetic, several authors confirmed that in fact this compound behaves as a Stoner-enhanced Pauli paramagnet (see e.g. [7.75] and references therein). When hydrogen is absorbed just once, its susceptibility is lowered by a factor of nearly 4. This has been attributed by *Schlapbach* [7.75] to a reduction of the enhancement factor and has been corroborated by magnetic measurements [7.77] as well as by the observation of the g-shift of the Gd ESR in Gd-doped $LaNi_5H_x$ [7.79]. At this point we want to stress that after just one absorption process we are still dealing with a pure bulk phenomenon, since the specific surface area of the Ni segregations may still be assumed to be neglegibly small.

Simple hydrogen absorption has been further examined by means of several Mössbauer experiments. By investigating Gd-doped samples with the ^{155}Gd transition ($E_\gamma = 86.5$ keV) *Bauminger* et al. [7.66] could confirm the different hydride phases which had already been detected by *Kuijpers* [7.14] by means of magnetic measurements. Furthermore a large reduction of the *s*-electron density at the Gd nuclei has been reported which, in our opinion has erroneously been interpreted as inconsistent with the ESR results of *Walsh* et al. [7.79]. The ^{57}Fe ($E_\gamma = 14.4$ keV) results obtained on Fe-doped $LaNi_5$ [7.80–82] show some discrepancies in the interpretation of the room temperature Mössbauer spectra which points we believe to the influence of the metallurgy on the hyperfine parameters. In case of the hydrides [7.83–85] the data again suggest the presence of Ni clusters, whereas no evidence for the formation of Fe clusters has been found. While for certain concentrations magnetic ordering temperatures were found to have been substantially reduced after hydrogen uptake, the Fe hyperfine field at 4 K was proved to remain almost unchanged. The small change in the Fe-isomer shift upon hydrogenation points to the predominance of the La–H interaction in this compound. The complete insensitivity of the ^{119}Sn ($E_\gamma = 23.8$ keV) hyperfine parameters in Sn doped $LaNi_5$ to hydrogenation, as reported by *Oliver* et al. [7.86], once more confirms this interpretation.

Apart from these studies, heavily cycled material has been investigated in order to examine the degree of reversibility of the absorption-desorption process after a large number of cycles [7.75, 79, 87–90]. Repeated cycling yields a distinct enhancement of the susceptibility which furthermore tends to become temperature and field dependent. As already mentioned, the bulk susceptibility is reduced after the absorption of hydrogen. Consequently the increase in χ has to be attributed to the formation of ferro- or at least superparamagnetic Ni precipitates, their specific area increasing with hydrogen content. The segregation and decomposition of the surface was confirmed by *Shaltiel* et al. [7.89] by applying ferromagnetic resonance (FMR). A signal which could be attributed to pure Ni metal clearly demonstrates the presence of clusters at the surface. With increasing number of cycles the magnetisation also increases and has been interpreted by *Schlapbach* [7.75] to consist of two superimposed terms, a linear one due to the bulk, the another field dependent term originating from the Ni precipitates at the surface. From a quantitative analysis of the *M* vs H curve Schlapbach was able to estimate that the Ni clusters contain about 6000 atoms.

Other Mössbauer transitions have been successfully applied to demonstrate the formation of the segregations in repeatedly cycled material. This shows that their volume must be of appreciable magnitude, since the clusters are visible for γ-rays with an energy of more than 20 keV. *Cohen* et al. [7.87, 88] used the ^{151}Eu spectroscopy ($E_\gamma = 21.6$ keV), while *Rummel* et al. [7.90] performed the only ^{61}Ni experiment ($E_\gamma = 67.4$ keV) known so far in the field of intermetallic hydrides. Identically treated samples were used and although the energy of

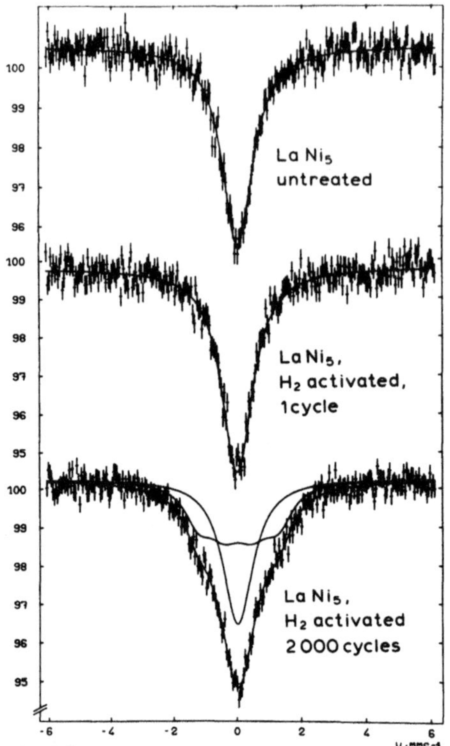

Fig. 7.2

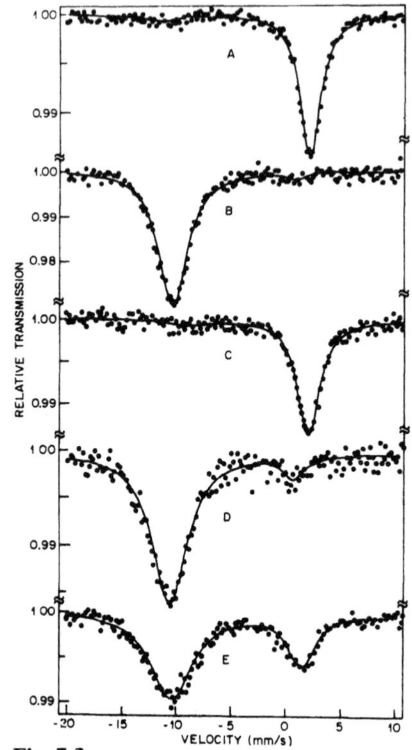

Fig. 7.3

Fig. 7.2. ^{61}Ni Mössbauer spectrum of $LaNi_5$ samples after various treatments: (*a*) no hydrogen exposure, (*b*) activated in hydrogen, (*c*) after 1584 thermally induced absorption-desorption cycles [7.90]

Fig. 7.3. Mössbauer spectra of ^{151}Eu in $LnTM_5$: *A*, a fresh sample; *B*, after hydriding at 150 bar; *C*, dehydrided after 10 cycles; *D*, hydrided after 1500 thermal cycles; *E*, dehydrided after 1500 cycles. The absorption lines near 2 mm/s arise from Eu^{3+} in the unhydrided host; those near − 11 mm/s arise from Eu^{2+} in the hydrided material [7.87]

radiation differs substantially, consistent results have been obtained. While Rummel et al. could not detect the second phase by x-ray diffraction, they could unequivocally ascertain it by the ^{61}Ni Mössbauer spectra. As can be seen by inspection from Fig. 7.2, no substantial effect is observed after one cycle. Repeated cycling (1584 times) leads, however, to the occurrence of a magnetic hyperfine pattern, its splitting agreeing perfectly with that of pure Ni metal. A careful quantitative analysis of the absorption areas yielded the result that some of the Ni clusters were still too small to develop long-range ferromagnetic order and thus remained superparamagnetic, even at liquid helium temperature.

Cohen et al. [7.87, 88] studied the degradation of thermally cycled Eu-doped $LaNi_5$. Eu was found to change its valency upon hydriding from 3+ to 2+ which fortunately is accompanied by large changes in isomer shift (up to

-13 mm/s). This makes an identification of the respective phases quite easy. The spectra in Fig. 7.3 demonstrate convincingly that the absorption-desorption process is by no means completely reversible. Full reversibility can only be achieved up to a few cycles (Fig. 7.3 curves A, B, C). After a larger number of cycles the material can no longer be completely dehydrided since stable EuH_2 has formed (curve E). On the assumption that La behaves identically to Eu, Cohen et al. explained the degradation as due to the formation of a stable hydride phase rather than via oxidation which has been only found to an extent of a few percent (curve D).

Oliver et al. [7.91] again made use of the favourable properties of the ^{151}Eu Mössbauer effect to study pure $EuNi_5$. Comparing several ternary Eu hydrides with different stoichiometries Oliver et al. concluded from the almost identical hyperfine field and isomer shift values (being close to that of EuH_2) that – as in $LaNi_5$ – the absorption process occurs primarily via the formation of the binary rare earth hydride.

In $CeNi_5$ *Malik* et al. [7.92] examined the substitution of Ni by Al and obtained an increase in stability of the hydride which on the other hand was accompanied by a drastic decrease of the absorption capacity. Susceptibility measurements yielded an increase of χ upon charging which has been attributed to the precipitation of Ni. The question about a change in valency of Ce from 4+ to 3+ upon hydrogenation remained unsolved. *Pedziwiatr* et al. [7.93] reported on a gradual change of the Ce valency from 4+ to 3+ in Cu substituted $CeNi_5$. Hydrogen uptake was found to further increase the Ce^{3+} content in the sample. In $CaNi_5$ *Yagisawa* and *Yoshikawa* [7.94] again observed a strong preference for the formation of Ni precipitates even in the parent compound. The behaviour of the bulk susceptibility upon hydrogen absorption is almost identical to that which has been described in detail for $LaNi_5$.

c) R_2TM_7 (Space Group $P6_3/mmc$ No. 194)

Magnetisation measurements for Y_2Ni_7 [7.95] and La_2Ni_7 [7.96] and their hydrides have been reported by Buschow. In Y_2Ni_7 surprisingly the Curie temperature rises upon hydriding. The magnetisation, however, decreases to about one half of its value in the parent compound. On the other hand, pressure experiments predict such a drop in T_c. Unexpectedly La_2Ni_7 is antiferromagnetic and becomes Pauli paramagnetic upon charging (Fig. 7.4) which Buschow explained by a reduced $5d$ density of states at E_F caused by the charge transfer between La and H. In contrast to La_2Ni_7, no broadening of the x-ray diffraction peaks after hydriding could be observed in the case of Y_2Ni_7. Thus a phase separation which could give rise to the enhanced value of T_c can be ruled out.

Buschow et al. [7.97] and *Buschow* [7.98] have also studied the hydrides of La_2Co_7 and of Ce_2Co_7. In contrast to the former case an increase in magnetic moment and Curie temperature has been obtained in the latter after hydrogen absorption. This was explained by Buschow in terms of a change in the valency of Ce from 4+ to 3+.

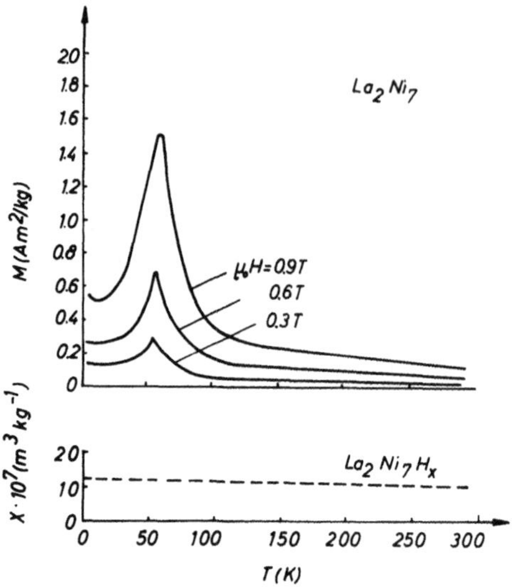

Fig. 7.4. Temperature dependence of the magnetisation (——) of pure La_2Ni_7 in different external fields and of the susceptibility (– – –) of the hydride $La_2Ni_7H_x$ [7.96]

d) R_2Fe_{14} B-Type Compounds

Very recently the development of an outstanding hard magnetic material based on Co-free *R*–Fe–B alloys was successfully demonstrated by rapidly quenched ribbons [7.99], by sintered magnets [7.100] and lastly by a ternary diffusion path [7.101]. Energy products of up to 45 MGOe have already been obtained. The principle source of the unique hard magnetic properties in Nd–Fe–B are the intrinsic properties of the tetragonal compound $Nd_2Fe_{14}B$ (space group $P4_2$/mnm, No. 136) which exhibits two Nd and six Fe sublattices [7.102–104]: the large saturation magnetisation and the high uniaxial crystal anisotropy ($\mu_0 M_S = 1.6$ T, $\mu_0 H_A = 7.5$ T) at room temperature. Moreover, the metallurgical possibility for the formation of small crystallites of the $Nd_2Fe_{14}B$ phase gives rise to a high coercivity $_IH_C$. The easy axis of magnetisation was determined to lie along the *c*-axis at room temperature, whereas below about 120 K a complex magnetic structure occurs. If another rare earth, with the exception of Eu and Yb, replaces Nd the crystal structure remains unchanged.

Due to the large number of inequivalent Fe lattice sites rather complicated Mössbauer spectra are obtained. In some cases, however, they could reasonably well be analyzed by considering the relative site occupancy of the Fe atoms in the unit cell [7.105–107].

Several of these compounds have already been examined with regard to their hydrogen absorption properties. For R = Nd up to 4 H atoms/f.u. were found to be dissolved, leading to a rise not only in the Fe moment [7.108, 109] but also in the Curie temperature [7.34, 108]. The Mössbauer spectra reveal not only an increase in B_{eff} but also the occurrence of a pattern characteristic of α-Fe upon charging with hydrogen. In cases where free metallic Fe is already present in the host compound this specific hyperfine pattern seems to exhibit an enlarged intensity after hydrogen absorption. For R = Dy, ^{161}Dy Mössbauer

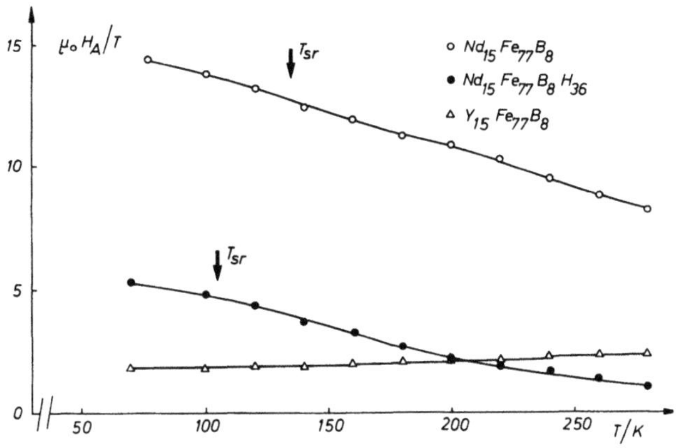

Fig. 7.5. Anisotropy field H_A vs temperature of $Nd_{15}Fe_{77}B_8$ (○), $Nd_{15}Fe_{77}B_8H_{36}$ (●), and $Y_{15}Fe_{77}B_8$ (△) [7.110]

studies have been carried out [7.106] which indicate a slight reduction of the Dy hyperfine field.

The temperature dependence of the anisotropy field H_A, measured by the so-called singular point detection technique, is displayed in Fig. 7.5 for $Y_{15}Fe_{77}B_8$, $Nd_{15}Fe_{77}B_8$, and $Nd_{15}Fe_{77}B_8H_{3.8}$ [7.110]. From comparing the data we deduce that for $T > 200$ K the contribution of the Nd sublattice to H_A is strongly reduced by hydrogen, leading to a drop of the global anisotropy of $Nd_{15}Fe_{77}B_8H_{36}$ even below that of $Y_{15}Fe_{77}B_8$. At low temperatures, however, the Nd contribution is estimated to still dominate the Fe sublattice contribution to H_A. A further indication for the persistence of the Nd anisotropy at low temperatures can be seen in the occurrence of the spin reorientation even in the hydride. It has been suggested that the shift of the reorientation temperature T_{SR} to lower values originates from the weakened Nd anisotropy upon hydrogen absorption.

Hydrogen absorption and desorption may provide an alternative to milling for disintegrating bulk Nd–Fe–*B* into a fine powder as required for the sintering process and also for the production of polymer bounded magnets. Since *Schlapbach* [7.34] has proved by means of XPS that $Nd_2Fe_{14}B$ shows a selective oxidation of Nd, it is suggested by analogy to $LaNi_5$ that in this case the surface decomposes into Nd_2O_3 and precipitates of Fe. These precipitations may even be large enough to be detected by ^{57}Fe Mössbauer spectroscopy. Up to now we have only obtained, in agreement with *Harris* et al. [7.111], coercivities comparable in magnitude with those from magnets produced by the standard milling and sintering technique. A further optimisation of the parameters during the hydriding process is required if hydrogen absorption is to supercede conventional methods of material pulverization.

e) R_6TM_{23} (TM = Mn, Fe)

These compounds, particularly the pseudobinary (Fe, Mn)-phases, exhibit several remarkable physical properties. Upon the absorption of hydrogen still

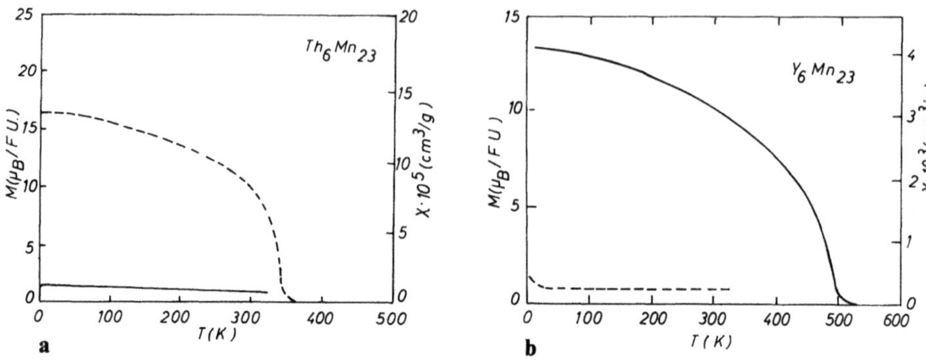

Fig. 7.6. (**a**) Temperature dependence of the magnetisation of Y_6Mn_{23} before (——, *left hand scale*) and after hydrogen absorption (– – –, *right hand scale*). (**b**) Temperature dependence of the magnetisation of Th_6Mn_{23} before (——, *right hand scale*) and after hydrogen absorption (– – –, *left hand scale*) [7.112]

further spectacular features appear (see e.g. [7.20]). All compounds crystallize in the fcc. Th_6Mn_{23} type of structure (space group Fm3m, No. 225), the unit cell consists of 116 atoms (1 *R* and 4 TM sublattices) and lattice constants of about 12 Å are commonly observed. Substantial amounts of hydrogen (deuterium) can be dissolved ranging from about 15 atoms/f.u. (Fe compounds) up to 30 (Th_6Mn_{23} [7.112–114]). Hydrogen pressures between 1 bar [7.112–113] and 130 bar [7.115] applied at temperatures between room temperature and 100 °C have been reported. No attempts are known to separate bulk from surface phenomena, as was done in case of $LaNi_5$.

The explanation of the fact that Y_6Mn_{23} apparently loses its magnetic order upon hydrogenation, while in the isostructural Pauli paramagnetic Th_6Mn_{23} magnetic ordering develops when hydrogen is dissolved (see Fig. 7.6), has been a matter of acute controversy. Various modifications of magnetic order have been claimed depending on the specific technique which has been applied i.e. magnetic measurements [7.112–114, 116, 117], neutron diffraction [7.118–125], or ^{57}Fe Mössbauer spectroscopy on Fe-doped samples [7.126].

According to the neutron diffraction results which have been reported most recently by *Hardman-Rhyne* et al. for $Y_6Mn_{23}D_{23}$ [7.121] and $Th_6Mn_{23}D_x$ ($x=16, 30$) [7.124], the current situation is the following: cubic Y_6Mn_{23} deuteride undergoes a crystallographic phase transition into a primitive tetragonal structure (P_4/mmm) at low temperature. In the parent intermetallic the Mn moments are ferrimagnetically coupled. Accompanied by the structural transition a magnetic transition also occurs at around 175 K. At still lower temperatures weak antiferromagnetic ordering of only some of the Mn moments is postulated which is in agreement with the interpretation of the Mössbauer study by *Stewart* et al. [7.126].

$Th_6Mn_{23}D_x$ only suffers a low-temperature distortion when $x=16$; $Th_6Mn_{23}D_{30}$ retains fcc. structure down to liquid helium temperature.

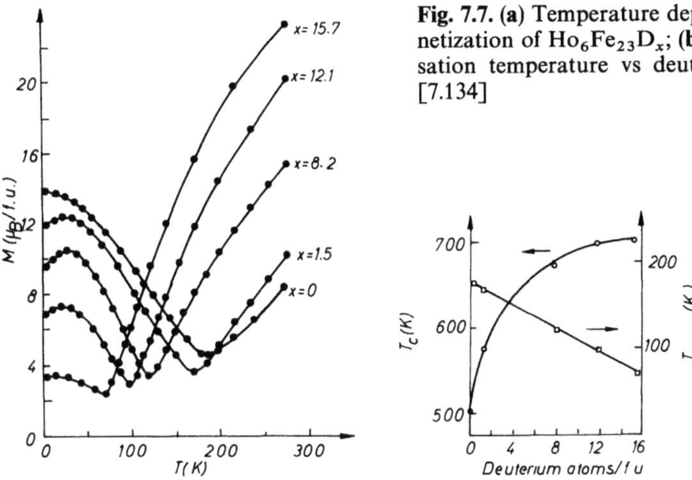

Fig. 7.7. (**a**) Temperature dependence of the magnetization of $Ho_6Fe_{23}D_x$; (**b**) Curie and compensation temperature vs deuterium concentration [7.134]

Whereas the former shows no long-range magnetic ordering even at 4 K, the latter exhibits almost ferromagnetism (only 4 out of 92 Mn spins point in the opposite direction). The difference in structural and magnetic properties of these two compounds is attributed by Hardman-Rhyne et al. to the altered arrangement of the deuterium atoms, to a deuterium induced charge transfer, and to a change in the band structure owing to the specific lattice expansion. The validity of a critical Mn–Mn distance for the formation of a magnetic moment, which has been introduced by *Buschow* and *Sherwood* [7.112], however, has been questioned. For details concerning the earlier investigations we refer to the reviews of *Buschow* [7.19, 20] and *Wallace* [7.3–7].

For the remaining R_6Mn_{23} compounds hydrogen absorption commonly leads to a drastic reduction of the Curie temperature and of the magnetisation [7.112, 127] which has sometimes even been claimed to have led to paramagnetism down to 4 K [7.128, 129]. Mössbauer studies on the rare earth nuclei R = Dy, Tm [7.130, 131], however, gave evidence that the small value of the hydride magnetisation had to be attributed to an antiferromagnetic arrangement of the R sublattice. From a Mössbauer study on ^{57}Fe-doped Er_6Mn_{23} *Stewart* et al. [7.132] concluded that only some of the Mn atoms had lost their moment after hydrogen absorption. The R moments were reported to have retained almost their free ion value upon hydrogenation, whereas the magnetic order of the Mn sublattice was substantially reduced. This suggests a pronounced weakening of the R–Mn exchange interaction in the R_6Mn_{23} hydrides.

Regarding the Fe compounds, the compensation points which are commonly observed for the parent intermetallics are shifted towards lower temperatures when hydrogen (deuterium) is dissolved [7.133, 134]. This is displayed in Fig. 7.7 taken from the study of *Pedziwiatr* et al. [7.134], where both the linear correlation between compensation temperature and absorbed deuterium as well as the substantial rise in T_c are also demonstrated. A reduction and an increase in the magnetisation can be observed below and above T_{comp} respectively [7.133–135]. From a neutron diffraction study *Rhyne* et al. [7.136] found that the R moments in $Ho_6Fe_{23}D_x$ remain close to their free ion value upon deuteration whereas the Fe moments strongly increase. The latter finding is confirmed by a recent Mössbauer study of *Pedziwiatr* et al. [7.137]. Similar results have been obtained for Tm_6Fe_{23} hydride by *Gubbens* et al. [7.138], while in the Lu case a decrease of the individual Fe hyperfine was reported by *Gubbens* et al. [7.139] to have occurred after hydrogen uptake. The reduction of B_{eff} was unexpectedly accompanied by a rise in the magnetic moment. This peculiar behaviour has been interpreted by the authors to arise from a hydrogen induced change in the conduction electron contribution to B_{eff}. At this point we want to recall that in these ferrimagnetic compounds, the R moment is predominant at low temperatures ($T < T_{comp}$) with the Fe moment dominating at higher temperatures. Thus a reduced magnetisation is not necessarily due to a diminished R moment. It can equally well be the result of an increase in the Fe moment. Besides the neutron study just mentioned [7.136] this has been ascertained by molecular field calculations, from which an enhancement of the Fe–Fe interaction and a pronounced weakening of the R–Fe interaction have been derived.

The pseudobinary system $(Y, Er)_6Fe_{23}H_x$ has also been examined by *Pedziwiatr* et al. [7.135] and was found to have similar properties to the two boundary compounds. More attention has been paid to the pseudobinaries $[Y_6(Fe, Mn)_{23}]$ where spectacular magnetic properties can be observed. Two critical concentrations for the onset of magnetic order have been obtained. While both end compounds exhibit Curie temperatures of about 500 K, there is no evidence for any long-range magnetic order in the concentration range $0.4 < x < 0.7$. From an analysis of their high pressure magnetisation data *Hilscher* et al. [7.140] suggested that these compounds form a spin glass system. This proposal was later confirmed by the neutron scattering experiments of *Lin* et al. [7.141]. An antiferromagnetic short-range order of atomic moments has been claimed with a correlation length of about 2.5 lattice constants. These results were ascertained by *Hardman-Rhyne* and *Rhyne* [7.120], but almost simultaneously questioned by *Crowder* and *James* [7.122]. In agreement with the neutron diffraction results, ^{57}Fe Mössbauer studies [7.142] revealed the complete absence of a magnetic moment on Fe for Mn concentrations exceeding $x = 0.3$. *Oesterreicher* and *Bittner* [7.143] were the first to perform magnetisation studies on various hydrides of this peculiar system. The general features of the host compounds remain almost unchanged, yet a shift to lower Mn concentrations is observed after hydrogen uptake. This has been interpreted as being due to the donation of electrons from hydrogen to the transition metals. By means of ^{166}Er and ^{57}Fe Mössbauer spectroscopy *Zukrowski* et al. [7.144] investigated $Er_6(Fe, Mn)_{23}H_x$ and came to the conclusion that the behaviour of the $3d$ magnetisation is almost identical to that of the corresponding Y series.

f) RTM$_3$ (TM = Fe, Co, Ni)

These compounds crystallize in the rhombohedral $PuNi_3$-type structure (space group $R\bar{3}m$, No. 166). *Buschow* [7.145] examined Ce-based compounds and found $CeCo_3$ and $CeNi_3$ to be Pauli paramagnetic. After hydrogen uptake $CeCo_3H_4$ had become ferromagnetic ($T_c = 80$ K) which is a unique behaviour for R–Co compounds and has been ascribed to a change in the valency of Ce from 4+ to 3+. This interpretation is suggested by the large hydrogen induced volume increase obtained by *van Essen* and *Buschow* [7.146]. The change in the valency of Ce upon hydrogenation can be more clearly observed in $CeNi_3$. The low value of the susceptibility and its temperature independence indicate tetravalent Ce. In the hydride, however, χ exhibits a pronounced temperature dependence, revealing an effective moment of 2.5 μ_B/Ce atom which corresponds to the 3+ state. In contrast to this, the ferromagnetic compound YNi_3 ($T_c = 85$ K) becomes Pauli paramagnetic after the absorption of about 4 H atoms/f.u. [7.147].

To date only compounds containing a heavy rare earth have been examined. In all cases ferrimagnetic behaviour has been detected, featuring compensation points at which the resultant of the rare earth and the Fe(Co) moments is equal to or near to zero. By studying the Co compounds *Malik* et al. [7.148] found for $GdCo_3$ an increase in magnetisation upon charging which they attributed to a decrease in the Co moment. In addition T_c was observed to be strongly reduced after the absorption of 4.6 H atoms/f.u. (a typical value for the maximum capacity of these Co-based compounds) which is also thought to be a consequence of the diminished Co moment, since T_c should be mainly determined by the Co–Co interaction.

For R = Dy, Ho [7.148], Er and Tm [7.149] hydrogen absorption leads to a reduction of magnetisation, Curie temperature and compensation point. In contrast to the host compounds, which can easily be saturated, the corresponding hydrides showed no sign of saturation in a field of $\mu_0 H = 2$ T (Fig. 7.8). A fanning out of the R moments as a function of temperature or applied field was given as an explanation for this lack of saturation and the steep variation in the magnetisation of the hydrides at low temperatures.

The magnetic properties of the parent RFe$_3$'s have been described in detail by *Herbst* and *Croat* [7.150] by applying a two-sublattice molecular field model. In the Mössbauer studies on the host intermetallics, due to the presence of a spin reorientation, the main effort was concentrated upon $ErFe_3$ [7.151–156]. When exposed to hydrogen up to 4 H atoms/f.u. are commonly absorbed [7.157–161]. The R moment decreases with the amount of hydrogen, while the Fe moment rises. As in case of the R_6Fe_{23} hydrides, the compensation points again decrease with x almost linearly. A weakening of the R–Fe exchange upon hydrogenation has thus been suggested as a likely explanation. Direct experimental evidence for this altered exchange interaction was later obtained by *Niarchos* et al. [7.160] by using both the ^{57}Fe and the ^{161}Dy (26 keV) Mössbauer effect in $DyFe_3$. While the Dy hyperfine field is reduced

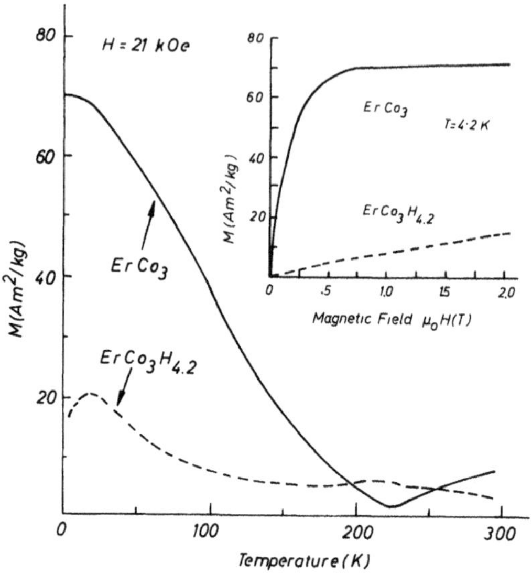

Fig. 7.8. Magnetisation vs temperature for $ErCo_3$ (——) and $ErCo_3H_{4.2}$ (- - -). In the insert the magnetisation at $T=4.2$ K is displayed as a function of the applied field [7.149]

with growing hydrogen concentration, the Fe hyperfine field increases as long as $x \lesssim 2.5$, but the changes are only rather small. On the other hand, a substantial reduction in T_{comp} has been obtained which showed a linear dependence on the volume expansion. This peculiar correlation led Niarchos et al. to the suggestion that the Dy–Fe exchange interaction is increasingly weakened by the presence of hydrogen in the lattice and moreover that it is primarily associated with the lattice expansion. Similar conclusions had already been drawn by *Niarchos* et al. [7.159] in an earlier ^{57}Fe and ^{166}Er (81 keV) Mössbauer study on some $ErFe_3$ hydrides.

Da Cunha and *Vasquez* [7.162] and *da Cunha* et al. [7.163] studied the influence of hydrogen uptake upon the spin reorientation process in $ErFe_3H_x$. The critical temperature was found to be drastically increased in the hydride. A value of 210 K has been given for $x=1.5$ which proved to be almost independent of further hydrogen absorption. These Mössbauer effect data are confirmed by magnetisation studies of *Malik* et al. [7.161] who observed a step in the magnetisation at a similar temperature.

In the case of the ferromagnetic compounds YFe_3 [7.164] and $ThFe_3$ [7.165] an increase of the Fe moment after hydriding was reported, whereas the Curie temperature remained unchanged in the former (545 K) but was found to decrease in the latter. The magnetisation anomaly at about 250 K observed in uncharged $ThFe_3$ [7.166] is again found in the hydride and has been attributed to the preservation of the coordination and of the Fe–Fe distances after the charging procedure. The results obtained by *Narasimhan* [7.158] for three (Er, Th)Fe_3 hydrides resembled the properties of the two end compounds.

Very recently *Höchst* et al. [7.167] reported on the investigation of YFe_2 and YFe_3 exposed to increasing doses of hydrogen, by means of synchrotron radiation; photoemission measurements revealed the absence of high *d*-like density of states at E_F for large exposures, which explains, according to the authors, the decrease of the magnetic moment and hyperfine field (*B*) which is commonly observed in the hydride for hydrogen concentrations larger than 3.5. The authors point out that their study concerns predominantly the surface. It should be mentioned that impurity segregation could have contributed significantly to the observed effects.

7.3.3 Laves-Phase-Type Compounds

These compounds occur predominantely either in the cubic $MgCu_2$ structure (space group *Fd*3m, No. 227) or in the hexagonal $MgZn_2$ structure (space group $P6_3$/mmc, No. 194).

a) Mn Compounds

RMn$_2$

Both types of structure are present here (see e.g. [7.44]). About 4 H atoms/f.u. can be absorbed whilst still preserving the crystal structure, thus leading to remarkable volume expansions of up to 40%. For still larger hydrogen concentrations the absence of x-ray diffraction peaks indicates at least microcrystallinity or even an amorphous nature of the sample [7.168]. *Buschow* [7.115] and *Buschow* and *Sherwood* [7.112] were the first to show that in $ScMn_2$ and in $LuMn_2$ a magnetic moment had developed on the Mn atoms after hydrogen absorption, leading to ferromagnetic order with a T_c of about 200 K. After cooling without an external magnetic field, the magnetisation vs temperature curves of $LuMn_2H_x$ exhibited a pronounced maximum which disappeared when the sample, prior to the measurement, was cooled in the presence of a magnetic field. This spin-glass-like behaviour was also obtained for YMn_2 hydride, whereas $ThMn_2$ was found to remain paramagnetic after having absorbed hydrogen. As is the case for the 6:23 compounds, here too a nonuniform behaviour is found upon hydrogen uptake when a nonmagnetic element is present. All the host compounds just mentioned were considered to be paramagnetic. Recently, however, *Nakamura* [7.169] found YMn_2 to be an antiferromagnet with a Mn moment of 2.7 μ_B aligned parallel to a [111] direction. This has been confirmed by a set of supplementary experiments: low temperature x-ray diffraction, magnetic and dilatometric measurements, neutron diffraction and spin-echo NMR. The Néel temperature was deduced to lie around 100 K and was found to be accompanied by an unusual large volume change of about 5%. This peculiar finding has been ascribed by Nakamura to the collapse of the Mn moments above T_N due to the strong spontaneous volume magnetostriction present in this material.

In $GdMn_2$ hydrogen uptake was found by *Buschow* and *Sherwood* [7.112] to reduce the magnetisation at low temperatures, whereas for temperatures above about 100 K an increase was obtained. This unique behaviour is explained by the authors in terms of an alignement of the Gd moments. This is thought to be caused by the Mn moments, which are assumed to order close to the Curie temperature of YMn_2 hydride ($T_c = 284$ K). From magnetisation studies on YMn_2H_x and $LuMn_2H_x$ with varying hydrogen content Buschow and Sherwood proposed the existence of a critical Mn–Mn distance for the formation of a local moment at a Mn site, as has already been suggested for the R_6Mn_{23} hydrides. A value of $d = (2.86 + 0.06)$ Å has been determined for both compounds, similar to that for $Th_6Mn_{23}H_x$, but substantially larger than that of $(Y, Lu)_6Mn_{23}H_x$. This was attributed to the different valence electron concentration in the respective compounds. However, as already mentioned in Sect. 7.3.3 the concept of critical distance has been questioned by neutron diffraction groups (see e.g. [7.124]. ^{161}Dy and ^{166}Er Mössbauer studies have been reported for $DyMn_2H_x$ [7.131] and $ErMn_2H_x$ [7.170] respectively. Magnetic ordering was found to have disappeared after hydrogen absorption. The magnetic hyperfine field at 1.5 K of about 500 T in $ErMn_2H_{4.6}$ was explained by intermediate relaxation rates. As has been mentioned earlier, the isomer shift values in the ternary hydrides are again found to lie close to those of the binary *R* hydrides.

$TiMn_2$

Hempelmann and *Hilscher* [7.171] found three Ti–Mn intermetallics to be stable at low temperature: the C14 Laves phase with a wide range of homogeneity reaching from $TiMn_2$ to $TiMn_{1.2}$, the ϱ-phase $TiMn_{1.17}$ and the ϕ-phase $TiMn_{1.08}$. While all parent compounds exhibit temperature independent paramagnetism, the three hydrides are ferromagnetic with Curie temperatures of about 212 K, 87 K, and 155 K respectively. The onset of ferromagnetism was attributed by Hempelmann and Hilscher to both volume expansion and an increase in $N(E_F)$ upon hydrogen absorption. In the context of the large homogeneity range of the Laves phase compound, an increase of the lattice constants was observed upon increasing the Ti concentration. Simultaneously the hydrogen capacity was found to grow, whereas stoichiometric $TiMn_2$ absorbs no hydrogen at all. Thus the existence of a critical range of the lattice dimensions for the absorption of hydrogen has been suggested which is in agreement with results obtained by *Oesterreicher* and *Bittner* [7.172] on $(Ti, Zr)Mn_2$. However, this may not be a general result.

Mössbauer studies performed by *Hempelmann* et al. [7.173] on $TiMn_{1.5}$ doped with ^{57}Fe yielded an asymmetric quadrupole doublet with a comparatively small splitting which, however, was strongly enhanced after hydrogenation (Fig. 7.9). From inspection of the figure it is obvious that upon hydrogenation the shape of the spectrum has changed to complete symmetry with narrow lines. The asymmetric doublet of the parent compound was attributed by Hempelmann et al. to a non-random distribution of the excess Ti atoms on the 2*a* sites, which was later confirmed by neutron diffraction experiments [7.174]. The presence of only one hyperfine pattern after

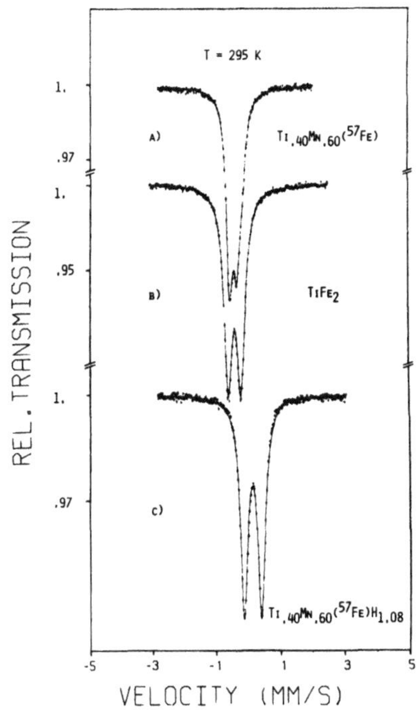

Fig. 7.9. Mössbauer spectra, recorded at room temperature for $Ti_{0.40}Mn_{0.60}(^{57}Fe)$ (*A*), $TiFe_2$ (*B*), and $Ti_{0.40}Mn_{0.60}(^{57}Fe)H_{1.08}$ (*C*) [7.173]

hydrogenation was explained by the authors as originating from the large lattice expansion ($\Delta V/V = 27\%$) which should release the space restriction for the (Ti, Mn) tetrahedra. Thus reasons for lattice distortions are no longer provided and all Fe atoms are now assumed to experience the same EFG, very similar to that one in the isostructural $TiFe_2$. It should be pointed out that an alternative explanation for the occurrence of narrow line widths after hydrogen absorption can be given by taking into account motional narrowing effects caused by rapid fluctuations of the hydrogen atoms.

In this work Hempelmann et al. also made an attempt to determine the type of ferromagnetism in the ternary Ti–Mn hydrides. They used the so-called Rhodes-Wohlfarth plot in order to classify ferromagnetic order into localized or itinerant type. We recall that the crucial parameter in this model is the ratio of magnetic carriers in the paramagnetic and in the ferromagnetic states, q_c/q_s as a function of the Curie temperature. In terms of this analysis the Ti–Mn Laves phase turned out to exhibit localized moments which is confirmed by the fact that the Arrott-plots (σ^2 vs B/σ) are not parallel. In the case of the ϱ- and the ϕ-phases however, the rather large values of q_c/q_s with the corresponding Curie temperatures fit well into the itinerant branch of the Rhodes-Wohlfarth plot, which led Hempelmann et al. to the conclusion that these compounds might exhibit pure band ferromagnetism.

The authors also commented on the onset of ferromagnetism upon hydrogenation of the Laves phase $TiMn_{1.5}$, by considering the large increase of the temperature-independent term of the paramagnetic susceptibility as well as of the specific heat coefficient γ observed after hydrogen absorption. This finding has been correlated with an increased density of states at the Fermi level $N(E_F)$ which continues to grow in the hydride when the amount of hydrogen is further increased. Within the framework of this simple picture the high value of $N(E_F)$ in the hydride was interpreted by Hempelmann et al. as one reason for the onset of ferromagnetism (Stoner criterion). Since the lattice expansion caused by hydrogen absorption brings about a narrowing of the *d*-band and thus an increase of $N(E_F)$, the interpretation of the YMn_2 results by *Buschow* and *Sherwood* [7.112] emphasizing this lattice expansion, was adopted by Hempelmann et al. within their model.

$ZrMn_2$

In a similar way to $TiMn_2$, the isostructural $ZrMn_2$ displays a wide homogeneity range, shifted however, to a larger Mn content, i.e. from $ZrMn_{1.8}$ [7.175] up to $ZrMn_{3.8}$ [7.176]. Such compounds exhibit favourable hydrogen absorption properties and compositions close to $ZrMn_{2.4}$ are particularly useful, since their equilibrium pressure is around 1 bar, whereas in stoichiometric $ZrMn_2$ it is two orders of magnitude lower [7.176].

Several authors [7.176–178] agree that upon hydrogen absorption Pauli paramagnetic $ZrMn_{2+x}$ becomes a spin glass; this was also found in the case of several multi-elemental storage compounds, where Mn had been partly replaced by Fe, Cr, or Co [7.179]. The reason for this probably lies in the presence of strong segregations which have been detected by means of XPS [7.56].

Substantial discrepancies in the magnetic ordering temperatures of the hydrides can be found in the literature which we attribute primarily to the different hydrogen pressures used during the charging procedure. Further reasons may be different stoichiometries and hydrogen concentrations. A recent neutron diffraction study performed by *Didisheim* and *Fischer* [7.180] on $ZrMn_2D_3$ demonstrated that the magnetically ordered state in this compound is in fact of great complexity.

b) Fe Compounds

Numerous investigations have been carried out on the RFe$_2$'s and their hydrides; this is due to their comparatively simple crystallographic and magnetic structure. Thus results from experiments such as neutron diffraction or Mössbauer spectroscopy can be interpreted more easily than those of transition-metal-rich compounds. Since for the a magnetic rare earths, magnetisation measurements are insufficient to separate the *R* from the Fe moment, the above experiments provide an ideal method to examine the influence on both kinds of atoms on the magnetic moment.

Apart from $ScFe_2$ (C14) all parent compounds display the cubic C15 structure. Stable hydride phases with ~ 2 and ~ 3.5 H(D) atoms/f.u. are known from x-ray and neutron diffraction to retain the parent structure with lattice parameters increased by ~5% and ~7% respectively. For a still larger hydrogen concentration (RFe_2H_4) a rhombohedral distortion [7.181] or even the distruction of the long-range lattice periodicity [7.97] has been detected. Thus the Mössbauer spectra of the hydrides sometimes exhibit a certain line broadening, predominantely when recorded at low temperatures which alternatively may be due to intermediate jump rates of the H atoms.

When the partner element of Fe does not carry a magnetic moment (Sc, Y, Ce, Lu) a substantial increase in magnetisation, ^{57}Fe hyperfine field and isomer shift has been observed [7.84, 97, 165, 182–185]. For YFe_2 *Buschow* and *van Diepen* [7.182] found the Curie temperature to have decreased upon hydrogenation, while in $CeFe_2$ hydride they observed a rise in T_c [7.165]. The largest increase in the magnetic moment has been detected in $ScFe_2H_x$ which was also found to exhibit an appreciable homogeneity range. *Grössinger* et al. [7.186] studied the influence of misplaced atoms upon the magnetic properties of non-stoichiometric parent compounds using pulsed fields and Mössbauer spectroscopy. An easy *c*-axis was reported and a drastic reduction of the anisotropy field with increasing Fe content in the sample. A strong increase in the Fe moment from 1.4 to 2.2 μ_B and in the ^{57}Fe hyperfine field from 24 to 30 T after hydriding has been observed by *Smit* and *Buschow* [7.184] which, however, was found to be less pronounced when an Fe excess is present in the sample [7.185]. This peculiar finding has been attributed to a different shielding of the Fe sublattice from the Sc sublattice by the H atoms in the case of an altered amount of Fe in "$ScFe_2$".

From the point of view of magnetic behaviour, the heavy RFe_2 hydrides have been particularly thoroughly investigated. The question about the influence of hydrogen upon the magnetic moments can be readily answered, because in this case decisive neutron diffraction results are now available. No such reliable conclusions, however, can be drawn from magnetisation or Mössbauer measurements. On the one hand, saturation is scarcely achieved [7.168, 187–189] and thus great discrepancies in the magnetic moments can be found in the literature; on the other hand particularly the ^{57}Fe Mössbauer spectra exhibit a substantial line broadening and wide hyperfine field distributions after hydrogenation [7.168, 181, 182, 188–192]. This brings about difficulties especially in determining the magnetic ordering temperature from the onset of the magnetic hyperfine splitting. One can, however, rely on the sublattice moments derived from liquid helium temperature spectra.

As a typical example from the neutron diffraction studies of *Fish* et al. [7.193] and *Rhyne* et al. [7.194] we choose $ErFe_2D_x$ (Fig. 7.10), although it has already been presented in the latest review of *Buschow* [7.20]. It represents, however, a case in which hydrogenation actually causes different ordering temperatures for the sublattice magnetisation. From an inspection of Fig. 7.10 the reduction of the overall Curie temperature on hydrogenation by as much as 50% of the host value can be deduced. The magnitude of the Fe sublattice magnetisation is sometimes observed to remain constant; in case of

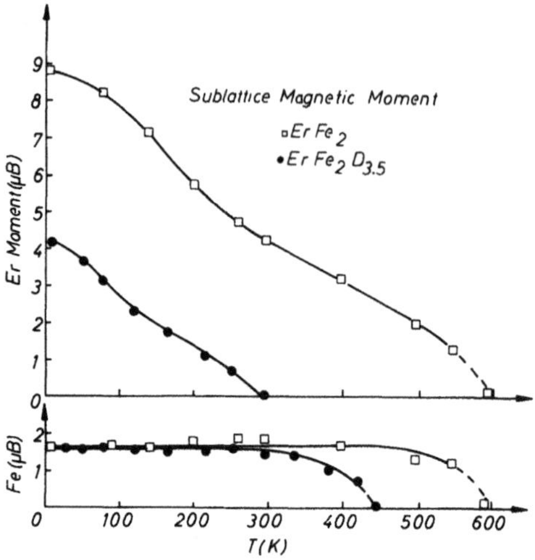

Fig. 7.10. Magnetic moment of Er and Fe vs temperature in $ErFe_2$ (□) and $ErFe_2D_{3.5}$ (○) [7.194]

$HoFe_2H(D)_x$ an increase in the Fe moment with x can be detected. Yet the flat temperature dependence of the Fe sublattice magnetisation is essentially unaffected by hydrogen absorption in each case. As already mentioned in the previous section, the rise in the magnetic moment of Fe upon hydrogen uptake is only observed for $x < 3.5$. This has likewise been demonstrated for some RFe_2 hydrides by *Dunlap* et al. [7.190] who obtained a breakdown of the ^{57}Fe hyperfine field in $ErFe_2H_{4.1}$.

The effect on the rare earths is more pronounced: the 0 K moment in the hydride is remarkably reduced from its free-ion value in the parent compound. Moreover the moment declines rapidly with elevated temperature, sometimes reaching zero well below the overall Curie point. This is in contrast to the behaviour in the pure intermetallics, where both sublattice moments disorder at identical temperatures. These features have led to the conclusion that a severe weakening of the *R*–Fe as well as of the *R*–*R* exchange interaction takes place upon hydrogen absorption. The hypothesis of a reduced *R*–Fe exchange is experimentally verified by the lowering of the compensation points with hydrogenation [7.188, 192] which has already been referred to for the cases of the R_6Fe_{23} and RFe_3. The relative insensitivity of the Fe–Fe exchange upon hydrogen uptake has been explained by a nearest-neighbour direct overlap exchange which should only be slightly influenced by the presence of hydrogen atoms. These suggestions, however, are inconsistent with certain results from magnetic measurements where the simple antiparallel arrangement of *R* and Fe spins is assumed to be retained in the hydrides as was been proposed prior to the neutron diffraction work [7.195]. More probably a "fanning" of the loosely coupled *R* moments takes place, arising from random local anisotropies [7.181, 188, 190, 194]. This too would explain the difficulties experienced in saturating the RFe_2 hydrides and the discrepancies in the magnitude of the Er moment as determined from neutron diffraction and from Mössbauer spectroscopy [7.188].

Very recently the question concerning the reduction of the R moment in the hydrides has be taken up again by *de Saxce* et al. [7.188]. They examined $ErFe_2H_x$ with variable hydrogen content by means of high field magnetisation measurements and ^{167}Er pulsed spin echo NMR. The R moment was found to be substantially reduced only for $x > 3$. For lower hydrogen concentrations the ferrimagnetic structure seemed to have been retained. For $x > 3.3$ no evidence for saturation at 4.2 K could be obtained even in a field of 16 T. This finding and the field dependence of the susceptibility has been attributed by de Saxce et al. to the tendency to ferromagnetic alignement of the Er and Fe sublattices rather than to the fanning of the R moments.

Cohen et al. [7.168] collected isomer shift data for several $DyTM_2$ hydrides. Once more the common trend could be shown: a considerable reduction of the Dy isomer shift with rising hydrogen content in the sample corresponding to a decrease in the *s*-electron density at the nuclear site. This reaches values which almost coincide with that of the binary Dy hydride.

c) Fe-Containing Pseudobinaries

The group of *Shaltiel* initiated the work on Zr-based pseudobinaries [7.196–199]. Various combinations have been examined in order to study the influence of alloying upon the formation and properties of the hydrides. Several concentration-dependent C15–C14 (and C14–C15) changes of structure have been observed which, after hydrogen absorption, were found to have been preserved.

More recently extensive investigations have been started by the Pittsburgh group on $Zr(Fe, Mn)_2$ like systems, where it has been established that particularly the hyperstoichiometric compounds can serve as hydrogen storage materials [7.178, 179, 200–205]. In pure $Zr(Fe, Mn)_2H_x$ the magnetic order is enhanced upon hydrogenation in the whole composition range. This result obtained from bulk magnetic measurements is confirmed by a recent ^{57}Fe Mössbauer study, where a substantial increase in B_{eff} has been observed [7.206]. Similar results have been reported on $Zr(Fe, Cr)_2$ [7.177, 179, 196]. In the case of $Zr(Fe, V)_2$ hydrogen absorption leads to an enhancement of ferromagnetism in the Fe-rich region only, whereas a suppression of superconductivity was found in the V-rich range [7.207] (see also Sect. 7.4). By contrast, in $Zr(Fe, Al)_2$ ferromagnetism is always strongly suppressed on hydrogenation [7.178, 204, 208]. Provided that the hydrides just mentioned exhibit magnetic ordering spin-glass-like behaviour is commonly observed.

^{59}Co and ^{55}Mn NMR and ^{57}Fe Mössbauer spectroscopy have been applied by *Fujii* et al. [7.209] and by *Okamoto* et al. [7.210] to examine $Y(Fe, Co)_2H_x$ and $Y(Fe, Mn)_2H_x$ respectively. Their results are summarized in Fig. 7.11. In the former case for Co concentrations $x > 0.1$ a common decrease of Fe and Co moment upon hydrogen absorption has been quoted. In the latter case a hydrogen induced enhancement of the Mn moment has been deduced from NMR frequency shifts. Competing coupling tendencies between the magnetic

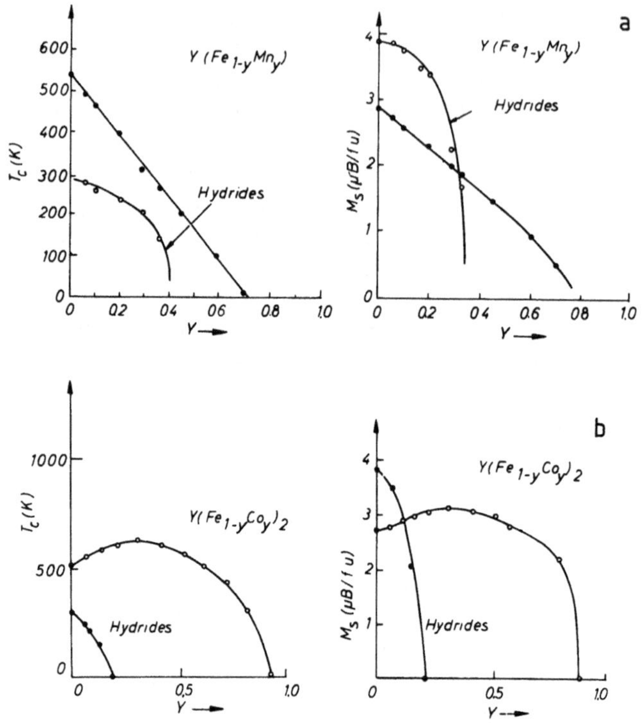

Fig. 7.11a, b. Magnetisation and Curie temperature vs concentration y of (**a**) $Y(Fe_{1-y}Co_y)_2$ (○) and their hydrides (●) [7.209]. (**b**) $Y(Fe_{1-y}Mn_y)_2$ (○) and their hydrides (●) [7.210]

atoms were suggested as an explanation for the spin-glass-like magnetism in the Mn-rich hydrides. *Oesterreicher* and *Bittner* [7.211] reported to have found some indications of partial amorphicity in this system.

Some $R(Fe, Co)_2$ hydrides have been studied by *Pourarian* et al. [7.212, 213], while *Sankar* et al. [7.214] reported on $Er(Fe, Mn)_2H_x$. The renowned arguments which have been applied to the ternary hydrides are again used on this occasion. The verification of the postulate by Pourarian et al. that the local moments residing at the Fe sites are retained in the Co-rich range should be taken as a challenge to Mössbauer spectroscopists. By means of EPR studies *Drulis* et al. [7.215] showed that in hexagonal GdFeAl the magnetic moment increases on hydrogen absorption up to the free-ion value of Gd^{3+}. This was ascribed to a complete randomization of the Fe moments due to the weakened Gd–Fe exchange. Their assumption of an ordered Gd sublattice at low temperatures seems to be questionable; the breakdown of the Fe–Fe exchange interaction, however, is strongly supported by Mössbauer studies on several *R*FeAl hydrides [7.216] which, even at 4.2 K, yielded almost no magnetic hyperfine splitting.

d) Compounds of Co, Ni and Others

In case of the Laves phases containing Co, which are all of the cubic C15 structure, the customary reduction of magnetic ordering after hydrogen absorption is obtained. This has been demonstrated by *Buschow* [7.217] for $GdCo_2H_x$ and by *de Jongh* et al. [7.218] for $PrCo_2H_x$. Clusters of free Co were found to give rise to spin-glass behaviour. This was shown by susceptibility measurements [7.218] and confirmed by *Buschow* and *van der Kraan* [7.219] in a study of ^{57}Fe-doped YCo_2H_4 and $HoCo_2H_4$. In the Mössbauer spectra a hyperfine pattern could be detected which almost coincided with that of Fe impurities in Co metal. For $GdNi_2$ *Malik* and *Wallace* [7.220] found a loss of the long-range atomic order upon hydrogen uptake which was accompanied by a drastic decrease in both the magnetisation and the Curie temperature. This trend, which commonly is observed in Ni compounds, has again been observed by *Drulis* et al. [7.215] who performed EPR measurements on $GdNiAlH_x$.

As already mentioned in Sect. 2.1, Gd compounds containing a non-magnetic metal make it possible to study directly the influence of hydrogen upon the *R–R* coupling. This has been done by using the 86.5 keV ^{155}Gd Mössbauer transition in $GdCu_2H_x$ [7.48] and in GdM_2H_x (M = Ru, Rh) [7.47]. Upon hydrogenation the Curie temperature behaves very differently in the two cases: a strong rise is observed in the former case and a decrease in the latter. This peculiarity has to be seen in the light of the influence of hydrogen on the oscillatory RKKY-type interaction present in these compounds. *Graaf* et al. collected numerous isomer shift and hyperfine field data for binary Gd compounds and some of their hydrides [7.221]. Hydrogen absorption commonly yields a substantial increase in isomer shift which is consistent with a reduction in *s*-electron density at the Gd nuclei. This indicates a charge transfer from Gd to hydrogen, a behaviour which is generally observed when rare earth

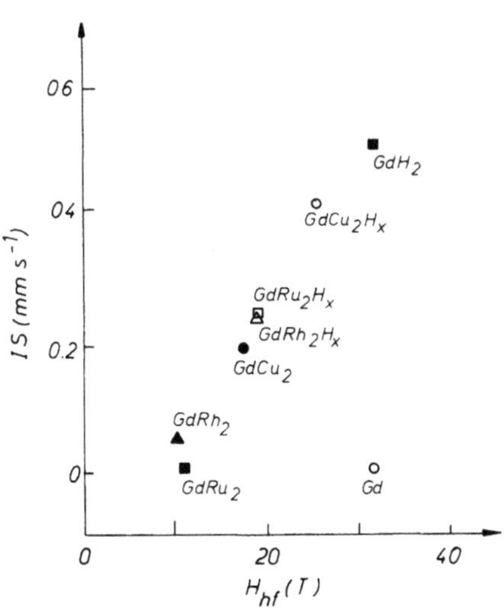

Fig. 7.12. Correlation between ^{155}Gd isomer shift (relative to ^{115}Eu : Pd) and hyperfine field in various intermetallic compounds GdM_2 and their hydrides [7.48]

Mössbauer data are considered. For several GdM_2 compounds and their hydrides Graaf et al. also claimed an almost linear correlation between Gd isomer shift and Gd hyperfine field (Fig. 7.12). The positive conduction electron contribution to the Gd hyperfine field is seen to be increasingly reduced with the (*s*-like) charge density at the nuclear site. This leads eventually to the semimetallic GdH_2, from which *Buschow* [7.20] derived the negative core contribution to B_{eff}.

Very dissimilar hydriding behaviour has been obtained for two Eu compounds: whereas in $EuRh_2$, Eu is found to change its valency ($3+ \rightarrow 2+$) upon hydrogen absorption [7.222, 223], it remains divalent in $EuMg_2H_x$ [7.91]. The hyperfine parameters of the hydrides were found to approach those of EuH_2; this has been attributed to significant clustering. *Huang* et al. [7.224] reported that the absorption of hydrogen had destroyed magnetic order in the weak itinerant ferromagnets $Ti(Be, Cu)_2$ and $ZrZn_{1.9}$. This has been explained by the influence of hydrogen upon the peculiar band structure present in these materials. By proton NMR studies combined with susceptibility measurements *Zogal* et al. [7.225] recently studied Fe_2P-type $ThNiAlH_x$ and $UNiAlH_x$. The existence of two antiferromagnetic phases has been claimed for the latter.

7.3.4 TiFe and Related CsCl-Type Compounds (Space Group Pm3m, No. 221)

Together with $LaNi_5$, TiFe belongs to the compounds which have most frequently been investigated with respect to their hydriding behaviour. When, more than ten years ago, *Reilly* and *Wiswall Jr.* [7.226] first identified the three different hydride phases, an enormous interest in TiFe hydrides developed. The main reason of course was the compound's excellent storage properties which promised useful applications.

Very soon Mössbauer spectroscopy was included in these investigations: by comparing transmission with back scattering spectra (using conversion electrons – CEMS) *Ron* et al. [7.227] were able to show that, after hydrogenation, Fe precipitates had formed close to the surface. Subsequently *Hempelmann* and co-workers [7.228, 229] studied TiFe during hydrogen treatment by means of susceptibility measurements. Although almost temperature independent to begin with, the susceptibility was found to increase substantially with rising numbers of charge-discharge cycles. Finally the sample behaved as a superparamagnet, a fact which has been attributed to Fe-rich precipitates that form magnetic clusters. *Hempelmann* and *Wicke* [7.229] supported this assumption by applying a simple band model from which they deduced that charged TiFe should not be stable in a well-ordered structure, but that the formation of Ti and Fe clusters is favoured.

Nevertheless the finding of Ron et al. was confirmed by *Shenoy* et al. [7.230] and by *Bläsius* and *Gonser* [7.231] who repeated the comparative transmission/CEMS study. From the latter work we have taken the surface emission spectrum of an activated TiFe disc which is displayed in Fig. 7.13; the α-Fe hyperfine pattern is clearly visible, whereas it is completely absent in a

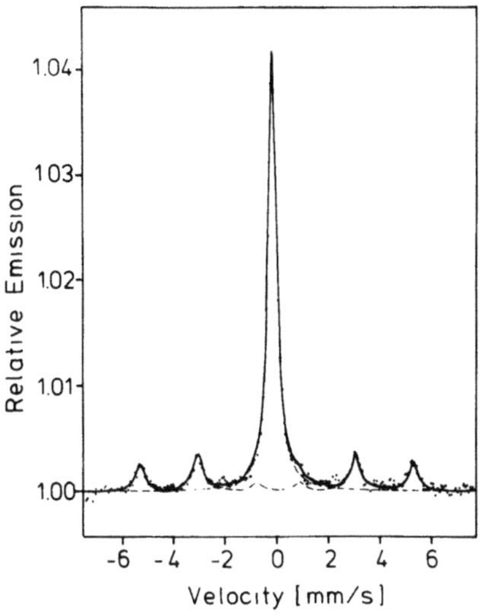

Fig. 7.13. Conversion electron Mössbauer spectrum of a hydrided TiFe disc recorded at room temperature [7.231]

transmission spectrum. This distinctly points to the fact that the clusters form predominantely at the surface ($d < 1000$ Å).

The group of *Schlapbach* performed several, chiefly surface-sensitive experiments on the activation behaviour of TiFe [7.73, 89, 232–237] described also in the comprehensive review of *Schlapbach* and *Riesterer* [7.238]. For uncharged TiFe, neutron scattering studies indicated complete CsCl order which was found to diminish only insignificantly upon charging. The increase in the magnetisation after hydrogen absorption was concluded to originate from ferromagnetic precipitates at the surface, in agreement with the Mössbauer results. Furthermore several electron microscopy studies have been carried out and have yielded somewhat contradictory results. *Von Waldkirch* et al. [7.239] reported Fe precipitates at the surface of TiFe, whereas *Schober* and *Westlake* [7.240] and *Khatamian* et al. [7.241, 242] claimed a decomposition into different oxides and $TiFe_2$. These discrepancies have been attributed by *Reilly* and *Reidinger* [7.243] to the different preparation techniques employed.

The different hydride phases detected by *Reilly* and *Wiswall, Jr.* [7.226] were found by *Swartzendruber* et al. [7.244] to give rise to specific hyperfine patterns: the cubic α-phase exhibits a single line spectrum and for the orthorhombic β- [7.233, 245, 246] and the monoclinic γ-phase [7.247] quadrupole split doublets were obtained. Later on this was confirmed by *Schäfer* et al. [7.246]. As commonly observed in Fe compounds, a distinct increase in isomer shift with hydrogen content is observed which recently has been interpreted by *Gupta* [7.55] using the augmented plane wave (APW) method. We recall that the negative isomer shift of TiFe relative to α-Fe (-0.14 mm/s) is cancelled in case of the β-hydride. For the γ-phase a further increase in isomer shift ($+0.27$ mm/s) is observed. Both values are in close agreement with the results derived by Gupta.

Mizuno and *Morozumi* [7.248] studied Ti_yFe ($0 \leq y \leq 1$) alloys and correlated a minimum in the Mössbauer line intensity with the maximum of the storage properties for $y \sim 1.15$ due to a weakening of the Ti–Fe binding forces. In our opinion the change in intensity is more likely to be due to the disturbed structure in the nonstoichiometric alloys which should strongly influence the f-factor.

If Co is alloyed to TiFe, a system is formed which exhibits unique magnetic properties thus and has been the subject of numerous investigations: while the boundary compounds TiFe ($y=0$) and TiCo ($y=1$) are Pauli paramagnets, ferromagnetism is observed in an intermediate concentration range ($0.25 < y < 0.65$) [7.41, 251]. Mössbauer measurements by *Bennett* and *Swartzendruber* [7.250], however, showed only a single line spectrum, giving no evidence for the existence of a localized moment at the iron site. Subsequently NMR studies on all nuclei present [7.251–253] were applied to elucidate this peculiar finding. The Knight shift results were explained by a greater s-admixture and a greater d-spin moment at the Fe sites than at the Co sites. A change of the sign of the hyperfine coupling constant was then presumed which should lead to a cancellation of B_{eff} in the intermediate concentration range [7.253]. On the other hand *Pickart* et al. [7.254] reported on polarized neutron measurements from which a moment of $0.15\,\mu_B$ for the (Fe, Co)-sublattice was derived for the Co concentrations $y=0.45$, 0.50, and 0.55.

More recently, several theoretical [7.255, 256] as well as experimental studies [7.41, 140, 257–261] on Ti(Fe, Co) have demonstrated the existence of so-called antistructure (AS) atoms, i.e. misplaced Fe(Co) atoms on the Ti sublattice. Since in such a case an Fe atom is surrounded by 8 Fe(Co) neighbours a local magnetic moment at these atoms may form. *Giner* and *Gautier* [7.255] were the first to point out the remarkable difference of the density of states (DOS) between atoms on correct sites and AS atoms; they showed this by solving the tight binding Hamiltonian using the coherent potential approximation (CPA). From the position of the Fermi level in the high-density region of the As–Fe and AS–Co DOS Giner and Gautier concluded that these atoms could indeed play an important role in the onset of ferromagnetism as observed by magnetic measurements. By performing KKR–CPA calculations *Schadler* and *Weinberger* [7.256] showed recently that Fe–AS atoms are responsible for the occurrence of magnetic order since the e_g and the t_{2g}-like virtual bound states of the Fe–AS atoms were found to lie very close to the Fermi energy with an exchange splitting much larger than that of the Co atoms.

By studying magnetovolume effects of some off-stoichiometric (Fe, Co) compounds *Buis* et al. [7.258] concluded that the ferromagnetic behaviour has to be ascribed to the presence of Fe(Co) atoms in a highly susceptible matrix. The pressure dependence of the spontaneous magnetic moment and of the Curie temperature are shown in Fig. 7.14, where the pressure derivatives of T_c and of σ_0 are plotted as a function of T_c. The diverging pressure derivative at decreasing T_c values obtained for the Co-rich stoichiometric and for the non-

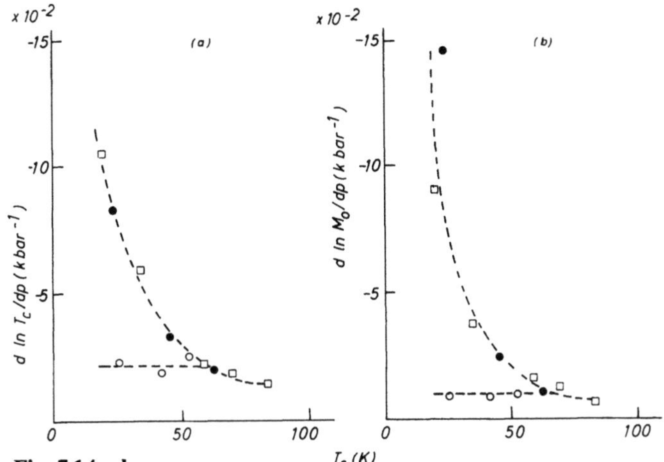

Fig. 7.14a, b

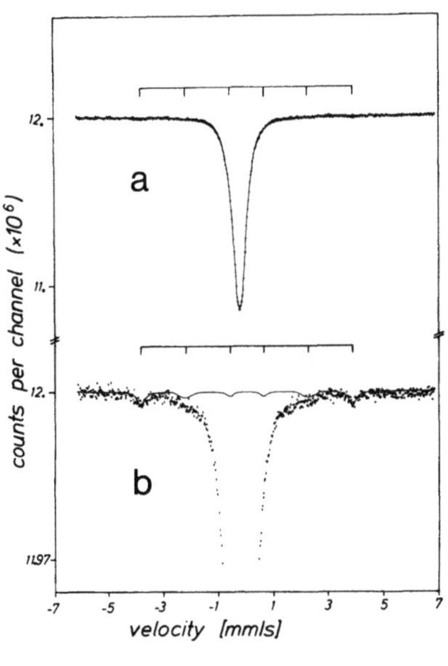

Fig. 7.14. Pressure derivatives of the Curie temperature (**a**) and magnetisation (**b**) of stoichiometric Ti(Fe, Co) compounds (●, Co-rich; ○, Fe-rich) and off-stoichiometric $Ti_z(Fe_{0.5}Co_{0.5})_{1-z}$ compounds (□) [7.260]

Fig. 7.15. Mössbauer spectrum recorded at 4.2 K for $Ti(Fe_{0.6}Co_{0.4})$; the bars on top of the spectrum indicate the line positions [7.41]

stoichiometric compounds clearly point to band-type magnetic behaviour. The constant values in case of the Fe-rich stoichiometric compounds can be explained by the presence of local magnetic moments at Fe–AS atoms.

Until recently, however, no direct experimental evidence for AS atoms has been presented; this is connected with the apparent puzzle regarding a zero hyperfine field within the ferromagnetic region. By Mössbauer measurements with an extended measuring time yielding count rates of about 12 million per channel *Hilscher* et al. [7.41] finally succeeded in resolving a weak magnetic hyperfine pattern which is otherwise buried in the background (see Fig. 7.15).

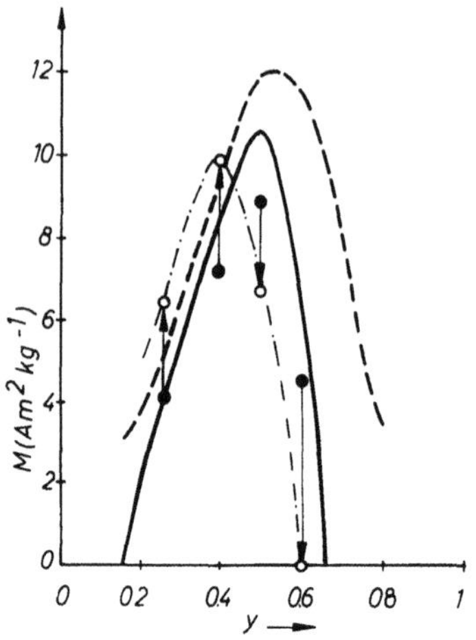

Fig. 7.16. Magnetisation vs concentration of $Ti(Fe_{1-y}Co_y)$ (●) and of some α-phase hydrides (○); the broken line represents the data obtained in a field of 7 T [7.41]

The intensity and splitting are in good agreement with the values predicted from bulk magnetic measurements for the Fe–AS atoms.

If the pseudobinary compounds are hydrided they behave very similarly to TiFe [7.41, 261, 262]. Again α-, β-, and γ-phases are found which exhibit equivalent Mössbauer spectra to those observed for TiFe by *Swartzendruber* et al. [7.244]. *Hilscher* et al. [7.41] tried to interpret the rise in T_c upon hydrogenation as a negative pressure effect by comparing their results with high pressure magnetisation data [7.140, 263, 264]. This, however, could successfully be done only for low Co concentrations (small changes of T_c). For higher Co content and larger increase in T_c ($\sim$40 K) no agreement could be obtained at all.

Here we want to recall that pressure experiments clearly point to the presence of localized moments in the Fe-rich regime, whereas for the Co-rich side of the system they point to itinerant ferromagnetism. The change of T_c and σ upon hydrogenation is positive for $y<0.45$ (Fe-rich) and negative for $y>0.45$ (Co-rich) (see Fig. 7.16). Therefore, not even a qualitative correlation between the hydrogen induced change of T_c and its pressure dependence can be expected for the Co-rich compounds. However, with the assumption that additional AS atoms are created upon hydrogenation the magnetic properties of the Ti(Fe, Co) hydrides could be explained at least qualitatively: the number of Fe–AS atoms carrying the local moments decreases, while $N(E_F)$ increases with the amount of Co. The occurrence of two critical concentrations for the onset of ferromagnetism ($y=0.25$ and $y=0.65$) is suggested to be a consequence of these two competing phenomena. The authors concluded that despite the presence of surface segregation the dominant feature responsible for the complex magnetic behaviour of this system is the presence of local moments which trigger the long-range magnetic order.

The paramagnetic behaviour of both the β- and the γ-phase was attributed by *Hilscher* et al. [7.41] to the fact that the formation of Fe–AS atoms is impossible in these non-cubic compounds. By performing proton spin-lattice relaxation studies on $TiCoH_x$ *Bowman* et al. [7.265] very recently emphasized the sensitivity of the position of E_F to hydrogen absorption. Additional studies of this kind in, combination with band structure calculations would give a deeper insight into this complex problem.

7.3.5 Compounds of Low Transition Metal Content

Mg_2Ni ($CuAl_2$ type of structure) belongs to the group of suitable hydrogen storage materials. By applying surface-sensitive techniques (AES, XPS, FMR) *Schlapbach* et al. [7.266], *Stucki* and *Schlapbach* [7.73] and *Shaltiel* et al. [7.89] could demonstrate that after continuous hydriding a decomposition at the surface takes place. As in $LaNi_5$ (see Sect. 7.3.2) this gives rise to the formation of superparamagnetic clusters. The bulk susceptibility, however, decreases after hydrogen absorption. Very recently *Aubertin* et al. [7.267] reported on Mössbauer studies on isostructural Fe-doped Zr_2Ni hydrides. The large hydrogen induced rise in isomer shift (+0.58 mm/s) has been attributed largely to charge transfer effects.

Eu_2IrH_5 has been assigned by *Moyer Jr.* and *Lindsay* [7.268] to the space group Fm3m (No. 225) and is claimed to order ferromagnetically below 20 K, as has recently has been corroborated by an ^{151}Eu Mössbauer study of *Stadnik* and *Moyer Jr.* [7.269]. Remarkable similarities to binary EuH_2 have been emphasized.

The remaining compounds with 2:1 stoichiometry which have been investigated with regard to their hydriding properties are all of the cubic Ti_2Ni type of structure (space group Fd3m, No. 227). Hf_2Fe was found by *Buschow* and *van Diepen* [7.270] and by *Tuscher* [7.271] to be converted from a Pauli paramagnet to a ferromagnet upon hydrogen absorption. Recently *Vulliet* et al. [7.272] extended this study to variable hydrogen content and detected the maximum magnetic moment at a concentration of 3 H atoms/f.u. The broadened Mössbauer spectra were interpreted to reflect a disordered magnetic phase.

Of the Ti-based compounds, Ti_2Ni, Ti_2Co and the pseudobinaries between these two and Fe have been examined. Pure Ti_2Fe, however, does not form [7.273, 274]. Nevertheless it can be stabilized by small amounts of oxygen (between 6% and 14% depending on the exact Ti:Fe ratio) [7.275–278]. The resulting ternary oxide (η-phase) is believed to play a significant role in the activation process of the storage compound TiFe [7.238, 279] which for large scale applications is frequently present as ferrotitanium, where substantial amounts of oxygen are dissolved.

The oxygen-free compounds Ti_2Co and Ti_2Ni were reported by *Tuscher* [7.274] and by *Hiebl* et al. [7.273] to be exchange-enhanced Pauli paramagnets displaying a complex temperature dependence. After hydrogen uptake a temperature independent susceptibility was obtained. Isostructural $Ti_2FeO_{0.5}$ shows a Curie-Weiss-like behaviour [7.275]. After hydrogenation an increase in the susceptibility was detected, its magnitude depending on the hydrogen content in the sample. For a hydrogen content exceeding 2 H atoms/f.u., magnetic ordering at low temperatures was inferred from the substantial broadening of the NMR lines.

Ternary oxides $Ti_{2-y}Fe_2O_z$ with varying Ti : Fe ratios and oxygen contents were studied earlier by *Mintz* et al. [7.276] and were more recently comprehensively reinvestigated by *Rupp* [7.278] and by *Rupp* and *Wiesinger* [7.280]. For a detailed reference list covering the field of hydrogen uptake by oxygen stabilized Ti_2Fe we refer to the paper of Rupp. From the Mössbauer spectra, a distribution of electric field gradients and isomer shifts due to incomplete occupation of lattice sites was deduced. At 4.2 K magnetic ordering upon hydrogenation has been observed only in the case of $Ti_2FeO_{0.2}H_{2.6}$; paramagnetism, i.e. a quadrupole split spectrum, has been obtained for all the remaining hydrides. For β-Ti (20 at.% Fe) as much as +0.66 mm/s was found for the hydrogen induced change in isomer shift, reflecting both a large increase in volume and a significant change of the electronic structure after charging. Additionally Rupp and Wiesinger could demonstrate that segregations of magnetically ordered Fe in the parent alloys are a consequence of an activation heat treatment under rough vacuum conditions. The hyperfine pattern characteristic for α-Fe which occurred in the transmission spectra indicated that those segregations had already achieved a volume large enough to be visible for 14.4 keV γ-rays. Very recently *Rogl* et al. [7.281] reported on a further ternary oxide, the so called $\varkappa$-phase (space group $P6_3/mmc$, No. 194). Only an insignificant rise in the susceptibility could be detected by *Rupp* and *Tuscher* [7.282] after hydrogen uptake.

When reinvestigating the Zr–Fe phase diagram *Aubertin* et al. studied hydrides of Zr-rich alloys [7.283] and of the η-phase $Zr_2FeO_{0.3}$ [7.284]. Regarding the latter, the Mössbauer spectrum differs from that of the Ti analogue presented by *Rupp* and *Wiesinger* [7.280] which is most probably due to the different amount of hydrogen in the two samples (2.4% and 2.6% respectively).

Th_7Fe_3-type compounds readily absorb remarkable amounts of hydrogen (up to about 30 H atoms/f.u.). By studying La_7Ni_3 hydride *Busch* et al. [7.285] and *Fischer* et al. [7.286] found that in fact it was metastable, disintegrating into LaH_3 and $LaNi_5$. Both the host compound and the hydride displayed Pauli paramagnetic behaviour. Roughly ten years ago *Buschow* et al. [7.287] investigated the hydrogen storage behaviour of a large number of Th–3*d* metal compounds. Only those with the 7 : 3 stoichiometry were found to absorb remarkable quantities of hydrogen, whereas the remaining ones exhibited rather poor storage capacities. More recently *Boltich* et al. [7.288] and *Malik* et al. [7.289] reported on the bulk magnetic properties of Th_7TM_3 (TM = Fe, Co, Ni). The parent compounds were found to be Pauli paramagnetic and below about 2 K superconductivity was detected. Hydrogen absorption led to a rise in the susceptibility for Th_7Co_3, while in Th_7Ni_3 it decreased. However, no evidence for magnetic order has been found in either hydride.

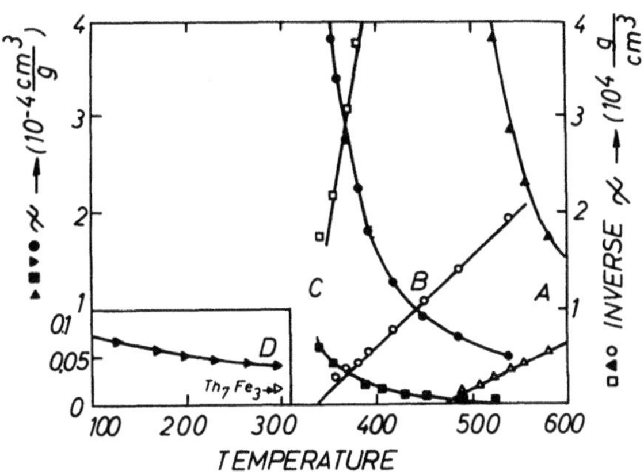

Fig. 7.17. Magnetic susceptibility of Th_7Fe_3 hydrides prepared at different hydrogen pressures [7.290]. The slowly hydrogenated sample (*D*) is not ferromagnetic. The ferromagnetism of the violently hydrogenated samples (*A*, *B*, *C*) is caused by a disproportionation reaction. (*A* −50 bar; *B* −10 bar; *C* −1 bar, 60°C; *D* −0.6 bar)

Th_7Fe_3, on the other hand, was claimed to have turned to ferromagnetism after hydrogen uptake with a Curie temperature of about 350 K. However, a comprehensive investigation by *Schlapbach* et al. [7.290] led to the result that ferromagnetism in these hydrides is obtained exclusively when violent charging conditions are applied. This leads to a disproportionation into Th_4H_{15} and a more Fe-rich compound which both exhibit Curie temperatures in the range cited and are moreover also known to absorb hydrogen [7.287]. After having been charged smoothly the resulting hydride was found to remain paramagnetic at least down to 80 K as displayed in Fig. 7.17. This confirms earlier Mössbauer studies of *Viccaro* et al. [7.291] who reported an almost vanishing ^{57}Fe hyperfine field.

7.3.6 Amorphous Alloys

Although the investigations started with Fe-metalloid (B)-type alloys, Fe-metal alloys, particularly $Zr_{1-y}Fe_y$, have received the most attention with regard to hydrogen absorption. Magnetisation and Mössbauer data yield a critical value around a concentration $y=0.4$, where a magnetic moment first appears on the Fe. For the ferromagnetic systems the average Fe moment as well as the Curie temperature T_c [7.292] frequently increase with increasing y-values. However, in Fe_yZr_{1-y} [7.293–295] these quantities pass though a maximum and then decline as y approaches 1 (Fig. 7.18). Other systems, notably sputtered Fe_yY_{1-y} [7.292, 296, 297] show no long-range magnetic order for any value of y because of a broad distribution of mainly ferromagnetic exchange with some antiferromagnetic interactions which leads to supermagnetic order. These varying

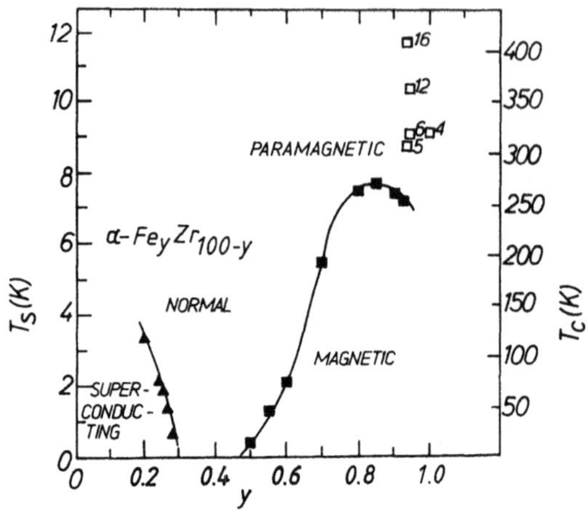

Fig. 7.18. Concentration dependence of the superconducting transition temperature (▲, *left-hand scale*) and of the magnetic ordering temperature (■, *right-hand scale*) of α-Fe_yZr_{1-y} [7.295] and of some hydrides (□) [7.292]; the numbers close to the open symbols indicate the hydrogen concentration

results show clearly the difficulty in extrapolating different alloy systems to $y=1$ in order to speculate about the magnetic state of pure amorphous iron.

In Fe_yZr_{1-y} below the critical Fe concentration ($y_c \sim 0.45$), superconductivity appears with transition temperatures of the order of 3 K. In the range where magnetic order occurs, the effective ^{57}Fe hyperfine field deduced from Mössbauer measurements increases monotonically with Fe (unlike T_c), indicating an increase in the Fe moment. All these characteristics were claimed to be consistent with a localized-moment description of magnetism in which an antiferromagnetic exchange component appears at high Fe concentrations. The concentration dependence of the electric quadrupole splitting shows a large drop between the Fe-rich and the Zr-rich region. The isomer shift increases with the Fe-concentration and is found to obey roughly the behaviour predicted by the model of *Miedema* and *van der Woude* [7.57].

In the case of the Fe concentrations $y \sim 0.9$, magnetisation studies with [7.294, 298] and without external pressure were interpreted on the basis of weak itinerant ferromagnetism. By studying thermomagnetisation curves and hysteresis loops, *Hiroyoshi* and *Fukamichi* [7.299, 300] obtained a phase diagram, where the Curie temperature and freezing points exhibit opposite behaviour. *Buschow* and *Smit* [7.301] reported a compositional short-range order to be present in melt-spun alloys but not in vapour-deposited ones. From magnetisation measurements *Kaul* [7.302] suggested the presence of two types of magnetic electrons: those exhibiting itinerant character, giving rise to ferromagnetism (single-particle contribution) and Invar anomalies, and those of a localized nature, being responsible for both the ferromagnetic (spin-wave contribution) and the micromagnetic behaviour. For the compound $Zr_{11}Fe_{89}$ two contradictary results have been reported. While *Boliang* et al. [7.303] claimed a reduction of the magnetisation below 60 K which they attributed to a change from ferro- to asperomagnetism, nothing of this kind was found by *Fries* et al. [7.304]. The complex conversion electron Mössbauer spectrum, consisting of an α-Fe component and an amorphous pattern of unknown composition,

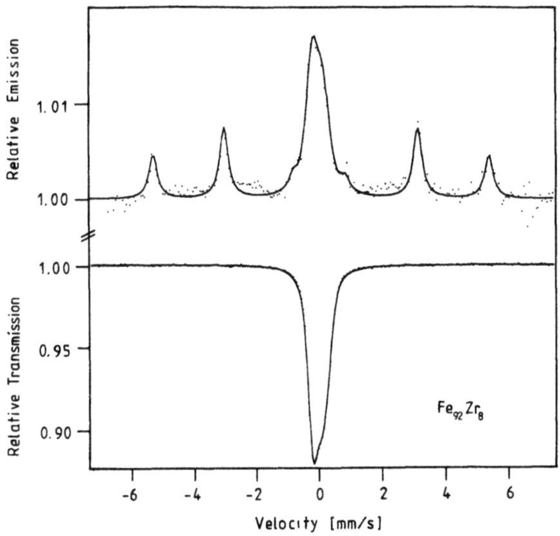

Fig. 7.19. Mössbauer emission *(upper)* and transmission spectrum *(lower)* of $Fe_{92}Zr_8$ [7.304]

led Fries et al. to the conclusion that these inhomogeneities in the ribbon surface might be the reason of differing results. In Fig. 7.19, taken from their work, an emission- (surface sensitive) and a transmission spectrum (bulk sensitive) which were recorded simultaneously can be compared. The Fe precipitates at the surface, detected by CEMS, are obvious.

So far, hydrogen absorption studies have mainly been performed on Fe-rich samples with a composition of or close to $y=0.9$ [7.40, 292, 303–308]. Substantial amounts of hydrogen can be absorbed. Generally an increase in both the Curie temperature and the magnetic Fe moment was reported (see Fig. 7.18). The magnetisation curves were found to resemble those of a collinear ferromagnet. *Fujimori* et al. [7.305, 306] pointed out that the large volume expansion upon hydrogenation might be mainly responsible for the enhancement of T_c which can be qualitatively explained in terms of the inverse effect of the large decrease of the Curie temperature under applied pressure. Since great differences between calculated and experimental values of T_c are found, *Boliang* et al. [7.303] suggested that, as in many similar cases, hydrogen does not act simply as a negative pressure, but directly alters the band structure, probably by forming chemical bonds with Zr, which has already been suggested by *Tanaka* et al. [7.309] for Zr–Ni and Zr–Pd in order to interpret their soft x-ray emission spectra.

Wronski et al. [7.307] noted the fact that drastic changes in magnetic ordering are accompanied by only a small change in the isomer shift. This finding they tentatively attributed to the formation of local deformations about the interstitial H atoms rather than to a simple charge transfer from hydrogen to the 3*d*-orbital of Fe.

In a thorough Mössbauer study with and without applied magnetic field *Fries* et al. [7.304] obtained the unique result that for $Fe_{89}Zr_{11}H_{50}$ the spins are

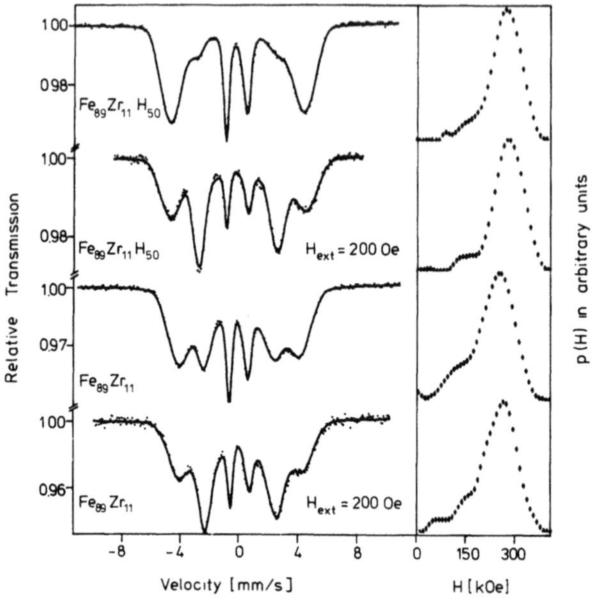

Fig. 7.20. Mössbauer spectra of $Fe_{89}Zr_{11}$ and of $Fe_{89}Zr_{11}H_{50}$ recorded with and without applied field; on the right-hand side of the spectra the hyperfine field distribution is displayed [7.304]

predominentely oriented perpendicular to the ribbon plane ($I_2/I_1=0.09$), whereas in the uncharged compound they were found to be more or less randomly distributed with a slight tendency to align in the ribbon plane (Fig. 7.20). An applied magnetic field of 200 Oe perpendicular to the γ-ray led to a nearly perfect alignement in the ribbon plane which is commonly observed in amorphous ribbons. Lowering the temperature also yielded a substantial rise in the ratio of the two outermost lines.

Very recently *Fries* et al. [7.308] published Mössbauer and XPS studies on a more Fe-rich Zr–Fe hydride. The hydrogen sites were found to resemble those of binary ZrH_2, and the crystallization temperature was substantially decreased upon charging. Furthermore a pronounced charge transfer of Zr 4*d*-electrons to hydrogen has been proposed. Compared to Zr–Fe only a few results have been reported for other glassy Zr transition metal (TM) hydrides [7.310, 311]. The TM=Ni system will be mentioned below in the chapter dealing with superconducting properties.

The influence of hydrogen uptake on the magnetic properties of amorphous Fe_yY_{1-y} has been extensively studied [7.292, 306]. A drastic change was observed in the Fe-rich region: whereas magnetisation and Mössbauer experiments in large applied fields clearly pointed to a non-collinear configuration of the Fe moments, these alloys were found to become excellent soft ferromagnets after hydrogen absorption.

Among other Fe-based amorphous alloys the system Fe_yGd_{1-y} was studied by *Forester* et al. [7.311]. They gained a sharp hydrogen induced reduction of compensation temperature and Curie temperature which is commonly observed in crystalline *R*–Fe compounds and is attributed to the weakening of the

R–Fe exchange upon hydrogen absorption. *Berry* and *Pritchet* [7.312] and *Chambron* et al. [7.313] reported a strong interaction of the hydrogen atoms with the Bloch walls which they observed by anisotropy measurements in (Fe, Ni)*B* type metallic glasses.

Finally several rare-earth-rich glassy hydrides have been studied by *Robbins* et al. [7.314] and by *Sellmyer* et al. [7.315]. Spin-glass behaviour was found for the host compounds which proved to be substantially influenced by hydrogenation.

7.4 Superconductivity

The discovery of superconductivity in Th_4H_{15} [7.316] and in PdH [7.317] together with the speculation that metallic hydrogen as a potential high temperature superconductor might be responsible for T_c values as high as 9 and 11 K for Pd–H and Pd–D respectively, was a stimulus for the development of a large variety of theories and experiments in this field. Moreover, large T_c values of 16.6 K, 15.6, and 13.6 K could be attained in Pd–Cu–H, Pd–Ag–H, and Pd–Au–H respectively, although the parent alloys as well Pd itself show no superconductivity [7.318].

Stritzker and *Wühl* [7.21] have already reviewed the effect of hydrogen upon the superconducting properties in metals and alloys with special emphasis laid on Th–H, Pd–H and related systems. Here hydrogen commonly contributes constructively to superconductivity, while in transition metals, gaseous impurities reduce the superconducting transition temperature T_c.

Before going into more detail we want to recall that the mechanism leading to superconductivity is based on the attractive interaction between conduction electrons formed via the lattice vibrations. This electron-phonon interaction enters via the electron-phonon enhancement factor into the so-called McMillan formula [7.319]:

$$T_c = \frac{\theta_D}{1.45} \exp\left(\frac{-1.04(1+\lambda)}{\lambda - \mu^*(1+0.62\lambda)}\right),$$

where θ_D is the Debye temperature and μ describes the repulsive electron-electron interaction. λ is given by

$$\lambda = \frac{N(E_F)\langle I^2 \rangle}{M\langle \omega^2 \rangle},$$

where $N(E_F)$ is the density of states (DOS) at the Fermi level, $\langle I^2 \rangle$ is the average value of the electron-phonon interaction over the Fermi surface, M denotes the atomic mass and $\langle \omega^2 \rangle$ is an averaged phonon frequency.

With regard to these relations the high superconducting transition temperatures experimentally observed may result either from an enlarged

$N(E_F)$ value caused by soft phonon mode contributions, or from additional new vibrational modes due to dissolved hydrogen in the host lattice. For the system Pd–H(D) *Stritzker* and *Wühl* [7.21] concluded that in the case of a suppression of spin fluctuations, the coupling of conduction electrons to the optical phonons becomes the main reason for the appearance of high T_c superconductivity. In Th–H, superconductivity is restricted to the stoichiometric compound Th_4H_{15} and may also be explained by low-lying optical H-phonon modes [7.21] rather than by describing Th_4H_{15} as an alloy of Th and metallic hydrogen [7.320]. For further discussions regarding superconducting binary metal hydrides such as Al–H and Nb–H we refer the reader to the review of *Stritzker* and *Wühl* [7.21].

We should, however, briefly mention superconducting amorphous alloys: Pd–Si [7.321, 322], Zr–Pd [7.323] and Zr–Ni [7.324] in which hydrogen absorption reduces T_c. For the last two systems in particular, the well-defined transition temperature becomes smeared out upon hydrogen uptake. This probably indicates that inhomogeneities appear in the amorphous hydrides. The superconducting properties of these amorphous alloys can be discussed on the same basis as those of their crystalline counterparts.

Turning to crystalline compounds, the pseudobinary cubic (C15) Laves phases $(Hf, Zr)V_2$ are high field superconducters with T_c reaching 10.1 K in the case of $Hf_{0.5}Zr_{0.5}V_2$ [7.325]. The transition temperature decreases upon absorption of hydrogen (deuterium) [7.326–328]. From specific heat and susceptibility measurements and from x-ray analysis *Geibel* et al. [7.328] proposed a phase diagram for ZrV_2H_x which for $x > 1.3$ shows a decomposition into $ZrV_2H_{1.3}$ and ZrV_2 on lowering the temperature. For this hydride specific heat measurements proved the absence of superconductivity above 2 K and an electronic specific heat coefficient which is drastically reduced to 3.2 mJ/K^2g-at. This decrease can be understood by considering recent band structure calculations [7.329] which were performed under the assumption that hydrogen delivers its electrons to the conduction band, thus shifting the Fermi energy into a valley of the DOS curve. Geibel et al. furthermore presented a reasonable interpretation of the isotope effect in $Hf_{0.5}Zr_{0.5}V_2$ in terms of the decomposition mentioned above.

Two examples are now given, in which upon hydrogen uptake superconductivity is suppressed, while magnetism occurs: cubic $CeRu_2$ is a superconductor with T_c ranging from 4.6 to 6.2 K, while $CeRu_2D_5$ is found to be magnetically ordered below 2.7 K. There pairs of adjacent [111] planes are ferromagnetically and antiferromagnetically coupled in the sequence $++--$. *Fruchart* et al. [7.330] explain the transition from superconductivity to magnetic order by a hydrogen induced valence change of Ce from 4+ to 3+. *Osterwalder* et al. [7.331] studied the effect of hydrogen absorption of the 4*f* localisation in $CeRu_2$ by means of XPS. The authors concluded that this intermetallic becomes a fully trivalent Ce compound upon hydriding and that the 4*f* conduction-band hybridisation decreases from 120 meV to about 60 meV.

Th_7Fe_3 is a Pauli paramagnet and is superconducting below 1.86 K [7.288, 291, 332] which indicates that the Fe 3*d*-band is completely filled. The hydride $Th_7Fe_3H_{30}$ has been claimed to order ferromagnetically with an Fe moment of 1.4 μ_B, Th being assumed to carry no magnetic moment. The Curie temperature was estimated to lie between 350 K and 400 K. *Malik* et al. [7.332] suggested hydrogen to be essentially anionic in this case too (see e.g. [7.3]). From this it is plausible that the Fe 3*d*-band is depopulated in the hydride and that the Fe moment destroys superconductivity. A re-investigation by *Schlapbach* et al. [7.290], however, showed that the hydride remains paramagnetic when the charging procedure is performed sufficiently smoothly, in order to avoid decomposition (see also Sect. 7.3.5 and Fig. 7.17).

The most important group of compounds for technical superconductivity are those with the A15 type of structure (Cr_3Si type) containing about 50 superconducting compounds [7.333]. The interest in the effect of hydrogenation upon the superconducting properties is of both scientific and technological nature, the latter in order to obtain a better performance in superconducting wires. Various authors have studied the influence of hydrogen uptake upon T_c in Nb_3Sn [7.334–337], Nb_3Ge [7.338], Ti_3Au [7.339], Ti_3Sb [7.340] and $Nb_3(Au, Ge)$ [7.341]. In the case of A15 compounds it is frequently found that upon hydrogenation the superconducting transition first increases ($\Delta T_c = +0.2$ K) as long as the hydrogen concentration $x \lesssim 0.03$. Subsequently a decrease is obtained to $T_c < 1.1$ K for $x \sim 1$ [7.336]. The compounds last mentioned are an exception, since upon hydrogen absorption T_c remains almost constant or increases only slightly. By neutron diffraction experiments *Vieland* et al. [7.335] investigated the crystal structure of some Nb_3Sn hydrides, and for hydrogen concentrations $x > 0.1$ confirmed the absence of the cubic-tetragonal lattice transformation occurring in the parent compound at about 50 K. The reduction of T_c by dissolving hydrogen in Nb_3Sn resembles the effect of alloying Nb_3Sn with Sb or Al, where T_c is altered in a similar way. Although this fact cannot be taken as direct evidence for the applicability of the rigid band approximation [7.335], the latter has been used by *Rama Rao* et al. [7.340] for the interpretation of the reduction of T_c in Ti_3SbH_x. The band structure of Ti_3Sb was assumed to resemble that of Nb_3Sn, where E_F lies in a maximum of the DOS curve of the *d*-band. Hydrogen absorption then increases the valence electron concentration by shifting the Fermi energy to a lower $N(E_F)$ value. Since T_c depends on the product $N(E_F) \cdot \langle I^2 \rangle$, a reduction of $N(E_F)$ upon hydrogenation may lead to a fall in T_c. A decrease in T_c is also observed for other Ti-based A15 superconductors in the sequence Ti_3Ir, Ti_3Pt, Ti_3Au. There too the valence electron concentration rises in going from Ir to Au. Thus T_c may be lowered by hydrogen absorption as well as by alloying.

In this connection we note that *Matacotta* et al. [7.337] reported an improvement of the current density of technical Nb_3SnH_x multifilamentary wires by low hydrogen admixture ($x < 0.05$), particularly in high magnetic fields around 12 T. *Ferdeghini* et al. [7.342] recently re-investigated the effect of hydrogen uptake on the superconducting performance of technical Nb_3Sn

wires. They obtained a maximum in the critical current density for a hydrogen content of $x=0.03$ at a magnetic field of $\mu_0 H=12$ T. They furthermore stated that the relative increase in performance of the wire was more pronounced at higher temperatures with a gain of J_c up to 215% at 10 K. After heat treating a Nb_3Sn wire in a hydrogen atmosphere *Wilhelm* and *Wohlleben* [7.343] observed an increase in both the upper critical field and the current density. Since T_c is slightly reduced, they attribute this enhancement in the hydrogen-treated samples to an increase in the normal-state resistivity.

Acknowledgement. This work was supported by the National Austrian Science Foundation (Fonds zur Förderung der wissenschaftlichen Forschung) under grant No. 4431.

References

7.1 G. Wortmann: J. de Phys. **12**, C6–333 (1976)
7.2 F.E. Wagner, G. Wortmann: "Mössbauer Studies of Metal-Hydrogen Systems", in *Hydrogen in Metals I*, ed. by G. Alefeld, J. Völkl (Springer, Berlin, Heidelberg 1978) p. 131
7.3 W.E. Wallace: "Magnetic and Electrical Properties of Rare Earth and Rare Earth Intermetallic Hydrides", in *Hydrides for Energy Storage*, ed. by A.F. Andresen, A.J. Maeland (Pergamon, Oxford 1978) p. 217
7.4 W.E. Wallace: "Magnetic Properties of Metal Hydrides and Hydrogenated Intermetallic Compounds" in *Hydrogen in Metals I*, ed. by G. Alefeld, J. Völkl (Springer, Berlin, Heidelberg 1978) p. 169
7.5 W.E. Wallace: Z. Phys. Chem. NF **115**, 219 (1979)
7.6 W.E. Wallace: "Studies of Rare Earth Intermetallic Compounds and Rare Earth Hydrides", in *The Rare Earths in Modern Science and Technology*, ed. by G.J. Mc Carthy, J.J. Rhyne, H.B. Silber (Plenum, New York 1980) p. 1–12
7.7 W.E. Wallace: "Magnetism of Lanthanide and Actinide Intermetallic Hydrides", in *Metal Hydrides*, ed. by G. Bambakidis (Plenum, New York 1981) p. 21
7.8 W.E. Wallace: J. Less-Common Met. **88**, 141 (1982)
7.9 G.K. Shenoy, B.D. Dunlap, P.J. Viccaro, D. Niarchos: Hyperf. Inter. **9**, 531 (1981)
7.10 G.K. Shenoy, B.D. Dunlap, P.J. Viccaro, D. Niarchos: "Hydrogen Storage Materials" in *Mössbauer Spectroscopy and Its Chemical Applications*, ed. by J.G. Stevens, G.K. Shenoy, Advances in Chemistry Series No. 194 (1981) p. 501
7.11 R.L. Cohen: J. de Phys. **41**, C1–333 (1980)
7.12 H. Oesterreicher: Appl. Phys. **24**, 169 (1981)
7.13 R.L. Cohen, J.H. Wernick: Science **214**, 1081 (1981)
7.14 F.A. Kuijpers: Philips Res. Repts. Suppl. 1973, No. 2, p. 1
7.15 K.H.J. Buschow: J. Magn. Magn. Mat. **29**, 91 (1982)
7.16 K.H.J. Buschow: "Change in Magnetic Properties of Rare Earth-Transition Metal Compounds upon H_2 Absorption", in *Hydrides for Energy Storage*, ed. by A.F. Andresen, A.J. Maeland (Pergamon, Oxford 1978) p. 273
7.17 K.H.J. Buschow, R.C. Sherwood: J. Appl. Phys. **49**, 1480 (1978)
7.18 K.H.J. Buschow, P.F. de Châtel: Pure & Appl. Chem. **52**, 135 (1979)
7.19 K.H.J. Buschow, P.C.P. Bouten, A.R. Miedema: Rept. Prog. Phys. **45**, 937 (1982)
7.20 K.H.J. Buschow: "Hydrogen Absorption in Intermetallic Compounds" in *Handbook on the Physics and Chemistry of Rare Earths*, ed. by K.A. Gschneidner, Jr., L. Eyring, Vol. 6 (North-Holland, Amsterdam 1984) p. 1
7.21 B. Stritzker, H. Wühl: "Superconductivity in Metal-Hydrogen Systems", In *Hydrogen in Metals II*, ed. by G. Alefeld, J. Völkl (Springer, Berlin, Heidelberg 1978) p. 243

7.22 E.P. Wohlfarth: "Iron, Cobalt and Nickel" in *Ferromagnetic Materials*, ed. by E.P. Wohlfarth, Vol. 1 (North-Holland, Amsterdam 1980) p. 3
7.23 B.L. Gyorffy, A.J. Pindor, J. Staunton, G.M. Stocks, H. Winter: J. Phys. F **15**, 1337 (1985)
7.24 T. Moriya (ed.): *Electron Correlation and Magnetism in Narrow-Band Systems*, Springer Ser. Solid-State Sci., Vol. 29 (Springer, Berlin, Heidelberg 1981)
7.25 G.G. Libowitz, A.J. Maeland: "Hydrides" In *Handbook on the Physics and Chemistry of Rare Earths*, Vol. 3, ed. by K.A. Gschneidner, Jr., L. Eyring (North-Holland, Amsterdam 1979) p. 299
7.26 A.C. Switendick: "The Change in Electronic Properties on Hydrogen Alloying and Hydride Formation" in *Hydrogen in Metals I*, ed. by G. Alefeld, J. Völkl (Springer, Berlin, Heidelberg 1978) p. 101
7.27 W.E. Wallace, K.H. Mader: J. Chem. Phys. **48**, 84 (1968)
7.28 W.E. Wallace, C. Deenadas, A.W. Thompson, R.S. Craig: J. Phys. Chem. Sol. **32**, 805 (1971)
7.29 A.M. van Diepen, R.S. Craig, W.E. Wallace: J. Phys. Chem. Sol. **32**, 1853 (1971)
7.30 Z. Bieganski, B. Stalinski: Phys. Stat. Sol. (a) **2**, K 161 (1970)
7.31 K. Knorr, B.E.F. Fender: "Crystal Fields in Pr-Hydrides" in *Crystal Field Effects in Metals and Alloys*, ed. by A. Furrer (Plenum, New York 1977) p. 42
7.32 J. Osterwalder, H.R. Ott, L. Schlapbach, J. Schefer, P. Fischer: J. Less-Common Met. **94**, 129 (1983)
7.33 J. Schefer, P. Fischer, W. Haelg, J. Osterwalder, L. Schlapbach, J.D. Jorgensen: J. Phys C **17**, 1575 (1984)
7.34 L. Schlapbach: J. Less. Common Met. **111**, 291 (1985)
7.35 F. Steglich, C.D. Bredl, W. Lieke, U. Rauchschwalbe, G. Sparn: Physica **126 B**, 82 (1984)
7.36 G.R. Stewart: Rev. Mod. Phys. **56**, 755 (1984)
7.37 J. Staunton, B.L. Gyorffy, A.J. Pindor, G.M. Stocks, H. Winter: J. Phys. F **15**, 1387 (1985)
7.38 R. Griessen, R. Feenstra: J. Phys. F **15**, 1013 (1985)
7.39 V. Heine: *The Physics of Metals*, Vol. 1, *Electrons* (University Press, Cambridge 1969) p. 1
7.40 J.M.D. Coey, D.H. Ryan, D. Gignoux, A. Lienard, J.P. Rebouillat: J. Appl. Phys. **53**, 7804 (1982)
7.41 G. Hilscher, G. Wiesinger, R. Hempelmann: J. Phys. F **11**, 2161 (1981)
7.42 K.H.J. Buschow: Rept. Prog. Phys. **40**, 1179 (1977)
7.43 K.H.J. Buschow: "Rare Earth Compounds" in *Ferromagnetic Materials*, ed. by E.P. Wohlfarth, Vol. 1 (North-Holland, Amsterdam 1980) p. 299
7.44 H. Kirchmayr, C. Poldy: "Magnetic Properties of Intermetallic Compounds of Rare Earth Metals" in *Handbook on the Physics and Chemistry of Rare Earths*, ed. by K.A. Gschneidner, L. Eyring, Vol. 2 (North-Holland, Amsterdam 1979) p. 55
7.45 W.E. Wallace: *Rare Earth Intermetallics* (Academic, New York 1973)
7.46 P. Vajda, J.N. Daou: J. Less-Common Met. **101**, 269 (1984)
7.47 I. Jacob, E.R. Bauminger, D. Davidov, I. Felner, S. Ofer, D. Shaltiel: J. Magn. Magn. Mat. **15–18**, 1269 (1980)
7.48 H. de Graaf, R.C. Thiel, K.H.J. Buschow: J. Phys. F **12**, 1239 (1982)
7.49 E. Wicke: J. Less-Common Met. **101**, 17 (1984)
7.50 P. Bennet, J.C. Fuggle: In *Electronic Structure and Properties of Hydrogen in Metals*, ed. by P. Jena, C.B. Satterthwaite (Plenum, New York 1983) p. 223
7.51 L. Schlapbach, J.P. Burger: "XPS/UPS Study of the Electronic Structure of $PdH_{0.6}$" in *Electronic Structure and Properties of Hydrogen in Metals*, ed. by P. Jena, C.B. Satterthwaite (Plenum, New York 1983) p. 229
7.52 U. Gonser (ed.): *Mössbauer Spectroscopy*, Topics Appl. Phys., Vol. 5 (Springer, Berlin, Heidelberg 1975)
7.53 D. Barb: *Grundlagen und Anwendungen der Mössbauerspektroskopie* (Akademie Verlag, Berlin 1980)

7.54 L.R. Walker, G.K. Wertheim, V. Jaccarino: Phys. Rev. Lett. **6**, 98 (1961)
7.55 M. Gupta: Solid State Commun. **42**, 501 (1982)
7.56 L. Schlapbach: Phys. Lett. **91**, 303 (1982)
7.57 A.R. Miedema, F. van der Woude: Physica **100B**, 145 (1980)
7.58 W. Steiner, R. Haferl: Phys. Stat. Sol. (a) **42**, 739 (1977)
7.59 P.C.M. Gubbens, K.H.J. Buschow: J. Phys. F **12**, 2715 (1982)
7.60 J. Zukrowski, A. Barnasik, K. Krop, R. Radwanski, J. Pszczola, J. Suwalski, Z. Kucharski, M. Lukasiak: Hyperf. Inter. **15/16**, 801 (1983)
7.61 P.C.M. Gubbens, A.M. van der Kraan: J. Magn. Magn. Mat. **9**, 349 (1978)
7.62 P.C.M. Gubbens, A.M. van der Kraan, K.H.J. Buschow: J. Appl. Phys. **56**, 2547 (1984)
7.63 F.A. Kuijpers, B.O. Loopstra: J. Phys. Chem. Sol. **35**, 301 (1974)
7.64 M. Yamaguchi, T. Katamune, T. Ohta: J. Appl. Phys. **53**, 2788 (1982)
7.65 M. Yamaguchi, T. Ohta, T. Katayama: J. Magn. Magn. Mat. **31–34**, 221 (1983)
M. Yamaguchi, D.K. Ross, T. Groto, T. Ohta: Z. Phys. Chem. NF **145**, 101 (1985)
7.66 E.R. Bauminger, D. Davidov, I. Felner, I. Nowik, S. Ofer, D. Shaltiel: Physica **86–88B**, 201 (1977)
7.67 M. Yamaguchi, T. Katamune, T. Ohta: J. Less-Common Met. **88**, 195 (1982)
7.68 M. Yamaguchi, S. Sasaki, T. Ohta: J. Less-Common Met. **73**, 201 (1980)
7.69 H. Figiel: J. Less-Common Met. **83**, L27 (1982)
7.70 G. Herbst, H. Kronmüller: Phys. Lett. **70A**, 341 (1979)
7.71 G. Herbst, H. Kronmüller: Z. Phys. Chem. NF **116**, 31 (1979)
7.72 F.T. Parker, H. Oesterreicher: J. Less-Common Met. **79**, 297 (1981)
7.73 F. Stucki, L. Schlapbach: J. Less-Common Met. **74**, 143 (1980)
7.74 L. Schlapbach, A. Seiler, F. Stucki, H.C. Siegmann: J. Less-Common Met. **73**, 145 (1980)
7.75 L. Schlapbach: J. Phys. F **10**, 2477 (1980)
7.76 J.H. Weaver, A. Franciosi, W.E. Wallace, H.K. Smith: J. Appl. Phys. **51**, 5847 (1980)
7.77 J. Palleau, G. Chouteau: J. de Phys. **41**, L227 (1980)
7.78 S.K. Malik, F.J. Arlinghaus, W.E. Wallace: Phys. Rev. B **25**, 6488 (1982)
7.79 W.M. Walsh, Jr., L.W. Rupp, Jr., P.H. Schmidt, L.D. Longinotti: "Magnetic Resonance of Gd in $LaNi_5$ and $LaNi_5$ Hydride" in AIP No. 29, 1975 Conf. Proc. p. 686
7.80 J. Lamloumi, C. Lartique, A. Percheron-Guegan, J.C. Achard, G. Jehanno: "Thermodynamic and Magnetic Properties of $LaNi_{5-x}Fe_xH_y$" in *The Rare Earths in Modern Science and Technology*, ed. by G.J. McCarthy, H.B. Silber, J.J. Rhyne (Plenum, New York 1982) p. 487
7.81 S.J. Campell, R.K. Day, J.B. Dunlop, A.M. Stewart: J. Magn. Magn. Mat. **31–34**, 167 (1983)
7.82 J. Lamloumi, A. Percheron-Guegan, J.C. Achard, G. Jehanno, D. Givord: J. de Phys. **45**, 1643 (1984)
7.83 U. Atzmony, D. Dayan, M.P. Dariel: Mat. Res. Bull. **16**, 793 (1981)
7.84 D. Niarchos, P.J. Viccaro, G.K. Shenoy, B.D. Dunlop, A.T. Aldred: Hyperf. Inter. **9**, 563 (1981)
7.85 F.W. Oliver, T. Kebede, K. Thompson, J. Gilchrist: Solid State Commun. **46**, 837 (1983)
7.86 F.W. Oliver, W. Morgan, E.C. Hammond, S. Wood, L. May: J. Appl. Phys. **57**, 3250 (1985)
7.87 R.L. Cohen, K.W. West, J.H. Wernick: J. Less-Common Met. **70**, 229 (1980)
7.88 R.L. Cohen, K.W. West, J.H. Wernick: J. Less-Common Met. **73**, 273 (1980)
7.89 D. Shaltiel, Th. v. Waldkirch, F. Stucki, L. Schlapbach: J. Phys. F **11**, 471 (1981)
7.90 H. Rummel, R.L. Cohen, P. Guetlich, K.W. West: Appl. Phys. Lett. **40**, 477 (1982)
7.91 F.W. Oliver, K.W. West, R.L. Cohen, K.H.J. Buschow: J. Phys. F **8**, 701 (1978)
7.92 S.K. Malik, E.B. Boltich, W.E. Wallace: J. Sol. State Chem. **33**, 263 (1980)
7.93 A.T. Pedziwiatr, F. Pourarian, W.E. Wallace: J. Appl. Phys. **55**, 1987 (1984)
7.94 K. Yagisawa, A. Yoshikawa: Z. Phys. Chem. NF **117**, 79 (1979)
7.95 K.H.J. Buschow: J. Less-Common Met. **97**, 185 (1984)
7.96 K.H.J. Buschow: J. Magn. Magn. Mat. **40**, 224 (1983)
7.97 K.H.J. Buschow, P.H. Smit, R.M. van Essen: J. Magn. Magn. Mat. **15–18**, 1261 (1980)

7.98 K.H.J. Buschow: J. Less-Common Met. **72**, 257 (1980)
7.99 J.J. Croat, J.F. Herbst, R.W. Lee, F.E. Pinkerton: J. Appl. Phys. **55**, 2078 (1984)
7.100 M. Sagawa, S. Fujimura, N. Togawa, H. Yamamoto, Y. Matsuura: J. Appl. Phys. **55**, 2083 (1984)
7.101 H.H. Stadelmaier, N.A. El Masry, S.R. Stallard: J. Appl. Phys. **57**, 4149 (1985)
7.102 J.F. Herbst, J.J. Croat, F.E. Pinkerton, W.B. Yelon: Phys. Rev. B **29**, 4176 (1984)
7.103 D. Givord, H.S. Li, J.M. Moreau: Solid State Commum. **50**, 497 (1984)
7.104 C.B. Shoemaker, D.P. Shoemaker, R. Fruchart: Acta Cryst. C **40**, 1665 (1984)
7.105 H. Onodera, Y. Yamaguchi, H. Yamamoto, M. Sagawa, Y. Matsuura, H. Yamamoto: J. Magn. Magn. Mat. **46**, 151 (1984)
7.106 J.M. Friedt, A. Vasquez, J.P. Sanchez, P. l'Heritier, R. Fruchart: J. Phys. F **16**, 651 (1986)
7.107 R. Grössinger, G. Hilscher, H. Kirchmayr, H. Sassik, R. Strnat, G. Wiesinger: Physica **130 B**, 307 (1985)
7.108 P. l'Heritier, P. Chaudouet, R. Madar, A. Rouault, J.-P. Senateur, R. Fruchart: C. R. Acad. Sc. Paris **299**, Serie II, 849 (1984)
7.109 K. Oesterreicher, H. Oesterreicher: Phys. Stat. Sol. (a) **85**, K61 (1984)
7.110 G. Wiesinger, G. Hilscher, R. Grössinger: J. Less-Common Met. **131**, 409 (1987)
7.111 I.R. Harris, C. Noble, T. Bailey: J. Less-Common Met. **106**, L1 (1985)
7.112 K.H.J. Buschow, R.C. Sherwood: J. Appl. Phys. **48**, 4643 (1977)
7.113 E.B. Boltich, W.E. Wallace, F. Pourarian, S.K. Malik: J. Phys. Chem. **86**, 524 (1982)
7.114 S.K. Malik, T. Takeshita, W.E. Wallace: Solid State Commun. **23**, 599 (1977)
7.115 P.C.M. Gubbens, A.M. van der Kraan, K.H.J. Buschow: J. Magn. Magn. Mat. **30**, 383 (1983)
7.116 K.H.J. Buschow: Solid State Commun. **21**, 1031 (1977)
7.117 E.B. Boltich, W.E. Wallace, F. Pourarian, S.K. Malik: J. Magn. Magn. Mat. **25**, 295 (1982)
7.118 M. Commandre, D. Fruchart, A. Rouault, D. Sauvage, C.B. Shoemaker, D.P. Shoemaker: J. de Phys. **40**, L639 (1979)
7.119 C. Crowder, B. Kebe, W.J. James, W. Yelon: "Magnetic and Structural Properties of $Y_6Mn_{23}D_{23}$" in *The Rare Earths in Modern Science and Technology*, ed. by G.J. McCarthy, H.B. Silber, J.J. Rhyne, Vol. 3 (Plenum, New York 1982) p. 473
7.120 K. Hardman-Rhyne, J.J. Rhyne: J. Less-Common Met. **94**, 23 (1983)
7.121 K. Hardman-Rhyne, J.J. Rhyne, E. Prince, C. Crowder, W.J. James: Phys. Rev. B **29**, 416 (1984)
7.122 C. Crowder, W.J. James: J. Less-Common Met. **95**, 1 (1983)
7.123 K. Hardman, J.J. Rhyne, H.K. Smith, W.E. Wallace: J. Less-Common Met. **74**, 97 (1980)
7.124 K. Hardman-Rhyne, H.K. Smith, W.E. Wallace: J. Less-Common Met. **96**, 201 (1984)
7.125 A. Delapalme, J. Deportes, R. Lemaire, K. Hardman, W.J. James: J. Appl. Phys. **50**, 1987 (1979)
7.126 G.A. Stewart, J. Zukrowski, G. Wortmann: J. Magn. Magn. Mat. **25**, 77 (1981)
7.127 K.H.J. Buschow: Solid State Commun. **40**, 207 (1981)
7.128 F. Pourarian, E.B. Boltich, W.E. Wallace, S.K. Malik: J. Less-Common Met. **74**, 153 (1980)
7.129 F. Pourarian, E.B. Boltich, W.E. Wallace, R.S. Craig, S.K. Malik: J. Magn. Magn. Mat. **21**, 128 (1980)
7.130 K.H.J. Buschow, P.C.M. Gubbens, W. Ras, A.M. van der Kraan: J. Appl. Phys. **53**, 8329 (1982)
7.131 P.C.M. Gubbens, W. Ras, A.M. van der Kraan, K.H.J. Buschow: Phys. Stat. Sol. (b) **117**, 277 (1983)
7.132 G.A. Stewart, J. Zukrowski, G. Wortmann: Solid State Commun. **39**, 1017 (1981)
7.133 E.B. Boltich, F. Pourarian, W.E. Wallace, H.K. Smith, S.K. Malik: Solid State Commun. **40**, 117 (1981)
7.134 A.T. Pedziwiatr, H.K. Smith, W.E. Wallace: J. Sol. State Chem. **47**, 41 (1983)

7.135 A.T. Pedziwiatr, E.B. Boltich, W.E. Wallace, R.S. Craig: J. Sol. State Chem. **46**, 342 (1983)
7.136 J.J. Rhyne, K. Hardman-Rhyne, H.K. Smith, W.E. Wallace: J. Less-Common Met. **94**, 95 (1983)
7.137 A.T. Pedziwiatr, Z.M. Stadnik, J. Zukrowski, H.K. Smith, W.E. Wallace: Solid State Commun. **55**, 455 (1985)
7.138 P.C.M. Gubbens, A.M. van der Kraan, K.H.J. Buschow: J. Phys. F **14**, 235 (1984)
7.139 P.C.M. Gubbens, A.M. van der Kraan, K.H.J. Buschow: Solid State Commun. **37**, 635 (1981)
7.140 G. Hilscher, N. Buis, J.J.M. Franse: Physica B **91**, 170 (1977)
7.141 C. Lin, P. v. Blanckenhagen, G. Hilscher, G. Wiesinger: J. Magn. Magn. Mat. **31–34**, 199 (1983)
7.142 G.J. Long, K. Hardman, W.J. James: Solid State Commun. **34**, 253 (1980)
7.143 H. Oesterreicher, H. Bittner: Phys-Stat. Sol. (a) **41**, K101 (1977)
7.144 J. Zukrowski, G.A. Stewart, G. Kalkowski, G. Wortmann, G. Wiesinger: "Mössbauer Studies of the $Er_6(Fe_{1-y}Mn_y)_{23}H_x$ System", in 1981 Conf. Cryst. Electric Field Effects Proc. p. 149 (Plenum, New York 1982)
7.145 K.H.J. Buschow: J. Less-Common Met. **72**, 257 (1980)
7.146 R.M. van Essen, K.H.J. Buschow: J. Less-Common Met. **70**, 189 (1980)
7.147 K.H.J. Buschow, R.M. van Essen: Solid State Commun. **32**, 1241 (1979)
7.148 S.K. Malik, W.E. Wallace, T. Takeshita: Solid State Commun. **28**, 977 (1978)
7.149 S.K. Malik, E.B. Boltich, W.E. Wallace: Solid State Commun. **37**, 329 (1981)
7.150 J.F. Herbst, J.J. Croat: J. Appl. Phys. **53**, 4304 (1982)
7.151 S.K. Arif, D.St.P. Bunbury, G.J. Bowden: J. Phys. F **5**, 1785 (1975)
7.152 A.M. van der Kraan, P.C.M. Gubbens, K.H.J. Buschow: Phys. Stat. Sol. (a) **31**, 495 (1975)
7.153 R.L. Davies, R.K. Day, J.B. Dunlop: J. Phys. F **7**, 1885 (1977)
7.154 G.J. Bowden, R.K. Day: J. Phys. F **7**, 181 (1977)
7.155 G.J. Bowden, R.K. Day: J. Phys. F **7**, 191 (1977)
7.156 G. Wiesinger, R. Haferl, H.R. Kirchmayr: Mikrochim. Acta, Suppl. **9**, 177 (1981)
7.157 S.K. Malik, T. Takeshita, W.E. Wallace: Magn. Lett. **1**, 33 (1976)
7.158 K.S.V.L. Narasimhan: Magnetisation Measurements on $Er_{1-x}Th_xFe_3$ and Some of Their Hydrides" in *The Rare Earths in Modern Science and Technology*, ed. by G.J. McCarthy, J.J. Rhyne (Plenum, New York 1978) p. 81
7.159 D. Niarchos, P.J. Viccaro, B.D. Dunlap, G.K. Shenoy: J. Appl. Phys. **50**, 7690 (1979)
7.160 D. Niarchos, P.J. Viccaro, B.D. Dunlap, G.K. Shenoy, A.T. Aldred: J. Less-Common Met. **73**, 283 (1980)
7.161 S.K. Malik, F. Pourarian, W.E. Wallace: J. Magn. Magn. Mat. **40**, 27 (1983)
7.162 J.B.M. da Cunha, A. Vasquez: Hyperf. Inter. **9**, 547 (1981)
7.163 J.B.M. da Cunha, P.J. Viccaro, A. Vasquez: Hyperf. Inter. **12**, 119 (1982)
7.164 K.H.J. Buschow: Solid State Commun. **19**, 421 (1976)
7.165 A.M. van Diepen, K.H.J. Buschow: Solid State Commun. **22**, 113 (1977)
7.166 A.M. van der Kraan, J.N.J. van der Velden, J.H.F. van Apeldoorn, P.C.M. Gubbens, K.H.J. Buschow: Phys. Stat. Sol. (a) **35**, 137 (1976)
7.167 H. Höchst, E. Colavita, K.H.J. Buschow: Phys. Rev. B **31**, 6167 (1985)
7.168 R.L. Cohen, K.W. West, F. Oliver, K.H.J. Buschow: Phys. Rev. B **21**, 941 (1980)
7.169 Y. Nakamura: J. Magn. Magn. Mat. **31–34**, 829 (1983)
7.170 P.J. Viccaro, G.K. Shenoy, D. Niarchos, B.D. Dunlap: J. Less-Common Met. **73**, 265 (1980)
7.171 R. Hempelmann, G. Hilscher: J. Less-Common Met. **74**, 103 (1980)
7.172 H. Oesterreicher, H. Bittner: Mat. Res. Bull. **13**, 83 (1978)
7.173 R. Hempelmann, E. Wicke, G. Hilscher, G. Wiesinger: Ber. Bunsenges. Phys. Chem. **87**, 48 (1983)
7.174 D. Fruchart, J.L. Soubeyroux, R. Hempelmann: J. Less-Common Met. **99**, 307 (1984)
7.175 R.M. van Essen, K.H.J. Buschow: Mat. Res. Bull. **15**, 1149 (1980)

7.176 F. Pourarian, H. Fujii, W.E. Wallace, V.K. Sinha, H.K. Smith: J. Phys. Chem. **85**, 3105 (1981)
7.177 I. Jacob, D. Davidov, D. Shaltiel: J. Magn. Magn. Mat. **20**, 226 (1980)
7.178 H. Fujii, F. Pourarian, W.E. Wallace: J. Less-Common Met. **88**, 187 (1982)
7.179 S. Hirosawa, F. Pourarian, W.E. Wallace: J. Magn. Magn. Mat. **43**, 187 (1984)
7.180 J.J. Didisheim, P. Fischer: J. Less. Common Met. **103**, 267 (1984)
7.181 P.J. Viccaro, G.K. Shenoy, B.D. Dunlap, D.G. Westlake, J.F. Miller: J. de Phys. **40**, C2–198 (1979)
7.182 K.H.J. Buschow, A.M. van Diepen: Solid State Commun. **19**, 79 (1976)
7.183 D. Niarchos, P.J. Viccaro, G.K. Shenoy, B.D. Dunlap, A.T. Aldred: Hyperf. Inter. **9**, 563 (1981)
7.184 P.H. Smit, K.H.J. Buschow: Phys. Rev. B **21**, 3839 (1980)
7.185 P.H. Smit, H.C. Donkersloot, K.H.J. Buschow: J. Appl. Phys. **53**, 2640 (1982)
7.186 R. Grössinger, R. Haferl, G. Hilscher, G. Wiesinger, K.H.J. Buschow, P.H. Smit: Inst. Phys. Conf. Ser. No. **55**, Chap. 4, 295 (1980)
7.187 Y. Berthier, T. de Saxce, D. Fruchart, P. Vulliet: Physica **130**B, 520 (1985)
7.188 T. de Saxce, Y. Berthier, D. Fruchart: J. Less-Common Met. **107**, 35 (1985)
7.189 F. Pourarian, W.E. Wallace, A. Elattar, J.F. Lakner: J. Less-Common Met. **74**, 161 (1980)
7.190 B.D. Dunlap, G.K. Shenoy, J.M. Friedt, P.J. Viccaro, D. Niarchos, H. Kierstead, A.T. Aldred, D.G. Westlake: J. Appl. Phys. **50**, 7682 (1979)
7.191 P.J. Viccaro, J.M. Friedt, D. Niarchos, D. Dunlap, G.K. Shenoy, A.T. Aldred, D.G. Westlake: J. Appl. Phys. **50**, 2051 (1979)
7.192 G.K. Shenoy, B. Schuttler, P.J. Viccaro, D. Niarchos: J. Less-Common Met. **94**, 37 (1983)
7.193 G.E. Fish, J.J. Rhyne, S.G. Sankar, W.E. Wallace: J. Appl. Phys. **50**, 2003 (1979)
7.194 J.J. Rhyne, G.E. Fish, S.G. Sankar, W.E. Wallace: J. de Phys. **40**, C5–209 (1979)
7.195 K.H.J. Buschow: Physica **86–88**B, 79 (1977)
7.196 D. Shaltiel, I. Jacob, D. Davidov: J. Less-Common Met. **53**, 117 (1977)
7.197 I. Jacob, D. Shaltiel, D. Davidov, I. Miloslavski: Solid State Commun. **23**, 669 (1977)
7.198 I. Jacob, D. Shaltiel: Solid State Commun. **27**, 175 (1978)
7.199 I. Jacob, J.M. Bloch, D. Shaltiel, D. Davidov: Solid State Commun. **35**, 155 (1980)
7.200 H. Fujii, V.K. Sinha, F. Pourarian, W.E. Wallace: J. Less-Common Met. **88**, 44 (1982)
7.201 F. Pourarian, W.E. Wallace: J. Less-Common Met. **91**, 223 (1983)
7.202 V.K. Sinha, W.E. Wallace: J. Less-Common Met. **91**, 229 (1983)
7.203 V.K. Sinha, W.E. Wallace: J. Less-Common Met. **91**, 239 (1983)
7.204 D. Rambabu, R. Nagarajan, S.K. Malik, R. Vijayaraghavan, F. Pourarian, W.E. Wallace: J. Magn. Magn. Mat. **31–34**, 759 (1983)
7.205 V.K. Sinha, F. Pouarian, W.E. Wallace: J. Less-Common Met. **87**, 283 (1982)
7.206 G. Wiesinger: Hyperf. Inter. **28**, 545 (1986)
7.207 H. Fujii, T. Okamoto, W.E. Wallace, F. Pourarian, T. Morisaki: J. Magn. Magn. Mat. **46**, 245 (1985)
7.208 F. Fujii, F. Pourarian, W.E. Wallace: J. Magn. Magn. Mat. **27**, 215 (1982)
7.209 H. Fujii, J. Fujimoto, S. Takeda, T. Hihara, T. Okamoto: J. Magn. Magn. Mat. **31–34**, 223 (1983)
7.210 T. Okamoto, H. Fujii, S. Takeda, T. Hihara: J. Less-Common Met. **88**, 181 (1982)
7.211 H. Oesterreicher, H. Bittner: J. Magn. Magn. Mat. **15–18**, 1264 (1980)
7.212 F. Pourarian, W.E. Wallace, S.K. Malik: J. Less-Common Met. **85**, 95 (1982)
7.213 F. Pourarian, W.E. Wallace, S.K. Malik: J. Magn. Magn. Mat. **25**, 299 (1982)
7.214 S.G. Sankar, D.M. Gualtieri, W.E. Wallace: "Low Temperature Magnetic Properties of the Hydrides and Deuterides of $Er(Fe, Mn)_2$", in *The Rare Earths in Modern Science and Technology*, ed. by G.J. Mc Carthy, J.J. Rhyne (Plenum, New York 1978) p. 69
7.215 H. Drulis, W. Petrynski, B. Stalinski: J. Less-Common Met. **101**, 229 (1984)
7.216 G. Wiesinger: J. Less-Common Met. **130**, 181 (1987)
7.217 K.H.J. Buschow: J. Less-Common Met. **51**, 173 (1977)

7.218 L.J. de Jongh, J. Bartolome, F.J.A.M. Greidanus, H.J.M. de Groot, H.L. Stipdonk, K.H.J. Buschow: J. Magn. Magn. Mat. **25**, 207 (1981)
7.219 K.H.J. Buschow, A.M. van der Kraan: J. Less-Common Met. **91**, 203 (1983)
7.220 S.K. Malik, W.E. Wallace: Solid State Commun. **24**, 283 (1977)
7.221 H. de Graaf, R.C. Thiel, K.H.J. Buschow: J. Phys. F **12**, 2079 (1982)
7.222 K.H.J. Buschow, R.L. Cohen, K.W. West: J. Appl. Phys. **48**, 5289 (1977)
7.223 R.L. Cohen, K.W. West, K.H.J. Buschow: Solid State Commun. **25**, 293 (1978)
7.224 S.Z. Huang, M.K. Wu, R.L. Meng, C.W. Chu, J.L. Smith: Solid State Commun. **38**, 1151 (1981)
7.225 O.J. Zogal, D.J. Lam, A. Zygmunt, H. Drulis, W. Petrynski, S. Stalinski: Phys. Rev. B **29**, 4837 (1984)
7.226 J.J. Reilly, R.H. Wiswall, Jr.: Inorg. Chem. **13**, 218 (1974)
7.227 M. Ron, R.S. Oswald, M. Ohring, G.M. Rothberg, M.R. Polcari: Bull. Amer. Phys. Soc. **21**, 273 (1976)
7.228 R. Hempelmann, D. Ohlendorf, E. Wicke: "Irreversible Change in Magnetic Properties of TiFe by Hydrogenation", in *Hydrides for Energy Storage*, ed. by A.F. Andresen, A.J. Maeland (Pergamon, Oxford 1978) p. 407
7.229 R. Hempelmann, E. Wicke: Ber. Bunsenges. Phys. Chem. **81**, 425 (1977)
7.230 G.K. Shenoy, D. Niarchos, P.J. Viccaro, B.D. Dunlap, A.T. Aldred, G.D. Sandrock: J. Less-Common Met. **73**, 171 (1980)
7.231 A. Bläsius, U. Gonser: Appl. Phys. **22**, 331 (1980)
7.232 L. Schlapbach, A. Seiler, F. Stucki: Mat. Res. Bull. **13**, 697 (1978)
7.233 G. Busch, L. Schlapbach, F. Stucki, P. Fischer, A.F. Andresen: Int. J. Hydrogen Energy **4**, 29 (1979)
7.234 L. Schlapbach, A. Seiler, F. Stucki, P. Zuercher, P. Fischer, J. Schefer: Z. Phys. Chem. NF **117**, 205 (1979)
7.235 F. Stucki, L. Schlapbach, A. Seiler: Ber. Bunsenges. Phys. Chem. **84**, 1067 (1980)
7.236 L. Schlapbach, F. Stucki, A. Seiler, H.C. Siegmann: Surf. Sci. **106**, 157 (1981)
7.237 F. Stucki: J. Less-Common Met. **83**, L37 (1982)
7.238 L. Schlapbach, T. Riesterer: Appl. Phys. A **32**, 169 (1983)
7.239 Th. v. Waldkirch, R. Wessicken, H.-U. Nissen, L. Schlapbach: "A TEM and SAD Study of TiFe and $LaNi_5$ Powder Particles: Evidence for Fe and Ni Precipitates", in 1982 Third Intern. Congress on Hydrogen and Materials Proc., p. 293, Paris
7.240 T. Schober, D.G. Westlake: Scripta Metall. **15**, 913 (1981)
7.241 D. Khatamian, G.C. Weatherly, F.D. Manchester, C.B. Alcock: J. Less-Common Met. **89**, 71 (1983)
7.242 D. Khatamian, G.C. Weatherly, F.D. Manchester: Acta Metall. **31**, 1771 (1983)
7.243 J.J. Reilly, F. Reidinger: J. Less-Common Met. **85**, 145 (1982)
7.244 L.J. Swartzendruber, L.H. Bennett, R.E. Watson: J. Phys. F **6**, L331 (1976)
7.245 P. Thompson, M.A. Pick, F. Reidinger, L.M. Corliss, J.M. Hastings, J.J. Reilly: J. Phys. F **8**, L75 (1978)
7.246 W. Schäfer, E. Lebsanft, A. Bläsius: Z. Phys. Chem. NF **115**, 201 (1979)
7.247 P. Thompson, J.J. Reilly, F. Reidinger, J.M. Hastings, L.M. Corliss: J. Phys. F **9**, L61 (1979)
7.248 T. Mizuno, T. Morozumi: J. Less-Common Met. **84**, 237 (1982)
7.249 Y. Asada, H. Nose: J. Phys. Soc. Jap. **35**, 409 (1973)
7.250 L.H. Bennett, L.J. Swartzendruber: Phys. Lett. **24**A, 359 (1967)
7.251 L.J. Swartzendruber, L.H. Bennett: J. Appl. Phys. **39**, 2215 (1968)
7.252 L.H. Bennett, L.J. Swartzendruber, R.E. Watson: Phys. Rev. **165**, 500 (1968)
7.253 J.C. Swartz, L.J. Swartzendruber, L.H. Bennett, R.E. Watson: Phys. Rev. B **1**, 146 (1970)
7.254 S.J. Pickart, R. Nathans, F. Menzinger: J. Appl. Phys. **39**, 2221 (1968)
7.255 J. Giner, F. Gautier: J. de Phys. **38**, C7–301 (1977)
7.256 G. Schadler, P. Weinberger: J. Phys. F. **16**, 42 (1986)
7.257 P.E. Brommer, J.J.M. Franse, H. Hoelscher: "The Occurence of Ferromagnetism in the $Ti(Fe_{1-x}Co_x)$ System", in Inst. Phys. Conf. Ser. 1981, No. **55**, Chap. 4, p. 279

7.258 N. Buis, P.E. Brommer, P. Disveld, M.S. Schalkwijk, J.J.M. Franse: J. Magn. Magn. Mat. **15–18**, 291 (1980)
7.259 H. Hoelscher, J.C.P. Klaasse, J.J.M. Franse, P.E. Brommer: "Thermal Expansion and Specific Heat in $TiFe_xCo_{1-x}$", in Inst. Phys. Conf. Ser. 1981, No. **55**, Chap. 4, p. 283
7.260 N. Buis, P. Disveld, P.E. Brommer, J.J.M. Franse: J. Phys. F **11**, 217 (1981)
7.261 G. Hilscher, G. Wiesinger, E. Lebsanft: J. Magn. Magn. Mat. **15–18**, 1273 (1980)
7.262 C.J. Koizumi, W.N. Cathey: J. Phys. F **10**, 497 (1980)
7.263 J. Beille, D. Bloch, F. Towfiq: Solid State Commun. **25**, 57 (1978)
7.264 J. Beille, D. Bloch, F. Towfiq, J. Voiron: J. Magn. Magn. Mat. **10**, 265 (1979)
7.265 R.C. Bowman, Jr., B.D. Craft, W.E. Tadlock, E.L. Venturini, J.S. Cantrell: J. Appl. Phys. **57**, 3036 (1985)
7.266 L. Schlapbach, D. Shaltiel, P. Oelhafen: Mat. Res. Bull. **14**, 1235 (1979)
7.267 F. Aubertin, S.J. Campbell, U. Gonser: Hyperf. Inter. **28**, 997 (1986)
7.268 R.O. Moyer, Jr., R. Lindsay: J. Less-Common Met. **70**, P57 (1980)
7.269 Z.M. Stadnik, R.O. Moyer, Jr.: J. Less-Common Met. **98**, 159 (1984)
7.270 K.H.J. Buschow, A.M. van Diepen: Solid State Commun. **31**, 469 (1979)
7.271 E. Tuscher: Monatsh. Chem. **110**, 1275 (1979)
7.272 P. Vulliet, G. Teisseron, J.L. Oddou, C. Jeandey, A. Yaouanc: J. Less-Common Met. **104**, 13 (1984)
7.273 K. Hiebl, E. Tuscher, H. Bittner: Monatsh. Chem. **110**, 869 (1979)
7.274 E. Tuscher: Monatsh. Chem. **111**, 535 (1980)
7.275 K. Hiebl, E. Tuscher, H. Bittner: Monatsh. Chem. **110**, 9 (1979)
7.276 M.H. Mintz, Z. Hadari, M.P. Dariel: J. Less-Common Met. **74**, 287 (1980)
7.277 C. Stioui, D. Fruchart, A. Rouault, R. Fruchart, E. Roudaut, J. Rebiere: Mat. Res. Bull. **16**, 869 (1981)
7.278 B. Rupp: J. Less-Common Met. **104**, 51 (1984)
7.279 A. Venkert, M.P. Dariel, M. Talianker: J. Less-Common Met. **103**, 361 (1984)
7.280 B. Rupp, G. Wiesinger: J. Less-Common Met. **104**, 65 (1984)
7.281 P. Rogl, B. Rupp, G. Wiesinger, J. Schefer, P. Fischer: J. Less-Common Met. **113**, 103 (1985)
7.282 B. Rupp, E. Tuscher: J. Less-Common Met. **104**, L9 (1984)
7.283 F. Aubertin, U. Gonser, S.J. Campbell: J. Less-Common Met. **101**, 437 (1984)
7.284 F. Aubertin, U. Gonser, S.J. Campbell: J. Phys. F **14**, 2213 (1984)
7.285 G. Busch, L. Schlapbach, Th. v. Waldkirch: J. Less-Common Met. **60**, 83 (1978)
7.286 P. Fischer, W. Haelg, L. Schlapbach, K. Yvon: J. Less-Common Met. **60**, 1 (1978)
7.287 K.H.J. Buschow, H.H. van Mal, A.R. Miedema: J. Less-Common Met. **42**, 163 (1975)
7.288 E.B. Boltich, S.K. Malik, W.E. Wallace: J. Less-Common Met. **74**, 111 (1980)
7.289 S.K. Malik, E.B. Boltich, W.E. Wallace: Solid State Commun. **33**, 921 (1980)
7.290 L. Schlapbach, C. Pina-Perez, T. Siegrist: Solid State Commun. **41**, 135 (1982)
7.291 P.J. Viccaro, G.K. Shenoy, B.D. Dunlap, D.G. Westlake, S.K. Malik, W.E. Wallace: J. de Phys. **40**, C2–157 (1979)
7.292 J.M.D. Coey, D.H. Ryan, Yu. Boliang: J. Appl. Phys. **55**, 1800 (1984)
7.293 T. Masumoto, S. Ohnuma, K. Shirakawa, M. Nose, K. Kobayashi: J. de Phys. **41**, C8–686 (1980)
7.294 K. Shirakawa, T. Kaneko, M. Nose, S. Ohnuma, H. Fujimori, T. Masumoto: J. Appl. Phys. **52**, 1829 (1981)
7.295 K.M. Unruh, C.L. Chien: Phys. Rev. B **30**, 4968 (1984)
7.296 J.M.D. Coey, D. Givord, A. Lienard, J.P. Rebouillat: J. Phys. F **11**, 2707 (1981)
7.297 J. Chappert, J.M.D. Coey, A. Lienard, J.P. Rebouillat: J. Phys. F **11**, 2727 (1981)
7.298 K. Shirakawa, K. Fukamichi, T. Kaneko, T. Masumoto: Physica **119** B, 192 (1983)
7.299 H. Hiroyoshi, K. Fukamichi: Phys. Lett. **85** A, 242 (1981)
7.300 H. Hiryoshi, K. Fukamichi: J. Appl. Phys. **53**, 2226 (1982)
7.301 K.H.J. Buschow, P.H. Smit: J. Magn. Magn. Mat. **23**, 85 (1981)
7.302 S.N. Kaul: Phys. Rev. B **27**, 6923 (1983)

7.303 Yu. Boliang, D.H. Ryan, J.M.D. Coey, Z. Altounian, J.O. Stroem-Olsen, F. Razavi: J. Phys. F **13**, L217 (1983)
7.304 S.M. Fries, H.-G. Wagner, U. Gonser, L. Schlapbach, R. Montiel-Montoya: J. Magn. Magn. Mat. **45**, 331 (1984)
7.305 H. Fujimori, K. Nakanishi, K. Shirakawa, T. Masumoto, T. Kaneko, N.S. Kazama: "Enhancement of Amorphous Fe–Zr Ferromagnetism by Hydrogen Absorption", in 1981 4th Int. Conf. on Rapidly Quenched Metals Proc., p. 1629 (Japan. Inst. of Metals, Sendai 1982)
7.306 H. Fujimori, K. Nakanishi, H. Hiroyoshi, N.S. Kazama: J. Appl. Phys. **53**, 7792 (1982); D.H. Ryan, J.M. Cadogan, E.J. Devlin, J.M.D. Coey: Z. Phys. Chem. NF **145**, 367 (1985)
7.307 Z.S. Wronski, A.H. Morrish, A.M. Stewart: Phys. Lett. **101** A, 294 (1984)
7.308 S.M. Fries, H.-G. Wagner, S.J. Campbell, U. Gonser, N. Blaes, P. Steiner: J. Phys. F **15**, 1179 (1985)
7.309 K. Tanaka, M. Higatani, K. Kai, K. Suzuki: J. Less-Common Met. **88**, 317 (1982)
7.310 K. Samwer, X.L. Yeh, W.L. Johnson: J. Non-Cryst. Sol. **61 & 62**, 631 (1984)
7.311 D.W. Forester, P. Lubitz, J.H. Schelleng, C. Vittoria: J. Non-Cryst. Sol. **61 & 62**, 685 (1984)
7.312 B.S. Berry, W.C. Pritchet: J. Appl. Phys. **52**, 1865 (1981)
7.313 W. Chambron, F. Lancon, A. Chamberod: Scr. Met. **18**, 29 (1984)
7.314 C.G. Robbins, Z.D. Chen, J.G. Zhao, M.J.O'Shea, D.J. Sellmyer: J. Appl. Phys. **53**, 7798 (1982)
7.315 D.J. Sellmyer, C.G. Robbins, M.J.O'Shea: J. Non-Cryst. Sol. **61 & 62**, 655 (1984)
7.316 C.B. Satterthwaite, I.L. Toepke: Phys. Rev. Lett. **25**, 741 (1970)
7.317 F. Skoskiewicz: Phys. Stat. Sol. (a) **11**, K123 (1972)
7.318 B. Stritzker: Z. Physik **268**, 261 (1974)
7.319 W.L. McMillan: Phys. Rev. **167**, 331 (1968)
7.320 W. Buckel: Z. Phys. Chem. NF **116**, 135 (1979)
7.321 B. Stritzker, H.L. Luo: Solid State Commun. **29**, 811 (1979)
7.322 A. Traverse, H. Bernas, J. Chaumont, X.-J. Fan, L. Mendoza-Zelis: Phys. Rev. B **30**, 6413 (1984)
7.323 K. Kai, M. Ikebe, K. Suzuki, T. Nomoto: J. Less-Common Met. **89**, 229 (1983)
7.324 E. Babic, B. Leontic, J. Lukatela, M. Miljak, M.G. Scott: "Transport Properties of some Hydrogen Doped Zr–Ni Metallic Glasses", in 1981 4th Intern. Conf. Rapidly Quenched Metals Proc., p. 1617 (Japan. Inst. of Metals, Sendai 1982)
7.325 K. Inoue, K. Tachikawa: "High Field Superconducting Properties of Laves Phases in V–Hf and V–Hf–Zr Alloys", in 1971, 12th Intern. Conf. Low Temp. Physics Proc., p. 483 (Acad. Press of Japan, Kyoto 1971)
7.326 P. Duffer, D.M. Gualtieri, V.U.S. Rao: Phys. Rev. Lett. **37**, 1410 (1976)
7.327 V.U.S. Rao, D.M. Gualtieri, S. Krishnamurthy, A. Patkin, P. Duffer: Phys. Lett. **67** A, 223 (1978)
7.328 C. Geibel, W. Goldacker, H. Keiber, V. Oestreich, H. Rietschel, H. Wühl: Phys. Rev. B **30**, 6363 (1984)
7.329 B.M. Klein, W.E. Pickett, D.A. Papaconstantopoulos, L.L. Boyer: Phys. Rev. B **27**, 6721 (1983)
7.330 D. Fruchart, F. Vaillant, A. Rouault, A. Benoit, J. Flouquet: J. Less-Common Met. **101**, 285 (1984)
7.331 J. Osterwalder, T. Riesterer, L. Schlapbach, F. Vaillant, D. Fruchart: Phys. Rev. B **31**, 8311 (1985)
7.332 S.K. Malik, W.E. Wallace, T. Takeshita: Solid State Commun. **28**, 359 (1978)
7.333 J. Muller: Pept. Progr. Phys. **43**, 641 (1980)
7.334 P.R. Sahm: Phys. Lett. **26** A, 459 (1968)
7.335 L.J. Vieland, A.W. Wicklund, J.G. White: Phys. Rev. B **11**, 3311 (1975)
7.336 C. Nölscher, P. Müller, H. Adrian, M. Lehmann, G. Saemann-Ischenko: Z. Phys. B **41**, 291 (1981)

7.337 F.C. Matacotta, C. Ferdeghini, M. Ferretti, G. Bruzzone: IEEE Trans. Mag. MAG-**19**, 897 (1983)
7.338 W.A. Lanford, P.H. Schmidt, J.M. Rowell, J.M. Poate, R.C. Dynes, P.D. Dernier: Appl. Phys. Lett. **32**, 339 (1978)
7.339 J.B. Vetrano, G.L. Guthrie, H.E. Kissinger: Phys. Lett. **26**A, 45 (1967)
7.340 K.V.S. Rama Rao, H. Sturm, B. Elschner, A. Weiss: Phys. Lett. **93**A, 492 (1983)
7.341 J. Elton, H. Oesterreicher: J. Less-Common Met. **90**, L37 (1983)
7.342 C. Ferdeghini, M. Ferretti, F.C. Matacotta, C. Rizzuto, A. Siri: Cryogenics **25**, 208 (1985)
7.343 M. Wilhelm, K. Wohlleben: IEEE Trans. Magn. MAG-**17**, 1625 (1981)

List of Tables

2.1 Crucibles 21

3.1 Thermodynamic data for $LaNi_5$-H 63

3.2 Thermodynamic data for FeTi-H 65

3.3 Thermodynamic data for $ZrMn_{2.8}$ 67

4.1 Ternary metal hydrides and deuterides studied by neutron diffraction: metal atom substructure, space group, lattice parameter and H(D) interstice 94–102

4.A Structure data for selected ternary metal hydrides and deuterides studied by neutron diffraction: space group, lattice parameters and atomic parameters 119–129

4.B Magnetic structures 130–131

5.1 Valence band and core level data for binary metal hydrides 152–155

5.2 Valence band and core level data for ternary metal hydrides 178–179

6.6.1 Heat of solution and formation of binary metal hydrides 266–269

6.6.2 Heat of formation of selected ternary metal hydrides 269–273

6.6.3 Heat of formation of some ternary metal hydrides calculated with the band-structure semi-empirical model 274–278

Subject Index

A 15 structure 176, 331
AB_2 Laves phase compounds, hydrides
electronic properties 201
heat of formation 67, 202; Tables 3.3, 6.6.2
magnetic properties 204, 309
pcT curves 38
phase diagram 15
preparation 37
structure, lattice parameters Tables 4.1, 4.A
substitutions 34
thermodynamic properties 66, 205
Activation 38, 60, 319; TAP 64 *
Amorphous alloys and hydrides 7, 28, 107, 316, 325
amorphisation 62
electronic properties 177, 205, 207
magnetic properties 207, 325
models 63
preparation 28, 205
resistivity 207
site occupancy 107
structural information 88
thermodynamic properties 62
Anionic model 114, 139, 290, 324
Anisotropy 303
Annealing 24, 60, 67
Antibonding band 158
Antiferromagnetic 175, 301, Chap. 7
Antistructure atoms 185, 190, 289, 320
Asperomagnetism 326
Atomic coordinates 87, 91
Augmented plane wave APW 142, 180
Augmented spherical wave ASW 143

Band structure 8; Chap. 5; 205, 249, 255
Band width 140, 150, 257; Table 5.1
Binary hydrides 106
electronic properties 150
phase diagram 169
structure 88

Bonding 105
Bonding band Tables 5.1, 2; 140, 150
Brillouin zone 142, 187
Bulk modulus 233

C 14 structure 201, 310, 312
C 15 structure 66, 201, 312
Calorimeter, -metric 58, 63
Catalysis 4
Cathodic charging 40
Capacity 87, 105, 208
reduction by impurities 42
Characterization 29
of hydrides 37
of intermetallics 29
Charge transfer Chap. 5; 143, 149, 181, 291, 317
Chemical potential 13, 53, 222, 226
Coercitivity 302
Coherent potential approximation CPA 146, 166, 255
Cohesion, cohesive energy 3, 144, 231
Compensation temperature 305, 307
Complex hydrides 88
Compton scattering 149
Configurational entropy 69, 71, 73, 115
Congruently 23
Contamination *see also* Impurity 32
by crucible 21
Conversion electron Mössbauer spectroscopy CEMS 318
Core level spectra, shift Chap. 5; Tables 5.1, 2; 158, 181, 203, 208
Correlation effects 107
Critical temperature, point 2, 224, 261
Crucible 20, 27, 33
ceramic crucible 20
cold crucible 20
corrosion 20
refractory metal crucible 20, 33
Crystal field 288
Crystallinity 89, 92, 309

* TAP 64: Topics in Applied Physics, Vol. 64 (*Hydrogen in Intermetallic Compounds II*).

Curie temperature T_C 93, 117; Table 4.B; Chap. 7
Czochralski technique 27

Debye temperature 173, 329
Decohesion 145
Decomposition 7, 28, 68, 289
Decrepitation 27, 37, 39
Degradation 300
De Haas van Alphen 149, 185
Density of states DOS, total, partial Chap. 5; 142, 150, 180, 255
Deoxidation 33, 35
Desintegration 303
Deuterium, deuteride 91, 295
Differential thermal analysis DTA 58
Diffractometer 90
Diffusion 7, 23; TAP 64
  diffusion path 110
Dingle temperature 160
Disorder 7, 87, 90, 108, 166, 177, 194
Disproportionation *see also* Decomposition 7, 37, 40, 62
Dissociation 2, 231

Einstein oscillator 55, 77
Einstein temperature 59, 162, 226, 229
Effective medium theory 145, 220, 251
Elastic energy 245
Electrical resistivity 148, 174
Electrochemical charging 40, 159
Electron energy loss spectroscopy EELS 156, 166
Electron field gradient EFG 294, 311
Electron paramagnetic resonance EPR 316
Electron-phonon interaction 147, 162, 165, 168, 173, 329
18-Electron rule 116
Electronic specific heat Chaps. 5, 7; 147, 151, 176, 231, 287
Electronic structure Chap. 5
  binary hydrides 150
  effect of hydrogen 140
  ternary hydrides 177
Ellingham diagram 16, 17
Embedding energy 251
Embrittlement 3, 145
Empirical correlations 88, 231
Energy bands 142, 150, 170, 187, 196
Energy distribution curve 149
Enthalpy 54, 225
Entropy 54
  hydride formation 54
  mixing 224

Equilibrium 50, 55, 80, 227
  partial 80
  pressure 2
  single condensed phase 50
  three condensed phases 56
  two condensed phases 50
Exchange energy 288

Fermi energy E_F 140, 145, 256; Chap. 5
Fermi surface FS 147, 160, 173
Ferrimagnetic 306
Ferromagnetic Chap. 7, 175
Ferromagnetic resonance 299
FeTi, FeTi hydrides
  electronic properties 185, 318
  heat of formation 65; Tables 3.2, 6.6.2
  magnetic properties 185, 318
  order parameter 186
  pcT curves 38
  phase diagram 15, 185
  preparation 27, 37
  resistivity 186
  single crystals 27
  structure, lattice parameters Chap. 4; Tables 4.1, 4.A; 185
  substitutions 41
  thermodynamic properties 64
Floating zone technique 27
Fractals 7
Free energy, formation, mixing, Gibbs 51, 57, 223
Friauf-phase 242

Gap 171, 200
Geometrical model 69, 88, 113, 242
Getter 21, 24
Glassy *see* Amorphous 28

Hard magnetic material 302
Haucke-phase 242
Heat capacity 59, 64
Heat of formation, solution Chaps. 3, 6; 2, 112, 144, 221, 230
  band-structure model, local 249, 258, 262
  effective medium theory 221
  Griessen-Driessen model 234, 249
  local heat of formation 242
  Miedema model 234, 237, 244
  partial 223
  semi-empirical models 234
  Shilov relation 248
  table, calculated values ternary hydrides Table 6.6.3

tabulated values binary hydrides Table 6.6.1
tabulated values ternary hydrides Tables 6.6.2, 3.1–3
volume effects 260
Heavy electron, -fermion 8, 175, 287
High pressure magnetisation 306
High pressure technology 7
Homogeneity range of intermetallics 14, 310
Hybridisation 175, 177, 182, 205, 251
Hydride carbides 88
Hydrogen bronze 88
Hydrogen site occupancy 91
Hydrogenation *see* Preparation of hydrides 37
Hyperfine interactions, -field 161, 292, 313
Hysteresis 44, 59, 81

Icosahedral 7
Imaginary binary hydride model 70, 116, 242
Implantation 6, 156, 158
Impurity/purity 18, 19, 35, 41
by crucible 21, 22
effect on capacity 42
effect on stability 41
hydrogen 141, 201
of hydrogen gas 37
of the intermetallic compounds 17, 35, 41
of the metallic constituents 17, 18, 19, 41
phase 92
segregation 309
Inhomogeneity *see also* Homogeneity 89
Interaction
attractive 116
energy 69
H–H 140
magnetic 288
repulsive 114
Interatomic distance 109
Interface 2
Interstices, interstitial sites Table 4.1; 257
Interstitial hole size 113, 231
Inverse photoemission BIS 149, 158
Ion implantation 156, 158
Isomer shift IS 190, 293, 299, 317
Isotherms *see* pcT curves
Isotope 1, 3, 295
isotope effect 76, 118, 160, 162
isotope mixture 91
isotope ratio IR 77
Itinerant 288

Jahn-Teller effect 163
Jellium model 141, 146

Kierstead model 73
Knight shift 148, 168, 174, 183
Korringa-Kohn-Rostocker KKR 142, 320

$LaNi_5$,$LaNi_5$ hydride
electronic properties 177, 182
electronic specific heat 181, 183
heat of formation 63, 185; Tables 3.1, 6.6.2
magnetic properties 181, 183, 298
pcT curves 35, 38
phase diagram 14
preparation 24, 38
resistivity 184
single crystal 27
structure, lattice parameters 108, 177, 182; Tables 4.1, 4.A
substitutions 24, 35, 42, 64, 184
thermodynamic properties 63
Lattice
contraction 103
distortion 162, 186
dynamics 76
expansion 103
parameter 91; Tables 4.1, 4.A
symmetry 92
Lattice gas 3, 6, 78, 79, 226
Laves phase 6, 201, 242, 257, 309, 315
Line broadening 60, 90, 301
Linear ordering 6, 107
Liquidus 23
Local band structure model 262, 264
Local density functional approximation LDF 141
Local heat of formation model 116

Magnetic
coupling 290
moment 93, 288
order 117; Table 4.B; 175
phase diagram 175
structure Table 4.B
susceptibility 31, 147, 148
Magnetism, concepts 286
Maximum hydrogen content 105
Mean field theory 288
Melting
arc melting 20
Levitation melting 20, 21
melting in crucibles 20
melting techniques 19, 21
Vacuum induction melting VIM 32

Metal atom substructure Table 4.1
Metal-semiconductor transition 8, 173
Metallography 29, 89
Matastable 5
Mg based compounds, hydrides
  electronic properties 193, 197
  heat of formation 68
  magnetic properties 200, 323
  pcT curves 38
  phase diagram 14
  preparation 20, 28, 37
  resistivity 200
  structure, lattice parameters Chap. 4; Tables 4.1, 4.A; 194
  substitutions 20
  thermodynamic properties 68
Microcrystallinity 309
Micrographs 30, 34
Microprobe 29, 89
Microtwinning 90
Miedema model 70, 146, 244
Minimum H-atom distance 110
Mischmetal 19
Miscibility gap 224
Mixed valence 8
Molar volume, partial 223, 253, 254
Mössbauer spectroscopy 148, 190; Chap. 7; 291
Mott transition 174
Muffin tin MT 142
Multi-step isotherms 75
Multiple plateau 73, 78
Multiple sublattice model 75
Muon spin rotation μSR 148, 183

Narrow band phenomena 8
Néel temperature T_N 93, 117; Table 4.B; 309
Neutron scattering, diffraction 60, 77, 90, 93, 182, 306, 312
  polarized 91
Nonstoichiometric hydrides 146
Nuclear density map 92
Nuclear magnetic resonance NMR 148, 161, 163, 174, 297

Octahedral 106, 107, 169, 170, 182, 264
Optic mode 76
Optical spectroscopy 150
Order, ordering 6, 80, 115, 169, 319
  linear ordering 6
Order-disorder transition 2, 78, 79, 166
Orthorhombic distortion 186
Oxide formation 16
Oxygen stabilized compounds 5, 111, 323

Pairing 6
Pairwise interaction 6, 69
Partial heat of solution 223
Partial molar properties 52
Pauli susceptibility 147, 151, 185, 298
Peritectic 24, 27
Phase diagrams of binary intermetallics 13
  La-Ni 14
  Mg-Ni 14
  of ternary intermetallics 16
  Ti-Cr 15
  Ti-Fe 15
Phase separation 72
Phase transition 78, 109, 297, 304
Phonons *see also* Electron-phonon coupling 162, 165
Photoelectron spectroscopy XPS, UPS Chap. 5; 149, 158, 167, 170, 177
Plateau
  pressure *see also* Pressure-composition isotherms 41, 53, 60, 80
  slope 7, 44, 62
Point group symmetry 93
Poisoning of the surface vy SO_2, CO, air 39, 89
Positron annihilation 149
Precipitations 25, 27
Preparation of hydrides Chap. 2; 37, 89
  electrochemically 40
  gas phase 37
  ion implantation 6
  via organometallic solutions 40
Preparation of intermetallics Chap. 2; 89
  chemical preparation 37
  gas phase 28
  melt, industrial scale 32
  melt, laboratory scale 17
  thin films, evaporation 28
Pressure-composition isotherms, pcT curves 2, 21, 57
  for characterization 29
  dynamic 81
  effect of annealing 25
  effect of substitutions 41
  for FeTi, $LaNi_5$, Mg_2Ni, $TiCr_2$ 38
  of industrially preparaed alloys 35
  measurements 57
  multi-step isotherms 73, 75
  slope 7, 44
Pressure-magnetisation isotherms 297
Price 18
Probe, H as a probe 7
Profile fitting 92
Protonic model 139

Pseudopotential 143
Pulverization 303
Purification
of constituent metals 19
intermetallic compounds 30
Purity *see* Impurity

Quasicrystals 2, 7

Raman spectra 173
Rapid solidification 28
Relaxation rate 161, 168, 294
Reorientation temperature 303
Repulsion 114, 117
Resistivity, electrical 148, 174, 290
Reviews 4, 88, 115, 139, 221, 285
Rietveld fitting method 91
Rigid band model 139, 161, 166, 201
Rocking curves 31
Ruderman-Kittel-Kasuya-Yosida
RKKY 287, 290, 317
Rule of reversed stability 112, 240, 247, 261

Schrödinger equation 141
Screening
core hole 149
hydrogen impurity 139
Sealing *see* Poisoning 39
Segregation 13, 22, 89, 299
Semiconducting 112, 168, 173
Semi-empirical models 234
Shear modulus 232
Sievert's law 64
Single crystal 27, 31
growth, FeTi, $LaNi_5$ 27
Site energy 69
Slater-Koster SK 143
Slope of plateau 7, 44
Solid-state reaction 28
Solidification 20, 22
rapid 28
unidirectional 26
Solidus 23
Solubility of hydrogen, solution 3, 13, 223
Space group 92; Table 4.1
Specific Heat
electronic *see* Electronic
lattice 59
partial molar 228
Specific surface area 39
Spin fluctuation 330
Spin glass 306, 309, 312, 316
Spin-lattice relaxation 161
Square-planar 106
Square-pyramidal 106, 107
Stability 112
Stoichiometry
of the intermetallic compounds 15
Stoner criterion 288
Stoner enhancement factor 181, 183, 298
Stoner-Wohlfarth model 288
Storage, storage capacity 4, 68, 87, 177
Strain 90
modulation 7
removal 59, 60
Structure Chap. 4
binary hydrides 88
magnetic Table 4.B
refinement 91
ternary hydrides Tables 4.1, 4.A
Sublattice substructure 3, 71, 74; Table 4.1
Sublattice magnetisation 313
Supercell 147
Superconductivity 8, 162; Chaps. 5, 7; 147, 166, 329
Superlattice 7
Superparamegnetism 299, 323, 325
Superstructure 80
Surface 3, TAP 64
area, specific 39
surface site 1

Ternary hydride
electronic properties 177
structure Table 4.1
thermodynamic properties Chaps. 3, 6; Tables 3.1–3, 6.6.2–3
Ternary intermetallic compounds 16
Tetragonal distortion 162, 169
Tetrahedral 106, 107, 264
Thermodynamic properties of alloy formation 12; Chaps. 3, 6
Thermodynamic properties of hydrides Chap. 3; 50
binary hydrides 237; Table 6.61
integral 50
multiple plateau 73
partial molar 52
plateau properties 53
quarternary hydrides 241
ternary hydrides 239; Tables 6.6.2–3
Thin films
characterization 32
preparation 28

Tight binding approximation 255
Time-of-flight method 90
Total Energy 143, 160, 167, 208

Unit cell volume 42, 231

Vacancy 265
Valence band, -spectra Chap. 5; 149, 158
Valence change 8, 118, 307, 330
Van't Hoff relation, plot 2, 58, 219, 229
Volatility 17

Westlake model 69, 113, 145
Work function 163, 256
Wyckoff positions 95

X-ray absorption near edge spectroscopy XANES 150, 159, 167, 186
X-ray absorption spectroscopy XAS 150
X-ray emission spectroscopy, soft SXES 150; Chap. 5; 164, 177, 200
X-ray photoelectron spectroscopy *see* Photoelectron spectroscopy Chap. 5; 149

GPSR Compliance

The European Union's (EU) General Product Safety Regulation (GPSR) is a set of rules that requires consumer products to be safe and our obligations to ensure this.

If you have any concerns about our products, you can contact us on ProductSafety@springernature.com

In case Publisher is established outside the EU, the EU authorized representative is:

Springer Nature Customer Service Center GmbH
Europaplatz 3
69115 Heidelberg, Germany

Batch number: 10682323

Printed by Printforce, the Netherlands